# Lecture Notes in Computer Science 4860

Commenced Publication in 1973
Founding and Former Series Editors:
Gerhard Goos, Juris Hartmanis, and Jan van Leeuwen

George Eleftherakis   Petros Kefalas
Gheorghe Păun   Grzegorz Rozenberg
Arto Salomaa (Eds.)

# Membrane Computing

8th International Workshop, WMC 2007
Thessaloniki, Greece, June 25-28, 2007
Revised Selected and Invited Papers

 Springer

Volume Editors

George Eleftherakis
Petros Kefalas
CITY College
Affiliated Institution of the University of Sheffield
Computer Science Department
13 Tsimiski St., 54624 Thessaloniki, Greece
E-mail: {eleftherakis, kefalas}@city.academic.gr

Gheorghe Păun
Institute of Mathematics of the Romanian Academy
PO Box 1-764, 014700 Bucharest, Romania
E-mail: george.paun@imar.ro

Grzegorz Rozenberg
Leiden University
Leiden Institute of Advanced Computer Science (LIACS)
Niels Bohrweg 1, 2333 CA Leiden, The Netherlands
E-mail: rozenber@liacs.nl

Arto Salomaa
Turku Centre for Computer Science (TUCS)
Leminkäisenkatu 14, 20520 Turku, Finland
E-mail: asalomaa@cs.utu.fi

Library of Congress Control Number: 2007941083

CR Subject Classification (1998): F.1, F.4, I.6, J.3

LNCS Sublibrary: SL 1 – Theoretical Computer Science and General Issues

ISSN    0302-9743
ISBN-10   3-540-77311-8 Springer Berlin Heidelberg New York
ISBN-13   978-3-540-77311-5 Springer Berlin Heidelberg New York

Springer is a part of Springer Science+Business Media

springer.com

© Springer-Verlag Berlin Heidelberg 2007
Printed in Germany

Typesetting: Camera-ready by author, data conversion by Scientific Publishing Services, Chennai, India
Printed on acid-free paper    SPIN: 12205752    06/3180    5 4 3 2 1 0

In Memory of Nadia Busi

# Preface

This volume contains a selection of papers presented at the Eighth Workshop on Membrane Computing, WMC8, which took place in Thessaloniki, Greece, during June 25–28, 2008. The first three workshops on membrane computing were organized in Curtea de Argeş, Romania – they took place in August 2000 (with the proceedings published in *Lecture Notes in Computer Science*, volume 2235), in August 2001 (with a selection of papers published as a special issue of *Fundamenta Informaticae*, volume 49, numbers 1–3, 2002), and in August 2002 (with the proceedings published in *Lecture Notes in Computer Science*, volume 2597). The next four workshops were organized in Tarragona, Spain, in July 2003, in Milan, Italy, in June 2004, in Vienna, Austria, in July 2005, and in Leiden, The Netherlands, in July 2006, with the proceedings published as volumes 2933, 3365, 3850, and 4361, respectively, of *Lecture Notes in Computer Science*.

The 2007 edition of WMC was organized by the South-East European Research Centre in Thessaloniki, under the auspices of the European Molecular Computing Consortium (EMCC). Special attention was paid to the interaction of membrane computing with biology and computer science, focusing on the biological roots of membrane computing, on applications of membrane computing in biology and medicine, and on possible electronically based implementations.

The pre-proceedings of WMC8 were published by the South-East European Research Centre, Thessaloniki, and they were available during the workshop. Each paper was refereed by two members of the Program Committee. Most of the papers selected for the present volume were significantly modified according to the discussions that took place during the workshop.

The volume includes all five invited talks as well as 22 regular papers, and so it presents a representative snapshot of current research in membrane computing (a comprehensive source of information about this fast-emerging area of natural computing is http://psystems.disco.unimib.it).

The Program Committee consisted of Gabriel Ciobanu (Iaşi, Romania), Erzsébeth Csuhaj-Varjú (Budapest, Hungary), Rudolf Freund (Vienna, Austria), Pierluigi Frisco (Edinburgh, UK), Marian Gheorghe (Sheffield, UK), Oscar H. Ibarra (Santa Barbara, USA), Petros Kefalas (Thessaloniki, Greece) – Co-chair, Vincenzo Manca (Verona, Italy), Giancarlo Mauri (Milan, Italy), Linqiang Pan (Wuhan, China), Gheorghe Păun (Bucharest, Romania) – Co-chair, Mario J. Pérez-Jiménez (Seville, Spain), and Athina Vakali (Thessaloniki, Greece).

The workshop was sponsored by City College, Thessaloniki, and the South-East European Research Centre (SEERC).

The editors are indebted to the members of the Program Committee, to all participants of WMC8, and in particular to the contributors to this volume.

Special thanks go to the organizers for their efficiency, and to Springer for the pleasant cooperation in the timely production of this volume.

At the beginning of September 2007, Nadia Busi passed away, after a short and severe illness. She was one of the most active researchers in membrane computing and the present volume includes the paper she presented at WMC8, one of the last conferences Nadia attended. Nadia will be remembered as a passionate researcher as well as a very nice person. As a token of respect and friendship that she enjoyed in our community, we devote this volume to her memory.

Rest in peace, Nadia. We all miss you.

October 2007

George Eleftherakis
Petros Kefalas
Gheorghe Păun
Grzegorz Rozenberg
Arto Salomaa

# Table of Contents

## Invited Lectures

## Regular Papers

# Psim: A Computational Platform for Metabolic P Systems

Luca Bianco[1] and Alberto Castellini[2]

[1] Cranfield University, Cranfield Health
Silsoe, Bedfordshire, MK45 4DT, UK
L.Bianco@cranfield.ac.uk
[2] Verona University, Computer Science Department
Strada Le Grazie 35, 37134 Verona, Italy
castellini@sci.univr.it

**Abstract.** Although born as unconventional models of computation, P systems can be conveniently adopted as modeling frameworks for biological systems simulations. This choice brings with it the advantage of producing easier to be devised and understood models than with other formalisms. Nevertheless, the employment of P systems for modeling purposes demands biologically meaningful evolution strategies as well as complete computational tools to run simulations on. In previous papers a strategy of evolution known as the *metabolic algorithm* has been presented; here a simulation tool called *Psim* (current version 2.4) is discussed and a case study of its application is also given.

## 1 Introduction

Membranes play a prominent role in living cells [1,20]. In fact, membranes not only act as a separation barrier indispensable to create different environments within cells boundaries, but they can also physically constitute a kind of "working board" whereby enzymes can activate and perform their duties on substrates. Other examples of the crucial role of membranes within cells are their ability to perform selective uptakes and expulsion of chemicals as well as being the interface of the cell with the surrounding environment allowing communication with neighboring cells.

P systems originate from the recognition of this important role of membranes and, by abstracting from the functioning and structure of living cells, they provide a novel computation model rooted in the context of formal language theory [34,36].

P systems investigations are nowadays focused on several research lines that make the field "a fast Emerging Research Front" in computer science (as stated by the Institute for Scientific Information). In particular, theoretical investigations on the power of the computational model have been carried on and important results have been achieved so far in order to characterize the computational power of many elements of P systems (such as objects and membranes) and,

G. Eleftherakis et al. (Eds.): WMC8 2007, LNCS 4860, pp. 1–20, 2007.

from a complexity viewpoint, P systems have been employed as well in the solution of NP hard problems. For a constant up to date bibliography of P systems we refer the reader to [39].

Parallel to these lines some more practical investigations are under way too. These studies exploit the resemblance of P systems to biological membranes in order to develop computational models of interesting biological systems. P systems seem to be particularly suitable to model biological systems, due to their direct correspondence of many elements (namely membranes, objects-chemicals and rules-reactions), even in their basic formulation, with real biological entities building the system to be modeled. Moreover, many extensions have been proposed to the standard formulation of P systems, such as some biologically relevant communication mechanisms [28,33,11], energy account [37] and active membranes [35] among others, which show the flexibility of the model. In this way, discrete mathematical tools can be used to represent interesting biological realities to be investigated. A further step is that of simulating all systems described in this way to get more information about their internal regulatory mechanisms and deeper insights into their underlying elements.

Although born as a non-conventional model of computation inspired by nature, P systems can therefore be employed as a simulation framework in which to embed the in silico simulation of interesting biological systems. The strength of this choice is, as said, the advantage that P systems share with biological systems many of their features and this leads to easy-to-devise and easy-to-understand descriptions of the studied realities. In fact, the membrane construct in P systems has a direct counterpart into biological membranes: objects correspond to all chemicals, proteins and macromolecules swimming in the aqueous solution within the cell and, eventually, rewriting rules represent biochemical reactions taking place in the controlled cellular environment. Other formalisms have been employed as modeling and simulation frameworks too, such as $\Pi$ calculus [29], Petri nets [38] and Ambient calculus [10], but in their case the very same notions of membranes, chemicals and reactions need to be reinterpreted and immersed in the particular representation formalism in a less immediate way.

Nevertheless, the employment of P systems as a modeling framework for biological systems posed, from a purely computational viewpoint, some new challenges to cope with, such as the identification of suitable, biologically meaningful, strategies for system evolution and the development of new automatic tools to describe, simulate and analyze the investigated system.

In previous works a novel strategy for systems evolution, called *metabolic algorithm* has been introduced [6,27,8], an hybrid (deterministic-stochastic) variant of which has been proposed as well [5]. Other strategies of evolution are known, such as *Dynamical Probabilistic P systems* [32,31] and *Multi-compartmental Gillespie's algorithm* [30,2].

Here we will focus on the metabolic algorithm in its deterministic version which has been confronted with the dynamics of several systems (a collection of case studies can also be found in [4]). Some examples of investigated systems by means of the metabolic algorithm are the Belousov-Zhabotinsky reaction (in the Brusselator formulation) [6,8], the Lotka-Volterra dynamics [6,27,7,14], the SIR (Susceptible-Infected-Recovered) epidemic [6], the leukocyte selective recruitment in the immune response [16,6], the Protein Kinase C activation [8], circadian rhythms [12] and mitotic cycles in early amphibian embryos [26]. In order to cope with the demand of computational tools to simulate the dynamics of P systems, we developed a first simulator called *Psim* [6], which has now been extended with several new features as will be explained later on. The new version of the simulator is freely available for download at [15].

The remaining of the discussion will firstly introduce (section 2) some theoretical aspects of the simulation framework we developed and some recent advances will be mentioned too. Section 3 will then be devoted to the newer version of the simulator itself and a practical case study will be given as well in such a way to show to the reader how to set up a simulation with the interface of Psim.

## 2   MP Systems

MP systems (Metabolic P systems) [21,26,24,23] are a special class of P systems [34,36], introduced for expressing the dynamics of metabolic (or, more generally speaking, biological) systems. Their dynamics is computed by means of a deterministic algorithm called *metabolic algorithm* which transforms populations of objects according to a *mass partition* principle, based on suitable generalizations of chemical laws.

A definition of MP systems follows, as given in [4].

**Definition 1 (MP system).** *An MP system of level $n-1$ (i.e., with $n \in \mathbb{N}$ membranes) is a construct*

$$\Pi = (T, \mu, Q, R, F, q_0)$$

*in which:*

- *$T$ is a finite set of symbols (or objects) called the alphabet;*
- *$\mu$ is the hierarchical membrane structure, constituted by $n$ membranes, labeled uniquely from 0 to $n-1$, or equivalently, associated in a one-to-one manner to labels from a set $L$ of $n-1$ distinct labels;*
- *$Q$ is the set of the possible states reachable by the MP system. Each element $q \in Q$ is a function $q : T \times L \longrightarrow \mathbb{R}$, from couples objects-membranes to real values. A value $q(X, l)$, with $X \in T$ and $l \in L$ gives the amount of objects $X$ inside a membrane labeled $l$, with respect to a conventional unit measure (grams, moles, individuals, ...);*
- *$R$ is the finite set of rewriting rules. Each $r \in R$ is specified according to the boundary notation [3]. In other words, each rule $r$ has the form $\alpha_r \longrightarrow \beta_r$, where $\alpha_r, \beta_r$ are strings defined over the alphabet $T$ enhanced with indexed parenthesis representing membranes. As an example, an hypothetical rule can have the form: $\alpha[_1\beta \longrightarrow \gamma[_1\delta$, with $\alpha, \beta, \gamma, \delta \in T^*$, meaning that $\alpha$ and $\beta$ are respectively changed in $\gamma$ and $\delta$, where all objects within $\alpha$ and $\gamma$ are outside membrane labeled 1, whereas elements of $\beta$ and $\delta$ are placed inside membrane 1;*

- $F$ *is the set of* reaction maps, *each* $f_r \in F$ *is a function uniquely associated to a rule* $r \in R$, *defined as* $f_r : Q \longrightarrow \mathbb{R}$ *and, given a certain state* $q$, *it produces* $f_r(q)$ *that is a real number specifying the strength of rule* $r$ *in acquiring objects;*
- $q_0 \in Q$ *is the* initial state *of the system. It specifies the initial amount of all the species throughout the various compartments of the system.*

Since encodings like that in [9] show that the membrane structure can be flattened by augmenting the alphabet size, the definition of the membrane structure $\mu$ is not very important in this context and the choice to employ *0-level* MP systems in the remaining of the discussion is not limiting from a theoretical point of view. Moreover, dealing with 0-level MP systems ends up in a easier discussion, in fact all states $q \in Q$ do not need the specification of a membrane label and in this way they have the simpler form $q : T \longrightarrow \mathbb{R}$. For this reason, in the following whenever the term MP system will be used, the more correct term *0-level MP system* has to be implicitly assumed.

The dynamics of MP systems has been calculated, starting from the initial state $q_0$ by means of an evolution strategy called *metabolic algorithm* [6,27,8], which is substantially different from the well known *non-deterministic and maximally parallel* paradigm followed by standard P systems. More precisely, the perspective of MP systems is to model systems at a population level rather than at an objects level. In this way, nothing can be precisely said about individuals but the investigation is focused on the macroscopic dynamics which assumes a deterministic flow in spite of individual behaviors.

## 2.1   The Metabolic Algorithm: Hints

Without entering into many details (which can be found in [6,27,8,25]), the metabolic algorithm is a deterministic strategy for MP systems evolution based on mass partition among rules of all elements in the alphabet $T$.

In very general terms, the metabolic algorithm can be summarized in the following main points [26]:

- Reactants are distributed among all the rules, as the system evolves, according to a "competition" strategy.
- If some rules compete for the same reactant, then each of them obtains a portion of the available substance that is proportional to its reaction strength (*reactivity*) in that state.
- The reactivity of a rule in a certain state measures the capability of the rule to acquire its reactants. It is calculated by the evaluation of the reaction map corresponding to the rule due and it depends on the state of the system, that is defined as the concentration and localization of all substances in the considered instant.
- The evolution strategy determines the reaction unit of all rules, that is, the unitary amount of substance which is dealt by the rule. The stoichiometry is used then to obtain the consumed and/or produced amount of substances for each rule.

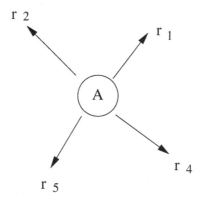

**Fig. 1.** Competition for object $A$ between rules $r_1, r_2, r_4$ and $r_5$

An example may be useful to clarify the concepts yet introduced. Let us suppose that in a given instant, four rules, namely $r_1, r_2, r_3$, and $r_4 \in R$, need molecules of a species $A$ (with $A$ belonging to the alphabet $T$) as reactant (see Figure 1), then a partition strategy for $A$ is necessary.

A real number called *reactivity* represents the strength of the rule (i.e., the rule's capability of obtaining matter to work on), given by the value assumed by a function uniquely associated to the rule called *reaction map*, in the considered state. For example, with respect to Figure 1, if we denote with $a, b$ and $c$ the concentrations of species $A, B, C$ respectively (in a state $q$ not specified for the sake of simplicity), then the reactivities of rules $r_1, r_2, r_4$ and $r_5$, which ask for $A$ molecules, can be:

$$f_1 = 200 \cdot a, \quad f_2 = 0.5 \cdot a^{1.25} \cdot b^{-1}, \quad f_4 = a^{1.25} \cdot (b+c)^{-1}, \text{ and } f_5 = 10,$$

where the choice of reaction maps $f_i$, $i = \{1, 2, 4, 5\}$ is completely arbitrary in this example.

We define the quantity

$$K_{A,q} = \sum_{i=1,2,4,5} f_i(q)$$

as the *total pressure* on $A$ in the state $q$ (the intuitive idea is that all reaction maps of rules competing for a certain species give the force that pushes that species to react).

Then, for each of the competing rules $r_j$ we consider the *partial pressure* (or *weight*) of $r_j$ on type $A$ as

$$w_{A,q}(r_j) = \frac{f_j(q)}{K_{A,q}}$$

(again the idea behind this is that the strongest the force pushing an element to follow a particular reaction channel, compared to other reaction channels, the more matter will follow that path).

Note that, in general, the quantity $K_{X,q}$ is defined for each couple $(X, q)$ where $X \in T$ and $q$ is a possible state of the system, moreover a weight $w_{X,q}(r)$ has to be calculated for each triple $(X, q, r)$ where $X$ and $q$ are respectively, an element of the biological alphabet and a state of the system, while $r \in R$ is a rule in which the element $X$ appears as a reactant (i.e., according to the terminology adopted above, $X \in \alpha_r$).

Getting back to the example discussed above, it should be easy to see that the partial pressure of $r_1$ on $A$ is

$$w_{A,q}(r_1) = \frac{200a}{200a + 0.5a^{1.25}b^{-1} + a^{1.25}(b+c)^{-1} + 10}$$

while the same pressure due to $r_2$ results to be equal to

$$w_{A,q}(r_2) = \frac{0.5a^{1.25}b^{-1}}{200a + 0.5a^{1.25}b^{-1} + a^{1.25}(b+c)^{-1} + 10}$$

and the other weights can be calculated analogously. The weights calculated so far determine the partition factors of the amount of species $A$, available in the state $q$, among the rules which need objects $A$ as reactants.

Now to calculate the reaction unit of a particular rule (i.e., the amount of reactant that can be dealt by the rule) we simply need to multiply the partial pressure of the rule on the reactant by the real amount of reactant present into the system at the considered state. For example, the reaction unit of rule $r_1$ (or equivalently, the amount of $A$ that that *can be* assigned to rule $r_1$) turns out to be $w_{A,q}(r_1) \cdot a = \frac{0.5a^{2.25}b^{-1}}{200a+0.5a^{1.25}b^{-1}+a^{1.25}(b+c)^{-1}+10}$.

In this way, if $r_1$ is a rule of the form $A \to X$, no matter what element is represented by $X \in T$, then the amount of $A$ associated to $r_1$ is exactly $u_1 = w_{A,q}(r_1) \cdot a$ and the effect of $r_1$'s application is the loss of $u_1$ units of $A$ and the acquisition of the same number of units of $X$.

In the case of cooperative rules (i.e., rules with more than just one reactant) things are a little bit more complicated since we need to take into account the real availability of all reactants taking part to the reaction. That is, for each $X$ belonging to the reactants of a certain rule $r$ we need firstly to compute the quantities $w_{X,q}(r) \cdot x$ and, since we have to respect species availability, the reaction unit associated to the rule is then computed as the minimum of those quantities. If we suppose that a rule $r_1$ has the form $AAB \to X$, then the reaction unit

$$u_1 = min(\frac{1}{2}w_{A,q}(r_1) \cdot a \, , \, w_{B,q}(r_1) \cdot b)$$

where the term $\frac{1}{2}$ in the first element of the minimum is due to the fact that $A$ appears twice in the stoichiometry of the rule.

In general terms, the metabolic algorithm is an effective procedure for calculating in each state of the system a reaction unit $u_i$ for all rules $r_i \in R$ by using a partition strategy that employs particular functions $f_i$ associated in a 1-1 manner to rules. After this calculation, the evolution of the system can be obtained in a straightforward way by consuming and producing species in a quantity given by rules' reaction units and by following the stoichiometry of the system.

Assuming a sorting on objects and on rules, let us denote with $M$ the $m \times n$ stoichiometric matrix associated to an MP system having $m$ symbols and $n$ rules (in which the $c_{i,j}$ element of $M$ denotes the gain or the loss of the $i$th object due to rule $j$ (it is the difference between the number of occurrences of the $i$th symbol among products and among reactants of $j$th rule) and with $U_q$ the $[u_1 \cdots u_n]^T$ vector of the reaction units in a state of the system $q$, calculated as mentioned above.

Then, as pointed out in [25] the transition from one state $q$ to the following one is done by means of the *delta operator* $(\Delta_x(q))$ which is a $m$-sized vector giving the variation of each species in the transition from state $q$ to the next state $q'$. In particular

$$\Delta_x(q) = M \times U_q$$

stating that the delta operator can be obtained as the product of the stoichiometric matrix $M$ by the reaction units vector of state $q$, $U_q$.

Since each row $i$ of $\Delta_x(q)$ gives the variation on the $i$th object, then if we think of a state $q$ as a vector containing the concentration of all the $m$ species at the corresponding instant, then the next state can easily be calculated as

$$q' = q + \Delta_x(q) = q + M \times U_q.$$

Just to exemplify the last concepts discussed, we can think about an alphabet $T = \{A, B, C\}$ and focus on a rule set comprising the following four rewriting rules:

$$
\begin{aligned}
r_1 &: A\ B \longrightarrow C \\
r_2 &: B\ B \longrightarrow A \\
r_3 &: C \quad\ \longrightarrow A \\
r_4 &: C \quad\ \longrightarrow B
\end{aligned}
$$

then, assuming the lexicographic order on elements of the alphabet, we can obtain the following stoichiometric matrix:

$$
M = \begin{bmatrix}
-1 & 1 & 1 & 0 \\
-1 & -2 & 0 & 1 \\
1 & 0 & -1 & -1
\end{bmatrix}
$$

in which, the first row corresponds to the object $A$ and states that we lose one conventional unit of $A$ due to rule $r_1$, we get one $A$ both from rule $r_2$ and $r_3$ and finally $r_4$ does not affect $A$ concentration at all.

Then, let us suppose to be in a state $q$, described by the vector of concentrations $q = [10\ 32\ 20]^T$ (i.e., we have 10 units of $A$, 32 of $B$ and 20 of $C$) and that the corresponding reaction units vector $U_q = [7\ 12\ 5\ 9]^T$ (i.e., reaction $r_1$ moves 7 mass units, $r_2$ 12, $r_3$ 5 and finally $r_4$ 9). In this way it is possible to calculate the next state $q'$ which turns out to be described by the following vector:

$$q' = q + M \times U_q = \begin{bmatrix} 10 \\ 32 \\ 20 \end{bmatrix} + \begin{bmatrix} -1 & 1 & 1 & 0 \\ -1 & -2 & 0 & 1 \\ 1 & 0 & -1 & -1 \end{bmatrix} \times \begin{bmatrix} 7 \\ 12 \\ 5 \\ 9 \end{bmatrix} = \begin{bmatrix} 20 \\ 10 \\ 13 \end{bmatrix}$$

describing the amount of all species at that particular instant.

In previous papers [26] a convenient and intuitive formalism for representing MP systems called *MP graphs* has been proposed. In particular, MP graphs are bipartite graphs describing both the stoichiometry (i.e., the shape of the rules) and the regulative part of MP systems that need to be effectively calculated in order to obtain the dynamics of the system (i.e., the reaction maps). According to what said above, MP graphs represent all the information needed to simulate MP systems by means of the metabolic algorithm. An example of MP graphs, as produced by the simulator we developed, will be shown later on.

## 2.2 Generalizing the Metabolic Algorithm

According to the formulation of the dynamics given in the previous section, the metabolic algorithm is a strategy that given a particular state $q$ provides the system with the corresponding reaction units vector $U_q$ which is used to calculate the transition to the state $q + 1$. As discussed in [25], other strategies can be considered whose aim is to produce a reasonable mass partition among all rules of an MP system, or in other words that give a different $U_q$ for each state $q$ of the system.

This view leads to the definition of several *metabolic algorithms* instead of a single one and the definition of MP systems can be generalized accordingly.

Based on the definition given in [22], Definition 1 can be easily generalized, in very general terms, in the following way:

**Definition 2 (Generalized MP systems).** *A 0-level (generalized) MP system is a 6-tuple*

$$\Pi = (T, Q, R, V, q_0, \phi)$$

*in which:*

- *$T$ is a finite set of symbols (or objects) called the* alphabet*;*
- *$V$ is a finite set of* variables*;*
- *$Q$ is the set of the possible states* reachable by the MP system. Each element *$q \in Q$ is a function $q : T \cup V \longrightarrow \mathbb{R}$;*
- *$R$ is the finite set of* rewriting rules*;*
- *$q_0 \in Q$ is the* initial state *of the system;*
- *$\phi$ is the strategy of evolution, $\phi : Q \longrightarrow \mathbb{R}^n$ with $|R| = n$.*

Note that nothing is said about the cardinality of the set of variables $V$ and they are not necessarily associated in a one-one manner to rules of $R$. Moreover, the strategy of evolution $\phi$, given a state $q$ has to be defined in such a way that it outputs the $n$-tuple providing the reaction unit vector of the system, or following the terminology used above, $\phi(q) = U_q$.

Complete freedom is left in the implementation of the strategy of evolution, whose only constraints are that given a state it has to provide the reaction unit vector corresponding to that state, which will be used to calculate the evolution of the system by means of the matrix product recalled in the previous section. As will be mentioned in the following, the specification of a fully customizable strategy of evolution will be one of the prominent features of the new version of the simulator Psim that has been implemented within the MNC Group of the University of Verona.

## 3   Psim

Based on the theoretical framework expressed in previous sections, a simulator called *Psim* (P systems simulator) has been developed to cope with the problem of calculating the dynamics of biological systems. An early version of Psim has been developed previously [6], with which the newer version shares the same philosophy, though extending some of its concepts and enhancing the simulation environment with many features.

The present release of Psim (version 2.4) has been developed in response to the need of an effective and easy to use tool to calculate the dynamics of MP systems by means of the metabolic algorithm. Its implementation has moreover followed some flexibility and extensibility principles which led to a tool that can be easily extended and integrated with other tools. In this way Psim, thanks to its immediate setup (nothing needs actually to be done provided a Java virtual machine is installed on the computer that is meant to run Psim) and to the easy user interface, can be used by people without a strong background in programming and a deep knowledge in the field of computer science. On the other hand, the extendability provided, by means of the plugin mechanism, allows people with stronger expertise in programming to build their own tools to complement and integrate the main core of Psim.

Some features of this tool, which is implemented by using the Java programming language, are listed below:

– Friendly user interface which is born to be easy-to-use and to interact with people not necessarily having a strong computer science background. Its immediacy can be found in the input side, which can be specified by means of a transposition of the concept of MP graphs into a point and click graphical interface. Moreover, the same simplicity principle holds for the output side as well, which is basically constituted by a graph containing the temporal evolution of all the species constituting the system (both on a temporal scale and on the phase plan space).
– Plugin architecture: the interaction with the system can either be done manually or by means of some specifically designed plugin which, thanks to the plugin support offered by the simulation engine, can interact with the simulation engine itself. More specifically, three different kinds of plugin can be devised and implemented in Java as well. *Input plugin* can be used to implement various sources for the data to run the simulation on

(let us think to some specific pathways databases like KEGG for instance); *output plugin* conversely, can be used to observe and analyze in various ways the results obtained from a simulation and can therefore give some meaningful insights into the simulated dynamics. Moreover, they can be used to export simulation data into particular formats. Finally, *experiment plugin* can directly control and intimately interact with the simulation engine, by controlling the execution flow, checking some properties and changing some experimental conditions. This kind of plugin can be very useful in tasks like model optimizations and stability analysis.

- Extreme flexibility. The simulation tool we propose is based on a simulation engine which is designed to accept the definition of new evolution strategies for the calculation of the systems dynamics. At the only price of the implementation of some specific interfaces, the developer has the chance to define his own simulation strategies and to design a customized library of *metabolic algorithms* to calculate the systems evolution.
- Models portability has been implemented by using the standard XML language and some extensions towards the SBML language are being considered too.
- Cross platform applicability, thanks to the choice of Java, Psim can be run on all platforms supporting the Java virtual machine architecture.

An aspect deserving a special emphasis here is the possibility offered by the simulation engine, to specify custom evolution strategies. Getting back to the definition of generalized MP systems, the architecture of the simulator allows the specification of a fully customized $\phi$ function. A set of evolution strategies can be devised by developing in Java a specific class implementing a particular interface provided by the main engine. Several different strategies can be handled simultaneously by the simulator that gives the chance to decide which simulation strategy employ in the simulation process. This gives the tool a very high level of flexibility and power as well as the plugin mechanism does. Since plugin can interact with the simulation engine at a different levels, such as input, output but also at the simulation level too, they can be used for various reasons within the simulator and this again gives users plenty of ways to improve the system and to extend its functionalities.

## 3.1   A Case Study

In this section we show an application of the Psim computational tool for the simulation of the well known mitotic oscillator as found in early amphibian embryos [18,19,26].

Mitotic oscillations are a mechanism exploited by nature to regulate the onset of mitosis, that is the process of cell division aimed at producing two identical daughter cells from a single parent. More precisely, mitotic oscillations concern the fluctuation in the activation state of a protein produced by cdc2 gene in fission yeasts or by homolog genes in other eukaryotes. The model here considered focuses on the simplest form of this mechanism, as it is found in early amphibian embryos. Here, the progressive accumulation of the cyclin protein leads to the activation of cdc2 kinase. This activation is achieved by a bound between cyclin and cdc2 kinase forming a complex known as M-phase promoting factor (or *MPF*). The complex triggers mitosis and degrades cyclin as well; the degradation of cyclin leads to the inactivation of the cdc2 kinase that brings the cell back to the initial conditions in which a new division cycle can take place.

Goldbeter [18,19] proposed a minimal structure for the mitotic oscillator in early amphibian embryos in which the two main entities are cyclin and cdc2 kinase. According to this model, depicted in Figure 2, the signalling protein cyclin is produced at a constant rate $v_i$ and it triggers the activation (by means of a dephosphorylation) of cdc2 kinase, passing from the inactive form labeled $M^+$ to the active one, denoted by $M$. This modification is reversible and the other way round is performed by the action of another kinase (not taken into account in the model) that brings $M$ back to its inactive form $M^+$. Moreover, active cdc2 kinase ($M$) elicits the activation of a protease $X^+$ that, when in the active (phosphorylated) form ($X$), is able to degrade the cyclin. This activation, as the previous one, is reversible as stated by the arrow connecting $X$ to $X^+$.

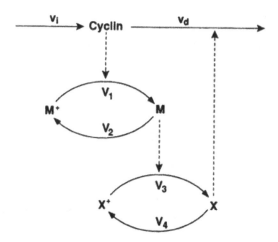

**Fig. 2.** The mitotic oscillator model by A. Goldbeter, from [18]

The set of differential equations devised by Goldbeter produces an oscillatory behavior in the activation of the three elements $M$, $C$, $X$ that repeatedly go through a state in which cells enter in a mitotic cycle (see Figure 3).

The goal of the case study showed here is to obtain a description and a simulation of the very same model of mitotic oscillations by means of the simulator Psim.

In general, there is no unique way to translate a differential equation system in terms of a metabolic P system, therefore we choose to obtain it by the application of the MP-ODE transformation [13]. The resulting MP system is reported here:

$$\Pi = (T, \mu, \mathcal{R}, F, q_0), \text{ where:}$$

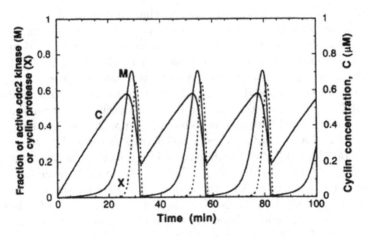

**Fig. 3.** Dynamics of the mitotic oscillator from [18]

- The alphabet: $T = \{A, C, X, X^+, M, M^+\}$
- The membrane structure: $\mu = [_0 \ ]_0$;
- The set of rules is $\mathcal{R} = \{r_1, r_2, ..., r_{10}\}$, where:

$$
\begin{array}{ll}
r_1 : A \to A\,C & r_6 : M^+ \to \lambda \\
r_2 : C \to X & r_7 : M \to M^+ \\
r_3 : X \to \lambda & r_8 : X^+ \to X\,M \\
r_4 : C \to \lambda & r_9 : M \to \lambda \\
r_5 : C \to M\,C & r_{10} : X \to X^+
\end{array}
$$

in which all symbols have the meaning described before (and $A$ is a kind of well to draw substance $C$ out from). Moreover, for every symbol in the system, we have introduced an inertia rule (i.e., a rule having the form $Y \to Y$, for each $Y \in T$ to model the inertia of the system), omitted in this set of rules.

- The set of reaction maps is $F = \{Fr_1, Fr_2, ..., Fr_{10}\}$, where:

$$
\begin{aligned}
Fr_1 &= k_1 \\
Fr_2 &= k_2 \frac{x}{k_3 + c} \\
Fr_3 &= k_2 \frac{c}{k_3 + c} \\
Fr_4 &= k_3 \\
Fr_5 &= k_5 \frac{m^+}{(k_6 + c)(k_7 + m^+)} \\
Fr_6 &= k_5 \frac{c}{(k_6 + c)(k_7 + m^+)} \\
Fr_7 &= k_8 \frac{1}{(k_9 + m)}
\end{aligned}
$$

$$Fr_8 = k_{10} \frac{m}{(k_{11} + x^+)}$$

$$Fr_9 = k_{10} \frac{x^+}{(k_{11} + x^+)}$$

$$Fr_{10} = k_{12} \frac{1}{(k_{13} + x)}$$

– The initial state $q_0$ of the single membrane system is defined by:
$q_0(A) = 1.3;$
$q_0(C) = 0.01;$
$q_0(X) = 0.01;$
$q_0(X^+) = 0.99;$
$q_0(M) = 0.01;$
$q_0(M^+) = 0.99;$
in which we deal with concentrations of species, rather than with objects, and in this way the initial amounts are real numbers.

For each element of $T$ the reaction map of inertia rules is set to 1600.

We start the modeling session by opening the Psim's main interface showed in Figure 4. This window allows the user to manage all the experiment's stages. In particular the main possible choices involve:

1. modeling the system, setting substances, initial conditions, reaction maps and rules;
2. starting the simulation;
3. displaying output charts.

The first step to consider while setting up a system's simulation is the specification of the corresponding MP graph. In what follows, some steps towards the creation of an MP graph modeling the mitotic oscillator are presented.

After clicking on the *New Experiment* label of the *File* menu, a window like the one depicted in the Figure 5 appears. This is the main window of the graphical interface allowing the user to input in a easy way the MP graph components by simply dragging them from the upper toolbar to the bottom white panel. This task is performed by using the following toolbar icons:

– The *blue circle*: adds a new *type node* that stores the name of a substance, its initial number of molar units and its inertia value (as explained in previous papers, inertias are a way to represent the fact that not all reactants need to react at a certain instant, they are a sort of resistance opposed to species to performing reactions).
– The *black circle*: adds a new *metabolic reaction node* that represents a reaction channel between interacting substances and stores the name of a reaction rule.
– The *red rectangle*: adds a new *reactivity node* building the regulatory part of MP graphs. In the simulator, reactivity nodes store the reactivity map function corresponding to the connected rule and, if necessary, a boolean guard function for the rule activation.

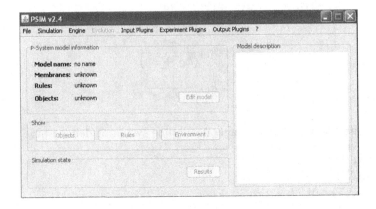

**Fig. 4.** The Psim's main interface

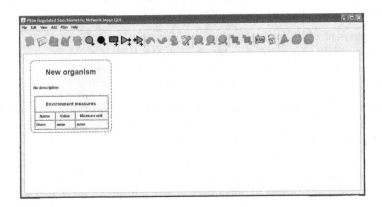

**Fig. 5.** Psim's input interface

- The *green triangles*: add *input gates* and *output gates* nodes that identify rules which respectively introduce new matter in the membrane or expel part of it from the system.

After the insertion of the nodes in the white panel the user can specify their internal parameters and connect the nodes by drawing arcs among them. The best way to accomplish this task is to start by defining the *type node* parameters and the *metabolic reaction node* parameters by double clicking on the corresponding nodes and filling in the window's field that automatically appears (Figure 6). Importantly enough, a parser has been implemented to check the consistency of inserted parameters and to alert the user if any irregularity arose.

At this point, one can connect the *type nodes* and the *metabolic reaction nodes* with each other by drawing arcs among them with the simple use of the mouse. This is a very important step because it allows to represent the stoichiometric part of the system by means of the MP graph topology (Figure 7).

**Fig. 6.** Insertion of the *type node* parameters

As an example, let us consider the reaction $r_2 : C \rightarrow X$. Within the input graphical interface it is represented by the $R2$ black circle that is connected by means of black arcs to the $C$ and the $X$ blue circles, representing the corresponding substances; the direction of the arrow represents the substance flow of the reaction.

A further modeling step is needed to add the *reactivity nodes* describing the regulatory part of the system. This can be done by first linking every *type node*, that affects a reaction map, with the corresponding *reactivity nodes* (as showed in Figure 8). Finally the reactivity map function of every reactivity node is specified by using the linked *type nodes* and the *environmental measures* as variables or constants (as reported in Figure 9). Figure 8 represents the final mitotic oscillator MP graph as produced by the Psim GUI.

This completes the modeling stage and the next logical step is to start the simulation of the specified system. This is done by clicking on the rightmost icon of the upper toolbar (the rightward arrow). The click causes a small window to pop out, in which it is possible to set the number of steps the simulation will span (Figure 10). A possible choice for this system is to run 150000 steps. By click on the *Start* button the dynamics computation begins.

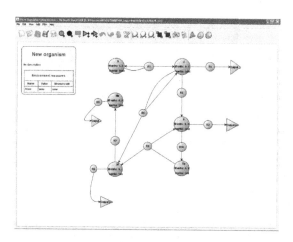

**Fig. 7.** Adding *type nodes, metabolic reaction nodes* and drawing arches among them

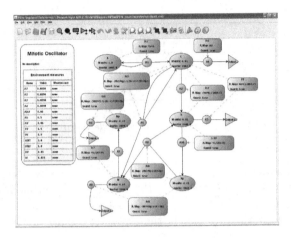

**Fig. 8.** An MP graph that models the mitotic oscillator

**Fig. 9.** Reactivity node input parameter window

**Fig. 10.** Set the number of steps for the simulation to 150000

When the simulation is finished the system prompts that results are available and ready to be visualized by the *Psim chart visualization form* (Figure 11). Using the bottom panel check boxes it is possible to decide elements to be displayed. In the considered case oscillations of cyclin $C$ (red line), active cdc2 kinase $M$ (blue line) and active protease $M$ (green line) are displayed but the phase space plot can be drawn as well.

We finally highlight an important mechanism of the Psim platform: plugin extensibility. As already mentioned, plugin allow the user to enhance the main Psim computing core with powerful functionalities for the data import, export, control and analysis. An example under development is the *experiment plugin* that stores the experiment state (concentration of the substances and environmental

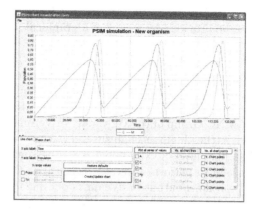

**Fig. 11.** Oscillations of the mitotic systems as calculated by Psim

measures) every $x$ steps, where $x$ is a parameter set by the user before the computation starts. This plugin could save, for instance, an XML file for every state, allowing the user to export the experiment samples in an standard way.

A software developer generates the plugin code (basically some Java classes) relying on the Psim's *JavaDoc* documentation obtainable at [15] which lists the *experiment plugin* methods to be mandatorily implemented. Plugin classes are meant to be archived in a Jar file and placed in a proper *plugin* directory. Provided this, at the following start up Psim will automatically find and load all the plugin contained in the *plugin* directory. The user can find all available experiment plugin statements in the main interface *Experiment Plugin* menu (Figure 4) in the form of a list. By clicking on the relative label is possible to activate the plugin that will be run at each step of the subsequent simulation and will save the state every $x$ steps chosen by the user filling in a plugin popup window.

The plugin just described yields a set of XML files but the same principles can be extended also to the other kinds of plugin (*input, output, engine plugin*).

A particular mention is deserved by *engine plugin* that allow to implement new simulation strategies which can be different from the *metabolic algorithm* described above. This gives the simulation tool a very high flexibility as well as extendability as discussed previously.

## 4   Conclusion and Further Work

P systems can be useful frameworks to embed biological systems models in. This demands for some modifications to the classical definition of P systems and particularly a biologically meaningful evolution strategy is needed. In previous papers an essentially deterministic strategy, called metabolic algorithm, for the calculation of biological systems dynamics has been provided as well as an extension of the classic model of P systems, known as MP systems, focused on the dynamics of bio-systems. Moreover, all data needed for the simulation of MP systems dynamics can be provided by means of a graphical formalism known as MP graphs.

The basics of MP systems have been briefly revisited in this paper and based on them a simulation tool called Psim has been highlighted, together with a case study of a well known and previously investigated model of mitotic cycles in early amphibians. Psim v. 2.4 is the latest release of the MP systems simulator developed within the MNC Group of the University of Verona and it has very interesting features such as the plugin mechanism and the meta-engine architecture which give the tool an high level of extendability and personalization. In particular, plugin can be useful to perform several tasks such as data import/export, control of the simulation flow, output of dynamics obtained and analysis of the results among others. Moreover, the meta-engine architecture of the simulator allows users to define their own evolution strategies by implementing some fixed interfaces of the simulator.

In the future we plan to enrich the core of this simulation tool by implementing a series of plugin such as the one described above to have a snapshot of the state of the system in particular instants. Other plugin under investigation involve some automatic procedures for parameter estimation given suitable observations of the reality to be modeled. Finally, we plan to employ this simulation tool for the calculation of the dynamics of systems not already modeled and in this respect the possibility to devise ad-hoc evolution strategies can be very important to tackle some specific issues related with the particular reality to be modeled.

## Acknowledgements

Authors would like to gratefully acknowledge the whole MNC Group of the University of Verona for the support in this work. Special thanks deserve the team leader, Prof. Vincenzo Manca, as well as Luca Marchetti and Michele Petterlini for their hard work while implementing the simulator Psim.

## References

1. Alberts, B., Johnson, A., Lewis, J., Raff, M., Roberts, K., Walter, P.: Biology of the Cell, 4th edn. Garland Science, New York (2002)
2. Bernardini, F., Gheorghe, M., Krasnogor, N., Muniyandi, R.C., Pérez-Jiménez, M.J., Romero-Campero, F.J.: On P systems as a modelling tool for biological systems. In: Freund, R., Păun, G., Rozenberg, G., Salomaa, A. (eds.) WMC 2005. LNCS, vol. 3850, pp. 193–213. Springer, Heidelberg (2006)
3. Bernardini, F., Manca, V.: Dynamical aspects of P systems. BioSystems 70, 85–93 (2002)
4. Bianco, L.: Membrane Models of Biological Systems. PhD thesis, Verona University (2007)
5. Bianco, L., Fontana, F.: Towards a hybrid metabolic algorithm. In: Hoogeboom, H.J., Păun, G., Rozenberg, G., Salomaa, A. (eds.) WMC 2006. LNCS, vol. 4361, pp. 183–196. Springer, Heidelberg (2006)

6. Bianco, L., Fontana, F., Franco, G., Manca, V.: P systems for biological dynamics. In: Ciobanu, G., Păun, G., Pérez-Jiménez, M.J. (eds.) Applications of Membrane Computing, Springer, Heidelberg (2006)
7. Bianco, L., Fontana, F., Manca, V.: Reaction-driven membrane systems. In: Wang, L., Chen, K., Ong, Y.S. (eds.) ICNC 2005. LNCS, vol. 3611, pp. 1155–1158. Springer, Heidelberg (2005)
8. Bianco, L., Fontana, F., Manca, V.: P systems with reation maps. International Journal of Fondations of Computer Science 16(1) (2006)
9. Bianco, L., Manca, V.: Encoding-decoding transitional systems for classes of P systems. In: Freund, R., Păun, G., Rozenberg, G., Salomaa, A. (eds.) WMC 2005. LNCS, vol. 3850, pp. 135–144. Springer, Heidelberg (2006)
10. Cardelli, L., Gordon, A.D.: Mobile ambients. In: Nivat, M. (ed.) ETAPS 1998 and FOSSACS 1998. LNCS, vol. 1378, pp. 140–155. Springer, Heidelberg (1998)
11. Cavaliere, M.: Evolution-communication P systems. In: Păun, G., Rozenberg, G., Salomaa, A., Zandron, C. (eds.) Membrane Computing. LNCS, vol. 2597, pp. 134–145. Springer, Heidelberg (2003)
12. Fontana, F., Bianco, L., Manca, V.: P systems and the modeling of biochemical oscillations. In: Freund, R., Păun, G., Rozenberg, G., Salomaa, A. (eds.) WMC 2005. LNCS, vol. 3850, pp. 199–208. Springer, Heidelberg (2006)
13. Fontana, F., Manca, V.: Discrete solution of differential equations by metabolic P systems. TCS (submitted)
14. Fontana, F., Manca, V.: Predator-prey dynamics in P dystems ruled by metabolic algorithm. BioSystems (accepted)
15. The Center for BioMedical Computing Web Site. Url: http://www.cbmc.it
16. Franco, G., Manca, V.: A membrane system for the leukocyte selective recruitment. In: Martín-Vide, C., Mauri, G., Păun, G., Rozenberg, G., Salomaa, A. (eds.) Membrane Computing. LNCS, vol. 2933, pp. 180–189. Springer, Heidelberg (2004)
17. Freund, R., Păun, G., Rozenberg, G., Salomaa, A. (eds.): WMC 2005. LNCS, vol. 3850, pp. 18–21. Springer, Heidelberg (2006)
18. Goldbeter, A.: A minimal cascade model for the mitotic oscillator involving cyclin and cdc2 kinase. PNAS 88(20), 9107–9111 (1991)
19. Goldbeter, A.: Biochemical Oscillations and Cellular Rhythms. The Molecular Bases of Periodic and Chaotic Behaviour. Cambridge University Press, New York (2004)
20. Lodish, H.F., Berk, A., Matsudaira, P., Kaiser, C., Krieger, M., Scott, M., Zipursky, L., Darnell, J.E.: Molecular Cell Biology, 5th edn. Scientific American Press, New York (2004)
21. Manca, V.: Topics and problems in metabolic P systems. In: BWMC4. Proc. of the Fourth Brainstorming Week on Membrane Computing (2006)
22. Manca, V.: Discrete simulations of biochemical dynamics. In: Garzon, M., Yan, H. (eds.) Preliminary proceedings of the 13th International Meeting on DNA Computing, June 4-8, 2007, University of Memphis, Memphis, USA (2007)
23. Manca, V.: Metabolic dynamics by MP systems. In: InterLink ERCIM Workshop, Eze, France (May 10-12, 2007)
24. Manca, V.: Metabolic P systems for biochemical dynamics. Progress in Natural Sciences, Invited Paper (2007)
25. Manca, V.: The metabolic algorithm for P systems principles and applications. Theoretical Computer Science (to appear, 2007)
26. Manca, V., Bianco, L.: Biological networks in metabolic P systems. BioSystems (to appear, 2007)

27. Manca, V., Bianco, L., Fontana, F.: Evolutions and oscillations of P systems: The-oretical considerations and applications to biochemical phenomena. In: Mauri, G., Păun, G., Pérez-Jiménez, M.J., Rozenberg, G., Salomaa, A. (eds.) WMC 2004. LNCS, vol. 3365, pp. 63–84. Springer, Heidelberg (2005)
28. Martin-Vide, C., Păun, G., Rozenberg, G.: Membrane systems with carriers. The-oretical Computer Science 270, 779–796 (2002)
29. Milner, R.: Communicating and Mobile Systems: The $\pi$ Calculus. Cambridge Uni-versity Press, Cambridge, England (1999)
30. Pérez-Jiménez, M.J., Romero-Campero, F.J.: P systems: a new computational modelling tool for systems biology. In: Priami, C., Plotkin, G. (eds.) Transactions on Computational Systems Biology VI. LNCS (LNBI), vol. 4220, pp. 176–197. Springer, Heidelberg (2006)
31. Pescini, D., Besozzi, D., Mauri, G.: Investigating local evolutions in dynamical probabilistic p systems. In: Ciobanu, G., Păun, G. (eds.) Pre-Proc. of First Inter-national Workshop on Theory and Application of P Systems, Timisoara, Romania, pp. 83–90 (September 26-27, 2005)
32. Pescini, D., Besozzi, D., Mauri, G., Zandron, C.: Dynamical probabilistic P sys-tems. Inter. Journal of Foundations of Computer Science 17(1), 183 (2006)
33. Păun, A., Păun, G.: The power of communication: P systems with sym-port/antiport. New Generation Computing 20(3), 295–306 (2002)
34. Păun, G.: Computing with membranes. J. Comput. System Sci. 61(1), 108–143 (2000)
35. Păun, G.: P systems with active membranes: Attacking NP-complete problems. Journal of Automata, Languages and Combinatorics 1(6), 75–90 (2001)
36. Păun, G.: Membrane Computing. An Introduction. Springer, Berlin (2002)
37. Păun, G., Suzuki, Y., Tanaka, H.: P systems with energy accounting. Int. J. Com-puter Math. 78(3), 343–364 (2001)
38. Reisig, W.: Petri Nets, An Introduction. EATCS, Monographs on Theoretical Com-puter Science (1985)
39. The P Systems Web Site. Url: http://psystems.disco.unimib.it

# Modeling the Dynamics of HIV Infection with Conformon-P Systems and Cellular Automata

Pierluigi Frisco and David Wolfe Corne

School of Mathematical and Computer Sciences
Heriot-Watt University, Edinburgh, EH14 4AS, UK
{pier,dwcorne}@macs.hw.ac.uk

**Abstract.** Further results on the study of the dynamics of HIV infection with grids of conformon-P systems are reported. This study clearly shows a subdivision in two phases of the mechanism at the base of the considered dynamics.

## 1   Introduction

The infection by the human immune-deficiency virus (HIV), the cause of acquired immunodeficiency syndrome (AIDS), has been widely studied both in the laboratory and with computer models in order to understand the different aspects that regulate the virus-host interaction.

Several mathematical models have been proposed (see, for example, [14,18,11]) but all of them fail to describe some aspects of the infection. The recent model reported by Dos Santos & Coutinho in [4], based on cellular automata (CA), clearly shows the different time scales of the infection and has a broad qualitative agreement to the density of healthy and infected cells observed *in vivo*. However, in [15] it is noted that this qualitative agreement is reached only if some parameters are chosen in a small interval. If some of the parameters are chosen outside this interval, then the cellular automata model of [4] does not follow the dynamics of what is observed *in vivo*.

In the present paper we continue our study on the modeling of the dynamics of HIV infection with grids of conformon-P systems started in [2]. There our model proved to be more robust than the CA model to a wide range of conditions and parameters, with more reproducible qualitative agreement to the overall dynamics and to the densities of healthy and infected cells observed *in vivo*.

## 2   The Modeling Platforms

### 2.1   Cellular Automata

Cellular automata are a regularly used platform for modeling, and are increasingly explored as modeling tools in the context of natural phenomena that exhibit characteristic spatiotemporal dynamics [16,3]. Of interest here, for example, are their use in modeling the spread of infection [1,12,4,11,17].

G. Eleftherakis et al. (Eds.): WMC8 2007, LNCS 4860, pp. 21–31, 2007.

A CA consists of a finite number of cells (invariably arranged in a regular spatial grid), each of which can be in one of a finite (typically small) number of specific states. In the usual approach, at each time step $t$ the status of the CA is characterized by its state vector; that is, the state of each of the cells. In the simplest type of CA, the state vector at time $t+1$ is obtained from that at time $t$ by the operation of a single rule applied in parallel (synchronously) to each cell. The rule specifies how the state of a cell will change as a function of its current state and the states of the cells in its neighborhood (see Figure 4).

In many applications, including that addressed here, it is appropriate for the rule to be probabilistic.

The straightforward nature of the time evolution of a CA, combined with its emphasis on local interactions, has made it an accessible and attractive tool for modeling many biological processes.

## 2.2  Conformon-P Systems

*Conformon-P systems* (cP systems) [6] have been introduced as a novel computational device (*P systems* are the chief systems arising in the emerging research area of *Membrane Computing* [13]) whose early inspiration comes from a theoretical model of the living cell.

CP systems are defined in an extremely simple way that does not limit either their computational power, or their modeling capabilities. As a variant of P systems, they capture the dynamics of interacting processes in a novel way, using constructs that characterize the flow of information between regions in a range of cell-like topological structures. Moreover, their definition allows them to model different kinds of process (a compartment defines locality in general, it is not necessarily a membrane compartment in a cell) and to integrate several degrees of abstraction in the same system.

P systems are well-defined models of parallel computational systems that have a rich and growing base [19] of theoretical understanding of their properties.

A cP system has conformons, a name-value pair, as objects. If $V$ is an alphabet (a finite set of letters) and $\mathbb{N}_0$ is the set of natural numbers (with 0 included), then we can define a conformon as $[\gamma, a]$, where $\gamma \in V$ and $a \in \mathbb{N}_0$, we will say that $\gamma$ is the *name* and $a$ is the *value* of the conformon $[\gamma, a]$. If, for instance, $V = A, B, C, \ldots, Z$, then $[A, 5], [C, 0], [Z, 14]$ are conformons, while $[AB, 21], [C, -15]$, and $[D, 0.5]$ are not.

Two conformons can interact according to an *interaction rule*. An interaction rule is of the form $\gamma \xrightarrow{n} \beta$, where $\gamma, \beta \in V$ and $n \in \mathbb{N}_0$, and it says that a conformon with name $\gamma$ can give $n$ from its value to the value of a conformon having name $\beta$. A rule can be applied only if the value of the conformon with name $\gamma$ is greater or equal to $n$. If, for instance, there are conformons $[G, 5]$ and $[R, 9]$ and the rule $G \xrightarrow{3} R$, the application of $r$ leads to $[G, 2]$ and $[R, 12]$.

The (membrane) compartments present in a cP system have a label (it is a name which makes it easier to refer to a compartment), every label being different. Compartments can be unidirectionally connected to each other and for each connection there is a *predicate*. A predicate is an element of the set

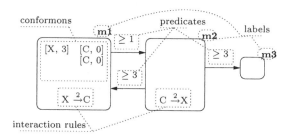

**Fig. 1.** A cP system

$\{\geq n, \leq n \mid n \in \mathbb{N}_0\}$. Examples of predicates are: $\geq 5, \leq 2$, etc. A connection and its predicate are referred as *passage rules*. If, for instance, there are two compartments (with labels) $m_1$ and $m_2$ and there is a passage rule from $m_1$ to $m_2$ having predicate $\geq 4$, then conformons having values greater than or equal to 4 can pass from $m_1$ to $m_2$. In a time unit any number of conformons can move between two connected membranes as long as the predicate of the passage rule is satisfied. Notice that we have *unidirectional passage rules* that is: $m_1$ connected to $m_2$ does not imply that $m_2$ is connected to $m_1$. Moreover, each passage rule has its own predicate. If, for instance, $m_1$ is connected to $m_2$ and $m_2$ is connected to $m_1$, the two connections can have different predicates.

A simple cP system is illustrated in Figure 1.

CP systems do not work under the requirement of maximal parallelism, typical to the majority of the models of P systems. When used as modeling platform cP systems can be classified as stochastic descriptive dynamic discrete models based on a discrete spatial heterogeneity. CP systems have been successfully used to model biological processes [8,2].

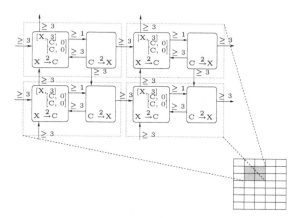

**Fig. 2.** A grid of cP systems

A *grid of cP systems* (Figure 2) is composed by *cells*, each cell being a simple conformon-P system connected to some other cells, the *neighborhood* of the cell.

Ongoing research is establishing the computational properties of (models of) cP systems [9,10,6,7,5].

CP systems can contain *modules*: groups of membranes with conformons and interaction rules able to perform a specific task. The task performed by a module can be considered atomic (i.e., completed in one time unit) in the context of the cP system containing it. Modules allow cP systems to be scalable.

Some modules are: *Splitter, Separator, Decreaser/Increaser* [6]. The combination of Separators and Decreaser/Increaser allows us to define *strict interaction* rule: $\gamma^{(a)} \xrightarrow{c} \beta_{(b)}$ where $\gamma, \beta \in V$, $a, b, c \in \mathbb{N}_0$, meaning that a conformon with name $\gamma$ can interact with $\beta$ passing just $c$ only if the value of $\gamma$ and $\beta$ before the interaction is $a$ and $b$ respectively. Notice that in a strict interaction just $c$ is passed even if the value of $\gamma$ could be decreased by any multiple of $c$. Interactions of the kind $\gamma \xrightarrow{c} \beta_{(b)}$ (before the interaction $\gamma$ can have any value while $\beta$ has $b$ as value) and $\gamma^{(a)} \xrightarrow{c} \beta$ (before the interaction $\gamma$ has $a$ as value while $\beta$ can have any value) can be defined, too.

## 3   The Process and the Models

We went on investigating the dynamics of HIV infection starting from the description of this process as present in [4,15].

The basic model of cP system we used is similar in the operation to the one reported in [2]. The differences are in the used probabilities and in the analysis we made.

The dynamics observed in HIV infections can be divided into three phases. Initially the amount of virus in the host grows in an exponential way, then the viral load drops to what is known as the "set point". Finally the amount of virus in the host increases slowly, accelerating near the onset of AIDS. The first two phases occur in the first weeks following the infection; the third phase can last years. This is plotted in Figure 3 where each unit in the $x$ axes represent a week in time.

In [4] this process was modeled with a CA in which each cell could be in any of four possible states: *healthy, A-infected, AA-infected*, and *dead*. In the (random) initial configuration a cell had probability $p_{HIV}$ to be *A-infected*, otherwise it is *healthy*.

The rules used in [4] are:

1. if an healthy cell has at least one A1-infected neighbor, then it becomes A1-infected;
2. if an healthy cell has not A1-infected neighbors but it has at least $2 < R < 8$ A2-infected neighbors, then it becomes A1-infected;
3. an A1-infected cell becomes A2-infected after $\tau$ time steps;
4. A2-infected cells become dead cells;
5. dead cells can become (are replaced by) healthy cells with probability $p_{repl}$;

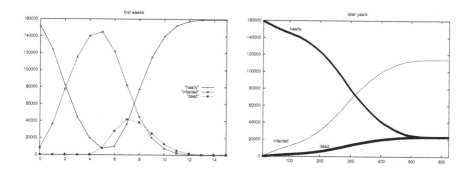

**Fig. 3.** Typical dynamics of HIV infection

6. newly introduced healthy cells can become A1-infected with probability $p_{infec}$.

The biological reasoning behind these rules is explained in [4]. Essentially, rules I and II model the basic spread of viral infection from cells to neighboring cells; rules III–VII model the short life of an infected cell, and the last rule models the body's continual replenishment of new healthy cells but maintaining a small probability of infection.

In [4] the following parameters were chosen: $p_{HIV} = 0.05$, $p_{repl} = 0.99$, $p_{infec} = 10^{-5}$, $R = 5$, and $\tau = 4$. They experimented with grids of size ranging from $300 \times 300$ to $1000 \times 1000$, and the averaged results of 500 simulations reported in [4] on toroidal grids ranging from $700 \times 700$ show a qualitative agreement to the density of healthy and infected cells observed *in vivo*.

In [15] it is shown that this qualitative agreement is reached only for values of the parameters close to the ones just indicated. If either $p_{HIV} < 10^{-2}$ or $p_{infec}$ is chosen in the range $10^{-2}$ to $10^{-4}$, then the CA model of [4] does not follow the dynamics of what is observed *in vivo*.

### 3.1  The CA Model

Our CA model followed the implementation of [4] with some minor differences. The main difference was that the state *A-infected* was represented by four separate states, *A-infected1*, *A-infected2*, *A-infected3*, and *A-infected4*. This enabled us to control the transition from *A-infected* to *AA-infected* over $\tau = 4$ time steps within a 'pure' CA framework.

The CA rules were as follows, in which we use *A-infected* as shorthand for the union of the four aforementioned states.

1. if a *healthy* cell has at least one *A-infected* neighbor, then it becomes *A-infected1* at the next time step;
2. if a *healthy* cell has no *A-infected* neighbors but at least $R$ *AA-infected* neighbors, then it becomes *A-infected1* at the next time step;

3. an *A-infected1* cell becomes *A-infected2* at the next time step;
4. an *A-infected2* cell becomes *A-infected3* at the next time step;
5. an *A-infected3* cell becomes *A-infected4* at the next time step;
6. an *A-infected4* cell becomes *AA-infected* at the next time step;
7. an *AA-infected* cell becomes *dead* at the next time step;
8. a *dead* cell becomes, at the next time step, either *healthy* (with probability $p_{repl} \times (1 - p_{infec})$), or *A-infected1* (with probability $p_{repl} \times p_{infec}$), or stays *dead* (with probability $1 - p_{repl}$).

As in [4], the first two model the basic spread of viral infection from cells to neighbouring cells, while most of the remainder model the short life of an infected cell, and the last rule models the continual replenishment of new healthy cells, but maintaining a small probability of infection.

In [4] a parameter $\tau$ is used for indicating the number of time-steps after which an *A-infected* cell became *AA-infected*, and maintained $\tau = 4$ throughout their work. This parameter setting is reflected in our use of four separate *A-infected* states.

## 3.2  The Grid of cP System Model

The main difference that our model has in respect to the one reported in [4] is that the interaction rules are divided in two subsets: *part 1* and *part 2* (see Appendix A). The rules in the two subsets differ in the probabilities associated to them.

Other differences as, for instance, the presence of *pre-dead* cells, were considered in order to simulate in terms of operations in a cP system some instructions of the CA presented in [4].

Each cell can be in one of five states: *1-healthy*, *A-infected*, *AA-infected*, *pre-dead*, and *dead* (in respect to the rules in *part 1*) identified by the presence of the conformons: $[H, 1], [A, 1], [AA, 1], [PD, 1]$, and $[D, 1]$ respectively. If, for instance, a cell is in an *healthy* state, then it will contain $[H, 1], [A, 0], [AA, 0]$, $[PD, 0]$, and $[D, 0]$ (similarly for the other cases). In the initial configuration, each cell contains the conformons $([R, 1], +\infty), ([V, 10], +\infty), ([E, 0], +\infty)$, and $([W, 0], +\infty)$.

In the following we consider and describe the rules in *part 1*.

If a cell is *A-infected*, then it can generate $[V, 11]$ (meaning: if a cell is *A-infected* it can generate a virus). This is performed by the rules:

$$\textbf{1: } R \xrightarrow{1} A_{(1)} \qquad\qquad \textbf{2: } A^{(2)} \xrightarrow{1} V_{(10)}$$

Notice that $[V, 10]$ does not represent a virus, but $[V, 11]$ does.

$[V, 11]$ conformons can pass from a cell to any other in its neighborhood (meaning: viruses can spread between cells).

An *1-healthy* cell can become *A-infected* if it contains a virus. This is performed by the rules:

$$\textbf{3: } V \xrightarrow{11} H_{(1)} \qquad \textbf{4: } H^{(12)} \xrightarrow{12} A_{(0)} \qquad \textbf{5: } A^{(12)} \xrightarrow{11} W_{(0)}$$

An *AA-infected* cell can generate $[E, 1]$ conformons. These conformons can pass to other cells and interact such that $[E, 4]$ conformons are created. When a $[E, 4]$ conformon is present in an *healthy* cell, then it can become *A-infected*. This process mimics rule II in Section 3 and it is performed by:

**6:** $R \xrightarrow{1} AA_{(1)}$    **7:** $AA^{(2)} \xrightarrow{1} E_{(0)}$    **8:** $E^{(1)} \xrightarrow{1} E_{(1)}$    **9:** $E^{(2)} \xrightarrow{2} E_{(2)}$

**10:** $E \xrightarrow{4} H_{(1)}$       **11:** $H^{(5)} \xrightarrow{5} A_{(0)}$       **12:** $A^{(5)} \xrightarrow{4} W_{(0)}$

and by the fact that $[E, 1]$ can pass from one cell to any other in its neighborhood. From rules 7, 8, and 9 it should be clear that only $[E, 1], [E, 2]$, and $[E, 4]$ can be present in the system. Because of rule 6 an *AA-infected* cell can generate $[E, 1]$. When two $[E, 1]$ are present in the same cell they can interact to create $[E, 2]$ (rule 8) and two $[E, 2]$ present in the same cell can interact to create $[E, 4]$ (rule 9). If the creation of $[E, 4]$ took place in an *healthy* cell, then this cell can become *A-infected* (rules 10, 11 and 12).

An *A-infected* cell can become *AA-infected* by the application of the rule:

**13:** $A^{(1)} \xrightarrow{1} AA_{(0)}$

An *AA-infected* cell can become *dead*. Before doing so it goes into the *pre-dead* state in which the $[V, 11], [E, 1], [E, 2]$, and $[E, 4]$ conformons present in it are removed. This is performed by the rules:

**14:** $AA^{(11)} \xrightarrow{1} PD_{(0)}$  **15:** $V^{(11)} \xrightarrow{1} PD_{(1)}$  **16:** $E \xrightarrow{1} PD_{(1)}$  **17:** $E \xrightarrow{2} PD_{(1)}$

**18:** $E \xrightarrow{4} PD_{(1)}$  **19:** $PD^{(1)} \xrightarrow{1} D_{(0)}$  **20:** $PD^{(2)} \xrightarrow{1} W_{(0)}$  **21:** $PD^{(3)} \xrightarrow{2} W_{(0)}$

**22:** $PD^{(5)} \xrightarrow{4} W_{(0)}$

A *dead* cell can become *2-healthy* cell by the application of the rule

**23:** $D^{(1)} \xrightarrow{1} H2_{(0)}$

The $R$ and $W$ conformons do not have a direct relationship with any aspect of HIV infection. In broad terms, the $R$ conformons can be regarded as 'food' molecules needed by a cell in a certain state to perform an action (for instance, if *A-infected* to generate a virus). The $W$ conformons can be regarded as 'waste' molecules, to which some conformons can pass part of their value. As $W$ conformons only receive values from other conformons, their initial value is not relevant for the simulation.

The state *2-healthy*, together with *A2-infected*, *AA2-infected*, *2-pre-dead*, and *2-dead* are managed by the rules in *part 2*. The rules in *part 2* are similar to the ones in *part 1* but they have $H2$ instead of $H$, $A2$ instead of $A$, $AA2$ instead of $AA$, $PD2$ instead of $PD$, and $D2$ instead of $D$.

In the diagrams related to the grid of cP systems the curve of *healthy* cells is obtained adding up the number of $H$ and $H2$ cells; the curve of *infected* cells is obtained adding up the number of $A$, $AA$, $A2$, and $AA2$ cells; the curve of *dead* cells is obtained adding up the number of $D$, $PD$, $D2$, and $PD2$ cells.

The interaction rules indicated in Appendix A can be logically divided in two sets: *state-change* and *internal dynamics*. The *state-change* rules allow the cells to pass from a state to another. For instance, rule 4 is a state change rule as when

it is applied in a cell the state of the cell passed from *1-healthy* to *A-infected*.
The *state-change* rules are: 4, 11, 13, 14, 19, 23, 27, 32, 34, 35, 40, and 44.

The remaining rules belong to *internal dynamics* as they do not directly effect
the state of a cell.

Differently from what was done in [2], in the present study the probabilities
associated to the *internal dynamics* rules in phase 1 are equal to the ones in
phase 2. The probabilities of the *state-change* rules in phase 1 are higher than
the ones in phase 2.

Considering what we said in Section 3, rules in *part 1* model the behavior of
the first two phases of the dynamics of HIV infection, while rules in *part 2* model
the behavior of the third phase.

## 4   Experiments and Results

The simulations performed with the cP system were based on a toroidal 50×50
grid, using a Moor neighborhood (considering Figure 4 a black cell can pass
conformons to any other of the grey cells) and with $p_{HIV} = 0.05$.

**Fig. 4.** The Moore neighborhood

All the 10 simulations (with different random number sequences) run for 16000
iterations and they all show a dynamics very similar to the one observed *in vivo*.
A typical outcome is depicted in Figure 5.

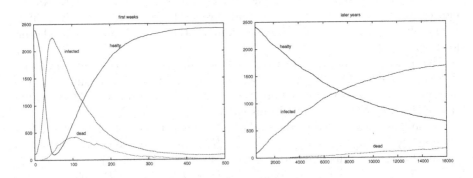

**Fig. 5.** Typical outcome for grids of cP systems

This outcome (even if run for a few configurations) fits the dynamics observed
*in vivo* better than the outcomes reported in [2]:

The tempo of the dynamics is constant during the simulation. In [2] the dynamics was 'too fast' in the later years (or 'too slow' in the first weeks). In the present study 1 year corresponds to 1560 iterations. This means that phase I and phase II (both taking place in at most 10 weeks) should correspond to 300 iterations. In this way the 16000 iterations of out tests correspond to a bit more than 10 years.

The percentage of *healthy* and *infected* cells in phase III is closer to what observed *in vivo* than what reported in [2].

The dynamics of *healthy* and *infected* cells in phase III is not flat as in [2] but shows a concavity similar to the one observed *in vivo*.

There are two major differences between the dynamics obtained by us and the one observed *in vivo*:

In phase III the number of *healthy* cells should become equal to the one of *dead* cells;

The curves followed by the number of *healthy* and *infected* cells in phase III do not change concavity.

## 5   Final Remarks

We consider the reported study still in its initial phases. In the future we will try to fit better the dynamics obtained with grids of cP systems to that observed *in vivo* and we will run the tests on different neighborhoods and different percentage of initially infected cells (as done in [2]).

Some results obtained by us indicates that the $E$ conformons do not play an important role in the whole dynamics, their effect is negligible. On this basis we will try to simplify our model in the number of interaction rules and conformons present in it.

## References

1. Ahmed, E., Agiza, H.N., Hassan, S.Z.: On modelling hepatitis B transmission using cellular automata. J. Stat. Phys. 92(3/4) (1998)
2. Corne, D.W., Frisco, P.: Dynamics of HIV infection studied with cellular automata and conformon-P systems. BioSystems (to appear)
3. Deutsch, A., Dormann, S.: Cellular Automaton Modeling of Biological Pattern Formation: Characterization, Applications, and Analysis, Birkäusen, Boston (2004)
4. Dos Santos, R.M., Coutinho, S.: Dynamics of HIV infection: a cellular automata approach. Physical Review Letters 87(16), 168102 (2001)
5. Frisco, P.: Conformon-p systems with negative values. In: Eleftherakis, G., Kefalas, P., Păun, G., Rozenberg, G., Salomaa, A. (eds.) WMC 2007. LNCS, vol. 4860, Springer, Heidelberg (this volume, 2007)
6. Frisco, P.: The conformon-P system: A molecular and cell biology-inspired computability model. Theoretical Computer Science 312(2-3), 295–319 (2004)
7. Frisco, P.: Infinite hierarchies of conformon-P systems. In: Hoogeboom, H.J., Păun, G., Rozenberg, G., Salomaa, A. (eds.) WMC 2006. LNCS, vol. 4361, pp. 395–408. Springer, Heidelberg (2006)

8. Frisco, P., Gibson, R.T.: A simulator and an evolution program for conformon-P systems. In: SYNASC 2005. 7th International Symposium on Simbolic and Numeric Algorithms for Scientific Computing, Workshop on Theory and Applications of P Systems, TAPS, Timisoara (Romania), September 26-27, 2005, pp. 427–430. IEEE Computer Society, Los Alamitos (2005)

9. Frisco, P., Ji, S.: Conformons-P systems. In: Hagiya, M., Ohuchi, A. (eds.) DNA Computing. LNCS, vol. 2568, pp. 291–301. Springer, Heidelberg (2003)

10. Frisco, P., Ji, S.: Towards a hierarchy of info-energy P systems. In: Păun, G., Rozenberg, G., Salomaa, A., Zandron, C. (eds.) Membrane Computing. LNCS, vol. 2597, pp. 302–318. Springer, Heidelberg (2003)

11. Kamp, C., Bornholdt, S.: From HIV infection to AIDS: a dynamically induced percolation transition? Proceedings of the Royal Society B: Biological Sciences 269(1504), 2035–2040 (2002)

12. Martins, M.L., Ceotto, G., Alves, S.G., Bufon, C.C.B., Silva, J.M., Laranjeira, F.F.: Cellular automata model for citrus variegated chlorosis. Phys. Rev. E 62(5), 7024–7030 (2000)

13. Păun, G.: Membrane Computing. An Introduction. Springer, Berlin (2002)

14. Perelson, A.S., Nelson, P.W.: Mathematical analysis of HIV-1 dynamics in vivo. SIAM Review 41(1), 3–44 (1999)

15. Strain, M.C., Levine, H.: Comment on Dynamics of HIV infection: a cellular automata approach. Physical review letters 89(21), 219805 (2002)

16. Toffoli, T., Margolus, N.: Cellular Automata Machines: A New Environment for Modeling. MIT press, Cambridge (1987)

17. Venkatachalam, S., Mikler, A.: Towards computational epidemiology: Using stochastic cellular automata in modeling spread of diseases. In: Proceedings of the 4th Annual International Conference on Statistics, Mathematics and Related Fields (2005)

18. Wodarz, D., Nowak, M.A.: Mathematical models of HIV pathogenesis and treatment. Bioessays 24(12), 1178–1187 (2002)

19. Zandron, C.: P-systems web page: http://psystems.disco.unimib.it

# A   Rules, Links, and Probabilities

| part 1 | | | part 2 | | |
| --- | --- | --- | --- | --- | --- |
| label | rule | prob. | label | rule | prob. |
| 1 | $R \xrightarrow{1} A_{(1)}$ | 0.7071 | 24 | $R \xrightarrow{1} A2_{(1)}$ | 0.7071 |
| 2 | $A^{(2)} \xrightarrow{1} V_{(10)}$ | 0.7071 | 25 | $A2^{(2)} \xrightarrow{1} V_{(10)}$ | 0.7071 |
| 3 | $V \xrightarrow{11} H_{(1)}$ | 0.79 | 26 | $V \xrightarrow{11} H2_{(1)}$ | 0.79 |
| 4 | $H^{(12)} \xrightarrow{12} A_{(0)}$ | 0.79 | 27 | $H2^{(12)} \xrightarrow{12} A2_{(0)}$ | 0.0001 |
| 5 | $A^{(12)} \xrightarrow{11} W_{(0)}$ | 0.79 | 28 | $A2^{(12)} \xrightarrow{11} W_{(0)}$ | 0.79 |
| 6 | $R \xrightarrow{1} AA_{(1)}$ | 0.7071 | 29 | $R \xrightarrow{1} AA2_{(1)}$ | 0.7071 |
| 7 | $AA^{(2)} \xrightarrow{1} E_{(0)}$ | 0.7071 | 30 | $AA2^{(2)} \xrightarrow{1} E_{(0)}$ | 0.7071 |
| 8 | $E^{(1)} \xrightarrow{1} E_{(1)}$ | 0.07071 | | | |
| 9 | $E^{(2)} \xrightarrow{2} E_{(2)}$ | 0.07071 | | | |
| 10 | $E \xrightarrow{4} H_{(1)}$ | 0.79 | 31 | $E \xrightarrow{4} H2_{(1)}$ | 0.79 |
| 11 | $H^{(5)} \xrightarrow{5} A_{(0)}$ | 0.79 | 32 | $H2^{(5)} \xrightarrow{5} A2_{(0)}$ | 0.0001 |
| 12 | $A^{(5)} \xrightarrow{4} W_{(0)}$ | 0.79 | 33 | $A2^{(5)} \xrightarrow{4} W_{(0)}$ | 0.79 |
| 13 | $A^{(1)} \xrightarrow{1} AA_{(0)}$ | 0.04 | 34 | $A2^{(1)} \xrightarrow{1} AA2_{(0)}$ | 0.0001 |
| 14 | $AA^{(11)} \xrightarrow{1} PD_{(0)}$ | 0.1 | 35 | $AA2^{(11)} \xrightarrow{1} PD2_{(0)}$ | 0.00075 |
| 15 | $V^{(11)} \xrightarrow{1} PD_{(1)}$ | 0.7071 | 36 | $V^{(11)} \xrightarrow{1} PD2_{(1)}$ | 0.7071 |
| 16 | $E \xrightarrow{1} PD_{(1)}$ | 0.7071 | 37 | $E \xrightarrow{1} PD2_{(1)}$ | 0.7071 |
| 17 | $E \xrightarrow{2} PD_{(1)}$ | 0.7071 | 38 | $E \xrightarrow{2} PD2_{(1)}$ | 0.7071 |
| 18 | $E \xrightarrow{4} PD_{(1)}$ | 0.7071 | 39 | $E \xrightarrow{4} PD2_{(1)}$ | 0.7071 |
| 19 | $PD^{(1)} \xrightarrow{1} D_{(0)}$ | 0.2 | 40 | $PD2^{(1)} \xrightarrow{1} D2_{(0)}$ | 0.001 |
| 20 | $PD^{(2)} \xrightarrow{1} W_{(0)}$ | 0.7071 | 41 | $PD2^{(2)} \xrightarrow{1} W_{(0)}$ | 0.7071 |
| 21 | $PD^{(3)} \xrightarrow{2} W_{(0)}$ | 0.7071 | 42 | $PD2^{(3)} \xrightarrow{2} W_{(0)}$ | 0.7071 |
| 22 | $PD^{(5)} \xrightarrow{4} W_{(0)}$ | 0.7071 | 43 | $PD2^{(5)} \xrightarrow{4} W_{(0)}$ | 0.7071 |
| 23 | $D^{(1)} \xrightarrow{1} H2_{(0)}$ | 0.1 | 44 | $D2^{(1)} \xrightarrow{1} H2_{(0)}$ | 0.001 |

**Links:**
[V, 11] can pass with probability 1 from any cell to any of its neighbors; [E, 1] can pass with probability 0.01 from any cell to any of its neighbors.

# (UREM) P Systems with a Quantum-Like Behavior: Background, Definition, and Computational Power

Alberto Leporati

Dipartimento di Informatica, Sistemistica e Comunicazione
Università degli Studi di Milano – Bicocca
Via Bicocca degli Arcimboldi 8, 20126 Milano, Italy
alberto.leporati@unimib.it

**Abstract.** Q-UREM P systems constitute an attempt to introduce and exploit in Membrane Computing notions and techniques deriving from quantum mechanics. As we will see, the approach we have adopted is different from what is usually done in Quantum Computing; in fact, we have been inspired by the functioning of creation and annihilation operators, that are sometimes used in quantum mechanics to exchange quanta of energy among physical systems. In this paper we will provide the background which has led to the current definition of Q-UREM P systems, and we will recall some results concerning their computational power.

## 1 The Quest for Quantum P Systems

*Membrane systems* (also known as P systems) have been introduced by Gheorghe Păun in 1998 [24] as a new class of distributed and parallel computing devices, inspired by the structure and functioning of living cells. The basic model consists of a hierarchical structure composed by several membranes, embedded into a main membrane called the *skin*. Membranes divide the Euclidean space into *regions*, that contain some *objects* (represented by symbols of an alphabet) and *evolution rules*. Using these rules, the objects may evolve and/or move from a region to a neighboring one. A *computation* starts from an initial configuration of the system and terminates when no evolution rule can be applied. Usually, the result of a computation is the multiset of objects contained into an *output membrane* or emitted from the skin of the system.

In what follows we assume the reader is already familiar with the basic notions and the terminology underlying P systems. For a layman–oriented introduction to P systems see [26], whereas for a systematic introduction we refer the reader to [25]. The latest information about P systems can be found in [29].

At the beginning of 2004, the Membrane Computing community started to query about the possibility to define a quantum version of P systems, and hence I started to work on the subject with some colleagues. A first paper [18] was presented in Palma de Mallorca in November 2004. There, we proposed two

G. Eleftherakis et al. (Eds.): WMC8 2007, LNCS 4860, pp. 32–53, 2007.

options: either to follow the steps usually performed in Quantum Computing to define the quantum version of a given computation device, or to propose a completely new computation device which is based on the most elementary operation which can be conceived in physics: the exchange of a quantum of energy among two quantum systems. In the former case we would have obtained yet another quantum computation device whose computation steps are defined as the action of unitary operators, whose computations are logically reversible, and in which there are severe constraints on the amount of information which can be extracted from the system by measuring its state. In the latter case, instead, we felt that a new and interesting computation device could be introduced. Indeed, after a long and careful investigation, we decided to adopt creation and annihilation operators as the most elementary operations which can be performed by our computation device.

It was since 2001 that several authors introduced the notion of energy in P systems [1,7,28,12,19,20]. Hence, we looked at the literature to find a model of P systems that could be easily transformed to a quantum computation device. Our first choice, explored in [18], was to focus on energy–based P systems, in which a given amount of energy is associated to each object of the system. Moreover, instances of a special symbol $e$ are used to denote free energy units occurring into the regions of the system. These energy units can be used to transform objects, using appropriate rules. The rules are defined according to conservativeness considerations. Indeed, in [18] we proposed two different versions of quantum P systems based on this classical model. Both versions were defined just like classical energy–based P systems, but for objects and rules. Objects were represented as pure states in the Hilbert space $\mathbb{C}^d$, for appropriate integers $d \geq 2$, whereas the definition of rules differs between the two models. In the former, rules are defined as bijective functions that operate on the objects of the alphabet; these functions are then implemented as unitary operators. In the latter, rules are defined as generic functions which map the alphabet into itself. Such functions are implemented using a generalization of the Conditional Quantum Control technique [2], and may yield to non unitary operators (a fact which is usually seen with suspicion in traditional Quantum Computing).

Several problems were also pointed out in [18], the most serious being that it is difficult to avoid undesired exchanges of energy among the objects, that yield the system to unintended states. Another difficulty was tied to the assignment of the amount of energy to every object of the system. In the original definition of energy–based P systems, every object incorporated a different amount of energy; in other words, the amount of energy uniquely determined the kind of object and, by acquiring or releasing energy from the environment, one object was transformed to another kind of object. Under this definition, we were able in [19] to simulate a single Fredkin gate. However, in order to simulate an entire Fredkin circuit [20,21] we were forced to relax the definition, and allow different kinds of objects to have the same amount of energy, otherwise the number of different kinds of objects would have become unmanageable.

Looking for some alternatives, we considered the model of P systems introduced in [8], in which a non–negative integer value is assigned to each membrane. Such a value can be conveniently interpreted as the *energy* of the membrane. In these P systems, rules are assigned to the membranes rather than to the regions of the system. Every rule has the form $(in_i : \alpha, \Delta e, \beta)$ or $(out_i : \alpha, \Delta e, \beta)$, where $i$ is the number of the membrane in a one-to-one labeling, $\alpha$ and $\beta$ are symbols of the alphabet and $\Delta e$ is a (possibly negative) integer number. The rule $(in_i : \alpha, \Delta e, \beta)$ is interpreted as follows: if a copy of $\alpha$ is in the region immediately surrounding membrane $i$, then this object crosses membrane $i$, is transformed to $\beta$, and modifies the energy of membrane $i$ from the current value $e_i$ to the new value $e_i + \Delta e$. Similarly, the rule $(out_i : \alpha, \Delta e, \beta)$ is interpreted as follows: if a copy of $\alpha$ is in the region surrounded by membrane $i$, then this object crosses membrane $i$, is transformed to $\beta$, and modifies the energy of membrane $i$ from the current value $e_i$ to the new value $e_i + \Delta e$. Both kinds of rules can be applied only if $e_i + \Delta e$ is non–negative. Since these rules transform one copy of an object to (one copy of) another object, in [8] they are referred to as *unit* rules. Hence, for conciseness, this model of P systems with unit rules and energy assigned to membranes has been abbreviated as *UREM P systems*. An important observation is that in [8] the rules of UREM P systems are applied in a *sequential* way: at each computation step, *one* rule is selected from the pool of currently active rules, and is applied. In [8] it has been proved that if we assign some local (that is, affecting only the membrane in which they are defined) priorities to the rules then UREM P systems are Turing complete, whereas if we omit the priorities then we do not get systems with universal computational power: indeed, we obtain a characterization of $PsMAT^\lambda$, the family of Parikh sets generated by context-free matrix grammars (without occurrence checking and with $\lambda$-rules).

Finally, in [17] a *quantum-like* version of UREM P systems (here referred to as Q-UREM P systems, for short) has been introduced, and it has been shown that such a model of computation is able to compute every partial recursive function (that is, it reaches the computational power of Turing machines) without the need to assign any priority between the rules of the system. In Q-UREM P systems, the rules $(in_i : \alpha, \Delta e, \beta)$ and $(out_i : \alpha, \Delta e, \beta)$ are realized through (not necessarily unitary) linear operators, which can be expressed as an appropriate composition of a truncated version of creation and annihilation operators. The operators which correspond to the rules have the form $|\beta\rangle \langle\alpha| \otimes O$, where $O$ is a linear operator which modifies the energy associated with the membrane (implemented as the state of a truncated quantum harmonic oscillator).

In [17] we also introduced a quantum-like version of register machines (QRMs, for short). It is our opinion that they could play the same role in proofs concerning the computational power of quantum computation devices as played by classical register machines for classical computing devices. Indeed, it has been shown in [17] that they are able to simulate any classical (deterministic) register machine, and hence they are (at least) Turing complete.

Subsequently, in [16] we have shown that, under the assumption that an external observer is able to discriminate a null vector from a non–null vector, the

**NP**–complete problem 3-SAT can be solved using quantum (Fredkin) circuits (built using the non–unitary creation and annihilation operators), QRMs and Q-UREM P systems. Precisely, for each type of computation device we have proposed a *brute force* technique that exploits quantum parallelism (as well as the ability to alter quantum states by using creation and annihilation operators) to explore the whole space of assignments to the boolean variables of any given instance $\phi$ of 3-SAT, in order to determine whether at least one of such assignments satisfies $\phi$. The solutions are presented in the so-called *semi–uniform* setting, which means that for every instance $\phi$ of 3-SAT a specific computation device (circuit, register machine or UREM P system) that solves it is built. Even if it is not formally proved, it is apparent that the proposed constructions can be performed in polynomial time by a classical deterministic Turing machine (whose output is a "reasonable" encoding of the machine, in the sense given in [13]).

In the rest of the paper we overview the basic notions of quantum mechanics which have led to the definition of Q-UREM P systems, and the results obtained so far about their computational power. Precisely, in section 2 we recall some basic notions on quantum computers, and we extend them to quantum systems which are able to assume a generic number $d \geq 2$ of base states. We also introduce creation and annihilation operators, by first giving a mathematical description and then illustrating two possible physical interpretations. In sections 3 and 4 we give the precise definitions of both classical and quantum-like register machines and UREM P systems, respectively. In section 5 we prove that Q-UREM P systems are able to compute any partial recursive function, and hence they are (at least) as powerful as Turing machines. In section 6 we show how to build two families of QRMs and Q-UREM P systems, respectively, that solve (in the semi–uniform setting) the **NP**–complete decision problem 3-SAT. Finally, section 7 contains some directions for future research.

## 2    Quantum Computers

From an abstract point of view, a quantum computer can be considered as made up of interacting parts. The elementary units (memory cells) that compose these parts are two–level quantum systems called *qubits*. A qubit is typically implemented using the energy levels of a two–level atom, or the two spin states of a spin–$\frac{1}{2}$ atomic nucleus, or a polarization photon. The mathematical description — independent of the practical realization — of a single qubit is based on the two–dimensional complex Hilbert space $\mathbb{C}^2$. The boolean truth values 0 and 1 are represented in this framework by the unit vectors of the canonical orthonormal basis, called the *computational basis* of $\mathbb{C}^2$:

$$|0\rangle = \begin{bmatrix} 1 \\ 0 \end{bmatrix} \qquad\qquad |1\rangle = \begin{bmatrix} 0 \\ 1 \end{bmatrix}$$

Qubits are thus the quantum extension of the classical notion of bit, but whereas bits can only take two different values, 0 and 1, qubits are not confined to their two basis (pure) states, $|0\rangle$ and $|1\rangle$, but can also exist in states which are coherent superpositions such as $\psi = c_0 \left|0\right\rangle + c_1 \left|1\right\rangle$, where $c_0$ and $c_1$ are complex numbers satisfying the condition $|c_0|^2 + |c_1|^2 = 1$. Performing a measurement of the state alters it. Indeed, performing a measurement on a qubit in the above superposition will return 0 with probability $|c_0|^2$ and 1 with probability $|c_1|^2$; the state of the qubit after the measurement (*post–measurement state*) will be $\left|0\right\rangle$ or $\left|1\right\rangle$, depending on the outcome.

A *quantum register* of size $n$ (also called an $n$–*register*) is mathematically described by the Hilbert space $\otimes^n \mathbb{C}^2 = \underbrace{\mathbb{C}^2 \otimes \ldots \otimes \mathbb{C}^2}_{n \text{ times}}$, representing a set of $n$ qubits labeled by the index $i \in \{1, \ldots, n\}$. An $n$–*configuration* (also *pattern*) is a vector $\left|x_1\right\rangle \otimes \ldots \otimes \left|x_n\right\rangle \in \otimes^n \mathbb{C}^2$, usually written as $\left|x_1, \ldots, x_n\right\rangle$, considered as a quantum realization of the boolean tuple $(x_1, \ldots, x_n)$. Let us recall that the dimension of $\otimes^n \mathbb{C}^2$ is $2^n$ and that $\{\left|x_1, \ldots, x_n\right\rangle : x_i \in \{0, 1\}\}$ is an orthonormal basis of this space called the $n$–register *computational basis*.

Computations are performed as follows. Each qubit of a given $n$–register is prepared in some particular pure state ($\left|0\right\rangle$ or $\left|1\right\rangle$) in order to realize the required $n$–configuration $\left|x_1, \ldots, x_n\right\rangle$, quantum realization of an input boolean tuple of length $n$. Then, a linear operator $G : \otimes^n \mathbb{C}^2 \rightarrow \otimes^n \mathbb{C}^2$ is applied to the $n$–register. The application of $G$ has the effect of transforming the $n$–configuration $\left|x_1, \ldots, x_n\right\rangle$ into a new $n$–configuration $G(\left|x_1, \ldots, x_n\right\rangle) = \left|y_1, \ldots, y_n\right\rangle$, which is the quantum realization of the output tuple of the computer. We interpret such modification as a computation step performed by the quantum computer. The action of the operator $G$ on a superposition $\Phi = \sum c^{i_1 \cdots i_n} \left|x_{i_1}, \ldots, x_{i_n}\right\rangle$, expressed as a linear combination of the elements of the $n$–register basis, is obtained by linearity: $G(\Phi) = \sum c^{i_1 \cdots i_n} G(\left|x_{i_1}, \ldots, x_{i_n}\right\rangle)$. We recall that linear operators which act on $n$–registers can be represented as order $2^n$ square matrices of complex entries. Usually (but not in this paper) such operators, as well as the corresponding matrices, are required to be unitary. In particular, this implies that the implemented operations are logically reversible (an operation is *logically reversible* if its inputs can always be deduced from its outputs).

All these notions can be easily extended to quantum systems which have $d > 2$ pure states. In this setting, the $d$–valued versions of qubits are usually called *qudits* [14]. As it happens with qubits, a qudit is typically implemented using the energy levels of an atom or a nuclear spin. The mathematical description — independent of the practical realization — of a single qudit is based on the $d$–dimensional complex Hilbert space $\mathbb{C}^d$. In particular, the pure states $\left|0\right\rangle, \left|\frac{1}{d-1}\right\rangle, \left|\frac{2}{d-1}\right\rangle, \ldots, \left|\frac{d-2}{d-1}\right\rangle, \left|1\right\rangle$ are represented by the unit vectors of the canonical orthonormal basis, called the *computational basis* of $\mathbb{C}^d$:

$$\left|0\right\rangle = \begin{bmatrix} 1 \\ 0 \\ \vdots \\ 0 \\ 0 \end{bmatrix}, \quad \left|\frac{1}{d-1}\right\rangle = \begin{bmatrix} 0 \\ 1 \\ \vdots \\ 0 \\ 0 \end{bmatrix}, \quad \cdots, \quad \left|\frac{d-2}{d-1}\right\rangle = \begin{bmatrix} 0 \\ 0 \\ \vdots \\ 1 \\ 0 \end{bmatrix}, \quad \left|1\right\rangle = \begin{bmatrix} 0 \\ 0 \\ \vdots \\ 0 \\ 1 \end{bmatrix}$$

As before, a *quantum register* of size $n$ can be defined as a collection of $n$ qudits. It is mathematically described by the Hilbert space $\otimes^n \mathbb{C}^d$. An $n$–*configuration* is now a vector $|x_1\rangle \otimes \ldots \otimes |x_n\rangle \in \otimes^n \mathbb{C}^d$, simply written as $|x_1, \ldots, x_n\rangle$, for $x_i$ running on $L_d = \left\{0, \frac{1}{d-1}, \frac{2}{d-1}, \ldots, \frac{d-2}{d-1}, 1\right\}$. An $n$–configuration can be viewed as the quantum realization of the "classical" tuple $(x_1, \ldots, x_n) \in L_d^n$. The dimension of $\otimes^n \mathbb{C}^d$ is $d^n$ and the set $\{|x_1, \ldots, x_n\rangle : x_i \in L_d\}$ of all $n$–configurations is an orthonormal basis of this space, called the $n$–*register computational basis*. Notice that the set $L_d$ can also be interpreted as a set of truth values, where 0 denotes falsity, 1 denotes truth and the other elements indicate different degrees of indefiniteness.

Let us now consider the set $\mathcal{E}_d = \left\{\varepsilon_0, \varepsilon_{\frac{1}{d-1}}, \varepsilon_{\frac{2}{d-1}}, \ldots, \varepsilon_{\frac{d-2}{d-1}}, \varepsilon_1\right\} \subseteq \mathbb{R}$ of real values; we can think to such quantities as energy values. To each element $v \in L_d$ we associate the energy level $\varepsilon_v$; moreover, let us assume that the values of $\mathcal{E}_d$ are all positive, equispaced, and ordered according to the corresponding objects: $0 < \varepsilon_0 < \varepsilon_{\frac{1}{d-1}} < \cdots < \varepsilon_{\frac{d-2}{d-1}} < \varepsilon_1$. If we denote by $\Delta\varepsilon$ the gap between two adjacent energy levels then the following linear relation holds:

$$\varepsilon_k = \varepsilon_0 + \Delta\varepsilon\,(d-1)\,k \qquad \forall\, k \in L_d \qquad (1)$$

Notice that it is not required that $\varepsilon_0 = \Delta\varepsilon$. As explained in [18,16], the values $\varepsilon_k$ can be thought of as the energy eigenvalues of the infinite dimensional quantum harmonic oscillator truncated at the $(d-1)$-th excited level, whose Hamiltonian on $\mathbb{C}^d$ is:

$$H = \begin{bmatrix} \varepsilon_0 & 0 & \cdots & & 0 \\ 0 & \varepsilon_0 + \Delta\varepsilon & \cdots & & 0 \\ \vdots & \vdots & \ddots & & \vdots \\ 0 & 0 & \cdots & & \varepsilon_0 + (d-1)\Delta\varepsilon \end{bmatrix} \qquad (2)$$

The unit vector $|H = \varepsilon_k\rangle = \left|\frac{k}{d-1}\right\rangle$, for $k \in \{0, 1, \ldots, d-1\}$, is the eigenvector of the state of energy $\varepsilon_0 + k\Delta\varepsilon$. To modify the state of a qudit we can use creation and annihilation operators on the Hilbert space $\mathbb{C}^d$, which are defined respectively as:

$$a^\dagger = \begin{bmatrix} 0 & 0 & \cdots & 0 & 0 \\ 1 & 0 & \cdots & 0 & 0 \\ 0 & \sqrt{2} & \cdots & 0 & 0 \\ \vdots & \vdots & \ddots & \vdots & \vdots \\ 0 & 0 & \cdots & \sqrt{d-1} & 0 \end{bmatrix} \qquad a = \begin{bmatrix} 0 & 1 & 0 & \cdots & 0 \\ 0 & 0 & \sqrt{2} & \cdots & 0 \\ \vdots & \vdots & \vdots & \ddots & \vdots \\ 0 & 0 & 0 & \cdots & \sqrt{d-1} \\ 0 & 0 & 0 & \cdots & 0 \end{bmatrix}$$

It is easily verified that the action of $a^\dagger$ on the vectors of the canonical orthonormal basis of $\mathbb{C}^d$ is the following:

$$a^\dagger \left|\frac{k}{d-1}\right\rangle = \sqrt{k+1}\left|\frac{k+1}{d-1}\right\rangle \qquad \text{for } k \in \{0, 1, \ldots, d-2\}$$
$$a^\dagger\,|1\rangle = \mathbf{0}$$

whereas the action of $a$ is:

$$a\left|\frac{k}{d-1}\right\rangle = \sqrt{k}\left|\frac{k-1}{d-1}\right\rangle \qquad \text{for } k \in \{1, 2, \ldots, d-1\}$$

$$a|0\rangle = \mathbf{0}$$

Using $a^\dagger$ and $a$ we can also introduce the following operators:

$$N = a^\dagger a = \begin{bmatrix} 0 & 0 & 0 & \cdots & 0 \\ 0 & 1 & 0 & \cdots & 0 \\ 0 & 0 & 2 & \cdots & 0 \\ \vdots & \vdots & \vdots & \ddots & \vdots \\ 0 & 0 & 0 & \cdots & d-1 \end{bmatrix} \qquad aa^\dagger = \begin{bmatrix} 1 & 0 & \cdots & 0 & 0 \\ 0 & 2 & \cdots & 0 & 0 \\ \vdots & \vdots & \ddots & \vdots & \vdots \\ 0 & 0 & \cdots & d-1 & 0 \\ 0 & 0 & \cdots & 0 & 0 \end{bmatrix}$$

The eigenvalues of the self–adjoint operator $N$ are $0, 1, 2, \ldots, d-1$, and the eigenvector corresponding to the generic eigenvalue $k$ is $|N = k\rangle = \left|\frac{k}{d-1}\right\rangle$.

One possible physical interpretation of $N$ is that it describes the *number of particles* of physical systems consisting of a maximum number of $d-1$ particles. In order to add a particle to the $k$ particles state $|N = k\rangle$ (thus making it switch to the "next" state $|N = k+1\rangle$) we apply the creation operator $a^\dagger$, while to remove a particle from this system (thus making it switch to the "previous" state $|N = k-1\rangle$) we apply the annihilation operator $a$. Since the maximum number of particles that can be simultaneously in the system is $d-1$, the application of the creation operator to a full $d-1$ particles system does not have any effect on the system, and returns as a result the null vector. Analogously, the application of the annihilation operator to an empty particle system does not affect the system and returns the null vector as a result.

Another physical interpretation of operators $a^\dagger$ and $a$, by operator $N$, follows from the possibility of expressing the Hamiltonian (2) as follows:

$$H = \varepsilon_0 \, \mathbb{I} + \Delta\varepsilon \, N = \varepsilon_0 \, \mathbb{I} + \Delta\varepsilon \, a^\dagger a$$

In this case $a^\dagger$ (resp., $a$) realizes the transition from the eigenstate of energy $\varepsilon_k = \varepsilon_0 + k\,\Delta\varepsilon$ to the "next" (resp., "previous") eigenstate of energy $\varepsilon_{k+1} = \varepsilon_0 + (k+1)\,\Delta\varepsilon$ (resp., $\varepsilon_{k-1} = \varepsilon_0 + (k-1)\,\Delta\varepsilon$) for any $0 \le k < d-1$ (resp., $0 < k \le d-1$), while it collapses the last excited (resp., ground) state of energy $\varepsilon_0 + (d-1)\,\Delta\varepsilon$ (resp., $\varepsilon_0$) to the null vector.

The collection of all linear operators on $\mathbb{C}^d$ is a $d^2$–dimensional linear space whose canonical basis is:

$$\{E_{x,y} = |y\rangle\langle x| \; : \; x, y \in L_d\}$$

Since $E_{x,y}|x\rangle = |y\rangle$ and $E_{x,y}|z\rangle = \mathbf{0}$ for every $z \in L_d$ such that $z \ne x$, this operator transforms the unit vector $|x\rangle$ into the unit vector $|y\rangle$, collapsing all the other vectors of the canonical orthonormal basis of $\mathbb{C}^d$ to the null vector. Each of the operators $E_{x,y}$ can be expressed, using the whole algebraic structure of the associative algebra of operators, as a suitable composition of creation and annihilation operators, as follows:

$$
E_{\frac{i}{d-1},\frac{j}{d-1}} = \begin{cases}
\frac{\sqrt{j!}}{(d-1)!} A_{a^\dagger,a^\dagger}^{d-2,d-1-j,0} & \text{if } i = 0 \\[2mm]
\frac{\sqrt{j!}}{(d-1)!} A_{a,a^\dagger}^{d-1,d-1-j,0} & \text{if } i = 1 \text{ and } j \geq 1 \\[2mm]
\frac{\sqrt{i!}}{(d-1)!\sqrt{j!}} A_{a^\dagger,a^\dagger}^{d-2-i,d-1,j} & \text{if } (i = 1,\ j = 0 \text{ and } d \geq 3) \text{ or} \\
& \quad (1 < i < d-2 \text{ and } j \leq i) \\[2mm]
\frac{\sqrt{j!}}{(d-1)!\sqrt{i!}} A_{a,a}^{i-1,d-1,d-1-j} & \text{if } (i = d-2,\ j = d-1 \text{ and } d \geq 3) \\
& \quad \text{or } (1 < i < d-2 \text{ and } j > i) \\[2mm]
\frac{1}{\sqrt{(d-1)!j!(d-1)}} A_{a^\dagger,a}^{d-1,j,0} & \text{if } i = d-2 \text{ and } j \leq d-2 \\[2mm]
\frac{1}{\sqrt{(d-1)!j!}} A_{a,a}^{d-2,j,0} & \text{if } i = d-1
\end{cases}
$$

Here we just recall, in order to keep the length of the paper under a reasonable size, that an alternative interpretation of qudits is possible, based on the values which can be assumed by the $z$ component of the angular momentum of semi–integer spin quantum systems. Also with this interpretation every linear operator, and in particular operators $E_{x,y}$, can be realized as appropriate compositions of *spin–rising* ($J_+$) and *spin–lowering* ($J_-$) operators, similarly to what we have done with creation and annihilation operators. For the details, we refer the reader to [18,16].

## 3   Classical and Quantum-Like Register Machines

A (classical, deterministic) *n–register machine* is a construct $M = (n, P, l_0, l_h)$, where $n$ is the number of registers, $P$ is a finite set of instructions injectively labeled with a given set $lab(M)$, $l_0$ is the label of the first instruction to be executed, and $l_h$ is the label of the last instruction of $P$. Registers contain non–negative integer values. Without loss of generality, we can assume $lab(M) = \{1, 2, \ldots, m\}$, $l_0 = 1$, and $l_h = m$. The instructions of $P$ have the following forms:

- $j : (INC(r), k)$, with $j, k \in lab(M)$
  Increment the value contained in register $r$, and then jump to instruction $k$.
- $j : (DEC(r), k, l)$, with $j, k, l \in lab(M)$
  If the value contained in register $r$ is positive then decrement it and jump to instruction $k$. If the value of $r$ is zero then jump to instruction $l$ (without altering the contents of the register).
- $m : Halt$
  Stop the machine. Note that this instruction can only be assigned to the final label $m$.

Register machines provide a simple universal computational model. Indeed, the results proved in [9] (based on the results established in [22]) as well as in [10] and [11] immediately lead to the following proposition.

**Proposition 1.** *For any partial recursive function $f : \mathbb{N}^{\alpha} \to \mathbb{N}^{\beta}$ there exists a deterministic $(\max\{\alpha, \beta\} + 2)$–register machine $M$ computing $f$ in such a way that, when starting with $(n_1, \ldots, n_{\alpha}) \in \mathbb{N}^{\alpha}$ in registers 1 to $\alpha$, $M$ has computed $f(n_1, \ldots, n_{\alpha}) = (r_1, \ldots, r_{\beta})$ if it halts in the final label $l_h$ with registers 1 to $\beta$ containing $r_1$ to $r_{\beta}$, and all other registers being empty; if the final label cannot be reached, then $f(n_1, \ldots, n_{\alpha})$ remains undefined.*

A *quantum-like $n$–register machine* is defined exactly as in the classical case, as a four–tuple $M = (n, P, l_0, l_h)$. Each register of the machine can be associated to an infinite dimensional quantum harmonic oscillator which is capable to assume the base states $|\varepsilon_0\rangle, |\varepsilon_1\rangle, |\varepsilon_2\rangle, \ldots$, corresponding to its energy levels, as described in section 2. The program counter of the machine is instead realized through a quantum system capable to assume $m$ different base states, from the set $\{|x\rangle : x \in L_m\}$. For simplicity, the instructions of $P$ are denoted in the usual way:

$$j : (INC(i), k) \qquad \text{and} \qquad j : (DEC(i), k, l)$$

This time, however, these instructions are appropriate linear operators acting on the Hilbert space whose vectors describe the (global) state of $M$. Precisely, the instruction $j : (INC(r), k)$ is defined as the operator

$$O_{j,r,k}^{INC} = |p_k\rangle \langle p_j| \otimes \left(\otimes^{r-1}\mathbb{I}\right) \otimes a^{\dagger} \otimes \left(\otimes^{n-r}\mathbb{I}\right)$$

with $\mathbb{I}$ the identity operator on $\mathcal{H}$ (the Hilbert space in which the state vectors of the infinite dimensional quantum harmonic oscillators associated with the registers exist), whereas the instruction $j : (DEC(r), k, l)$ is defined as the operator

$$O_{j,r,k,l}^{DEC} = |p_l\rangle \langle p_j| \otimes \left(\otimes^{r-1}\mathbb{I}\right) \otimes |\varepsilon_0\rangle \langle \varepsilon_0| \otimes \left(\otimes^{n-r}\mathbb{I}\right) + \\ |p_k\rangle \langle p_j| \otimes \left(\otimes^{r-1}\mathbb{I}\right) \otimes a \otimes \left(\otimes^{n-r}\mathbb{I}\right)$$

Hence the program $P$ can be formally defined as the sum $O_P$ of all these operators:

$$O_P = \sum_{j,r,k} O_{j,r,k}^{INC} + \sum_{j,r,k,l} O_{j,r,k,l}^{DEC}$$

Thus $O_P$ is the global operator which describes a computation step of $M$. When the program counter assumes the value $|p_m\rangle$, the application of $O_P$ would produce the null vector as a result. This would stop the execution of the machine, but it would also destroy the result of the computation. For this reason, in what follows we will add a term to $O_P$ that will allow us to extract the solution of the problem from a prefixed register when the program counter assumes the value $|p_m\rangle$.

A *configuration* of $M$ is given by the value of the program counter and the values contained in the registers. From a mathematical point of view, a configuration of $M$ is a vector of the Hilbert space $\mathbb{C}^m \otimes (\otimes^n \mathcal{H})$. A transition between two configurations is obtained by executing one instruction of $P$ (the one pointed at by the program counter), that is, by applying the operator $O_P$ to the current configuration of $M$.

As shown in [17], QRMs can simulate any (classical, deterministic) register machine, and thus they are (at least) computationally complete.

## 4  Classical and Quantum-Like UREM P Systems

We are now ready to focus our attention to P systems. As stated in the introduction, Q-UREM P systems have been introduced in [17] as a quantum-like version of UREM P systems. Hence, let us start by recalling the definition of the classical model of computation.

A UREM P system [8] of degree $d + 1$ is a construct $\Pi$ of the form:

$$\Pi = (A, \mu, e_0, \ldots, e_d, w_0, \ldots, w_d, R_0, \ldots, R_d)$$

where:

- $A$ is an alphabet of *objects*;
- $\mu$ is a *membrane structure*, with the membranes labeled by numbers $0, \ldots, d$ in a one-to-one manner;
- $e_0, \ldots, e_d$ are the initial energy values assigned to the membranes $0, \ldots, d$. In what follows we assume that $e_0, \ldots, e_d$ are non–negative integers;
- $w_0, \ldots, w_d$ are multisets over $A$ associated with the regions $0, \ldots, d$ of $\mu$;
- $R_0, \ldots, R_d$ are finite sets of *unit rules* associated with the membranes $0, \ldots, d$. Each rule has the form $(\alpha : a, \Delta e, b)$, where $\alpha \in \{in, out\}$, $a, b \in A$, and $|\Delta e|$ is the amount of energy that — for $\Delta e \geq 0$ — is added to or — for $\Delta e < 0$ — is subtracted from $e_i$ (the energy assigned to membrane $i$) by the application of the rule.

By writing $(\alpha_i : a, \Delta e, b)$ instead of $(\alpha : a, \Delta e, b) \in R_i$, we can specify only one set of rules $R$ with

$$R = \{(\alpha_i : a, \Delta e, b) \ : \ (\alpha : a, \Delta e, b) \in R_i, \ 0 \leq i \leq d\}$$

The *initial configuration* of $\Pi$ consists of $e_0, \ldots, e_d$ and $w_0, \ldots, w_d$. The transition from a configuration to another one is performed by nondeterministically choosing one rule from some $R_i$ and applying it (observe that here we consider a *sequential* model of applying the rules instead of choosing rules in a maximally parallel way, as it is often required in P systems). Applying $(in_i : a, \Delta e, b)$ means that an object $a$ (being in the membrane immediately outside of $i$) is changed into $b$ while entering membrane $i$, thereby changing the energy value $e_i$ of membrane $i$ by $\Delta e$. On the other hand, the application of a rule $(out_i : a, \Delta e, b)$ changes object $a$ into $b$ while leaving membrane $i$, and changes the energy value $e_i$ by $\Delta e$. The rules can be applied only if the amount $e_i$ of energy assigned to membrane $i$ fulfills the requirement $e_i + \Delta e \geq 0$. Moreover, we use some sort of local priorities: if there are two or more applicable rules in membrane $i$, then one of the rules with $\max |\Delta e|$ has to be used.

A sequence of transitions is called a *computation*; it is *successful* if and only if it halts. The *result* of a successful computation is considered to be the distribution of energies among the membranes (a non–halting computation does not produce a result). If we consider the energy distribution of the membrane structure as the input to be analyzed, we obtain a model for accepting sets of (vectors of) non–negative integers.

The following result, proved in [8], establishes computational completeness for this model of P systems.

**Proposition 2.** *Every partial recursive function $f : \mathbb{N}^\alpha \to \mathbb{N}^\beta$ can be computed by a UREM P system with (at most) $\max\{\alpha, \beta\} + 3$ membranes.*

It is interesting to note that the proof of this proposition is obtained by simulating register machines. In the simulation, a P system is defined which contains one subsystem for each register of the simulated machine. The contents of the register are expressed as the energy value $e_i$ assigned to the $i$-th subsystem. A single object is present in the system at every computation step, which stores the label of the instruction of the program $P$ currently simulated. Increment instructions are simulated in two steps by using the rules $(in_i : p_j, 1, \widetilde{p}_j)$ and $(out_i : \widetilde{p}_j, 0, p_k)$. Decrement instructions are also simulated in two steps, by using the rules $(in_i : p_j, 0, \widetilde{p}_j)$ and $(out_i : \widetilde{p}_j, -1, p_k)$ or $(out_i : \widetilde{p}_j, 0, p_l)$. The use of priorities associated to these last rules is crucial to correctly simulate a decrement instruction. For the details of the proof we refer the reader to [8].

On the other hand, by omitting the priority feature we do not get systems with universal computational power. Precisely, in [8] it is proved that P systems with unit rules and energy assigned to membranes without priorities and with an arbitrary number of membranes characterize the family $PsMAT^\lambda$ of Parikh sets generated by context–free matrix grammars (without occurrence checking and with $\lambda$-rules).

In Q-UREM P systems, all the elements of the model (multisets, the membrane hierarchy, configurations, and computations) are defined just like the corresponding elements of the classical P systems, but for objects and rules. The objects of $A$ are represented as pure states of a quantum system. If the alphabet contains $d \geq 2$ elements then, recalling the notation introduced in section 2, without loss of generality we can put $A = \left\{ |0\rangle, \left|\frac{1}{d-1}\right\rangle, \left|\frac{2}{d-1}\right\rangle, \dots, \left|\frac{d-2}{d-1}\right\rangle, |1\rangle \right\}$, that is, $A = \{|a\rangle : a \in L_d\}$. As stated above, the quantum system will also be able to assume as a state any superposition of the kind:

$$c_0 |0\rangle + c_{\frac{1}{d-1}} \left|\frac{1}{d-1}\right\rangle + \dots + c_{\frac{d-2}{d-1}} \left|\frac{d-2}{d-1}\right\rangle + c_1 |1\rangle$$

with $c_0, c_{\frac{1}{d-1}}, \dots, c_{\frac{d-2}{d-1}}, c_1 \in \mathbb{C}$ such that $\sum_{i=0}^{d-1} \left|c_{\frac{i}{d-1}}\right|^2 = 1$. A multiset is simply a collection of quantum systems, each in its own state.

In order to represent the energy values assigned to membranes we must use quantum systems which can exist in an infinite (countable) number of states. Hence we assume that every membrane of the quantum-like P system has an associated infinite dimensional quantum harmonic oscillator whose state represents the energy value assigned to the membrane. To modify the state of such harmonic oscillator we can use the infinite dimensional version of creation ($a^\dagger$) and annihilation ($a$) operators[1] described in section 2, which are commonly used in quantum mechanics. The actions of $a^\dagger$ and $a$ on the state of an infinite dimensional harmonic oscillator are analogous to the actions on the states of truncated harmonic oscillators; the only difference is that in the former case there is no state with maximum energy, and hence the creation operator never produces the null vector. Also in this case it is possible to express operators $E_{x,y} = |y\rangle \langle x|$ as appropriate compositions of $a^\dagger$ and $a$.

As in the classical case, rules are associated to the membranes rather than to the regions enclosed by them. Each rule of $R_i$ is an operator of the form

$$|y\rangle \langle x| \otimes O, \qquad \text{with } x, y \in L_d \tag{3}$$

where $O$ is a linear operator which can be expressed by an appropriate composition of operators $a^\dagger$ and $a$. The part $|y\rangle \langle x|$ is the *guard* of the rule: it makes the rule "active" (that is, the rule produces an effect) if and only if a quantum system in the basis state $|x\rangle$ is present. The semantics of rule (3) is the following: If an object in state $|x\rangle$ is present in the region immediately outside membrane $i$, then the state of the object is changed to $|y\rangle$ and the operator $O$ is applied to the state of the harmonic oscillator associated with the membrane. Notice that the application of $O$ can result in the null vector, so that the rule has no effect even if its guard is satisfied; this fact is equivalent to the condition $e_i + \Delta e \geq 0$ on the energy of membrane $i$ required in the classical case. Differently from the classical case, no local priorities are assigned to the rules. If two or more rules are associated to membrane $i$, then they are summed. This means that, indeed, we can think to each membrane as having only one rule with many guards. When an object is present, the inactive parts of the rule (those for which the guard is not satisfied) produce the null vector as a result. If the region in which the object occurs contains two or more membranes, then all their rules are applied to the object. Observe that the object which activates the rules never crosses the membranes. This means that the objects specified in the initial configuration can change their state but never move to a different region. Notwithstanding, transmission of information between different membranes is possible, since different objects may modify in different ways the energy state of the harmonic oscillators associated with the membranes.

The application of one or more rules determines a *transition* between two configurations. A *halting configuration* is a configuration in which no rule can be applied. A sequence of transitions is a *computation*. A computation is *successful* if and only if it *halts*, that is, reaches a halting configuration. The *result* of a successful computation is considered to be the distribution of energies among the membranes in the halting configuration. A non–halting computation does not produce a result. Just like in the classical case, if we consider the energy

---

[1] We recall that an alternative formulation that uses spin–rising ($J_+$) and spin–lowering ($J_-$) operators instead of creation and annihilation is also possible.

distribution of the membrane structure as the input to be analyzed, we obtain a model for accepting sets of (vectors of) non–negative integers.

In [17] it has been proved that Q-UREM P systems are computationally complete. Precisely, we can state the following theorem, whose proof is once again obtained by simulating register machines.

**Theorem 1.** *Every partial recursive function $f : \mathbb{N}^\alpha \to \mathbb{N}^\beta$ can be computed by a Q-UREM P system with (at most) $\max\{\alpha, \beta\} + 3$ membranes.*

*Proof.* Let $M = (n, P, 1, m)$ be a deterministic $n$–register machine that computes $f$. Let $m$ be the number of instructions of $P$. The initial instruction of $P$ has the label 1, and the halting instruction has the label $m$. Observe that, according to Proposition 1, $n = \max\{\alpha, \beta\} + 2$ is enough.

The input values $x_1, \ldots, x_\alpha$ are expected to be in the first $\alpha$ registers, and the output values are expected to be in registers 1 to $\beta$ at the end of a successful computation. Moreover, without loss of generality, we may assume that at the beginning of a computation all the registers except (eventually) the registers from 1 to $\alpha$ contain zero.

We construct the quantum P system

$$\Pi = (A, \mu, e_0, \ldots, e_n, w_0, \ldots, w_n, R_0, \ldots, R_n)$$

where:

- $A = \{|j\rangle \mid j \in L_m\}$
- $\mu = [_0 [_1 ]_1 \cdots [_\alpha ]_\alpha \cdots [_n ]_n ]_0$
- $e_i = \begin{cases} |\varepsilon_{x_i}\rangle & \text{for } 1 \leq i \leq \alpha \\ |\varepsilon_0\rangle & \text{for } \alpha + 1 \leq i \leq n \\ \mathbf{0} & \text{(the null vector) for } i = 0 \end{cases}$
- $w_0 = |0\rangle$
- $w_i = \emptyset \qquad \text{for } 1 \leq i \leq n$
- $R_0 = \emptyset$
- $R_i = \sum_{j=1}^m O_{i_j} \qquad \text{for } 1 \leq i \leq n$
  where the $O_{i_j}$'s are local operators which simulate instructions of the kind $j : (INC(i), k)$ and $j : (DEC(i), k, l)$ (one local operator for each increment or decrement operation which affects register $i$). The details on how the $O_{i_j}$'s are defined are given below.

The value contained into register $i$, $1 \leq i \leq n$, is represented by the energy value $e_i = |\varepsilon_{x_i}\rangle$ of the infinite dimensional quantum harmonic oscillator associated with membrane $i$. Figure 1 depicts a typical configuration of $\Pi$. The skin contains one object of the kind $|j\rangle$, $j \in L_m$, which mimics the program counter of machine $M$. Precisely, if the program counter of $M$ has the value $k \in \{1, 2, \ldots, m\}$ then the object present in region 0 is $\left|\frac{k-1}{m-1}\right\rangle$. In order to avoid cumbersome notation, in what follows we denote by $|p_k\rangle$ the state $\left|\frac{k-1}{m-1}\right\rangle$ of the quantum system which mimics the program counter.

The sets of rules $R_i$ depend upon the instructions of $P$. Precisely, the simulation works as follows.

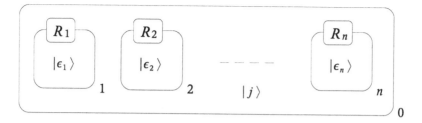

**Fig. 1.** A configuration of the simulating P system

1. Increment instructions $j : (INC(i), k)$ are simulated by a guarded rule of the kind $|p_k\rangle \langle p_j| \otimes a^\dagger \in R_i$.
   If the object $|p_j\rangle$ is present in region 0, then the rule transforms it into object $|p_k\rangle$ and increments the energy level of the harmonic oscillator contained into membrane $i$.
2. Decrement instructions $j : (DEC(i), k, l)$ are simulated by a guarded rule of the kind:

$$|p_l\rangle \langle p_j| \otimes |\varepsilon_0\rangle \langle \varepsilon_0| + |p_k\rangle \langle p_j| \otimes a \in R_i$$

In fact, let us assume that the object $|p_j\rangle$ is present in region 0 (if $|p_j\rangle$ is not present then the above rule produces the null operator), and let us denote by $O$ the above rule. The harmonic oscillator may be in the base state $|\varepsilon_0\rangle$ or in a base state $|\varepsilon_x\rangle$ with $x$ a positive integer.
   If the state of the harmonic oscillator is $|\varepsilon_0\rangle$ then the rule produces:

$$
\begin{aligned}
O(|p_j\rangle \otimes |\varepsilon_0\rangle) &= \\
&= (|p_l\rangle \langle p_j| \otimes |\varepsilon_0\rangle \langle \varepsilon_0|)(|p_j\rangle \otimes |\varepsilon_0\rangle) + (|p_k\rangle \langle p_j| \otimes a)(|p_j\rangle \otimes |\varepsilon_0\rangle) = \\
&= |p_l\rangle \otimes |\varepsilon_0\rangle + |p_k\rangle \otimes \mathbf{0} = |p_l\rangle \otimes |\varepsilon_0\rangle
\end{aligned}
$$

that is, the state of the oscillator is unaltered and the program counter is set to $|p_l\rangle$.
   If the state of the harmonic oscillator is $|\varepsilon_x\rangle$, for a positive integer $x$, then the rule produces:

$$
\begin{aligned}
O(|p_j\rangle \otimes |\varepsilon_x\rangle) &= \\
&= (|p_l\rangle \langle p_j| \otimes |\varepsilon_0\rangle \langle \varepsilon_0|)(|p_j\rangle \otimes |\varepsilon_x\rangle) + (|p_k\rangle \langle p_j| \otimes a)(|p_j\rangle \otimes |\varepsilon_x\rangle) = \\
&= |p_l\rangle \otimes \mathbf{0} + |p_k\rangle \otimes a |\varepsilon_x\rangle = |p_k\rangle \otimes |\varepsilon_{x-1}\rangle
\end{aligned}
$$

that is, the energy level of the harmonic oscillator is decremented and the program counter is set to $|p_k\rangle$.

The set $R_i$ of rules is obtained by summing all the operators which affect (increment or decrement) register $i$. The Halt instruction is simply simulated by doing nothing when the object $|p_m\rangle$ appears in region 0.

It is apparent from the description given above that after the simulation of each instruction each energy value $e_i$ equals the value contained into register $i$, with $1 \leq i \leq m$. Hence, when the halting symbol $|p_m\rangle$ appears in region 0, the energy values $e_1, \ldots, e_\beta$ equal the output of program $P$.    ∎

Let us conclude this section by observing that, in order to obtain computational completeness, it is not necessary that the objects cross the membranes. This fact avoids one of the problems raised in [18]: the existence of a "magic" quantum transportation mechanism which is able to move objects according to the target contained into the rule. In quantum P systems with unit rules and energy assigned to membranes, the only problem is to keep the object $|p_j\rangle$ localized in region 0, so that it never enters into the other regions. In other words, the major problem of this kind of quantum P systems is to oppose the tunnel effect. Let us also note that with Q-UREM P systems we do not need to exploit the membrane hierarchy to obtain computational completeness: all the subsystems contained into the region enclosed by the skin are at the same hierarchy level.

## 5    Solving 3-SAT with QRMs and with Q-UREM P Systems

Q-UREM P systems are not only able to compute all partial recursive functions, like Turing machines, but they can also be very efficient computation devices. Indeed, in this section we show how we can solve in polynomial time the **NP**–complete decision problem 3-SAT by using QRMs and Q-UREM P systems. As we will see, the solution provided by Q-UREM P systems will be even more efficient that the one obtained with QRMs.

It is important to stress that our solutions assume that a specific non–unitary operator, built using the truncated version of creation and annihilation operators, can be realized as an instruction of QRMs and as a rule of Q-UREM P systems, respectively. The construction relies also upon the assumption that an external observer is able to discriminate, as the result of a measurement, a null vector from a non–null vector.

### 5.1    The 3-SAT Problem

A *boolean* variable is a variable which can assume one of two possible truth values: TRUE and FALSE. As usually done in the literature, we will denote TRUE by 1 and FALSE by 0. A *literal* is either a directed or a negated boolean variable. A *clause* is a disjunction of literals, whereas a *3-clause* is a disjunction of exactly three literals. Given a set $X = \{x_1, x_2, \ldots, x_n\}$ of boolean variables, an *assignment* is a mapping $a : X \rightarrow \{0, 1\}$ that associates to each variable a truth value. The number of all possible assignments to the variables of $X$ is $2^n$. We say that an assignment *satisfies* the clause $C$ if, assigned the truth values to all the variables which occur in $C$, the evaluation of $C$ (considered as a boolean formula) gives 1 as a result.

The 3-SAT decision problem is defined as follows.

*Problem 1.* NAME: 3-SAT.

- INSTANCE: a set $C = \{C_1, C_2, \ldots, C_m\}$ of 3-clauses, built on a finite set $\{x_1, x_2, \ldots, x_n\}$ of boolean variables.
- QUESTION: is there an assignment of the variables $x_1, x_2, \ldots, x_n$ that satisfies all the clauses in $C$?

Notice that the number $m$ of possible 3-clauses is polynomially bounded with respect to $n$: in fact, since each clause contains exactly three literals, we can have at most $(2n)^3 = 8n^3$ clauses.

In what follows we will equivalently say that an instance of 3-SAT is a boolean formula $\phi_n$, built on $n$ free variables and expressed in conjunctive normal form, with each clause containing exactly three literals. The formula $\phi_n$ is thus the conjunction of the above clauses.

It is well known [13] that 3-SAT is an **NP**–complete problem.

## 5.2   Solving 3-SAT with QRMs

Let $\phi_n$ be an instance of 3-SAT containing $n$ free variables. We will first show how to evaluate $\phi_n$ with a classical register machine; then, we will initialize the input registers with a superposition of all possible assignments, we will compute the corresponding superposition of output values into an output register, and finally we will apply the linear operator $2^n |1\rangle \langle 1|$ to the output register to check whether $\phi_n$ is a positive instance of 3-SAT.

The register machine that we use to evaluate $\phi_n$ is composed by $n+1$ registers. The first $n$ registers correspond (in a one-to-one manner) to the free variables of $\phi_n$, while the last register is used to compute the output value. The *structure* of the program used to evaluate $\phi_n$ is the following:

$$\phi = 0$$
$$\textbf{if } C_1 = 0 \textbf{ then goto } \text{end}$$
$$\textbf{if } C_2 = 0 \textbf{ then goto } \text{end}$$
$$\vdots$$
$$\textbf{if } C_m = 0 \textbf{ then goto } \text{end}$$
$$\phi = 1$$
$$\text{end:}$$

where $\phi$ denotes the output register, and $C_1, C_2, \ldots, C_m$ are the clauses of $\phi_n$. Let $X_{i,j}$, with $j \in \{1, 2, 3\}$, be the literals (directed or negated variables) which occur in the clause $C_i$ (hence $C_i = X_{i,1} \vee X_{i,2} \vee X_{i,3}$). We can thus write the above structure of the program, at a finer grain, as follows:

$$\phi = 0$$
$$\textbf{if } X_{1,1} = 1 \textbf{ then goto } \text{end}_1$$
$$\textbf{if } X_{1,2} = 1 \textbf{ then goto } \text{end}_1$$

$$\begin{aligned}
&\quad\quad \textbf{if } X_{1,3} = 1 \textbf{ then goto } \text{end}_1 \\
&\quad\quad \textbf{goto } \text{end} \\
&\text{end}_1\text{: } \textbf{if } X_{2,1} = 1 \textbf{ then goto } \text{end}_2 \\
&\quad\quad \textbf{if } X_{2,2} = 1 \textbf{ then goto } \text{end}_2 \\
&\quad\quad \textbf{if } X_{2,3} = 1 \textbf{ then goto } \text{end}_2 \\
&\quad\quad \textbf{goto } \text{end} \\
&\text{end}_2\text{: } \cdots\cdots \\
&\quad\quad\quad \vdots \\
&\text{end}_{m-1}\text{: } \textbf{if } X_{m,1} = 1 \textbf{ then goto } \text{end} \\
&\quad\quad \textbf{if } X_{m,2} = 1 \textbf{ then goto } \text{end} \\
&\quad\quad \textbf{if } X_{m,3} = 1 \textbf{ then goto } \text{end} \\
&\quad\quad \phi = 1 \\
&\text{end:}
\end{aligned} \tag{4}$$

In the above structure it is assumed that each literal $X_{i,j}$, with $1 \leq i \leq m$ and $j \in \{1, 2, 3\}$, is substituted with the corresponding variable which occurs in it; moreover, if the variable occurs negated into the literal then the comparison must be done with 0 instead of 1:

$$\textbf{if } X_{i,j} = 0 \textbf{ then goto } \text{end}_i$$

Since the free variables of $\phi_n$ are bijectively associated with the first $n$ registers of the machine, in order to evaluate $\phi_n$ we need a method to check whether a given register contains 0 (or 1) without destroying its value. Let us assume that, when the program counter of the machine reaches the value $k$, we have to execute the following instruction:

$$k\text{: } \textbf{if } X_{i,j} = 1 \textbf{ then goto } \text{end}_i$$

We translate such instruction as follows (where, instead of $X_{i,j}$, we specify the register which corresponds to the variable indicated in $X_{i,j}$):

$$\begin{aligned}
k\text{: } & DEC(X_{i,j}), k+1, k+2 \\
k+1\text{: } & INC(X_{i,j}), \text{end}_i
\end{aligned}$$

The instruction:

$$k\text{: } \textbf{if } X_{i,j} = 0 \textbf{ then goto } \text{end}_i$$

is instead translated as follows:

$$\begin{aligned}
k\text{: } & DEC(X_{i,j}), k+1, \text{end}_i \\
k+1\text{: } & INC(X_{i,j}), k+2
\end{aligned}$$

Notice that the only difference with the previous sequence of instructions is in the position of "$\text{end}_i$" and "$k+2$". Moreover, the structure of the program is always the same. As a consequence, given an instance $\phi_n$ of 3-SAT, the program $P$ of a register machine which evaluates $\phi_n$ can be obtained in a very straightforward (mechanical) way.

On a *classical* register machine, this program computes the value of $\phi_n$ for a given assignment to its variables $x_1, x_2, \ldots, x_n$. On a QRM we can initialize the registers with the following state (where $H_1$ is the Hadamard operator, applied to a single qubit):

$$\otimes^n H_1 |0\rangle \otimes |0\rangle$$

which sets the output register $\phi$ to 0 and the registers corresponding to $x_1, x_2, \ldots, x_n$ to a superposition of all possible assignments. Then, we apply the global operator $O_P$ which corresponds to the program $P$ until the program counter reaches the value $|p_{\text{end}}\rangle$, thus computing in the output register a superposition of all classical results. The operator $O_P$ is built as described in section 3, with the only difference that now it contains also the term:

$$|p_{\text{end}}\rangle \langle p_{\text{end}}| \otimes \text{ID}_n \otimes 2^n |1\rangle \langle 1| =$$
$$|p_{\text{end}}\rangle \langle p_{\text{end}}| \otimes \text{ID}_n \otimes \underbrace{\left[ (|1\rangle \langle 1| + |1\rangle \langle 1|) \circ \ldots \circ (|1\rangle \langle 1| + |1\rangle \langle 1|) \right]}_{n \text{ times}}$$

which extracts the result from the output register when the program counter assumes the value $|p_{\text{end}}\rangle$. The number of times we have to apply $O_P$ is equal to the length of $P$, that is, $2 \cdot 3m + 2 = 6m + 2$: two instructions for each literal in every clause, plus two final instructions.

Now, if $\phi_n$ is not satisfiable then the contents of the output register are $|0\rangle$, and when the program counter reaches the value $|p_{\text{end}}\rangle$ the operator $O_P$ transforms it to the null vector. On the other hand, if $\phi_n$ is satisfiable then the contents of the output register will be a superposition $\alpha_0 |0\rangle + \alpha_1 |1\rangle$, with $\alpha_1 \neq 0$. By applying the operator $O_P$ we obtain (here $|\psi_n\rangle$ denotes the state of the $n$ input registers):

$$O_P \big( |p_{\text{end}}\rangle \otimes |\psi_n\rangle \otimes (\alpha_0 |0\rangle + \alpha_1 |1\rangle) \big) =$$
$$= \big( |p_{\text{end}}\rangle \langle p_{\text{end}}| \otimes \text{ID}_n \otimes 2^n |1\rangle \langle 1| \big) \cdot$$
$$\cdot \big( |p_{\text{end}}\rangle \otimes |\psi_n\rangle \otimes (\alpha_0 |0\rangle + \alpha_1 |1\rangle) \big) =$$
$$= |p_{\text{end}}\rangle \langle p_{\text{end}} | p_{\text{end}}\rangle \otimes \text{ID}_n |\psi_n\rangle \otimes 2^n |1\rangle \langle 1| (\alpha_0 |0\rangle + \alpha_1 |1\rangle) =$$
$$= |p_{\text{end}}\rangle \otimes |\psi_n\rangle \otimes (2^n \alpha_0 |1\rangle \langle 1|0\rangle + 2^n \alpha_1 |1\rangle \langle 1|1\rangle) =$$
$$= |p_{\text{end}}\rangle \otimes |\psi_n\rangle \otimes (\mathbf{0} + 2^n \alpha_1 |1\rangle) =$$
$$= |p_{\text{end}}\rangle \otimes |\psi_n\rangle \otimes 2^n \alpha_1 |1\rangle$$

that is, a non–null vector.

We can thus conclude that if an external observer is able to discriminate between a null vector and a non–null vector, and it is possible to build and apply the operator $2^n |1\rangle \langle 1|$ to the output register of a QRM, then we have a family of QRMs that solve 3-SAT in polynomial time. This solution is given in a *semi–uniform* setting: in particular, the program $P$ executed by the QRM depends upon the instance $\phi_n$ of 3-SAT we want to solve.

## 5.3   Solving 3-SAT with Q-UREM P Systems

In this section we finally show how to build a (semi–uniform) family of Q-UREM P systems that solves 3-SAT. Let $\phi_n$ be an instance of 3-SAT containing $n$ free variables. The structure and the initial configuration of the P system that determines whether $\phi_n$ is satisfiable is similar to what shown in Figure 1, the only difference being that there are $n+1$ subsystems instead of $n$.

As we have done with QRMs, let us start by showing how to evaluate $\phi_n$ for a given assignment of truth values to its variables $x_1, \ldots, x_n$. The input values are set as the energies $|\varepsilon_{x_i}\rangle$ of the harmonic oscillators associated with the membranes from 1 to $n$. The energy (eventually) associated with the skin membrane is not used. The $(n+1)$-th membrane, whose harmonic oscillator will contain the output at the end of the computation, is initialized with $|\varepsilon_0\rangle$. The alphabet $A$ consists of all the possible values which can be assumed by the program counter of the QRM that evaluates $\phi_n$. In the initial configuration the P system contains only one copy of the object $|p_1\rangle$, corresponding to the initial value of the program counter, in the region enclosed by the skin membrane.

The evaluation of $\phi_n$ could be performed by simulating the QRM obtained from $\phi_n$ as explained in the previous section. However, we can obtain a slightly more efficient P system as follows. We start from the program structure (4), which can be obtained from $\phi_n$ in a straightforward way. Now, let us suppose we must execute the following instruction:

$$k: \textbf{if } X_{i,j} = 1 \textbf{ then goto } \text{end}_i$$

As told above, this instruction is performed as follows in a register machine:

$$k: DEC(X_{i,j}), k+1, k+2$$
$$k+1: INC(X_{i,j}), \text{end}_i$$

If we had to simulate these two instructions using a Q-UREM P system, we should use the following sum of rules:

$$\underbrace{\left(|p_{\text{end}_i}\rangle \langle p_{k+1}| \otimes a^\dagger\right)}_{k+1:\ INC(X_{i,j}),\ \text{end}_i} + \underbrace{\left(|p_{k+2}\rangle \langle p_k| \otimes |\varepsilon_0\rangle \langle \varepsilon_0| + |p_{k+1}\rangle \langle p_k| \otimes a\right)}_{k:\ DEC(X_{i,j}),\ k+1,\ k+2} \in R_\ell$$

where $\ell = \langle i, j \rangle$ is the index of the variable (in the set $\{x_1, x_2, \ldots, x_n\}$) which occurs in literal $X_{i,j}$. As we can see, this operator produces the vector $|p_{k+2}\rangle \otimes |\varepsilon_0\rangle$ if the harmonic oscillator of membrane $\ell$ is in state $|\varepsilon_0\rangle$; otherwise, it produces the vector $|p_{\text{end}_i}\rangle \otimes |\varepsilon_1\rangle$. Hence we can simplify the above expression as follows:

$$|p_{\text{end}_i}\rangle \langle p_k| \otimes |\varepsilon_1\rangle \langle \varepsilon_1| + |p_{k+2}\rangle \langle p_k| \otimes |\varepsilon_0\rangle \langle \varepsilon_0| =$$
$$= |p_{\text{end}_i}\rangle \langle p_k| \otimes a^\dagger a + |p_{k+2}\rangle \langle p_k| \otimes aa^\dagger$$

We denote this operator by $O^{(1)}_{i,j,k}$. Similarly, if the instruction to be executed is:

$$k: \textbf{if } X_{i,j} = 0 \textbf{ then goto } \text{end}_i$$

then we use the operator

$$O_{i,j,k}^{(0)} = |p_{\text{end}_i}\rangle \langle p_k| \otimes aa^\dagger + |p_{k+2}\rangle \langle p_k| \otimes a^\dagger a \in R_\ell$$

which produces the vector $|p_{k+2}\rangle \otimes |\varepsilon_1\rangle$ if the harmonic oscillator of membrane $\ell$ is in state $|\varepsilon_1\rangle$, otherwise it produces the vector $|p_{\text{end}_i}\rangle \otimes |\varepsilon_0\rangle$.

Since the value $|p_{k+1}\rangle$ is no longer used, we can "compact" the program by redefining the operators $O_{i,j,k}^{(0)}$ and $O_{i,j,k}^{(1)}$ respectively as:

$$O_{i,j,k}^{(0)} = |p_{\text{end}_i}\rangle \langle p_k| \otimes aa^\dagger + |p_{k+1}\rangle \langle p_k| \otimes a^\dagger a$$
$$O_{i,j,k}^{(1)} = |p_{\text{end}_i}\rangle \langle p_k| \otimes a^\dagger a + |p_{k+1}\rangle \langle p_k| \otimes aa^\dagger$$

The "**goto end**" instructions in (4) can be executed as if they were **if** statements whose condition is the negation of the condition given in the previous **if**. Hence the two instructions:

> 7: **if** $X_{2,3} = 1$ **then goto** $\text{end}_2$
> 8: **goto end**

can be translated to:

> 7: **if** $X_{2,3} = 1$ **then goto** $\text{end}_2$
> 8: **if** $X_{2,3} = 0$ **then goto end**

which are realized by the operators $O_{2,3,7}^{(1)}$ and $O_{2,3,8}^{(0)}$ (to be added to membrane $\langle 2, 3 \rangle$). The last instruction ($\phi = 1$) of the program can be implemented as:

$$|p_{\text{end}}\rangle \langle p_{\text{end}-1}| \otimes a^\dagger$$

to be added to membrane $n + 1$.

For each membrane $i \in \{1, 2, \ldots, n\}$, the set of rules $R_i$ is obtained by summing all the operators which concern variable $x_i$.

Note that the formulation given in terms of Q-UREM P systems is simpler than the one obtained with QRMs. As usual, if we consider a single assignment to the variables of $\phi_n$ then at the end of the computation we will obtain the result of the evaluation of $\phi_n$ as the energy of the output membrane. Instead, if we initialize the harmonic oscillators of the $n$ input membranes with a uniform superposition of all possible classical assignments to $x_1, x_2, \ldots, x_n$, then at the end of the computation the harmonic oscillator of membrane $n+1$ will be in one of the following states:

- $|0\rangle$, if $\phi_n$ is not satisfiable;
- a superposition $\alpha_0 |0\rangle + \alpha_1 |1\rangle$, with $\alpha_1 \neq 0$, if $\phi_n$ is satisfiable.

Once again, we add the rule:

$$|p_{\text{end}}\rangle \langle p_{\text{end}}| \otimes 2^n |1\rangle \langle 1| \in R_{n+1}$$

to membrane $n + 1$ to extract the result.

We have thus obtained a family of Q-UREM P systems which solves 3-SAT in polynomial time. Also this scheme works in the *semi–uniform* setting: in fact, it is immediately verified that the rules of the system depend upon the instance $\phi_n$ of 3-SAT to be solved.

# 6   Directions for Future Research

In this paper we have overviewed the state of the art concerning Q-UREM P systems. As we have seen, Q-UREM P systems seem to be a very powerful model of computation. Hence, one possible direction for future research is to further study their computational properties, for example assuming the presence of entangled objects. Another line of research is to study the limits of the computational power of Q-UREM P systems by attacking harder than NP-complete problems. On this front, we conjecture that Q-UREM P systems could be able to solve (in semi–uniform way) EXP-complete problems, such as SUCCINT CIRCUIT VALUE.

## Acknowledgments

This research was partially funded by Università degli Studi di Milano–Bicocca — FIAR 2006.

## References

1. Alford, G.: Membrane systems with heat control. In: Păun, G., Rozenberg, G., Salomaa, A., Zandron, C. (eds.) Membrane Computing. LNCS, vol. 2597, Springer, Heidelberg (2003)
2. Barenco, A., Deutsch, D., Ekert, A., Jozsa, R.: Conditional quantum control and logic gates. Physical Review Letters 74, 4083–4086 (1995)
3. Benioff, P.: Quantum mechanical Hamiltonian models of computers. Annals of the New York Academy of Science 480, 475–486 (1986)
4. Deutsch, D.: Quantum theory, the Church–Turing principle, and the universal quantum computer. Proceedings of the Royal Society of London A 400, 97–117 (1985)
5. Feynman, R.P.: Simulating physics with computers. International Journal of Theoretical Physics 21(6-7), 467–488 (1982)
6. Feynman, R.P.: Quantum mechanical computers. Optics News 11, 11–20 (1985)
7. Freund, R.: Energy-controlled P systems. In: Păun, G., Rozenberg, G., Salomaa, A., Zandron, C. (eds.) Membrane Computing. LNCS, vol. 2597, pp. 247–260. Springer, Heidelberg (2003)
8. Freund, R., Leporati, A., Oswald, M., Zandron, C.: Sequential P systems with unit rules and energy assigned to membranes. In: Margenstern, M. (ed.) MCU 2004. LNCS, vol. 3354, pp. 200–210. Springer, Heidelberg (2005)
9. Freund, R., Oswald, M.: GP systems with forbidding context. Fundamenta Informaticae 49(1-3), 81–102 (2002)
10. Freund, R., Păun, G.: On the number of non-terminals in graph-controlled, programmed, and matrix grammars. In: Margenstern, M., Rogozhin, Y. (eds.) MCU 2001. LNCS, vol. 2055, pp. 214–225. Springer, Heidelberg (2001)
11. Freund, R., Păun, G.: From regulated rewriting to computing with membranes: Collapsing hierarchies. Theoretical Computer Science 312, 143–188 (2004)
12. Frisco, P.: The conformon–P system: a molecular and cell biology–inspired computability model. Theoretical Computer Science 312, 295–319 (2004)
13. Garey, M.R., Johnson, D.S.: Computers and Intractability. A Guide to the Theory on NP–Completeness. W.H. Freeman and Company (1979)

14. Gottesman, D.: Fault-tolerant quantum computation with higher-dimensional systems. Chaos, Solitons, and Fractals 10, 1749–1758 (1999)
15. Gruska, J.: Quantum Computing. McGraw-Hill, New York (1999)
16. Leporati, A., Felloni, S.: Three "quantum" algorithms to solve 3-SAT. Theoretical Computer Science 372, 218–241 (2007)
17. Leporati, A., Mauri, G., Zandron, C.: Quantum sequential P systems with unit rules and energy assigned to membranes. In: Freund, R., Păun, G., Rozenberg, G., Salomaa, A. (eds.) WMC 2005. LNCS, vol. 3850, pp. 310–325. Springer, Heidelberg (2006)
18. Leporati, A., Pescini, D., Zandron, C.: Quantum energy–based P systems. In: Proceedings of the First Brainstorming Workshop on Uncertainty in Membrane Computing, Palma de Mallorca, Spain, November 8–10, 2004, pp. 145–167 (2004)
19. Leporati, A., Zandron, C., Mauri, G.: Simulating the Fredkin gate with energy–based P systems. Journal of Universal Computer Science 10(5), 600–619 (2004)
20. Leporati, A., Zandron, C., Mauri, G.: Universal families of reversible P systems. In: Margenstern, M. (ed.) MCU 2004. LNCS, vol. 3354, pp. 257–268. Springer, Heidelberg (2005)
21. Leporati, A., Zandron, C., Mauri, G.: Reversible P systems to simulate Fredkin circuits. Fundamenta Informaticae 74, 529–548 (2006)
22. Minsky, M.L.: Finite and Infinite Machines. Prentice Hall, Englewood Cliffs, New Jersey (1967)
23. Nielsen, M.A., Chuang, I.L.: Quantum Computation and Quantum Information. Cambridge University Press, Cambridge (2000)
24. Păun, G.: Computing with membranes. Journal of Computer and System Sciences 1(61), 108–143 (2000), see also Turku Centre for Computer Science – TUCS Report No. 208 (1998)
25. Păun, G.: Membrane Computing. An Introduction. Springer, Heidelberg (2002)
26. Păun, G., Pérez-Jiménez, M.J.: Recent computing models inspired from biology: DNA and membrane computing. Theoria 18, 72–84 (2003)
27. Păun, G., Rozenberg, G.: A guide to membrane computing. Theoretical Computer Science 287(1), 73–100 (2002)
28. Păun, G., Suzuki, Y., Tanaka, H.: P systems with energy accounting. International Journal Computer Math. 78(3), 343–364 (2001)
29. The P systems Web page: http://psystems.disco.unimib.it/

# The Calculus of Looping Sequences for Modeling Biological Membranes

Roberto Barbuti[1], Andrea Maggiolo–Schettini[1],
Paolo Milazzo[1], and Angelo Troina[2,3]

[1] Dip. di Informatica, Università di Pisa, Largo B. Pontecorvo 3, 56127 - Pisa, Italy
[2] LIX - École Polytechnique Rue de Saclay, 91128 - Palaiseau, France
[3] LSV - ENS Cachan 61 Avenue du Président Wilson, 94235 - Cachan, France
{barbuti,maggiolo,milazzo,troina}@di.unipi.it

**Abstract.** We survey the formalism Calculus of Looping Sequences (CLS) and a number of its variants from the point of view of their use for describing biological membranes. The CLS formalism is based on term rewriting and allows describing biomolecular systems. A first variant of CLS, called Stochastic CLS, extends the formalism with stochastic time, another variant, called LCLS (CLS with links), allows describing proteins interaction at the domain level. A third variant is introduced for easier description of biological membranes. This extension can be encoded into CLS as well as other formalisms capable of membrane description such as Brane Calculi and P Systems. Such encodings allow verifying and simulating descriptions in Brane Calculi and P Systems by means of verifiers and simulators developed for CLS.

## 1 Introduction

Cell biology, the study of the morphological and functional organization of cells, is now an established field in biochemical research. Computer Science can help the research in cell biology in several ways. For instance, it can provide biologists with models and formalisms capable of describing and analyzing complex systems such as cells. In the last few years many formalisms originally developed by computer scientists to model systems of interacting components have been applied to Biology. Among these, there are Petri Nets [16], Hybrid Systems [1], and the π-calculus [9,25]. Moreover, new formalisms have been defined for describing biomolecular and membrane interactions [2,7,8,11,21,23]. Others, such as P Systems [17,18], have been proposed as biologically inspired computational models and have been later applied to the description of biological systems.

The π–calculus and new calculi based on it [21,23] have been particularly successful in the description of biological systems, as they allow describing systems in a compositional manner. Interactions of biological components are modeled as communications on channels whose names can be passed; sharing names of private channels allows describing biological compartments. However, these calculi offer very low–level interaction primitives, and this causes models to become

G. Eleftherakis et al. (Eds.): WMC8 2007, LNCS 4860, pp. 54–76, 2007.

very large and difficult to read. Calculi such as those proposed in [7,8,11] give a more abstract description of systems and offer special biologically motivated operators. However, they are often specialized to the description of some particular kinds of phenomena such as membrane interactions or protein interactions. Finally, P Systems have a simple notation and are not specialized to the description of a particular class of systems, but they are still not completely general. For instance, it is possible to describe biological membranes and the movement of molecules across membranes, and there are some variants able to describe also more complex membrane activities. However, the formalism is not so flexible to allow describing easily new activities observed on membranes without extending the formalism to model such activities.

Therefore, we conclude that there is a need for a formalism having a simple notation, having the ability of describing biological systems at different levels of abstraction, having some notions of compositionality and being flexible enough to allow describing new kinds of phenomena as they are discovered, without being specialized to the description of a particular class of systems. For this reason in [3] we have introduced the Calculus of Looping Sequences (CLS).

CLS is a formalism based on term rewriting with some features, such as a commutative parallel composition operator, and some semantic means, such as bisimulations, which are common in process calculi. This permits to combine the simplicity of notation of rewriting systems with the advantage of a form of compositionality. Actually, in [4] we have defined bisimilarity relations on CLS terms which are congruences with respect to the operators. The bisimilarity relation may be used to verify a property of a system by assessing its bisimilarity with a system one knows to enjoy that property. The fact that bisimilarity is a congruence is very important for a compositional account of behavioral equivalence. In [5,6], we have defined two extensions of CLS. The first, Stochastic CLS, allows describing quantitative aspects of the modeled systems such as the time spent by occurrences of chemical reactions. The second, CLS with links, allows describing protein interaction more precisely at a lower level of abstraction, namely at the domain level.

In this paper, after recalling CLS and the two mentioned extensions, we focus on the modeling of biological membranes by means of CLS. Now, CLS does not offer an easy representation for membranes whose nature is fluid and for

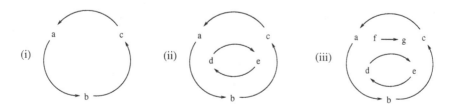

**Fig. 1.** (i) represents $(a \cdot b \cdot c)^L$; (ii) represents $(a \cdot b \cdot c)^L \rfloor (d \cdot e)^L$; (iii) represents $(a \cdot b \cdot c)^L \rfloor ((d \cdot e)^L \mid f \cdot g)$.

proteins which consequently move freely on membrane surfaces. For this reason, in [15] we have defined a CLS variant, called CLS+, which introduces a new operator allowing commutativity on membrane surfaces. We show how CLS+ can be encoded into CLS. In [3,15] we have shown how Brane Calculi [7] and P Systems [18] can be translated into CLS. Here we recall the ideas on which the translations are based.

CLS appears to allow description and manipulation of biological membranes and, moreover, offers, via translations, verification and simulation tools to other formalisms for membrane description.

# 2   The Calculus of Looping Sequences (CLS)

In this section we recall the Calculus of Looping Sequences (CLS) and we give some guidelines for the modeling of biological systems. CLS is essentially based on term rewriting, hence a CLS model consists of a term and a set of rewrite rules. The term is intended to represent the structure of the modeled system, and the rewrite rules to represent the events that may cause the system to evolve.

## 2.1   Formal Definition

We start with defining the syntax of terms. We assume a possibly infinite alphabet $\mathcal{E}$ of symbols ranged over by $a, b, c, \ldots$.

**Definition 1 (Terms).** *Terms $T$ and sequences $S$ of CLS are given by the following grammar:*

$$T ::= S \mid (S)^L \rfloor T \mid T \mid T$$
$$S ::= \epsilon \mid a \mid S \cdot S$$

*where $a$ is a generic element of $\mathcal{E}$, and $\epsilon$ represents the empty sequence. We denote with $\mathcal{T}$ the infinite set of terms, and with $\mathcal{S}$ the infinite set of sequences.*

In CLS we have a sequencing operator $\_ \cdot \_$, a looping operator $(\_)^L$, a parallel composition operator $\_ \mid \_$ and a containment operator $\_ \rfloor \_$. Sequencing can be used to concatenate elements of the alphabet $\mathcal{E}$. The empty sequence $\epsilon$ denotes the concatenation of zero symbols. A term can be either a sequence or a looping sequence (that is the application of the looping operator to a sequence) containing another term, or the parallel composition of two terms. By definition, looping and containment are always applied together, hence we can consider them as a single binary operator $(\_)^L \rfloor \_$ which applies to one sequence and one term.

Brackets can be used to indicate the order of application of the operators, and we assume $(\_)^L \rfloor \_$ to have precedence over $\_ \mid \_$. In Figure 1 we show some examples of CLS terms and their visual representation.

In CLS we may have syntactically different terms representing the same structure. We introduce a structural congruence relation to identify such terms.

**Definition 2 (Structural Congruence).** *The structural congruence relations* $\equiv_S$ *and* $\equiv_T$ *are the least congruence relations on sequences and on terms, respectively, satisfying the following rules:*

$$S_1 \cdot (S_2 \cdot S_3) \equiv_S (S_1 \cdot S_2) \cdot S_3 \qquad S \cdot \epsilon \equiv_S \epsilon \cdot S \equiv_S S$$

$$S_1 \equiv_S S_2 \text{ implies } S_1 \equiv_T S_2 \text{ and } \left(S_1\right)^L \rfloor T \equiv_T \left(S_2\right)^L \rfloor T$$

$$T_1 \mid T_2 \equiv_T T_2 \mid T_1 \qquad T_1 \mid (T_2 \mid T_3) \equiv_T (T_1 \mid T_2) \mid T_3 \qquad T \mid \epsilon \equiv_T T$$

$$\left(\epsilon\right)^L \rfloor \epsilon \equiv_T \epsilon \qquad \left(S_1 \cdot S_2\right)^L \rfloor T \equiv_T \left(S_2 \cdot S_1\right)^L \rfloor T$$

Rules of the structural congruence state the associativity of $\cdot$ and $\mid$, the commutativity of the latter and the neutral role of $\epsilon$. Moreover, axiom $\left(S_1 \cdot S_2\right)^L \rfloor T \equiv_T \left(S_2 \cdot S_1\right)^L \rfloor T$ says that looping sequences can rotate. In the following, for simplicity, we will use $\equiv$ in place of $\equiv_T$.

Rewrite rules will be defined essentially as pairs of terms, with the first term describing the portion of the system in which the event modeled by the rule may occur, and the second term describing how that portion of the system changes when the event occurs. In the terms of a rewrite rule we allow the use of variables. As a consequence, a rule will be applicable to all terms which can be obtained by properly instantiating its variables. Variables can be of three kinds: two of these are associated with the two different syntactic categories of terms and sequences, and one is associated with single alphabet elements. We assume a set of term variables $TV$ ranged over by $X, Y, Z, \ldots$, a set of sequence variables $SV$ ranged over by $\tilde{x}, \tilde{y}, \tilde{z}, \ldots$, and a set of element variables $\mathcal{X}$ ranged over by $x, y, z, \ldots$. All these sets are possibly infinite and pairwise disjoint. We denote by $\mathcal{V}$ the set of all variables, $\mathcal{V} = TV \cup SV \cup \mathcal{X}$, and with $\rho$ a generic variable of $\mathcal{V}$. Hence, a pattern is a term that may include variables.

**Definition 3 (Patterns).** *Patterns* $P$ *and* sequence patterns $SP$ *of CLS are given by the following grammar:*

$$P \quad ::= \quad SP \quad \mid \quad \left(SP\right)^L \rfloor P \quad \mid \quad P \mid P \quad \mid \quad X$$
$$SP \quad ::= \quad \epsilon \quad \mid \quad a \quad \mid \quad SP \cdot SP \quad \mid \quad \tilde{x} \quad \mid \quad x$$

*where* $a$ *is a generic element of* $\mathcal{E}$, *and* $X, \tilde{x}$ *and* $x$ *are generic elements of* $TV, SV$ *and* $\mathcal{X}$, *respectively. We denote with* $\mathcal{P}$ *the infinite set of patterns.*

We assume the structural congruence relation to be trivially extended to patterns. An *instantiation* is a partial function $\sigma : \mathcal{V} \to \mathcal{T}$. An instantiation must preserve the type of variables, thus for $X \in TV, \tilde{x} \in SV$ and $x \in \mathcal{X}$ we have $\sigma(X) \in \mathcal{T}, \sigma(\tilde{x}) \in \mathcal{S}$ and $\sigma(x) \in \mathcal{E}$, respectively. Given $P \in \mathcal{P}$, with $P\sigma$ we denote the term obtained by replacing each occurrence of each variable $\rho \in \mathcal{V}$ appearing in $P$ with the corresponding term $\sigma(\rho)$. With $\Sigma$ we denote the set of all the possible instantiations and, given $P \in \mathcal{P}$, with $Var(P)$ we denote the set of variables appearing in $P$. Now we define rewrite rules.

**Definition 4 (Rewrite Rules).** *A rewrite rule is a pair of patterns* $(P_1, P_2)$, *denoted with* $P_1 \mapsto P_2$, *where* $P_1, P_2 \in \mathcal{P}$, $P_1 \not\equiv \epsilon$ *and such that* $Var(P_2) \subseteq Var(P_1)$. *We denote with* $\mathfrak{R}$ *the infinite set of all the possible rewrite rules.*

**Table 1.** Guidelines for the abstraction of biomolecular entities into CLS

| Biomolecular Entity | CLS Term |
|---|---|
| Elementary object (genes, domains, other molecules, etc...) | Alphabet symbol |
| DNA strand | Sequence of elements repr. genes |
| RNA strand | Sequence of elements repr. transcribed genes |
| Protein | Sequence of elements repr. domains or single alphabet symbol |
| Molecular population | Parallel composition of molecules |
| Membrane | Looping sequence |

A rewrite rule $P_1 \mapsto P_2$ states that a term $P_1\sigma$, obtained by instantiating variables in $P_1$ by some instantiation function $\sigma$, can be transformed into the term $P_2\sigma$. We define the semantics of CLS as a transition system, in which states correspond to terms, and transitions correspond to rule applications.

**Definition 5 (Semantics).** *Given a set of rewrite rules $\mathcal{R} \subseteq \Re$, the semantics of CLS is the least transition relation $\rightarrow$ on terms closed under $\equiv$, and satisfying the following inference rules:*

$$\frac{P_1 \mapsto P_2 \in \mathcal{R} \quad P_1\sigma \not\equiv \epsilon \quad \sigma \in \Sigma}{P_1\sigma \rightarrow P_2\sigma} \qquad \frac{T_1 \rightarrow T_2}{T \mid T_1 \rightarrow T \mid T_2} \qquad \frac{T_1 \rightarrow T_2}{(S)^L \rfloor T_1 \rightarrow (S)^L \rfloor T_2}$$

*where the symmetric rule for the parallel composition is omitted.*

A *model* in CLS is given by a term describing the initial state of the system and by a set of rewrite rules describing all the events that may occur.

## 2.2 Modeling Guidelines

We describe how CLS can be used to model biomolecular systems analogously to what done by Regev and Shapiro in [24] for the $\pi$–calculus. An abstraction is a mapping from a real–world domain to a mathematical domain, which may allow highlighting some essential properties of a system while ignoring other, complicating, ones. In [24], Regev and Shapiro show how to abstract biomolecular systems as concurrent computations by identifying the biomolecular entities and events of interest and by associating them with concepts of concurrent computations such as concurrent processes and communications. In particular, they give some guidelines for the abstraction of biomolecular systems to the $\pi$–calculus, and give some simple examples.

The use of rewrite systems, such as CLS, to describe biological systems is founded on a different abstraction. Usually, entities (and their structures) are abstracted by terms of the rewrite system, and events by rewriting rules. We have already introduced the biological interpretation of CLS operators in the previous section. Here we want to give more general guidelines.

**Table 2.** Guidelines for the abstraction of biomolecular events into CLS

| Biomolecular Event | Examples of CLS Rewrite Rule |
|---|---|
| State change | $a \mapsto b$ <br> $\widetilde{x} \cdot a \cdot \widetilde{y} \mapsto \widetilde{x} \cdot b \cdot \widetilde{y}$ |
| Complexation | $a \mid b \mapsto c$ <br> $\widetilde{x} \cdot a \cdot \widetilde{y} \mid b \mapsto \widetilde{x} \cdot c \cdot \widetilde{y}$ |
| Decomplexation | $c \mapsto a \mid b$ <br> $\widetilde{x} \cdot c \cdot \widetilde{y} \mapsto \widetilde{x} \cdot a \cdot \widetilde{y} \mid b$ |
| Catalysis | $c \mid P_1 \mapsto c \mid P_2$ <br> where $P_1 \mapsto P_2$ is the catalyzed event |
| State change <br> on membrane | $\left(a \cdot \widetilde{x}\right)^L \rfloor X \mapsto \left(b \cdot \widetilde{x}\right)^L \rfloor X$ |
| Complexation <br> on membrane | $\left(a \cdot \widetilde{x} \cdot b \cdot \widetilde{y}\right)^L \rfloor X \mapsto \left(c \cdot \widetilde{x} \cdot \widetilde{y}\right)^L \rfloor X$ <br> $a \mid \left(b \cdot \widetilde{x}\right)^L \rfloor X \mapsto \left(c \cdot \widetilde{x}\right)^L \rfloor X$ <br> $\left(b \cdot \widetilde{x}\right)^L \rfloor (a \mid X) \mapsto \left(c \cdot \widetilde{x}\right)^L \rfloor X$ |
| Decomplexation <br> on membrane | $\left(c \cdot \widetilde{x}\right)^L \rfloor X \mapsto \left(a \cdot b \cdot \widetilde{x}\right)^L \rfloor X$ <br> $\left(c \cdot \widetilde{x}\right)^L \rfloor X \mapsto a \mid \left(b \cdot \widetilde{x}\right)^L \rfloor X$ <br> $\left(c \cdot \widetilde{x}\right)^L \rfloor X \mapsto \left(b \cdot \widetilde{x}\right)^L \rfloor (a \mid X)$ |
| Catalysis <br> on membrane | $\left(c \cdot \widetilde{x} \cdot SP_1 \cdot \widetilde{y}\right)^L \rfloor X \mapsto \left(c \cdot \widetilde{x} \cdot SP_2 \cdot \widetilde{y}\right)^L \rfloor X$ <br> where $SP_1 \mapsto SP_2$ is the catalyzed event |
| Membrane crossing | $a \mid \left(\widetilde{x}\right)^L \rfloor X \mapsto \left(\widetilde{x}\right)^L \rfloor (a \mid X)$ <br> $\left(\widetilde{x}\right)^L \rfloor (a \mid X) \mapsto a \mid \left(\widetilde{x}\right)^L \rfloor X$ <br> $\widetilde{x} \cdot a \cdot \widetilde{y} \mid \left(\widetilde{z}\right)^L \rfloor X \mapsto \left(\widetilde{z}\right)^L \rfloor (\widetilde{x} \cdot a \cdot \widetilde{y} \mid X)$ <br> $\left(\widetilde{z}\right)^L \rfloor (\widetilde{x} \cdot a \cdot \widetilde{y} \mid X) \mapsto \widetilde{x} \cdot a \cdot \widetilde{y} \mid \left(\widetilde{z}\right)^L \rfloor X$ |
| Catalyzed <br> membrane crossing | $a \mid \left(b \cdot \widetilde{x}\right)^L \rfloor X \mapsto \left(b \cdot \widetilde{x}\right)^L \rfloor (a \mid X)$ <br> $\left(b \cdot \widetilde{x}\right)^L \rfloor (a \mid X) \mapsto a \mid \left(b \cdot \widetilde{x}\right)^L \rfloor X$ <br> $\widetilde{x} \cdot a \cdot \widetilde{y} \mid \left(b \cdot \widetilde{z}\right)^L \rfloor X \mapsto \left(b \cdot \widetilde{z}\right)^L \rfloor (\widetilde{x} \cdot a \cdot \widetilde{y} \mid X)$ <br> $\left(b \cdot \widetilde{z}\right)^L \rfloor (\widetilde{x} \cdot a \cdot \widetilde{y} \mid X) \mapsto \widetilde{x} \cdot a \cdot \widetilde{y} \mid \left(b \cdot \widetilde{z}\right)^L \rfloor X$ |
| Membrane joining | $\left(\widetilde{x}\right)^L \rfloor (a \mid X) \mapsto \left(a \cdot \widetilde{x}\right)^L \rfloor X$ <br> $\left(\widetilde{x}\right)^L \rfloor (\widetilde{y} \cdot a \cdot \widetilde{z} \mid X) \mapsto \left(\widetilde{y} \cdot a \cdot \widetilde{z} \cdot \widetilde{x}\right)^L \rfloor X$ |
| Catalyzed <br> membrane joining | $\left(b \cdot \widetilde{x}\right)^L \rfloor (a \mid X) \mapsto \left(a \cdot b \cdot \widetilde{x}\right)^L \rfloor X$ <br> $\left(\widetilde{x}\right)^L \rfloor (a \mid b \mid X) \mapsto \left(a \cdot \widetilde{x}\right)^L \rfloor (b \mid X)$ <br> $\left(b \cdot \widetilde{x}\right)^L \rfloor (\widetilde{y} \cdot a \cdot \widetilde{z} \mid X) \mapsto \left(\widetilde{y} \cdot a \cdot \widetilde{z} \cdot \widetilde{x}\right)^L \rfloor X$ <br> $\left(\widetilde{x}\right)^L \rfloor (\widetilde{y} \cdot a \cdot \widetilde{z} \mid b \mid X) \mapsto \left(\widetilde{y} \cdot a \cdot \widetilde{z} \cdot \widetilde{x}\right)^L \rfloor (b \mid X)$ |
| Membrane fusion | $\left(\widetilde{x}\right)^L \rfloor (X) \mid \left(\widetilde{y}\right)^L \rfloor (Y) \mapsto \left(\widetilde{x} \cdot \widetilde{y}\right)^L \rfloor (X \mid Y)$ |
| Catalyzed <br> membrane fusion | $\left(a \cdot \widetilde{x}\right)^L \rfloor (X) \mid \left(b \cdot \widetilde{y}\right)^L \rfloor (Y) \mapsto$ <br> $\left(a \cdot \widetilde{x} \cdot b \cdot \widetilde{y}\right)^L \rfloor (X \mid Y)$ |
| Membrane division | $\left(\widetilde{x} \cdot \widetilde{y}\right)^L \rfloor (X \mid Y) \mapsto \left(\widetilde{x}\right)^L \rfloor (X) \mid \left(\widetilde{y}\right)^L \rfloor (Y)$ |
| Catalyzed <br> membrane division | $\left(a \cdot \widetilde{x} \cdot b \cdot \widetilde{y}\right)^L \rfloor (X \mid Y) \mapsto$ <br> $\left(a \cdot \widetilde{x}\right)^L \rfloor (X) \mid \left(b \cdot \widetilde{y}\right)^L \rfloor (Y)$ |

First of all, we select the biomolecular entities of interest. Since we want to describe cells, we consider molecular populations and membranes. Molecular populations are groups of molecules that are in the same compartment of the cell. Molecules can be of many types: we classify them as DNA and RNA strands, proteins, and other molecules. DNA and RNA strands and proteins can be seen as non–elementary objects. DNA strands are composed by genes, RNA strands are composed by parts corresponding to the transcription of individual genes, and proteins are composed by parts having the role of interaction sites (or domains). Other molecules are considered as elementary objects, even if they are complexes. Membranes are considered as elementary objects, in the sense that we do not describe them at the level of the lipids they are made of. The only interesting properties of a membrane are that it may contain something (hence, create a compartment) and that it may have molecules on its surface.

Now, we select the biomolecular events of interest. The simplest kind of event is the change of state of an elementary object. Then, we may have interactions between molecules: in particular complexation, decomplexation and catalysis. These interactions may involve single elements of non–elementary molecules (DNA and RNA strands, and proteins). Moreover, we may have interactions between membranes and molecules: in particular a molecule may cross or join a membrane. Finally, we may have interactions between membranes: in this case there may be many kinds of interactions (fusion, division, etc. . . ).

The guidelines for the abstraction of biomolecular entities and events into CLS are given in Table 1 and Table 2, respectively. Entities are associated with CLS terms: elementary objects are modeled as alphabet symbols, non–elementary objects as CLS sequences and membranes as looping sequences. Biomolecular events are associated with CLS rewrite rules. In the figure we give some examples of rewrite rules for each type of event. The list of examples is not complete: one could imagine also rewrite rules for the description of complexation/decomplexation events involving more than two molecules, or catalysis events in which the catalyzing molecule is on a membrane and the catalyzed event occurs in its content, or more complex interactions between membranes. We remark that in the second example of rewrite rule associated with the complexation event we have that one of the two molecules which are involved should be either an elementary object or a protein modeled as a single alphabet symbol. As before, this is caused by the problem of modeling protein interaction at the domain level. This problem is solved by the extension of CLS with links, called LCLS, we shall describe in the following.

## 2.3   Examples

A well–known example of biomolecular system is the epidermal growth factor (EGF) signal transduction pathway[26,19]. If EGF proteins are present in the environment of a cell, they should be interpreted as a proliferation signal from the environment, and hence the cell should react by synthesizing proteins which stimulate its proliferation. A cell recognizes the EGF signal because it has on its membrane some EGF receptor proteins (EGFR), which are transmembrane proteins (they have some intra–cellular and some extra–cellular domains). One of the extra–cellular domains binds to one EGF protein in the environment, forming a signal–receptor complex on the membrane. This causes a conformational change on the receptor protein that enables it to bind to another one signal–receptor complex. The formation of the binding of the two signal–receptor complexes (called dimerization) causes the phosphorylation of some intra–cellular domains of the dimer. This, in turn, causes the internal domains of the dimer to be recognized by a protein that is inside the cell (in the cytoplasm), called SHC. The protein SHC binds to the dimer, enabling a long chain of protein–protein interactions, which finally activate some proteins, such as one called ERK, which bind to the DNA and stimulate synthesis of proteins for cell proliferation.

Now, we use CLS to build a model of the first steps of the EGF signaling pathway up to the binding of the signal-receptor dimer to the SHC protein. In the following we shall refine the model by using the LCLS extension to describe interactions at the domain level.

We model the EGFR, EGF, and SHC proteins as the alphabet symbols $EGFR$, $EGF$ and $SHC$, respectively. The cell is modeled as a looping sequence (representing its external membrane), initially composed only by $EGFR$ symbols, containing $SHC$ symbols and surrounded by $EGF$ symbols. The rewrite rules modeling the first steps of the pathway are the following:

$$EGF \mid \left(EGFR \cdot \widetilde{x}\right)^L \rfloor X \;\mapsto\; \left(CMPLX \cdot \widetilde{x}\right)^L \rfloor X \qquad (R1)$$

$$\left(CMPLX \cdot \widetilde{x} \cdot CMPLX \cdot \widetilde{y}\right)^L \rfloor X \;\mapsto\; \left(DIM \cdot \widetilde{x} \cdot \widetilde{y}\right)^L \rfloor X \qquad (R2)$$

$$\left(DIM \cdot \widetilde{x}\right)^L \rfloor X \;\mapsto\; \left(DIMp \cdot \widetilde{x}\right)^L \rfloor X \qquad (R3)$$

$$\left(DIMp \cdot \widetilde{x}\right)^L \rfloor (SHC \mid X) \;\mapsto\; \left(DIMpSHC \cdot \widetilde{x}\right)^L \rfloor X \qquad (R4)$$

Rule R1 describes the binding of a EGF protein to a EGFR receptor protein on the membrane surface. The result of the binding is a signal-receptor complex denoted $CMPLX$. Rule R2 describes the dimerization of two signal-receptor complex, the result is denoted $DIM$. Rule R3 describes the phosphorylation (and activation) of a signal-receptor dimer, that is the replacement of a $DIM$ symbol with a $DIMp$ symbol. Finally, rule R4 describes the binding of an active dimer $DIMp$ with a SHC protein contained in the cytoplasm. The result is a $DIMpSHC$ symbol placed on the membrane surface.

A possible initial term for the model in this example is given by a looping sequence composed by some $EGFR$ symbols, containing some $SHC$ symbols and with some $EGF$ symbols outside. A possible evolution of such a term by means of application of the given rewrite rules is the following (we write on each transition the name of the rewrite rule applied):

$$EGF \mid EGF \mid \left(EGFR \cdot EGFR \cdot EGFR \cdot EGFR\right)^L \rfloor (SHC \mid SHC)$$

$$\xrightarrow{(R1)} EGF \mid \left(EGFR \cdot CMPLX \cdot EGFR \cdot EGFR\right)^L \rfloor (SHC \mid SHC)$$

$$\xrightarrow{(R1)} \left(EGFR \cdot CMPLX \cdot EGFR \cdot CMPLX\right)^L \rfloor (SHC \mid SHC)$$

$$\xrightarrow{(R2)} \ \left(EGFR \cdot DIM \cdot EGFR\right)^L \rfloor (SHC \mid SHC)$$

$$\xrightarrow{(R3)} \ \left(EGFR \cdot DIMp \cdot EGFR\right)^L \rfloor (SHC \mid SHC)$$

$$\xrightarrow{(R4)} \ \left(EGFR \cdot DIMpSHC \cdot EGFR\right)^L \rfloor SHC$$

We show another example of modeling of a biomolecular system with CLS, that is the modeling of a simple gene regulation process. This kind of processes are essential for cell life as they allow a cell to regulate the production of proteins that may have important roles for instance in metabolism, growth, proliferation and differentiation.

The example we consider is as follows: we have a simple DNA fragment consisting of a sequence of three genes. The first, denoted $p$, is called *promoter* and is the place where a *RNA polymerase* enzyme (responsible for translation of DNA into RNA) binds to the DNA. The second, denoted $o$, is called *operator* and it is the place where a *repressor* protein (responsible for regulating the activity of the RNA polymerase) binds to the DNA. The third, denoted as $g$, is the gene that encodes for the protein whose production is regulated by this process.

When the repressor is not bound to the DNA, the RNA polymerase can scan the sequence of genes and transcribe gene $g$ into a piece of RNA that will be later translated into the protein encoded by $g$. When the repressor is bound to the DNA, it becomes an obstacle for the RNA polymerase that cannot scan any more the sequence of genes.

The CLS model of this simple regulation process is a follows. The sequence of genes is represented as the CLS sequence $p \cdot o \cdot g$, the RNA polymerase enzyme as $polym$, the repressor protein as $repr$, and the piece of RNA obtained by the translation of gene $g$ as $rna$. The rewrite rules describing the process are the following:

$$polym \mid p \cdot \widetilde{x} \ \mapsto \ pp \cdot \widetilde{x} \tag{R1}$$

$$repr \mid \widetilde{x} \cdot o \cdot \widetilde{y} \ \mapsto \ \widetilde{x} \cdot ro \cdot \widetilde{y} \tag{R2}$$

$$pp \cdot o \cdot \widetilde{x} \ \mapsto \ p \cdot po \cdot \widetilde{x} \tag{R3}$$

$$\widetilde{x} \cdot po \cdot g \ \mapsto \ \widetilde{x} \cdot o \cdot pg \tag{R4}$$

$$\widetilde{x} \cdot pg \ \mapsto \ polym \mid rna \mid \widetilde{x} \cdot g \tag{R5}$$

Rules R1 and R2 describe the binding of the RNA polymerase and of the repressor to the corresponding genes in the DNA sequences. The results of these bindings are that the symbols representing the two genes are replaced by $pp$ and $ro$, respectively. Rules R3, R4 and R5 describe the activity of the RNA polymerase enzyme in the absence of the repressor: it moves from gene $p$ to gene $o$ in rule R3, then it moves from gene $o$ to gene $g$ in rule R4, and finally it produces the RNA fragment and leaves the DNA in rule R5. Note that, in order to apply rule R3, the repressor must be not bound to the DNA.

The only possible evolution of a term representing an initial situation in which
no repressors are present is

$$polym \mid p \cdot o \cdot g \xrightarrow{(R1)} pp \cdot o \cdot g \xrightarrow{(R3)} p \cdot po \cdot g$$
$$\xrightarrow{(R4)} p \cdot o \cdot pg \xrightarrow{(R5)} polym \mid rna \mid p \cdot o \cdot g$$

that represent the case in which the RNA polymerase enzyme can scan the
DNA sequence and transcribe gene $g$ into a piece of RNA. When the repressor
is present, instead, a possible evolution is

$$repr \mid polym \mid p \cdot o \cdot g \xrightarrow{(R1)} repr \mid pp \cdot o \cdot g \xrightarrow{(R2)} pp \cdot ro \cdot g$$

and it corresponds to a situation in which the repressor stops the transcription
of the gene by hampering the activity of the RNA polymerase.

## 3    Two Extensions of CLS

In this section we describe two extensions of CLS. The first, Stochastic CLS,
allows describing quantitative aspects of the modeled systems, such as the time
spent by occurrences of chemical reactions. The second, CLS with links, allows
describing protein interaction more precisely at a lower level of abstraction,
namely at the domain level.

### 3.1    Stochastic CLS

In CLS only qualitative aspects of biological systems are considered, such as
their structure and the presence (or the absence) of certain molecules. As a
consequence, on CLS models it is only possible to verify properties such as
the reachability of particular states or causality relationships between events.
It would be interesting to verify also properties such as the time spent to
reach a particular state, or the probability of reaching it. To face this problem, in [6] we have developed a stochastic extension of CLS, called *Stochastic
CLS*, in which quantitative aspects, such as time and probability are taken into
account.

The standard way of extending a formalism to model quantitative aspects of
biological systems is by incorporating the stochastic framework developed by
Gillespie with its simulation algorithm for chemical reactions [12] in the semantics of the formalism. This has been done, for instance, for the $\pi$–calculus [20,22].
The idea of Gillespie's algorithm is that a rate constant is associated with each
chemical reaction that may occur in the system. Such a constant is obtained by
multiplying the kinetic constant of the reaction by the number of possible combinations of reactants that may occur in the system. The resulting rate constant
is then used as the parameter of an exponential distribution modeling the time
spent between two occurrences of the considered chemical reaction.

The use of exponential distributions to represent the (stochastic) time spent
between two occurrences of chemical reactions allows describing the system as a

Continuous Time Markov Chain (CTMC), and consequently it allows verifying properties of the described system by means of analytic means and by means of stochastic model checkers.

In Stochastic CLS, incorporating Gillespie's stochastic framework is not a simple exercise. The main difficulty is counting the number of possible reactant combinations of the chemical reaction described by a rewrite rule. This means counting the number of different positions where the rewrite rule can be applied, by taking into account that rules may contain variables. We have defined the Stochastic CLS in [6], and showed how to derive a CTMC from the semantics of a system modeled in Stochastic CLS. This allows performing simulation and verification of properties of the described systems, for instance by using stochastic model checkers, such as PRISM [13].

Let us consider the simple regulation process we modeled with CLS in Section 2.3. We now extend the CLS model by including a kinetic constant in each rewrite rule. The result is a Stochastic CLS model. In order to make the model a little more realistic we add two rewrite rules describing the unbinding of the RNA polymerase and of the repressor from the DNA. Hence, the rewrite rules of the Stochastic CLS model are the following:

$$polym \mid p \cdot \widetilde{x} \stackrel{0.1}{\longmapsto} pp \cdot \widetilde{x} \tag{R1}$$

$$pp \cdot \widetilde{x} \stackrel{2}{\longmapsto} polym \mid p \cdot \widetilde{x} \tag{R1'}$$

$$repr \mid \widetilde{x} \cdot o \cdot \widetilde{y} \stackrel{1}{\longmapsto} \widetilde{x} \cdot ro \cdot \widetilde{y} \tag{R2}$$

$$\widetilde{x} \cdot ro \cdot \widetilde{y} \stackrel{10}{\longmapsto} repr \mid \widetilde{x} \cdot o \cdot \widetilde{y} \tag{R2'}$$

$$pp \cdot o \cdot \widetilde{x} \stackrel{100}{\longmapsto} p \cdot po \cdot \widetilde{x} \tag{R3}$$

$$\widetilde{x} \cdot po \cdot g \stackrel{100}{\longmapsto} \widetilde{x} \cdot o \cdot pg \tag{R4}$$

$$\widetilde{x} \cdot pg \stackrel{30}{\longmapsto} polym \mid rna \mid \widetilde{x} \cdot g \tag{R5}$$

We developed a simulator based on Stochastic CLS, and we used it to study the behaviour of the regulation process. In particular, we performed simulations by varying the quantity of repressors and we observed the production of RNA fragments in each case. The initial configuration of the system is given by the following term

$$\underbrace{repr \mid \ldots \mid repr}_{n} \mid \underbrace{polym \mid \ldots \mid polym}_{100} \mid p \cdot o \cdot g$$

and we performed simulations with $n = 0, 10, 25$ and $50$. The results of the simulations are shown in Figure 2. By varying the number of repressors from 0 to 50 the rate of transcription of the DNA into RNA molecules decreases.

## 3.2   CLS with Links (LCLS)

A formalism for modeling proteins interactions at the domain level was developed in the seminal paper by Danos and Laneve [11], and extended in [14]. This

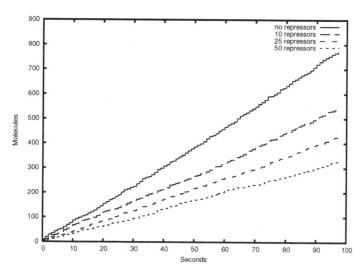

**Fig. 2.** Simulation result of the regulation process: number of RNA molecules over time

formalism allows expressing proteins by a node with a fixed number of domains; binding between domains allow complexating proteins. In this section we extend CLS to represent proteins interaction at the domain level. Such an extension, called Calculus of Linked Looping Sequences (LCLS), is obtained by labelling elementary components of sequences. Two elements with the same label are considered to be linked.

To model a protein at the domain level in CLS it would be natural to use a sequence with one symbol for each domain. However, the binding between two domains of two different proteins, that is the linking between two elements of two different sequences, cannot be expressed in CLS. To represent this, we extend CLS by labels on basic symbols. If in a term two symbols appear having the same label, we intend that they represent domains which are bound to each other. If in a term there is a symbol with a label and no other symbol with the same label, we intend that the term represents only a part of a system we model, and that the symbol will be linked to some other symbol in another part of the term representing the full model.

As membranes create compartments, elements inside a looping sequence cannot be linked to elements outside. Elements inside a membrane can be linked either to other elements inside the membrane or to elements of the membrane itself. An element can be linked at most to another element.

As an example, we model in LCLS the first steps of the EGF pathway described before. We model the EGFR protein as the sequence $R_{E1} \cdot R_{E2} \cdot R_{I1} \cdot R_{I2}$ where $R_{E1}$ and $R_{E2}$ are two extra–cellular domains and $R_{I1}$ and $R_{I2}$ are two intra–cellular domains. The membrane of the cell is modeled as a looping sequence which could contain EGFR proteins. Outside the looping sequence (i.e. in the environment) there could be EGF proteins, and inside (i.e. in the cytoplasm) there could be SHC proteins. Rewrite rules modeling the pathway are the following:

$$EGF \mid \left(R_{E1} \cdot \widetilde{x}\right)^{L} \rfloor X \;\mapsto\; \left(sR_{E1} \cdot \widetilde{x}\right)^{L} \rfloor X \tag{R1}$$

$$\left(sR_{E1} \cdot R_{E2} \cdot x \cdot y \cdot \widetilde{x} \cdot sR_{E1} \cdot R_{E2} \cdot z \cdot w \cdot \widetilde{y}\right)^{L} \rfloor X \;\mapsto\;$$
$$\left(sR_{E1} \cdot R_{E2}^{1} \cdot x \cdot y \cdot sR_{E1} \cdot R_{E2}^{1} \cdot z \cdot w \cdot \widetilde{x} \cdot \widetilde{y}\right)^{L} \rfloor X \tag{R2}$$

$$\left(R_{E2}^{1} \cdot R_{I1} \cdot \widetilde{x} \cdot R_{E2}^{1} \cdot R_{I1} \cdot \widetilde{y}\right)^{L} \rfloor X \;\mapsto\;$$
$$\left(R_{E2}^{1} \cdot PR_{I1} \cdot \widetilde{x} \cdot R_{E2}^{1} \cdot R_{I1} \cdot \widetilde{y}\right)^{L} \rfloor X \tag{R3}$$

$$\left(R_{E2}^{1} \cdot PR_{I1} \cdot \widetilde{x} \cdot R_{E2}^{1} \cdot R_{I1} \cdot \widetilde{y}\right)^{L} \rfloor X \;\mapsto\;$$
$$\left(R_{E2}^{1} \cdot PR_{I1} \cdot \widetilde{x} \cdot R_{E2}^{1} \cdot PR_{I1} \cdot \widetilde{y}\right)^{L} \rfloor X \tag{R4}$$

$$\left(R_{E2}^{1} \cdot PR_{I1} \cdot R_{I2} \cdot \widetilde{x} \cdot R_{E2}^{1} \cdot PR_{I1} \cdot R_{I2} \cdot \widetilde{y}\right)^{L} \rfloor (SHC \mid X) \;\mapsto\;$$
$$\left(R_{E2}^{1} \cdot PR_{I1} \cdot R_{I2}^{2} \cdot \widetilde{x} \cdot R_{E2}^{1} \cdot PR_{I1} \cdot R_{I2} \cdot \widetilde{y}\right)^{L} \rfloor (SHC^{2} \mid X) \tag{R5}$$

Rule R1 represents the binding of the EGF protein to the receptor domain $R_{E1}$ with $sR_{E1}$ as a result. Rule R2 represents that when two EGFR proteins activated by proteins EGF occur on the membrane, they may bind to each other to form a dimer (shown by the link 1). Rule R3 represents the phosphorylation of one of the internal domains $R_{I1}$ of the dimer, and rule R4 represents the phosphorylation of the other internal domain $R_{I1}$ of the dimer. The result of each phosphorylation is $pR_{I1}$. Rule R5 represents the binding of the protein SHC in the cytoplasm to an internal domain $R_{I2}$ of the dimer. Remark that the binding of SHC to the dimer is represented by the link 2, allowing the protein SHC to continue the interactions to stimulate cell proliferation.

Let us denote the sequence $R_{E1} \cdot R_{E2} \cdot R_{I1} \cdot R_{I2}$ by EGFR. By starting from a cell with some EGFR proteins on its membrane, some SHC proteins in the cytoplasm and some EGF proteins in the environment, a possible evolution is the following:

$$EGF \mid EGF \mid \left(EGFR \cdot EGFR \cdot EGFR\right)^{L} \rfloor (SHC \mid SHC)$$
$$\xrightarrow{(R1)} EGF \mid \left(sR_{E1} \cdot R_{E2} \cdot R_{I1} \cdot R_{I2} \cdot EGFR \cdot EGFR\right)^{L} \rfloor (SHC \mid SHC)$$
$$\xrightarrow{(R1)} \left(sR_{E1} \cdot R_{E2} \cdot R_{I1} \cdot R_{I2} \cdot EGFR \cdot sR_{E1} \cdot R_{E2} \cdot R_{I1} \cdot R_{I2}\right)^{L} \rfloor (SHC \mid SHC)$$
$$\xrightarrow{(R2)} \left(sR_{E1} \cdot R_{E2}^{1} \cdot R_{I1} \cdot R_{I2} \cdot sR_{E1} \cdot R_{E2}^{1} \cdot R_{I1} \cdot R_{I2} \cdot EGFR\right)^{L} \rfloor (SHC \mid SHC)$$

$$\xrightarrow{(R3)} \left(sR_{E1}\cdot R_{E2}^1\cdot pR_{I1}\cdot R_{I2}\cdot sR_{E1}\cdot R_{E2}^1\cdot R_{I1}\cdot R_{I2}\cdot EGFR\right)^L \rfloor (SHC \mid SHC)$$

$$\xrightarrow{(R4)} \left(sR_{E1}\cdot R_{E2}^1\cdot pR_{I1}\cdot R_{I2}\cdot sR_{E1}\cdot R_{E2}^1\cdot pR_{I1}\cdot R_{I2}\cdot EGFR\right)^L \rfloor (SHC \mid SHC)$$

$$\xrightarrow{(R5)} \left(sR_{E1}\cdot R_{E2}^1\cdot pR_{I1}\cdot R_{I2}^2\cdot sR_{E1}\cdot R_{E2}^1\cdot pR_{I1}\cdot R_{I2}\cdot EGFR\right)^L \rfloor (SHC^2 \mid SHC)$$

## 4  CLS and Membranes

What could seem strange in CLS is the use of looping sequences for the description of membranes, as sequencing is not a commutative operation and this do not correspond to the usual fluid representation of membranes in which objects can move freely. What one would expect is to have a multiset or a parallel composition of objects on a membrane. In the case of CLS, what could be used is a parallel composition of sequences. To address this problem, we define an extension of CLS, called CLS+, in which the looping operator can be applied to a parallel composition of sequences, and we show that we can translate quite easily CLS+ models into CLS ones.

### 4.1  Definition of CLS+

Terms in CLS+ are defined as follows.

**Definition 6 (Terms).** *Terms $T$, branes $B$, and sequences $S$ of CLS+ are given by the following grammar:*

$$
\begin{aligned}
T &::= S \mid (B)^L \rfloor T \mid T \mid T \\
B &::= S \mid B \mid B \\
S &::= \epsilon \mid a \mid S \cdot S
\end{aligned}
$$

*where $a$ is a generic element of $\mathcal{E}$. We denote with $\mathcal{T}$ the infinite set of terms, with $\mathcal{B}$ the infinite set of branes, and with $\mathcal{S}$ the infinite set of sequences.*

The structural congruence relation of CLS+ is a trivial extension of the one of CLS. The only difference is that commutativity of branes replaces rotation of looping sequences.

**Definition 7 (Structural Congruence).** *The structural congruence relations $\equiv_S$, $\equiv_B$ and $\equiv_T$ are the least congruence relations on sequences, on branes and on terms, respectively, satisfying the following rules:*

$$S_1 \cdot (S_2 \cdot S_3) \equiv_S (S_1 \cdot S_2) \cdot S_3 \qquad S \cdot \epsilon \equiv_S \epsilon \cdot S \equiv_S S$$

$$S_1 \equiv_S S_2 \text{ implies } S_1 \equiv_B S_2$$

$$B_1 \mid B_2 \equiv_B B_2 \mid B_1 \qquad B_1 \mid (B_2 \mid B_3) \equiv_B (B_1 \mid B_2) \mid B_3 \qquad B \mid \epsilon \equiv_B B$$

$$S_1 \equiv_S S_2 \; implies \; S_1 \equiv_T S_2$$

$$B_1 \equiv_B B_2 \; implies \; (B_1)^L \rfloor T \equiv_T (B_2)^L \rfloor T$$

$$T_1 \mid T_2 \equiv_T T_2 \mid T_1 \quad T_1 \mid (T_2 \mid T_3) \equiv_T (T_1 \mid T_2) \mid T_3 \quad T \mid \epsilon \equiv_T T \quad (\epsilon)^L \rfloor \epsilon \equiv \epsilon$$

Now, to define patterns in CLS+ we consider an additional type of variables with respect of CLS, namely brane variables. We assume a set of brane variables $BV$ ranged over by $\overline{x}, \overline{y}, \overline{z}, \ldots$.

**Definition 8 (Patterns).** Patterns $P$, brane patterns $BP$ and sequence patterns $SP$ of CLS+ are given by the following grammar:

$$P \; ::= \; SP \quad | \quad (BP)^L \rfloor P \quad | \quad P \mid P \quad | \quad X$$
$$BP \; ::= \; SP \quad | \quad BP \mid BP \quad | \quad \overline{x}$$
$$SP \; ::= \; \epsilon \quad | \quad a \quad | \quad SP \cdot SP \quad | \quad \widetilde{x} \quad | \quad x$$

where $a$ is a generic element of $\mathcal{E}$, and $X, \overline{x}, \widetilde{x}$ and $x$ are generic elements of $TV, BV, SV$ and $\mathcal{X}$, respectively. We denote with $\mathcal{P}$ the infinite set of patterns.

**Definition 9 (Rewrite Rules).** A rewrite rule is a pair of patterns $(P_1, P_2)$, denoted with $P_1 \mapsto P_2$, where $P_1, P_2 \in PP$, $P_1 \not\equiv \epsilon$ and such that $Var(P_2) \subseteq Var(P_1)$. We denote with $\Re$ the infinite set of all the possible rewrite rules.

Now, differently from CLS, we have that a rule such as $a \mid b \mapsto c$ could be applied to elements of a looping sequence. For instance, $a \mid b \mapsto c$ can be applied to the term $(a \mid b)^L \rfloor d$ so to obtain the term $(c)^L \rfloor d$. However, a rule such as $(a)^L \rfloor b \mapsto c$ still cannot be applied to elements of a looping sequences, as $((a)^L \rfloor b)^L \rfloor c$ is not a CLS+ term.

The rules that can be applied to elements of a looping sequence are those having the form $(B_1, B_2)$ with $B_1, B_2 \in \mathcal{B}$. We call these rules *brane rules* and we denote as $\Re_\mathcal{B} \subset \Re$ their infinite set. Now, in the semantics of CLS+ we have to take into account brane rules and allow them to be applied also to elements of looping sequences. Hence, we define the semantics as follows.

**Definition 10 (Semantics).** Given a set of rewrite rules $\mathcal{R} \subseteq \Re$, and a set of brane rules $\mathcal{R}_\mathcal{B} \subseteq \mathcal{R}$, such that $(\mathcal{R} \setminus \mathcal{R}_\mathcal{B}) \cap \Re_\mathcal{B} = \varnothing$, the semantics of CLS is the least transition relation $\rightarrow$ on terms closed under $\equiv$, and satisfying the following inference rules:

$$\frac{(P_1, P_2) \in \mathcal{R} \quad P_1\sigma \not\equiv \epsilon \quad \sigma \in \Sigma}{P_1\sigma \rightarrow P_2\sigma} \qquad \frac{T_1 \rightarrow T_2}{T \mid T_1 \rightarrow T \mid T_2} \qquad \frac{T_1 \rightarrow T_2}{(B)^L \rfloor T_1 \rightarrow (B)^L \rfloor T_2}$$

$$\frac{(BP_1, BP_2) \in \mathcal{R}_\mathcal{B} \quad BP_1\sigma \not\equiv \epsilon \quad \sigma \in \Sigma}{BP_1\sigma \rightarrow_\mathcal{B} BP_2\sigma}$$

$$\frac{B_1 \rightarrow_\mathcal{B} B_2}{B \mid B_1 \rightarrow_\mathcal{B} B \mid B_2} \qquad \frac{B_1 \rightarrow_\mathcal{B} B_2}{(B_1)^L \rfloor T \rightarrow (B_2)^L \rfloor T}$$

*where $\to_\mathcal{B}$ is a transition relation on branes, and where the symmetric rules for the parallel composition of terms and of branes are omitted.*

In the definition of the semantics of CLS+ we use an additional transition relation $\to_\mathcal{B}$ on branes. This relation is used to describe the application of a brane rule to elements of a looping sequence. As usual, a CLS+ model is composed by a term, representing the initial state of the modeled system, and a set of rewrite rules.

In the following section we show that CLS+ models can be translated into CLS models. The translation into CLS is compositional and preserves the semantics of the model.

## 4.2   Translating CLS+ into CLS

The first step of the translation of a CLS+ model into CLS is a preprocessing procedure. For each brane rule $(BP_1, BP_2)$ in the CLS+ model, we add to the set of rules of the model a new rule, namely $((BP_1 \mid \overline{x})^L \rfloor X, (BP_2 \mid \overline{x})^L \rfloor X)$. This new rule is redundant in the model, as every time it can be applied to a CLS+ term, also the original one can be applied with the same result. However, the translation we are going to define will translate the original rule into a CLS rule that will be applicable only inside looping sequences, or at the top level of the term, and will translate the new rule into a CLS rule applicable only to elements that compose a looping sequence.

Now, the translation of CLS+ into CLS consists mainly of an encoding function, denoted $\{\!|\cdot|\!\}$, which maps CLS+ patterns into CLS patterns. This encoding function will be used to translate each rewrite rule of the CLS+ model into a rewrite rule for the corresponding CLS model, and to translate the term representing the initial state of the system in the CLS+ model into a CLS term for the corresponding CLS model.

The encoding function for CLS+ patterns is defined as follows. We assume a total and injective function from brane variables into a subset of term variables that are never used in CLS models. More easily, we assume brane variables to be a subset of the term variables of CLS. Moreover, we assume *in* and *out* to be symbols of the alphabet $\mathcal{E}$ never used in CLS models.

The encoding follows the "ball–bearing" technique described by Cardelli in [7]. Intuitively, every CLS+ looping sequence is translated into a couple of CLS looping sequences, one contained in the other, with the brane patterns of the CLS+ looping sequence between the two corresponding CLS looping sequences.

**Definition 11 (Encoding Function).** *The encoding function $\{\!|\cdot|\!\}$ maps CLS+ patterns into CLS patterns, and is given by the following recursive definition:*

$$\{\!|SP|\!\} = SP$$

$$\{\!|X|\!\} = X$$

$$\{\!|(BP)^L \rfloor P|\!\} = (out)^L \rfloor (BP \mid (in)^L \rfloor \{\!|P|\!\})$$

$$\{\!|P_1 \mid P_2|\!\} = \{\!|P_1|\!\} \mid \{\!|P_2|\!\}$$

A CLS rewrite rule is obtained from each CLS+ rewrite rule of the translated model by applying the encoding function to the two patterns of the rule. More precisely, given a CLS+ rule $P_1 \mapsto P_2$, the corresponding CLS rule is $(in)^L \rfloor (\{[P_1]\} \mid X) \mapsto (in)^L \rfloor (\{[P_2]\} \mid X)$ where $X$ is a term variable that does not occur in $P_1$ and $P_2$. For example, by applying the encoding to the two patterns of the CLS+ rewrite rule

$$R = b \cdot x \mid c \mapsto b \cdot x$$

we obtain

$$R_{\{[\cdot]\}} = (in)^L \rfloor (b \cdot x \mid c \mid X) \mapsto (in)^L \rfloor (b \cdot x \mid X).$$

The encoding of a CLS+ term into a CLS term is as follows: given a CLS+ term $T$ the corresponding CLS term is $(in)^L \rfloor \{[T]\}$. In this case we have that the encoding function never encounters variables. Consider, as an example, the following CLS+ term:

$$T = a \mid \left(c \mid d \mid b \cdot b \mid d\right)^L \rfloor d$$

the corresponding CLS term is as follows:

$$T_{\{[\cdot]\}} = (in)^L \rfloor \left(a \mid (out)^L \rfloor \left(c \mid d \mid b \cdot b \mid d \mid (in)^L \rfloor d\right)\right)$$

Now, it is easy to see that $R$ can be applied to $T$, because parallel components in the looping sequence can be commuted, and the result of the application is

$$T' = a \mid \left(b \cdot b \mid d \mid d\right)^L \rfloor d$$

but the corresponding CLS rewrite rule $R_{\{[\cdot]\}}$ cannot be applied to $T_{\{[\cdot]\}}$. However, we have that $R \in \mathcal{R}_\mathcal{B}$, and hence, by the preprocessing phase described above, we have that also

$$R' = \left(b \cdot x \mid c \mid \overline{x}\right)^L \rfloor X \mapsto \left(b \cdot x \mid \overline{x}\right)^L \rfloor X$$

is a rule of the CLS+ model. By translating rule $R'$ we obtain

$$R'_{\{[\cdot]\}} = (in)^L \rfloor \left((out)^L \rfloor (b \cdot x \mid c \mid \overline{x} \mid (in)^L \rfloor X) \mid Y\right) \mapsto$$
$$(in)^L \rfloor \left((out)^L \rfloor (b \cdot x \mid \overline{x} \mid (in)^L \rfloor X) \mid Y\right)$$

that can be applied to $T_{\{[\cdot]\}}$. The result of the application is

$$(in)^L \rfloor \left(a \mid (out)^L \rfloor (b \cdot b \mid d \mid d \mid (in)^L \rfloor d)\right)$$

that corresponds exactly to the encoding of $T'$.

### 4.3   CLS, Brane Calculi and P Systems

Brane Calculi are a family of process calculi specialized in the description of membrane activity, and they allow associating processes with membranes of a membrane structure. Each process is composed by actions whose execution has an effect on the membrane structure. Some examples of actions are phagocytosis (a membrane engulfs another one), exocytosis (a membrane expels another one), and pinocytosis (a new membrane is created inside another one). These three actions are enough to define the simplest of Brane Calculi, namely the PEP calculus. Other actions, such as fusion of membranes and mitosis can be used to define different calculi of the family. Moreover, extensions of Brane Calculi allow describing interactions with molecules and complexes, such as letting them enter and exit membranes.

We have given a sound and complete encoding of the PEP calculus into CLS in [3,15]. Here, to recall shortly the encoding technique, we give a very simple example of PEP system and we show its translation into CLS. The PEP system we consider is the following

$$\phi(\!|\diamond|\!) \circ \phi^{\perp}(\mathbf{0})(\!|\diamond|\!)$$

representing two adjacent membranes $\phi_n(\!|\diamond|\!)$ and $\phi_n^{\perp}(\mathbf{0})(\!|\diamond|\!)$ ($\circ$ denotes juxtaposition) both containing nothing of relevant (what is between brackets $(\!|\ |\!)$ is the content of the membrane and $\diamond$ is the null system). The processes associated with the two membranes are $\phi$ and $\phi^{\perp}(\mathbf{0})$, respectively, representing two complementary phagocytosis actions: the first says that the membrane it is associated with can be engulfed by another membrane, and the second that the membrane it is associated with can engulf another membrane, that will be sourrounded by another new membrane whose associated process is the parameter of the action (in this case it is the idle process $\mathbf{0}$). Hence, in accordance with the semantics of the PEP calculus, we have that the only transition that can be performed by the system is the following, leading to a system that is equivalent to the null system $\diamond$:

$$\phi(\!|\diamond|\!) \circ \phi^{\perp}(\mathbf{0})(\!|\diamond|\!) \quad \rightarrow \quad \mathbf{0}(\!|\mathbf{0}(\!|\mathbf{0}(\!|\diamond|\!)|\!)|\!) \quad \equiv \quad \diamond$$

By applying the encoding to the system we obtain the following CLS term $T$:

$$act \cdot circ \cdot e \cdot brane \cdot b \cdot \phi \cdot a \cdot \mathbf{0} \cdot a \cdot b \cdot \mathbf{0} \cdot e \cdot brane \cdot d \cdot \phi^{\perp} \cdot c \cdot \mathbf{0} \cdot c \cdot \mathbf{0} \cdot c \cdot d \cdot \mathbf{0}$$

where $act$ is a sort of program counter that precedes the symbol representing the next action to be executed, symbol $circ$ represents $\circ$, symbol $brane$ represents a membrane $(\!|\ |\!)$, symbols $\phi$ and $\phi^{\perp}$ represent the corresponding actions, symbol $\mathbf{0}$ represents the idle process and symbols $a, b, c, d$ and $e$ are used as separators of actions and parameters. The translation consists also of a set of CLS rewrite rules to be applied to terms obtained by the encoding of PEP systems. Such a set of rewrite rules does not depend on the encoded PEP system, hence it is always the same. By applying rewrite rules, the long sequence obtained from the encoding is transformed into a hierarchy of looping sequences corresponding to the membrane hierarchy in the original PEP system, then rewrite rules are applied that correspond to the semantics of the actions occurring in the processes associated with membranes.

Hence, by means of application of rewrite rules, the result of the encoding of the PEP system may evolve as follows (where $\rightarrow^*$ represents a sequence of rewrite rule applications):

$$
\begin{aligned}
T \;\rightarrow\; & act \cdot brane \cdot b \cdot \phi \cdot a \cdot \mathbf{0} \cdot a \cdot b \cdot \mathbf{0} \mid act \cdot brane \cdot d \cdot \phi^{\perp} \cdot c \cdot \mathbf{0} \cdot c \cdot \mathbf{0} \cdot c \cdot d \\
\rightarrow^* \; & \left( act \cdot \phi \cdot a \cdot \mathbf{0} \cdot a \right)^{L} \rfloor \; act \cdot \mathbf{0} \mid \left( act \cdot \phi^{\perp} \cdot c \cdot \mathbf{0} \cdot c \cdot \mathbf{0} \cdot c \right)^{L} \rfloor \; act \cdot \mathbf{0} \\
\rightarrow\; & \left( act \cdot \mathbf{0} \right)^{L} \rfloor \; (act \cdot \mathbf{0} \mid \left( act \cdot \mathbf{0} \right)^{L} \rfloor \; \left( act \cdot \mathbf{0} \right)^{L} \rfloor \; act \cdot \mathbf{0}) \\
\rightarrow^* \; & act \cdot \mathbf{0}
\end{aligned}
$$

Differently from Brane Calculi, P Systems (in their most common formulation) do not allow describing complex membrane activities such as phagocytosis and exocytosis. However, they are specialized in the description of reactions between molecules which are placed in a compartment of a complex membrane structure.

A P System is a membrane structure (a nesting of membranes) in which there could be multisets of objects representing molecules. A set of multiset rewrite rules is associated with each membrane, and describe the reactions that may occur between the molecules contained in the membrane. The result of the application of a rewrite rule can either remain in the same membrane, or exit the membrane, or enter an inner membrane. Priorities can be imposed on rewrite rules, meaning that some rules can be applied only if some others cannot, and it is possible for a membrane to dissolve and release its content to in the environment.

A peculiarity of P Systems is that rewrite rules are applied in a fully–parallel manner, namely in one step of evolution of the system all rules are applied as many times as possible (to different molecules), and this is one of the main differences with respect to CLS in which at each step one only rewrite rule is applied. We show that P Systems can be translated into CLS, and that the execution of a (fully parallel) step of a P System is simulated by a sequence of steps in CLS. A variant of P Systems, called Sequential P Systems, in which rules are applied sequentially is described in [10]. We do not consider the translation of this variant into CLS as it would be quite trivial and of little interest.

To recall the encoding technique, we give a simple example of P System and we show its translation into CLS. We focus on the translation of multiple parallelism, hence we consider a P System (depicted in Figure 3) consisting of a single membrane with only two rules, without priorities and without membrane dissolutions. We give a simplified translation: more details can be found in [15].

The alphabet of objects in the considered P System is $\{a, b, c\}$. A multiset of objects from this alphabet is represented by a CLS term as follows: let $n_a, n_b$ and $n_c$ be the number of occurrences of $a, b$ and $c$ in the multiset, respectively, then

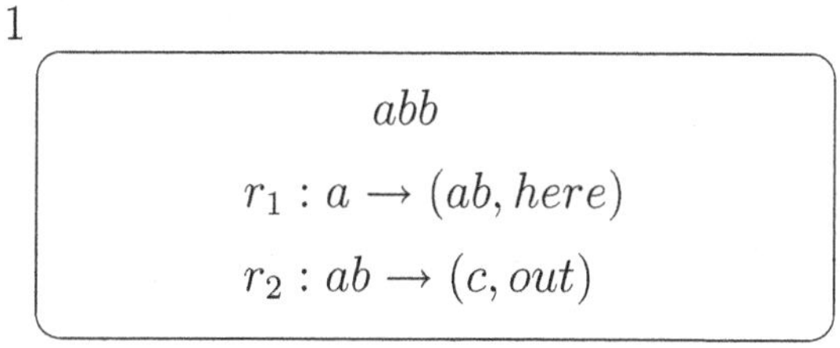

1

$$abb$$
$$r_1 : a \rightarrow (ab, here)$$
$$r_2 : ab \rightarrow (c, out)$$

**Fig. 3.** A simple example of a P system

$$a \cdot \overbrace{1 \cdot \ldots \cdot 1}^{n_a} \mid b \cdot \overbrace{1 \cdot \ldots \cdot 1}^{n_b} \mid c \cdot \overbrace{1 \cdot \ldots \cdot 1}^{n_c}$$

is the term representing the multiset. We choose this representation as it allows us checking whether an object is absent, by checking whether the corresponding symbol if followed by zero 1s. An empty multiset is represented as $a \mid b \mid c$.

The CLS term obtained by the translation of the considered P System is the following:

$$(1)^L \rfloor (Check \mid a \cdot 1 \mid b \cdot 1 \cdot 1 \mid c \mid r_1 \mid r_2 \mid (next)^L \rfloor (a \mid b \mid c))$$

where the membrane of the P System is represented by a looping sequence composed by the membrane label (in this case 1). Inside the looping sequence there is a *Check* symbol representing the current state of the system, the translation of the multiset of objects of the membrane, two symbols $r_1$ and $r_2$ corresponding to the evolutionary rules of the membrane, and an empty multiset surrounded by a looping sequence *next*. This empty multiset is used to store temporary information on the result of the application of evolutionary rules.

The CLS rewrite rules obtained by the encoding of the considered P System are the following:

$$(1)^L \rfloor (X \mid Check \mid a \cdot 1 \cdot \tilde{x} \mid r_1) \mapsto (1)^L \rfloor (X \mid Check \mid a \cdot 1 \cdot \tilde{x} \mid r_1 \cdot 1) \quad \text{(C1)}$$

$$(1)^L \rfloor (X \mid Check \mid a \mid r_1) \mapsto (1)^L \rfloor (X \mid Check \mid a \mid r_1 \cdot 0) \quad \text{(C2)}$$

$$(1)^L \rfloor (X \mid Check \mid a \cdot 1 \cdot \tilde{x} \mid b \cdot 1 \cdot \tilde{y} \mid r_1 \cdot z \mid r_2) \mapsto$$
$$(1)^L \rfloor (X \mid Check \mid a \cdot 1 \cdot \tilde{x} \mid b \cdot 1 \cdot \tilde{y} \mid r_1 \cdot z \mid r_2 \cdot 1) \quad \text{(C3)}$$

$$(1)^L \rfloor (X \mid Check \mid a \mid r_1 \cdot z \mid r_2) \mapsto (1)^L \rfloor (X \mid Run \mid a \mid r_1 \cdot z \mid r_2 \cdot 0) \quad \text{(C4)}$$

$$(1)^{L} \rfloor (X \mid Check \mid b \mid r_1 \cdot z \mid r_2) \mapsto (1)^{L} \rfloor (X \mid Run \mid b \mid r_1 \cdot z \mid r_2 \cdot 0) \quad \text{(C5)}$$

$$(1)^{L} \rfloor (X \mid Run \mid a \cdot 1 \cdot \tilde{x} \mid r_1 \cdot 1 \mid (next)^{L} \rfloor (Y \mid a \cdot \tilde{y} \mid b \cdot \tilde{z})) \mapsto$$
$$(1)^{L} \rfloor (X \mid Run \mid a \cdot \tilde{x} \mid r_1 \cdot 1 \mid (next)^{L} \rfloor (Y \mid a \cdot 1 \cdot \tilde{y} \mid b \cdot 1 \cdot \tilde{z})) \quad \text{(R1)}$$

$$(1)^{L} \rfloor (X \mid Run \mid a \cdot 1 \cdot \tilde{x} \mid b \cdot 1 \cdot \tilde{y} \mid r_2 \cdot 1 \mid (next)^{L} \rfloor (Y \mid c \cdot \tilde{z})) \mapsto$$
$$(1)^{L} \rfloor (X \mid Run \mid a \cdot \tilde{x} \mid b \cdot \tilde{y} \mid r_2 \cdot 1 \mid (next)^{L} \rfloor (Y \mid c \cdot 1 \cdot \tilde{z})) \quad \text{(R2)}$$

$$(1)^{L} \rfloor (X \mid Run \mid a \mid r_1 \cdot 1) \mapsto (1)^{L} \rfloor (X \mid Run \mid a \mid r_1 \cdot 0) \quad \text{(R3)}$$

$$(1)^{L} \rfloor (X \mid Run \mid a \mid r_2 \cdot 1) \mapsto (1)^{L} \rfloor (X \mid Run \mid a \mid r_2 \cdot 0) \quad \text{(R4)}$$

$$(1)^{L} \rfloor (X \mid Run \mid b \mid r_2 \cdot 1) \mapsto (1)^{L} \rfloor (X \mid Run \mid b \mid r_2 \cdot 0) \quad \text{(R5)}$$

$$(1)^{L} \rfloor (X \mid Run \mid r_1 \cdot 0 \mid r_2 \cdot 0) \mapsto (1)^{L} \rfloor (X \mid Update \mid r_1 \cdot 0 \mid r_2 \cdot 0) \quad \text{(R6)}$$

$$(1)^{L} \rfloor (X \mid Update \mid x \cdot \tilde{x} \mid (next)^{L} \rfloor (Y \mid x \cdot 1 \cdot \tilde{y})) \mapsto$$
$$(1)^{L} \rfloor (X \mid Update \mid x \cdot 1 \cdot \tilde{y} \cdot \tilde{x} \mid (next)^{L} \rfloor (Y \mid x)) \quad \text{(U1)}$$

$$(1)^{L} \rfloor (X \mid Update \mid (next)^{L} \rfloor (a \mid b \mid c)) \mapsto$$
$$(1)^{L} \rfloor (X \mid Check \mid (next)^{L} \rfloor (a \mid b \mid c)) \quad \text{(U2)}$$

Rules (C1)–(C5) describe the steps performed by the system while it is in *Check* state: the objective of this phase is to test whether each evolutionary rule is applicable or not. When all rules have been tested, the systems moves into a state called *Run*, whose steps are given by the application of rules (R1)–(R6). In this second phase, evolutionary rules previously identified as applicable are actually applied, and the result of the application is stored inside the looping sequence *next*. Finally, when no evolutionary rule is further applicable, the system moves into a state called *Update*, in which the content of the looping sequence *next* is used to reset the multiset of objects of the membrane by applying rule (U1)–(U2). When this update operation has been performed, the system moves back to the *Check* state.

## 5   Conclusions

We have surveyed the formalism CLS and a number of its variants from the point of view of its use for describing biological membranes. Verification and simulation tools have been developed for CLS and its variants and can be used to study properties of membrane systems. Via translations, these tools can be used to study systems described by other formalisms such as Brane Calculi and P Systems, capable of describing biological membranes.

# References

1. Alur, R., Belta, C., Ivancic, F., Kumar, V., Mintz, M., Pappas, G.J., Rubin, H., Schug, J.: Hybrid Modeling and Simulation of Biomolecular Networks. In: Di Benedetto, M.D., Sangiovanni-Vincentelli, A.L. (eds.) HSCC 2001. LNCS, vol. 2034, pp. 19–32. Springer, Heidelberg (2001)
2. Barbuti, R., Cataudella, S., Maggiolo-Schettini, A., Milazzo, P., Troina, A.: A Probabilistic Model for Molecular Systems. Fundamenta Informaticae 67, 13–27 (2005)
3. Barbuti, R., Maggiolo-Schettini, A., Milazzo, P., Troina, A.: A Calculus of Looping Sequences for Modelling Microbiological Systems. Fundamenta Informaticae 72, 21–35 (2006)
4. Barbuti, R., Maggiolo-Schettini, A., Milazzo, P., Troina, A.: Bisimulation Congruences in the Calculus of Looping Sequences. In: Barkaoui, K., Cavalcanti, A., Cerone, A. (eds.) ICTAC 2006. LNCS, vol. 4281, pp. 93–107. Springer, Heidelberg (2006)
5. Barbuti, R., Maggiolo-Schettini, A., Milazzo, P.: Extending the Calculus of Looping Sequences to Model Protein Interaction at the Domain Level. In: Mandoiu, I., Zelikovsky, A. (eds.) ISBRA 2007. LNCS (LNBI), vol. 4463, pp. 638–649. Springer, Heidelberg (2007)
6. Barbuti, R., Maggiolo-Schettini, A., Milazzo, P., Tiberi, P., Troina, A.: Stochastic CLS for the Modeling and Simulation of Biological Systems (submitted for publication), draft available at: http://www.di.unipi.it/~milazzo/
7. Cardelli, L.: Brane Calculi. Interactions of Biological Membranes. In: Danos, V., Schachter, V. (eds.) CMSB 2004. LNCS (LNBI), vol. 3082, pp. 257–280. Springer, Heidelberg (2005)
8. Chabrier-Rivier, N., Chiaverini, M., Danos, V., Fages, F., Schachter, V.: Modeling and Querying Biomolecular Interaction Networks. Theoretical Computer Science 325(1), 25–44 (2004)
9. Curti, M., Degano, P., Priami, C., Baldari, C.T.: Modelling Biochemical Pathways through Enhanced pi-calculus. Theoretical Computer Science 325, 111–140 (2004)
10. Dang, Z., Ibarra, O.H.: On P Systems Operating in Sequential and Limited Parallel Modes. In: Workshop on Descriptional Complexity of Formal Systems, pp. 164–177 (2004)
11. Danos, V., Laneve, C.: Formal Molecular Biology. Theoretical Computer Science 325, 69–110 (2004)
12. Gillespie, D.: Exact Stochastic Simulation of Coupled Chemical Reactions. Journal of Physical Chemistry 81, 2340–2361 (1977)
13. Kwiatkowska, M., Norman, G., Parker, D.: Probabilistic Symbolic Model Checking with PRISM: a Hybrid Approach. Int. Journal on Software Tools for Technology Transfer 6, 128–142 (2004)
14. Laneve, C., Tarissan, F.: A Simple Calculus for Proteins and Cells. In: MeCBIC 2006. Workshop on Membrane Computing and Biological Inspired Process Calculi (to appear in ENTCS)
15. Milazzo, P.: Qualitative and Quantitative Formal Modeling of Biological Systems. PhD Thesis, Università di Pisa (2007)
16. Matsuno, H., Doi, A., Nagasaki, M., Miyano, S.: Hybrid Petri Net Representation of Gene Regulatory Network. In: Pacific Symposium on Biocomputing, pp. 341–352. World Scientific Press, Singapore (2000)

17. Păun, G.: Computing with Membranes. Journal of Computer and System Sciences 61, 108–143 (2000)

18. Păun, G.: Membrane Computing. An Introduction. Springer, Heidelberg (2002)

19. Pérez–Jiménez, M.J., Romero–Campero, F.J.: A Study of the Robustness of the EGFR Signalling Cascade Using Continuous Membrane Systems. In: Mira, J.M., Álvarez, J.R. (eds.) IWINAC 2005. LNCS, vol. 3561, pp. 268–278. Springer, Heidelberg (2005)

20. Priami, C.: Stochastic π–Calculus. The Computer Journal 38, 578–589 (1995)

21. Priami, C., Quaglia, P.: Beta Binders for Biological Interactions. In: Danos, V., Schachter, V. (eds.) CMSB 2004. LNCS (LNBI), vol. 3082, pp. 20–33. Springer, Heidelberg (2005)

22. Priami, C., Regev, A., Silvermann, W., Shapiro, E.: Application of a Stochastic Name–Passing Calculus to Representation and Simulation of Molecular Processes. Information Processing Letters 80, 25–31 (2001)

23. Regev, A., Panina, E.M., Silverman, W., Cardelli, L., Shapiro, E.: BioAmbients: An Abstraction for Biological Compartments. Theoretical Computer Science 325, 141–167 (2004)

24. Regev, A., Shapiro, E.: The π–Calculus as an Abstraction for Biomolecular Systems. In: Modelling in Molecular Biology. Natural Computing Series, pp. 219–266. Springer, Heidelberg (2004)

25. Regev, A., Silverman, W., Shapiro, E.Y.: Representation and Simulation of Biochemical Processes Using the pi-calculus Process Algebra. In: Pacific Symposium on Biocomputing, pp. 459–470. World Scientific Press, Singapore (2001)

26. Wiley, H.S., Shvartsman, S.Y., Lauffenburger, D.A.: Computational Modeling of the EGF–Receptor System: a Paradigm for Systems Biology. Trends in Cell Biology 13, 43–50 (2003)

# Membrane Computing in Connex Environment

Mihaela Maliţa[1] and Gheorghe Ştefan[2]

[1] Anselm College, Manchester, N.H., U.S.A.
mmalita@anselm.edu
http://www.anselm.edu/homepage/mmalita/
[2] Polytechnical University of Bucharest, Romania,
and BrightScale, Inc., Sunnyvale, CA, U.S.A.
gstefan@brightscale.com
http://arh.pub.ro/gstefan/

**Abstract.** The Connex technology is presented as a possible way to implement efficiently membrane computations in silicon environment. The opportunity is offered by the recent trend of promoting the parallel computation as a real competitor on the consumer market. The Connex environment has an integral parallel architecture, which is introduced here and its main performances are presented. Some suggestions are provided about how to use the Connex environment as accelerator for membrane computation.

## 1 Introduction

The computation model of membrane computing can be supported by a specific physical environment or by non-specific, on-silicon parallel architectures. The second trend is investigated from the view point of the Connex technology: a highly integrated parallel machine.

**Membrane computing summary:** The membrane computing model is based on multiset rewriting rules applied on a membrane structure populated with objects belonging to a finite alphabet. The potential degree of parallelism is very high in P systems because in each step all possible rules are applied (see details in [10]).

**Connex environment summary:** The Connex technology has an intensive integral parallel architecture [15]. The first embodiment of this technology (see [14]) targets the high definition TV market, but the chip CA1024 can be used also as a general purpose machine for data intensive computing. Application in graphics, data mining, neural network [1], and communication are efficiently supported by the Connex technology. Then, why not for membrane computing!?

**The Intel study:** Since 2002 the clock speed of the processor has improved less than 20%/year, after a long period characterized by around 50%/year. That is why the promise of parallel computing starts to fascinate in a special way. *Intel Inc.* published seminal studies (see [4], [3]) about the next generation of parallel computers. The future processors will contain multi- or maybe many–processors

G. Eleftherakis et al. (Eds.): WMC8 2007, LNCS 4860, pp. 77–96, 2007.
© Springer-Verlag Berlin Heidelberg 2007

optimized for the magic triad of **Recognition – Mining – Synthesis** (RMS). The main problem for this promised development is to find the way to program efficiently the next generation of parallel machines. New programming languages or more sophisticated computation models are needed to fructify the opportunities offered by the new coming parallel computation technologies. In this context membrane computing could play a very promising role.

**The Berkeley study:** Rather than starting from the market opportunities, as Intel did with the RMS domains, the Berkeley approach [2] starts from their "13 dwarfs" (dense linear algebra, sparse linear algebra, spectral methods, ... finite state machines) identified as **parallel computational patterns** able to cover almost all the applications for the next few decades. While Intel takes into consideration a continuous transition from multi- to many-processors, the Berkeley approach is oriented from the start toward the many-processor systems working on data-intensive computation applications. Here is also the place for membrane computation if a good representation will be developed for it.

**Application oriented vs. functionally oriented parallel architectures:** A complex, intense and general purpose application requires usually a multi-threaded approach. In contrast with it, there are functions involving data intense computations. By the rule, multi-processors are involved in the first case (because they are able to exploit thread-level parallelism), and many-processors are needed in the second case. A multi-processor has usually, according to Flynn, a MIMD (multiple instructions - multiple data) architecture, rarely a SIMD (single instructions - multiple data) architecture, and never a MISD (multiple instructions - single data) type one. For a many-processor machine the architecture must be shaped starting from a functional approach, and usually involves all the special forms of possible parallelism.

**Our functional approach:** The **integral parallel architecture** (IPA) is a parallel architecture derived starting from the computational model of *partial recursive functions* [7]. The Turing machine model has been successfully used to ground various sequential computing architectures. Because the functional approach of Kleene is more related with circuits (which are intrinsic parallel structures) we consider there is a better fit between the functional recursive model and the parallel computation. The composition rule provides the best starting point to develop parallel architectures able to support efficiently the other two rules: the primitive recursion and the minimalization. If the *13 dwarfs* will be able to cover the RMS domains, maybe then an IPA will be enough powerful to cover efficiently the 13 computational patterns emphasized by the seminal work done at Berkeley. *A three level hierarchy results. It is topped by application domains (RMS), mediated by computational patterns (the 13 dwarfs), and grounded on various IPAs.*

In the following sections the idea of IPA and the Connex environment are introduced in order to offer various suggestions for a *membrane computing accelerator*. Membrane computing being an intrinsic parallel computational model has the chance to open new ways toward the efficient use of parallel machines.

# 2   Integral Parallel Architecture (IPA)

Various taxonomies were proposed for parallel computations (see [5] [18]). All
of them tell us about different forms of parallelism. We can discuss about many
forms only when we use the parallel approach to accelerate specific computations.
But, when a real complex and intensive computation must pe done, sometimes
we can not use only one form of parallelism. Actual computations involve usually
all possible forms. For example, using Flynn's taxonomy, MIMD or SIMD machines can be defined, but it is not so easy to define MIMD or SIMD application
domains.

General purpose or even application domain oriented parallel machines must
be able to perform all the forms of parallelism, no matter how these forms are
segregated. We propose in the following a new taxonomy and a way to put
together all the resulting forms of parallelism in order to solve efficiently data
intensive computations.

## 2.1   Parallelism and Partial Recursiveness

We claim that the most suggestive classic computational model for defining
parallel architectures is the model of partial recursive functions, because the
rules defining it have direct correspondences in circuits – the intrinsic parallel
support for computation.

**Composition and basic parallel structures.** The first rule, of composition,
provides the basic parallel structures to be used in defining all the forms of
parallelism. Let be $m$ $n$-ary functions $h_i(x_0, \ldots x_{n-1})$, for $i = 0, 1, \ldots m - 1$, and
a $m$-ary function $g(y_0, \ldots y_{m-1})$. In this case, the composition rule is defined as
computing the function

$$f(x_0, \ldots x_{n-1}) = g(h_0(x_0, \ldots x_{n-1}), \ldots h_{m-1}(x_0, \ldots x_{n-1}))$$

The associated physical structure (containing simple circuits or simple program-
mable machines) is shown in Figure 1.

The following four particular, but meaningful forms (see Figure 2) can be
emphasized:

1. **data parallel composition:** with $n = m$, each function $h_i = h$ depends on
   a single input variable $x_i$, for $i = 0, 1, \ldots n - 1$, and $g$ performs the identity
   function (see Figure 2a). Given an input **vector** containing $n$ scalars,

$$\mathbf{X} = \{x_0, x_1, \ldots, x_{n-1}\},$$

   the result is another vector:

$$\{h(x_0), h(x_1), \ldots, h(x_{n-1})\}.$$

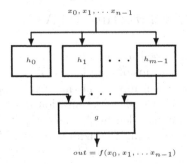

**Fig. 1. The physical structure associated to the composition rule.** The composition of the function $g$ with the functions $h_0, \ldots, h_{m-1}$ implies a two-level system. The first level, performing in **parallel** $m$ computations, is **serially** connected with the second level which performs a reduction function.

2. **speculative composition:** with $n = 1$, i.e., $x_0 = x$, (see Figure 2b), and $g$ performs the identity function. It computes a **vector of functions**

$$\mathbf{H} = [h_0(x), \ldots, h_{n-1}(x)]$$

on the same **scalar** input $x$, generating a vector of results:

$$\mathbf{H}(x) = \{h_0(x), h_1(x), \ldots, h_{n-1}(x)\}$$

3. **serial composition:** with $n = m = 1$ (see Figure 2c). A *"pipe"* of two different machines receives a **stream** of $n$ scalars as input:

$$< \mathbf{X} >=< x_0, x_1, \ldots, x_{n-1} >$$

and provides another stream of scalars

$$< f(x_0), f(x_1), \ldots, f(x_{n-1}) > .$$

In the general case the function $f(x)$ is a composition of more than two functions $h$ and $g$. Thus, the function $f$ can be expressed as a vector of functions $\mathbf{F}$ receiving as input a data stream $< \mathbf{X} >$:

$$\mathbf{F} = [f_0, \ldots f_{p-1}]$$

(in Figure 2c $\mathbf{F} = [h, g]$, and $p = 2$).

4. **reduction composition:** each $h_i$ performing the identity function (see Figure 2d), receives a vector $\{x_0, \ldots, x_{n-1}\}$ as input and provides the scalar, $g(x_0, \ldots, x_{n-1})$ (it transforms a stream of vectors into a stream of scalars).

Concluding, the composition rule provides the context of defining computation using the following basic concepts:

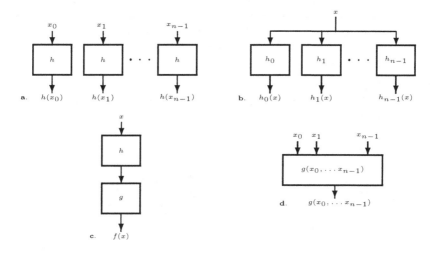

**Fig. 2. The four simple forms of composition. a.** Data parallel composition. **b.** Speculative composition. **c.** Serial composition. **d.** Reduction composition.

**scalar** : $x$
**vector** : $\mathbf{X} = \{x_0, x_1, \ldots, x_{n-1}\}$
**stream** : $<\mathbf{X}> = <x_0, x_1, \ldots, x_{n-1}>$
**function** : $f(x)$
**vector of functions** :
    – $\mathbf{F} = [f_0, \ldots f_{p-1}]$ applied on streams
    – $\mathbf{F}(x) = [f_0(x), \ldots f_{p-1}(x)]$ applied on scalars.

Using the previously defined forms all the requirements for the next two rules (primitive recursion, minimalization) are fulfilled.

**The primitive recursive rule.** There are two ways to implement in parallel the primitive recursive rule. In both cases a lot of data is supposed available to be computed, i.e., there are vectors or streams of data as inputs for the primitive recursive function.

The primitive recursive rule computes the function $f(x, y)$ using the rule:

$$f(x, y) = h(x, f(x, y - 1)),$$

where $f(x, 0) = g(x)$. This rule can be translated in the following serial composition:

$$f(x, y) = h(x, h(x, h(x, \ldots, h(x, g(x)) \ldots)))$$

If the function $f(x, y)$ must be computed for the vector of scalars $\mathbf{X} = \{y_0, y_1, \ldots, y_{n-1}\}$, then a data parallel structure is used. Each machine will compute, using a local *data loop*, the function $f(x, y_i)$ in $max(y_0, y_1, \ldots, y_{n-1})$ "cycles".

If the function $f(x, y)$ must be computed for a stream of scalars, a time parallel structure is used. A "pipe" of $n$ machines will receive in each "cycle" a new scalar from the stream of scalars. If $y > n$, then a *data loop* can be closed from the output of the pipe to its input.

**Minimalization.** Minimalization has also two kinds of parallel solutions: one using data parallel structures and another using time parallel structures.

The minimalization rule assumes

$$f(x) = min(y)[m(x, y) = 0]$$

i.e., the value of $f(x)$ is the minimum $y$ for which $m(x, y) = 0$.

The first, "brute force" implementation uses the speculative structure represented in Figure 2b, where each block computes a function which returns a pair containing a predicate and a scalar:

$$h_i = \{(m(x, i) = 0), i\}$$

after that, the reduction step (using a structure belonging to the class represented in Figure 2d) selects the smallest $i$ from all pairs having the form $\{1, i\}$, if any, that were generated on the previous speculative composition level (all pairs of the form $\{0, i\}$ are ignored).

The second implementation occurs in time-parallel environments where speculation can be used to speed-up the pipe processing. **Reconfigurable pipes** can be conceived and implemented using special reduction features distributed along a pipe. Let be a pipe of functions described by the function vector:

$$\mathbf{P} = [f_0(x), \dots, f_{p-1}(x)]$$

where $y_i = f_i(x)$, for $i = 0, \dots, p - 1$. The associated reconfigurable pipe means to transform the original pipe characterized by:

$$\mathbf{P} = [\dots, f_i(y_{i-1}), \dots]$$

into a pipe characterized by:

$$\mathbf{P} = [\dots, f_i(y_{i-1}, \dots, y_{i-s}), \dots]$$

where $f_i(y_{i-1}, \dots, y_{i-s})$ is a function or a program which decides in each step the variable to be involved in the current computation, selecting (which is one of the simplest reduction functions) one variable of $\{y_{i-1}, \dots y_{i-s}\}$. The maximum *degree of speculation* is $s$.

## 2.2  Functional Taxonomy of Parallel Computing

According to the previously identified simple form of compositions (see Figure 2) we propose a functional taxonomy of parallel computation. We will consider the following types of parallel processing:

**data-parallel computing:** uses operators that take vectors as arguments and returns vectors, scalars (by reduction operations) or streams (input values for time-parallel computations); it is very similar to a SIMD machine.

**time-parallel computing:** uses operators that take streams as arguments and returns streams, scalars, or vectors (input values for data-parallel computations): it is a kind of MIMD machine which works to compute only one function (while a true MIMD performs multi-threading).

**speculative-parallel computing:** with operators that take scalars as arguments and return vectors reduced to scalars using selection (used mainly to speed up time-parallel computations); this contains a true MISD-like structure (completely ignored in the current multi-processing environments).

An IPA is a parallel architecture featured with all kinds of parallelism.

### 2.3   IPA and Market Tendencies

The market tendencies emphasized in the Intel approach and based on Berkeley's dwarfs demand for an IPA. IPA is a many-core (not multi-core) architecture designed to support data intensive computations. It is supposed to work as an accelerator in a mono- or multi-core environment. For all the computational patterns emphasized in the Berkeley's view an IPA provides efficient solutions. Even for the 13th dwarf – Finite State Machine – the speculative & time-parallel aspects of an IPA provides a solution. (Berkeley's view claims that "nothing helps".)

The need for solving real hard applications promotes IPA as an efficient actual solution.

## 3   The Connex System

### 3.1   Structural Description

The first embodiment of an IPA is the Connex System. It is part of CA1024 chip produced by Connex Technology Inc[1]. The Connex System contains mainly an array of 1024 PEs working as a **data parallel sub-system**, DPS, a stream accelerator machine containing 8 PEs (the **time parallel sub-system**, TPS). DPS is driven by an instruction *sequencer*, S, used to broadcast in each clock cycle the same instruction toward each PE from DPS. An *input output controller*, IOC, feeds DPS with data and sends out the results from it. An interconnection fabric allows DPS and TPS to communicate with each other and with the other components of the chip. S and IOC interact using interrupts. They are both simple stack machines with their own data and program memory.

The Connex System uses also other components on the chip to be interfaced with the external world. They are: a MIPS processor acting as a local host, PCI interface to the external host, and a DDR interface to the external memory.

---

[1] From the moment the title of this paper was announced the name of the company changed in **BrightScale Inc.**

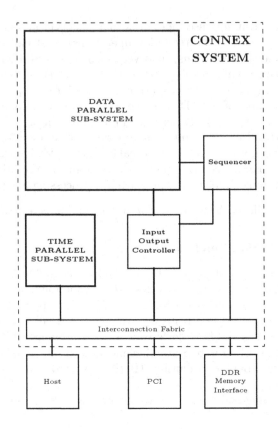

**Fig. 3.** The Connex System

TPS receives streams of data under the control of the local host, and sends the results into the external memory. DPS receives the data vectors from the external memory and sends back the results in the same place. Thus, the two parallel machines communicate usually through the content of the external memory. A data stream is converted into a vector of data, and vice versa, by the programs, run by *Host* and IOC, used to control the buffers organized in the external memory.

## 3.2 General Performances

The first embodiment of the Connex Architecture is designed for $130nm$ standard process technology, and has the following general performances:

- clock frequency: $f_{CK} = 200\ MHz$
- area for the Connex System: $\sim 70\ mm^2$
- 200 *GOPS* (OP is a 16-bit simple operation; no multiplication, division or floating point)

- $> 60\ GOPS/Watt$
- $> 2\ GOPS/mm^2$
- internal bandwidth: $400\ GB/sec$
- external bandwidth: $3.2\ GB/sec$, involving an additional $2\ Watt$.

### 3.3   Specific Performances

The first application domain investigated in the Connex environment is of **High Definition TV** (HDTV). We estimated 80% of the computational power of the Connex System is necessary to decode in real time two H.264 HDTV streams. Some figures referring to specific functions in HDTV domains follow:

- $8 \times 8$ DCT: $4.2\ clock\_cycle$ ($0.066\ clock\_cycle/pixel$)
- $8 \times 8$ IDCT: $4.9\ clock\_cycle$ ($0.077\ clock\_cycle/pixel$)
- $4 \times 4$ SAD: $0.04\ clock\_cycle$ ($0.0025\ clock\_cycle/pixel$)

**Graphics** is another application domain. A preliminary investigation for an image having the complexity characterized by:

- dynamic images having 10,000 triangles, each covering an average of 100 pixels, one-half being obscured by other triangles
- ambient and diffuze illumination model
- 1920 x 1080 display screen, at 30 frames per second

provides the following figures:

- uses $6.6\ GOPS = 3.3\%$ of the total computational power of the Connex System
- and $390\ MB/sec = 12.2\%$ of the total external bandwidth of the CA1024 chip.

For **linear algebra** domain we present here only the computation of the dot product for vectors of up to 1024 components. Two cases are estimated:

- for vectors having as components 32-bit floats:
  $150\ clock\_cycle$ ($> 1.3\ MDot\_Product/sec$)
- for vectors having as components 16-bit signed integer:
  $28\ clock\_cycle$ ($\sim 7\ MDot\_Product/sec$)

The **neural network** domain is also targeted as an application domain. A preliminary estimation is done in [1]: 5 Giga Connection Updates per Second (about 17 times faster than the fastest *specialized* chip on the market: *Hitachi WSI*).

All these estimations are very encouraging for those who are looking for using the Connex environment as an accelerator for membrane computation.

## 4   An IPA: The Connex Architecture

The IPA of the Connex System is described in the following two subsections. The vector section describes the architecture of the data parallel sub-system, and the stream section is devoted to describe the time parallel sub-system.

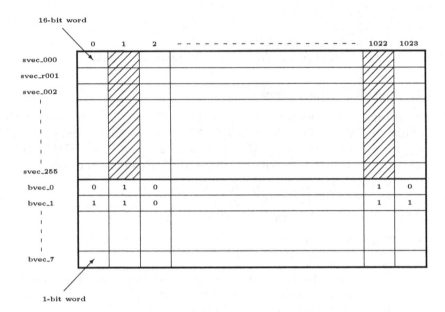

**Fig. 4. The vector variables of the data parallel subsystem.** If the execution is conditioned by AND(bvector0, bvector1), then only column1, ... column1022 of scalars can be involved in computation.

### 4.1   Vector Section

The main physical resources of the Connex System are represented in Figure 4 and are described also in the following pseudo-Verilog form:

```
// 256 16-bit scalar vectors & the hardwired index vector
   reg [15:0]   svec_000[0:1023],
                svec_001[0:1023],
                ...
                svec_255[0:1023],
                ixVect[0:1023]   ;
   initial
      ixVect = {0, 1, 2, ... 1023}; // 16-bit scalars

// 8 Boolean vectors
   reg          bvec_0[0:1023]   ,
                bvec_1[0:1023]   ,
                ...
                bvec_7[0:1023]   ;

// 1024 32-bit scalars stored in the scalar memory
   reg [31:0]   scalar_memory[0:1023]   ;
```

```
// Flag vectors used to predicate
   wire         cryFlag[0:1023] ,
                zeroFlag[0:1023],
                eqFlag[0:1023]  ,
                gtFlag[0:1023]  ,
                ...             ;
```

The Boolean vectors are used to select the active components of the scalar vectors.

```
   ScalarOP: ADD, SUB, INC, BWAND, BWOR, BWXOR, ...
   BooleanOP: AND, OR, ...
```

The **where** construct is a sort of "spatial if".

```
// 'where' construct
   where BooleanOP(booleanVect_i, booleanVect_j, ...) {
       svect_k = ScalarOP(svect_p, svect_q, ...),
       bvect_r = xxxFlag;
   elsew {
       ...
       }
       }
```

Here is an example of how this construct can be used:

```
   where AND(bvect_2, OR(bvect_0, bvect_5)) {
       svect_034 = ADD(svect_012, svect_078, svect_002),
       bvect_3 = cryVect;
       }
   elsew {
       svect_034 = ADD(svect_022, svect_222),
       bvect_3 = cryFlag;
       }
```

It is executed by the Connex System as follows:

```
   selVect = OR(bvect_0, bvect_5); // a temporary variable
   selVect = AND(bvect_2, selVect);
   for(i=0; i<1024; i=i+1)
     if (selVect[i]) {
       svect_034[i] = ADD(svect_012[i], svect_078[i]),
           bvect_3[i] = cryVect[i];
       svect_034[i] = ADD(svect_034[i], svect_002[i]),
           bvect_3[i] = OR(cryVect[i], bvect_3[i]);
       }
     else {svect_034[i] = ADD(svect_022[i], svect_222[i]),
           bvect_3[i] = cryVect[i];
         }
```

There are two distinct ways to generate selections. One is pattern based. It starts from the index vector. Here is an example:

```
/* Pattern based selection example (each other of 4 bit in selVect
will be set on 1)*/
   svect_000 = ixVect;
   svect_000 = AND(svect_000, 16'b11);
   svect_000 = XOR(svect_000, 16'b11), selVect = zeroFlag;
```

The second way to make a selection is to start from the data contained in the scalar vector.

```
// Patternless (data dependent) selection example:
   svect_070 = SUB(svect_070, 16'b10011001), selVect = gtFlag;
```

Usually, any operation specified by one line, having the form:

```
svect_xyz = ScalarOp(...), bvect_q = BooleanOP(...);
```

is executeable in one clock cycle. (Exceptions are specified. For example MULT (...) is executed in 9 clock cycles for 16-bit signed integers, and in 10 clock cycles for unsigned integers.)

## 4.2   Stream Section

The stream section of the Connex System receives the input stream $< X >$ and sends back the output stream $< Z >$, where:

```
<X> = <x_0, ... x_(p-1)>
<Z> = <z_0, ... z_(q-1)>
```

with $p = q$ or $p \neq q$.

The function of the two-dimension pipe $(n \times w)$ is specified by the function vector F, as follows:

```
F = [func_0, ... func_7];
func_i(y_(i-1), y_(i-2), ... y_(i-w)) = y_i;
```

where: func_i is the program executed by $PE_i$. It could be a one instruction looping program, if the pipe "advances" in each clock cycle, or a $s$-instruction loop for pipe propagation executed at each $s$ clock cycles. Each PE can have the associated program using variables generated by the previous $w$ PEs. The degree of speculation is $w$.

Let be, as an example, the following partially defined computation:

```
   ...
   x = ...
   y = y[15] ? y + (x + c1) : y + (x + c2);
   ...
```

Where c1 and c2 are constants. The associated function vector is:

```
F = [... func_i(...),
         func_(i+1)(y_i),
         func_(i+2)(y_i),
         func_(i+3)(y_(i-1), y_(i-2)), ...];
```

where:

```
...
y_i = ...;
y_(i+1) = y_i + c1;
y_(i+2) = y_i + c2;
y_(i+3) = y_(i+3)[15] ? y_(i+3) + y_(i+1) : y_(i+3) + y_(i+2);
...
```

The output of the processing element $PE_i$ works as input for both, $PE_{i+1}$ and $PE_{i+2}$. The processing element $PE_{i+3}$ receives the input variables from the previous two machines $PE_{i+1}$ and $PE_{i+2}$. The second constant dimension of the pipe allows these "shortcuts" which accelerate the computation.

### 4.3   Putting Together the Vector Section and the Stream Section

The two sections of the IPA interact through the content of the external memory. In the external memory a vector or a stream have the same representation. Thus, depending on the source or on the destination, an array of data can be interpreted as a vector or as a stream.

Data exchange between the vector section (DPS) and the stream section (TPS) is done by executing one of the two operation:

```
X <= <Y>; // stream to vector transfer
<X> <= Y; // vector to stream transfer
```

where:

```
X = {x_0, ... x_(n-1)};
<X> = <x_0, ... x_(n-1)>;
Y = {y_0, ... y_(n-1)};
<Y> = <y_0, ... y_(n-1)>;
```

because the destination and the source must have the same dimension $n$.

## 5   How to Use Connex to Accelerate Membrane Computing

The key is the representation. The big amount of parallel resources of the Connex architecture can be activated only if an appropriate representation of membrane is adopted. Follow some simple suggestions. The functionality used in these proposals are described in **Appendix: About VectorC**.

**The first suggestion:** Using the formal definition from [11] (see pag. 11), the content of a membrane system can have associated an $n$-component vector which contains an $m$-component list (with $n \geq m$). For example (see Fig. 3 in [Păun '0x]):

```
[[[<w_3>]<w_2>]<w_1>]... =
[[[a f c] ] ]...
```

where each symbol is represented by a 2-byte word, (**index**, **ASCII_code**), as follows:

(1,[) (2,[) (3,[) (0,a) (0,f) (0,c) (3,]) (2,]) (1,]) ...

For this first suggestion, only the square parenthesis are indexed, and all the objects are represented with the index having the value 0.

The sets of rules $(R_1, R_2, \ldots)$ are represented inside the program run by the sequencer S. Thus, the list representing the membrane system will evolve as follows:

```
(00) [[[a f c] ] ]... =>
(01) [[[a b f f c] ] ]... =>            // in 11 clock cycles
(02) [[[a b b f f f f c] ] ]... =>      // in 15 clock cycles
(03) [[b b b f f f f f f f c ] ]... => // in 27 clock cycles
(04) [[d d d f f f f c ] ]... =>        // in 10 clock cycles
(05) [[d e d e d e f f c ] ]... =>      // in 10 clock cycles
(06) [d e d e d e d f c ]... =>         // in 10 clock cycles
(07) [d d d d f c ] e e e... =>         // in 15 clock cycles
```

The degree of parallelism is not big enough in each cycle during the previously described computation. Only in step (04) all the three $b$s were substituted by $d$s in parallel (in 2 clock cycles). The parallelism is also involved in searching different symbols such as [, ], a, f. On the other hand, all the insertions asked by the evolution rules are performed sequentially.

The performance of the implementation can be increased only by changing the representation of the membrane system.

**The second suggestion:** Another way to represent the membrane system is to use indexes also for objects. The most significant byte of each vector component is used to tell us how many objects of the kind indicated by the other byte are represented. The same membrane system have now the following content:

(1,[) (2,[) (3,[) (1,a) (1,f) (1,c) (3,]) (2,]) (1,]) ...

For the same rules applied results the following evolution of the system:

```
[[[1a 1f 1c] ] ]... =>
[[[1a 1b 2f 1c] ] ]... =>   // in 5 clock cycles
[[[1a 2b 4f 1c] ] ]... =>   // in 5 clock cycles
[[3b 8f 1c ] ]... =>        // in 10 clock cycles
[[3d 4f 1c ] ]... =>        // in 7 clock cycles
```

```
[[3d 3e 2f 1c ] ]... =>        // in 8 clock cycles
[4d 3e 1f 1c ]... =>           // in 8 clock cycles
[4d 1f 1c ] 3e... =>           // in 5 clock cycles
```

Now applying the rule $f \rightarrow ff$ is executed by simply doubling the index associated to f. The same for the rule $d \rightarrow de$. For the rule $ff \rightarrow f$ the index is divided. The main effects are: the representation is kept smaller and the execution time is reduced more than two times.

The degree of parallelism remains small because the application supposes to work only in one membrane at a time. It will be improved if many membranes having the same rules are processed in the same time.

**The third suggestion:** The degree of parallelism increases if on the lowest level more similar membranes are defined. Let us make a little more complex the example presented in [11] (see Fig. 3). Suppose on the lowest level there are two membranes ([1a 1f 1c] and [2a 1f 1c]) with the initial content a little different, but working governed by the same rules. Results the following evolution:

```
[[[1a 1f 1c] [2a 1f 1c] ] ]... =>
[[[1a 1b 2f 1c] [2a 2b 2f 1c] ] ]... => // in 5 clock cycles
[[[1a 2b 4f 1c] [2a 4b 4f 1c] ] ]... => // in 5 clock cycles
[[3b 8f 1c 6b 8f 1c ] ]... =>           // in 10 clock cycles
[[3d 4f 1c 6d 4f 1c ] ]... =>           // in 7 clock cycles
[[3d 3e 2f 1c 6d 6e 2f 1c ] ]... =>     // in 8 clock cycles
[4d 3e 1f 1c 7d 6e 1f 1c]... =>         // in 8 clock cycles
[4d 1f 1c 7d 1f 1c ] 9e... =>           // in 10 clock cycles
```

The execution time has very little increased (only in the last step). It is obvious that having 3 or more low level membranes the degree of parallelism will increase correspondingly.

The performance can be increased more if the rules are integrated in or as a vector representation. In the previous examples the rules were applied sequentially because they were "known" only by the program issued by the sequencer S. The sequencer must know only to apply rules defined inside the Connex Array in an appropriate manner.

# 6 Concluding Remarks

**Functional vs. Flynn's taxonomy.** Our functional taxonomy works better in many-processor environment, while Flynn's taxonomy fits better the multiprocessor environment. The functional taxonomy supposes three different types equally involved in defining a high performance architecture, while Flynn's taxonomy proposes also three kinds of parallel machines, only one of them being (MIMD) considered as an effective efficient solution for real machines (see [6]).

**Limited non-determinism.** The physical resources added for the speculative mechanism are used to support a sort of limited non-deterministic computation inside an IPA.

**Can we accelerate molecular computing in vector environment?** The vector section of an IPA can be used to accelerate molecular computation if appropriate representations are imagined. Molecular computing has a huge potential for data parallelism and vector processing is a special kind of data parallel computation. The main problem is to reformulate the molecular approach to fit with the restrictions and promises imposed/offered by the vector computation. The Connex System has also some additional features helping the implementation of specific search functions, very helpful for rewriting rule based processing. Various insert and delete capabilities can be used for the same purpose.

**An efficient P-Architecture is slightly different from the current Connex Architecture.** Although the Connex environment is helpful for investigating molecular computing based applications, there are needed few specific features in order to obtain a market efficient environment.

**Why not a P-language?** A very useful intermediary step toward the definition of a marketable environment for this new computation model is providing a P-language. Working with the basic definition of P-systems is not enough flexible for solving real and complex problems. Using a high level type language and developing for it a specific environment will speed-up the work for a specific membrane platform.

## Acknowledgments

I would like to thank Emanuele Altieri, Frank Ho, Bogdan Mîţu, Tom Thomson, Dominique Thiébaut, and Dan Tomescu for their technical contributions in developing the Connex System.

## References

1. Andonie, R., Maliţa, M.: The Connex Array as a Neural Network Accelerator. In: Third IASTED International Conference on Computational Intelligence, 2007, Bannf, Alberta, Canada, July 2-4, 2007 (accepted, 2007)
2. Asanovic, K., et al.: The Landscape of Parallel Computing Research: A View from Berkeley, Technical Report No. UCB/EECS-2006-183 (December 18, 2006)
3. Borkar, S.Y., et al.: Platform 2015: Intel Processor and Platform Evolution for the Next decade, Intel Corporation (2005)
4. Dubey, P.: A Platform 2015 Workload Model: Recognition, Mining and Synthesis Moves Computers to the Era of Tera, Intel Corporation (2005)
5. Flynn, M.J.: Some computer organization and their affectiveness. IEEE Trans. Comp. C21(9), 948–960 (1972)
6. Hennessy, J.L., Patterson, D.A.: Computer Architecture. A Quantitative Approach, 4th edn. Morgan Kaufmann, San Francisco (2007)

7. Kleene, S.C.: General Recursive Functions of Natural Numbers. Math. Ann. 112 (1936)
8. Maliţa, M., Ştefan, G., Stoian, M.: Complex vs. Intensive in Parallel Computation. In: International Multi-Conference on Computing in the Global Information Technology - Challenges for the Next Generation of IT&C - ICCGI 2006, Bucharest, Romania (August 1-3, 2006)
9. Mîţu, B.: private communication
10. Păun, G.: Membrane Computing. An Introduction. Springer, Berlin (2002)
11. Păun, G.: Introduction to Membrane Computing. In: Ciobanu, G., Păun, G., Pérez-Jiménez, M.J. (eds.) Applications of Membrane Computing, ch. 1, Springer, Heidelberg (2006)
12. Ştefan, G.: The CA1024: A Massively Parallel Processor for Cost-Effective HDTV. In: SPRING PROCESSOR FORUM: Power-Efficient Design, Doubletree Hotel, San Jose, CA, May 15-17, 2006 and in SPRING PROCESSOR FORUM JAPAN, June 8-9, 2006, Tokyo (2006)
13. Ştefan, G., Sheel, A., Mîţu, B., Thomson, T., Tomescu, D.: The CA1024: A Fully Programable System-On-Chip for Cost-Effective HDTV Media Processing. In: Hot Chips: A Symposium on High Performance Chips, Memorial Auditorium, Stanford University (August 20-22, 2006)
14. Ştefan, G.: The CA1024: SoC with Integral Parallel Architecture for HDTV Processin. In: 4th International System-on-Chip (SoC) Conference & Exhibit, Radisson Hotel Newport Beach, CA (November 1-2, 2006)
15. Ştefan, G.: Integral Parallel Computation. In: Proceedings of the Romanian Academy. Series A: Mathematics, Physics, Technical Sciences, Information Science, vol. 7(3) (September-December 2006)
16. Thiébaut, D., Ştefan, G., Maliţa, M.: DNA search and the Connex technology. In: International Multi-Conference on Computing in the Global Information Technology - Challenges for the Next Generation of IT&C - ICCGI 2006, Bucharest, Romania (August 1-3, 2006)
17. Thiébaut, D., Maliţa, M.: Pipelining the Connex array. In: BARC 2007, Boston (January 2007)
18. Xavier, C., Iyengar, S.S.: Introduction to Parallel Algorithms. John Wiley & Sons, Inc., Chichester (1998)

# APPENDIX: About VectorC

**VectorC** is a simulator for the Connex Architecture written in C++. In order to program the user must know how Connex Architecture works. VectorC is defined by BrightScale Inc. (www.brightscale.com) and is used to explore different possible applications for the Connex environment. This helps C++ developers to write code for Connex Architecture. Writing Code using C++ provides the best in class development/debug tools, along with a familiar language to program. VectorC consists of a set of libraries that extend the language.

Sources and examples are found on the Saint Anselm College website of Mihaela Malita:

`http://www.anselm.edu/homepage/mmalita/ResearchS07/WebsiteS07/`

The Connex machine is an array of vectors where different operations can be performed in parallel. The elements of Vectors can be processed in parallel. The machine keeps track of a selection vector called Selection. This selection is used as a mask to put over any vector. Only the marked elements will be processed.

The class vector is the most important class. Here are some definitions that extend the classic vector from C++.

Main files are selection.cpp, vector.cpp, vectorio.cpp, vectorlib. cpp, mm_print.cpp with their header files.

The C vector class is overloaded over the STl vector class.

**Example 1.** *Let's see how we can add two vectors in parallel.*

```
vector X, Y;
X += Y;      // X[i] += Y[i], for all i
X += 5;      // X[i] += 5 add with a scalar, for all i
```

*In order to compare vectors in parallel we use:*

```
vector  A, B;     //declare two vectors A and B
selection C;      //declare a selection vector
C = A > b;        // c[i] = a[i] > b[i] for all i
C = A > 5;        // c[i] = a[i] > 5  for all i
```

*Selection* C *is a Boolean vector with values 1 for all the positions* i, *where* $A[i] > b$ *and 0 for the rest. Same for all the comparison operators* $(<, <=, >=, ==)$. *As we see the input parameter can be either a vector or scalar.* ◇

A selection vector is a boolean vector. All the vectors and selections have the same size, i.e., SIZEOF_VECTOR=1024, for this version of the chip.

The selected elements for a vector can be also moved right or left. See below the definition for getleft(). All the selected positions take the value of the left element.

```
vector vector::getLeft() {
  vector v;
  for (int i = (SIZEOF_VECTOR - 1); i > 0; i--)   {
    v.cntnt[i] = vector::cntnt[i-1];
  }
  v.cntnt[0] = vector::cntnt[SIZEOF_VECTOR - 1];
  return v;
}
```

**Example 2.** *Let be* V= 0 1 2 **3** 4 5 **3** 7 8 9 10 *and all 3's are selected. Then the effect of* getletf(V,1) *means all selected are shifted to the right (or the selected take the value of the left neighbor) Result is* V= 0 1 2 **2** 4 5 **5** 7 8 9 10

*The piece of code for this is:*

```
.. initialize vector V..;
```

```
selection Sel = (V == 3); // select positions with value =3
WHERE (Sel) {              // shift right everything selected
    V = getLeft(V, 1);
}
ENDW
```

◇

Below we shall give two examples to show how we can program in parallel some operations with Connex.

**Example 3. Parallel substitution.** *Substitute all 5's with 0 in a vector* V. *We initialize first a vector which we call* V *with all the integers starting from 0. That is* V *will be* 0,1,2,3,4,5 .... *Then we choose 3 indexes, 3, 10 and 15, and set them to 5 (*V[3] = V[10] = V[15] = 5*). In order to change only the elements that are 5 we use a WHILE structure where* V == 5 *actually marks (selects) all the elements in* V *that are 5. The next operation, that is* V = 0, *takes place only for those selected.*
*Initial* V :

0 1 2 **5** 4 **5** 6 7 8 9 **5** 11 12 13 14 **5** 16 17 18 19

V *after substitution:*

0 1 2 **0** 4 **0** 6 7 8 9 **0** 11 12 13 14 **0** 16 17 18 19

    *The size of the vectors is contained in a constant variable*

```
// uses example use of WHERE - ENDW block
 #include "mm_print.h"        // for printing the vector
 int main () {
    vector V = indexvector(); // parallel is 1 clock cycle
    V[3] = 5;
    V[10] = 5;
    V[15] = 5;
    WHERE (V == 5) {   // select the 5's
        V = 0;         // only 5's become 0
    }
    ENDW
    mm_print_vector(V);
    return 0;
}
```

◇

**Example 4. Count the differences between two vectors**, *that is count the number of elements that are different on the same position. We initialize two vectors* V1 *and* V2 *with* indexvector(), *the function that gives all the elements their index value. Then we modify vector* V2 *in two positions (we modify* V1[4], *and* V1[10] *). Here is the data:*

V1 = 0 1 2 3 4 5 6 7 8 9 **10** 11 12 13 14 15 16 17 18 19
V2 = 0 1 2 3 **5** 5 6 7 8 9 **9** 11 12 13 14 15 16 17 18 19

*The algorithm goes as follows:*

− *we subtract the vectors in parallel (one clock cycle)*
− *we mark (select) the non zero positions (this type of search is done in 1 cycle)*
− *we count the selections ( counting the marked positions is again 1 cycle (the function* getCount(Sel) *from VectorC is doing this).*

*The output is the difference* **V2** − **V1**:
0 0 0 0 -1 0 0 0 0 0 1 0 0 0 0 0 0 0 0 0.
*The selection vector* Sel *(where* getcount(Sel) *is applied):*
0 0 0 0 1 0 0 0 0 0 1 0 0 0 0 0 0 0 0 0.
*Number of different cells: 2.*

```
#include "mm_print.h"
int main () {
    vector V1 = indexvector();  // initialization
    V1[4] = 5;
    V1[10] = 9;// V1 = [0, 1, 2, 3, 5, 5, 6, 7, 8, 9, 9, 10, ...]
    vector V2 = indexvector();  // V2 = [0, 1, 2, 3, ...]

    selection Sel;
    V2 -= V1;               // V2 = V2 - V1
    mm_print_vector(V2); mm_print_cycles();
    Sel = (V2 != 0);  // Select any non-zero differences
    mm_print_vector(Sel);
    cout << "Number of different cells: "
            << getCount(Sel);
    return 0;
}
```

◇

The VectorC approach is very flexible we can extend at any point the libraries with more operations for the vector class.

# Skin Output in P Systems with Minimal Symport/Antiport and Two Membranes

Artiom Alhazov[1,2] and Yurii Rogozhin[1,3]

[1] Institute of Mathematics and Computer Science
Academy of Sciences of Moldova
Str. Academiei 5, Chişinău, MD-2028 Moldova
{artiom,rogozhin}@math.md
[2] Åbo Akademi University
Department of Information Technologies
Turku Center for Computer Science, FIN-20520 Turku, Finland
aalhazov@abo.fi
[3] Rovira i Virgili University,
Research Group on Mathematical Linguistics,
Pl. Imperial Tàrraco 1, 43005 Tarragona, Spain

**Abstract.** It is known that symport/antiport P systems with two membranes and minimal cooperation can generate any recursively enumerable sets of natural numbers using exactly one superfluous object in the output membrane, where the output membrane is an elementary membrane. In this paper we consider symport/antiport P systems where the output membrane is the skin membrane. In this case we prove an unexpected characterization: symport/antiport P systems (and purely symport P systems) with two membranes and minimal cooperation generate exactly the recursively enumerable sets of natural numbers.

## 1 Introduction

P systems with *symport/antiport* rules, i.e., P systems with *pure communication rules assigned to membranes*, first were introduced in [24]; symport rules move objects across a membrane together in one direction, whereas antiport rules move objects across a membrane in opposite directions. These operations are very powerful, i.e., P systems with symport/antiport rules have universal computational power with only one membrane, e.g., see [12], [15], [13].

A comprehensive overview of the most important results obtained in the area of P systems and tissue P systems with *symport/antiport* rules, with respect to the development of computational completeness results improving descriptional complexity parameters as the number of membranes and cells, the size of the rules, and the number of objects can be found in [1].

For instance, in [3] one obtains the exact characterization of $NRE$ for symport/antiport P systems with three membranes and minimal cooperation and for corresponding purely symport P systems.

In [5] one shows that if some P system with two membranes and with minimal cooperation, i.e., a P system with symport/antiport rules of size one or a

G. Eleftherakis et al. (Eds.): WMC8 2007, LNCS 4860, pp. 97–112, 2007.

P system with symport rules of size two, generates a set of numbers *containing zero*, then this set is **finite**. After that one proves that P systems with symport/antiport rules of size one can generate any *recursively enumerable* set of natural numbers without zero (i.e., they are computationally complete with just **one superfluous object** remaining in the output membrane at the end of a halting computation). The same result is true also for purely symport P systems of size two. Therefore, one superfluous object is both necessary and sufficient in case of two membranes.

The question about precise characterization of computational power of symport/antiport P systems (purely symport P systems) with two membranes and minimal cooperation is still open.

Interpreting the result of the computation as the sequence of terminal symbols sent to the environment, one shows that P systems with two membranes and symport rules of size two or symport/antiport rules of size one generate all recursively enumerable languages [6].

In this paper we show that P systems with minimal symport/antiport (and purely symport P systems) with two membranes characterize $NRE$ when we consider the **output in the skin membrane** rather than the elementary one.

## 2    Basic Notations and Definitions

For the basic elements of formal language theory needed in the following, we refer to [30]. We just list a few notions and notations: $\mathbb{N}$ denotes the set of natural numbers (i.e., of non-negative integers). $V^*$ is the free monoid generated by the alphabet $V$ under the operation of concatenation and the empty string, denoted by $\lambda$, as unit element; by $NRE$, $NREG$, and $NFIN$ we denote the family of recursively enumerable sets, regular sets, and finite sets of natural numbers, respectively. For $k \geq 1$, by $\mathbb{N}_k RE$ we denote the family of recursively enumerable sets of natural numbers excluding the initial segment 0 to $k-1$. Particularly, $\mathbb{N}_1 RE = \{N \in NRE \mid 0 \notin N\}$. The families of recursively enumerable sets of vectors of natural numbers are denoted by $PsRE$.

### 2.1    Counter Automata

A non-deterministic *counter automaton* (see [11], [1]) is a construct

$$M = (d, Q, q_0, q_f, P), \text{ where:}$$

- $d$ is the number of counters, and we denote $D = \{1, ..., d\}$;
- $Q$ is a finite set of states, and without loss of generality, we use the notation $Q = \{q_i \mid 0 \leq i \leq f\}$ and $F = \{0, 1, ..., f\}$,
- $q_0 \in Q$ is the initial state,
- $q_f \in Q$ is the final state, and
- $P$ is a finite set of instructions of the following form:

1. $(q_i \rightarrow q_l, k+)$, with $i, l \in F$, $i \neq f$, $k \in D$ ("increment" -instruction). Add one to counter $k$ and change the state of the system from $q_i$ to $q_l$.

2. $(q_i \rightarrow q_l, k-)$, with $i, l \in F$, $i \neq f$, $k \in D$ ("decrement" -instruction). If the value of counter $k$ is greater than zero, then this instruction decrements it by 1 and changes the state of the system from $q_i$ to $q_l$. Otherwise (when the value of counter $k$ is zero) the computation is blocked in state $q_i$.

3. $(q_i \rightarrow q_l, k = 0)$, with $i, l \in F$, $i \neq f$, $k \in D$ ("test for zero" -instruction). If the value of counter $k$ is zero, then this instruction changes the state of the system from $q_i$ to $q_l$. Otherwise (the value stored in counter $k$ is greater than zero) the computation is blocked in state $q_i$.

4. *halt*. This instruction stops the computation of the counter automaton, and it has to be only assigned to the final state $q_f$.

A transition of the counter automaton consists in updating/checking the value of a counter according to an instruction of one of the types described above and by changing the current state to another one. The computation starts in state $q_0$ with all counters being equal to zero. The result of the computation of a counter automaton is the value of the first $k$ counters when the automaton halts in state $q_f \in Q$ (without loss of generality we may assume that in this case all other counters are empty). A counter automaton thus (by means of all computations) generates a set of $k$-vectors of natural numbers. If $k = 1$, then by $N(M)$ we denote the corresponding numeric set generated by $M$.

## 2.2   P Systems with Symport/Antiport Rules

The reader is supposed to be familiar with basics of membrane computing, e.g., from [26]; see the P systems web page, [34], for the comprehensive information.

A *P system with symport/antiport rules* is a construct

$$\Pi = (O, \mu, w_1, \ldots, w_k, E, R_1, \ldots, R_k, i_0), \text{ where:}$$

1. $O$ is a finite alphabet of symbols called *objects*;
2. $\mu$ is a *membrane structure* consisting of $k$ membranes that are labeled in a one-to-one manner by $1, 2, \ldots, k$;
3. $w_i \in O^*$, for each $1 \leq i \leq k$, is a finite multiset of objects associated with the region $i$ (delimited by membrane $i$);
4. $E \subseteq O$ is the set of objects that appear in the environment in an infinite number of copies;
5. $R_i$, for each $1 \leq i \leq k$, is a finite set of symport/antiport rules associated with membrane $i$; these rules are of the forms $(x, in)$ and $(y, out)$ *(symport rules)* and $(y, out; x, in)$ *(antiport rules)*, respectively, where $x, y \in O^+$;
6. $i_0$ is the label of a membrane of $\mu$ that identifies the output region.

A P system with symport/antiport rules is defined as a computational device consisting of $k$ hierarchically nested membranes identifying $k$ distinct regions (the membrane structure $\mu$), where to each membrane $i$ there are assigned a multiset of objects $w_i$ and a finite set of symport/antiport rules $R_i$, $1 \leq i \leq k$. A rule $(x, in) \in R_i$ permits the objects specified by $x$ to be moved into region $i$ from the immediately outer region. Notice that for P systems with symport rules the rules in the skin membrane of the form $(x, in)$, where $x \in E^*$, are forbidden. A rule $(x, out) \in R_i$ permits the multiset $x$ to be moved from region $i$ into the outer region. A rule $(y, out; x, in)$ permits the multisets $y$ and $x$, which are situated in region $i$ and the outer region of $i$, respectively, to be exchanged. It is clear that a rule can be applied if and only if the multisets involved by this rule are present in the corresponding regions. The size of a symport rule $(x, in)$ or $(x, out)$ is $|x|$, while the size of an antiport rule $(y, out; x, in)$ is $max\{|x|, |y|\}$.

As usual, a computation in a P system with symport/antiport rules is obtained by applying the rules in a non-deterministic maximally parallel manner. Specifically, in this variant, a computation is restricted to moving objects through membranes, since symport/antiport rules do not allow the system to modify the objects placed inside the regions. Initially, each region $i$ contains the corresponding finite multiset $w_i$, whereas the environment contains only objects from $E$ that appear in infinitely many copies.

A computation is successful if starting from the initial configuration, the P system reaches a configuration where no rule can be applied anymore. The result of a successful computation is a natural number obtained by counting all objects in region $i_0$. The set of natural numbers computed in this way by a P system $\Pi$ is denoted by $N(\Pi)$. If the multiplicity of each object is counted separately, then a vector of natural numbers is obtained, denoted by $Ps(\Pi)$, see [26].

By $NOP_m(sym_s, anti_t)$ we denote the family of sets of natural numbers generated by P systems with symport/antiport rules with at most $m > 0$ membranes, symport rules of size at most $s \geq 0$, and antiport rules of size at most $t \geq 0$. In the papers on P systems, following [26], $i_0$ is assumed to be an elementary membrane. In this paper we will write $\mathbb{N}^{skin}OP_m(sym_s, anti_t)$ if $i_0$ is the skin membrane. If $t = 0$, then we may omit $anti_t$.

## 3   Main Results

**Theorem 1.** $\mathbb{N}^{skin}OP_2(sym_1, anti_1) = \mathbb{N}RE$.

*Proof.* We simulate a counter automaton $M = (d, Q, q_0, q_f, P)$. Recall that $M$ starts with empty counters. We also suppose that all instructions from $P$ are labeled in a one-to-one manner with elements of $\{1, \dots, n\} = I$, $n$ is a label of the *halt* instruction and $I' = I \setminus \{n\}$. We denote by $I_+$, $I_-$, and $I_{=0}$ the set of labels for the "increment" -, "decrement" -, and "test for zero" - instructions, respectively. We also use the following notation: $C = \{c_k\}, k \in D$ and $Q' = Q \setminus \{q_0\}$. We construct the P system $\Pi_1$ as follows:

$$\Pi_1 = (O, [_1 [_2 \ ]_2 \ ]_1, w_1, w_2, E, R_1, R_2, 1),$$
$$O = E \cup \{q_0, \#, L, T_1, T_2, P_2, J_1, J_2, J_3\} \cup \{b_j \mid j \in I\} \cup \{d_j \mid j \in I'\},$$
$$E = Q' \cup C \cup \{a_j \mid j \in I\} \cup \{a'_j, e_j \mid j \in I'\} \cup \{J_0, P_1\} \cup \{F_i \mid 0 \leq i \leq 9\},$$
$$w_1 = q_0 L J_1 J_2 J_3,$$
$$w_2 = \# T_1 T_2 P_2 \prod_{j \in I} b_j \prod_{j \in I'} d_j,$$
$$R_i = R_{i,s} \cup R_{i,r} \cup R_{i,f}, \quad i = 1, 2.$$

We code the counter automaton as follows. Region 1 will hold the current state of the automaton, represented by a symbol $q_i \in Q$ and also the value of all counters, represented by the number of occurrences of symbols $c_k \in C$, $k \in D$, where $D = \{1, ..., d\}$.

We split our proof into several parts that depend on the logical separation of the behavior of the system. We will present the rules and the initial symbols for each part, but we remark that the system we present is the union of all these parts. The rules $R_i$ are given by three phases:

1. START: preparation of the system for the computation.
2. RUN: simulation of instructions of the counter automaton.
3. END: terminating the computation.

The parts of the computations illustrated in the following describe different phases of the evolution of the P system. For simplicity, we focus on explaining a particular phase and omit the objects that do not participate in the evolution at that time. Each rectangle represents a membrane, each variable represents a copy of an object in a corresponding membrane (symbols outside of the outermost rectangle are found in the environment). In each step, the symbols that will evolve (will be moved) are written in **boldface**. The labels of the applied rules are written above the symbol $\Rightarrow$.

## 1. START

We use the following idea: in our system we have a symbol $L$ which moves from region 1 to the environment and back in an infinite loop. This loop may be stopped only if all stages are completed correctly.

$$R_{1,s} = \{ \mathbf{1s1} : (L, out), \quad \mathbf{1s2} : (L, in) \}.$$
$$R_{2,s} = \emptyset.$$

Notice that some rules are never executed during a correct simulation: applying them would lead to an infinite computation. To help the reader, we will underline the labels of such rules in the description below.

## 2. RUN

$$R_{1,r} = \{ \mathbf{1r1} : (q_i, out; a_j, in) \mid (j : q_i \rightarrow q_l, c_k \gamma) \in P, \gamma \in \{+, -, = 0\} \}$$
$$\cup \{ \mathbf{1r2} : (q_f, out; a_n, in) \}$$
$$\cup \{ \mathbf{1r3} : (b_j, out; a'_j, in) \mid j \in I' \}$$
$$\cup \{ \mathbf{1r4} : (a_j, out; J_0, in), \quad \mathbf{1r5} : (J_1, out; b_j, in) \mid j \in I \}$$
$$\cup \{ \mathbf{1r6} : (J_0, out; J_1, in) \}$$
$$\cup \{ \mathbf{1r7} : (a'_j, out; c_k, in) \mid (j : q_i \rightarrow q_l, c_k +) \in P \}$$
$$\cup \{ \mathbf{1r8} : (a'_j, out) \mid j \in I_- \cup I_{=0} \}$$
$$\cup \{ \mathbf{1r9} : (d_j, in) \mid j \in I_+ \cup I_{=0} \}$$

$\cup \{\mathbf{1r10} : (c_k, out; d_j, in) \mid (j : q_i \to q_l, c_k-) \in P\}$

$\cup \{\mathbf{1r11} : (J_3, out; d_j, in) \mid j \in I_-\}$

$\cup \{\mathbf{1r12} : (J_3, out; J_1, in)\}$

$\cup \{\mathbf{1r13} : (d_j, out; e_j, in) \mid j \in I'\}$

$\cup \{\mathbf{1r14} : (e_j, out, q_l, in) \mid (j : q_i \to q_l, c_k\gamma) \in P, \gamma \in \{+, -, = 0\}\}$

$\cup \{\mathbf{1r15} : (b_n, out; F_0, in)\}$

$\cup \{\mathbf{1r16} : (\#, out), \mathbf{1r17} : (\#, in)\}.$

$R_{2,r} = \{\mathbf{2r1} : (b_j, out; a_j, in), \quad \mathbf{2r2} : (a_j, out; J_2, in) \mid j \in I\}$

$\cup \{\mathbf{2r3} : (a_j, out; J_1, in) \mid j \in I\}$

$\cup \{\mathbf{2r4} : (d_j, out; a'_j, in) \mid j \in I'\}$

$\cup \{\mathbf{2r5} : (a'_j, out; c_k, in) \mid (j : q_i \to q_l, c_k = 0) \in P\}$

$\cup \{\mathbf{2r6} : (a'_j, out; e_j, in) \mid j \in I_{=0}\}$

$\cup \{\mathbf{2r7} : (a'_j, out; J_1, in) \mid j \in I_{=0}\}$

$\cup \{\mathbf{2r8} : (e_j, out; d_j, in) \mid j \in I_{=0}\}$

$\cup \{\mathbf{2r9} : (e_j, out; J_1, in) \mid j \in I_{=0}\}$

$\cup \{\mathbf{2r10} : (d_j, in) \mid j \in I_+ \cup I_-\}$

$\cup \{\mathbf{2r11} : (a'_j, out) \mid j \in I_+ \cup I_-\}$

$\cup \{\mathbf{2r12} : (J_2, out; b_j, in) \mid j \in I'\}$

$\cup \{\mathbf{2r13} : (J_2, out; J_1, in), \quad \mathbf{2r14} : (\#, out; J_0, in)\}.$

First of all, we mention that if during the phase RUN object $J_3$ comes to the environment by rules $\mathbf{1r11}$, $\mathbf{1r12}$ (**Scenario 0**), it remains there forever and cannot move object $L$ to region 2 (during the phase END), thus to stop the infinite loop. So, the computation never halts.

Let us explain the synchronization of $a_j$ coming to the environment and $b_j$ leaving the environment: the first one brings $J_0$ into region 1 while the latter brings $J_1$ into the environment; then rule $\mathbf{1r6}$ moves $J_0$ and $J_1$ back.

If $a_j$ comes to the environment without $b_j$ leaving it or $b_j$ is in region 1 or 2 at that moment (it is possible after applying rules $\mathbf{2r3}$, $\mathbf{2r7}$, $\mathbf{2r13}$), $J_1$ remains in region 1 (or 2) and $J_0$ comes to region 1 and after that in region 2 by rules $\mathbf{1r4}$, $\mathbf{2r14}$ (**Scenario 1**), thus causing an endless computation since $\mathbf{1r16}$ and $\mathbf{1r17}$ are always applicable.

If $b_j$ leaves the environment without $a_j$ coming there, $J_0$ remains in the environment and $J_1$ comes there (**Scenario 2**), so $\mathbf{1r12}$ is applied and $J_3$ comes to the environment. The computation never halts, see scenario 0.

**Scenario 3** takes place when two symbols $a_j$ and symbol $b_j, j \in I$ appear in region 1 and in the environment accordingly. In this case rules $\mathbf{1r4}, \mathbf{1r5}$ will be applied, and rule $\mathbf{1r4}$ two times. Thus, two symbols $J_0$ appear in region 1 and rule $\mathbf{2r14}$ will be applied eventually. The computation never halts, see scenario 1.

We also mention that applying rule $\mathbf{1r11}$ causes scenario 0 (modeling a "decrement"-instruction, with no $c_k$ in region 1); applying $\mathbf{2r5}$ leads to scenario 3 (modeling a "test for zero"-instruction, with some $c_k$ in region 1), and applying **2r7** and **2r9** eventually causing scenario 1. Therefore, in order for a computation to halt, no underlined rules should be applied.

We will now consider the "main" line of computation. We explain the behavior of simulating the instruction $(j : q_i \to q_l, c_k \gamma)$. Index $s$ stands for any possible instruction associated to state $q_l$.

**"Increment"** -instruction:

$$q_l \mathbf{a}_j a_s a'_j e_j c_k J_0 \boxed{\mathbf{q_i}\, J_1 J_2 J_3 \boxed{b_j d_j \#}} \Rightarrow^{1r1} q_l q_i a_s a'_j e_j c_k J_0 \boxed{\mathbf{a_j}\, J_1 J_2 J_3 \boxed{b_j d_j \#}} \Rightarrow^{2r1}$$

$$q_l q_i a_s \mathbf{a'_j} e_j c_k J_0 \boxed{\mathbf{b_j J_1 J_2 J_3} \boxed{\mathbf{a_j} d_j \#}} \Rightarrow^{1r3,2r2} q_l q_i a_s \mathbf{b_j} e_j c_k J_0 \boxed{\mathbf{a'_j J_1 a_j} J_3 \boxed{\mathbf{J_2 d_j} \#}}$$

$$\Rightarrow^{1r4,1r5,2r4} q_l q_i a_j a_s \mathbf{e_j} c_k \mathbf{J_1} \boxed{\mathbf{b_j d_j J_0} J_3 \boxed{\mathbf{J_2 a'_j} \#}} \qquad (A)$$

$$\Rightarrow^{1r6,1r13,2r11,2r12} q_l q_i a_j a_s \mathbf{d_j c_k} J_0 \boxed{J_1 J_2 \mathbf{a'_j} e_j J_3 \boxed{b_j \#}} \Rightarrow^{1r7,1r9,1r14}$$

$$q_i a_j \mathbf{a_s} a'_j e_j J_0 \boxed{\mathbf{q_l d_j} J_1 J_2 J_3 c_k \boxed{b_j \#}} \Rightarrow^{1r1,2r10} q_l q_i a_j a'_j e_j J_0 \boxed{\mathbf{a_s} J_1 J_2 J_3 c_k \boxed{b_j d_j \#}}$$

In that way, $q_i$ is replaced by $q_l$ and $c_k$ is moved from the environment into region 1. Notice that symbols $a_j$, $b_j$, $a'_j$, $d_j$, $e_j$, $J_0$, $J_1$, $J_2$ have returned to their original positions. Symbol $d_j$ returns to region 2 in the first step of the simulation of the next instruction (the last step of the illustration).

**"Decrement"** -instruction:

(i) *There is some $c_k$ in region 1:*

We consider configuration (A) above with symbol $c_k$ in region 1.

$$q_l q_i a_j a_s \mathbf{e_j J_1} \boxed{\mathbf{b_j d_j J_0} J_3 c_k \boxed{\mathbf{J_2 a'_j} \#}} \Rightarrow^{1r6,1r13,2r11,2r12}$$

$$\mathbf{q_l} q_i a_j a_s \mathbf{d_j} J_0 \boxed{J_1 J_2 \mathbf{a'_j} e_j J_3 c_k \boxed{b_j \#}} \Rightarrow^{1r8,1r10,1r14}$$

$$q_i a_j \mathbf{a_s} a'_j e_j c_k J_0 \boxed{\mathbf{q_l} J_1 J_2 J_3 \mathbf{d_j} \boxed{b_j \#}} \Rightarrow^{1r1,2r10} q_l q_i a_j a'_j e_j c_k J_0 \boxed{\mathbf{a_s} J_1 J_2 J_3 \boxed{b_j d_j \#}}$$

In the way described above, $q_i$ is replaced by $q_l$ and $c_k$ is removed from region 1 to the environment. Notice that symbols $a_j$, $a'_j$, $b_j$, $d_j$, $e_j$, $J_0$, $J_1$, $J_2$ have returned to their original positions. Symbol $d_j$ returns to region 2 in the first step of the simulation of the next instruction (the last step of the illustration).

(ii) *There is no $c_k$ in region 1:*

Again we start with configuration (A).

$$q_l q_i a_j a_s \mathbf{e_j J_1} \boxed{\mathbf{b_j d_j J_0} J_3 \boxed{\mathbf{J_2 a'_j} \#}}$$

$$\Rightarrow^{1r6,1r13,2r11,2r12} \mathbf{q_l} q_i a_j a_s \mathbf{d_j} J_0 \boxed{J_1 J_2 \mathbf{a'_j} e_j J_3 \boxed{b_j \#}} \Rightarrow^{1r8,\underline{1r11},1r14}$$

Now rule **1r11** will be applied, leading to an infinite computation (see scenario 0).

**"Test for zero"** -instruction:

$q_i$ is replaced by $q_l$ if there is no $c_k$ in region 1, otherwise $a'_j$ in region 2 exchanges with $c_k$ in region 1 and the computation will never stop.

(i) *There is no $c_k$ in region 1:*

We consider configuration (A) above.

$$q_l q_i a_j a_s e_j \mathbf{J_1} \boxed{\mathbf{b_j d_j J_0} J_3 \boxed{\mathbf{J_2 a'_j} \#}} \Rightarrow^{1r6, 1r13, 2r12} q_l q_i a_j a_s \mathbf{d_j} J_0 \boxed{J_1 J_2 e_j J_3 \boxed{\mathbf{a'_j b_j} \#}}$$

$$\Rightarrow^{1r9, 2r6} q_l q_i a_j a_s J_0 \boxed{\mathbf{d_j} J_1 J_2 J_3 \mathbf{a'_j} \boxed{e_j b_j \#}} \Rightarrow^{1r8, 2r8} \mathbf{q_1} q_i a_j a_s a'_j J_0 \boxed{e_j J_1 J_2 J_3 \boxed{b_j d_j \#}}$$

$$\Rightarrow^{1r14} q_i a_j \mathbf{a_s} a'_j e_j J_0 \boxed{\mathbf{q_l} J_1 J_2 J_3 \boxed{b_j d_j \#}}$$

In this case, $q_i$ is replaced by $q_l$. Notice that symbols $a_j$, $a'_j$, $b_j$, $d_j$, $e_j$, $J_0$, $J_1$, $J_2$ have returned to their original positions.

(ii) *There is some $c_k$ in region 1:*

Consider configuration (A) with object $c_k$ in region 1:

$$q_l q_i a_j a_s e_j \mathbf{J_1} \boxed{\mathbf{b_j d_j J_0} J_3 \mathbf{c_k} \boxed{\mathbf{J_2 a'_j} \#}} \Rightarrow^{1r6, 1r13, \underline{2r5}, 2r12}$$

Now applying rule **2r5** leads to an infinite computation.

$$\mathbf{q_1} q_i a_j a_s a_s a'_s \mathbf{d_j} J_0 J_0 \boxed{J_1 J_2 \mathbf{a'_j} e_j J_3 \boxed{b_j b_s c_k \#}} \Rightarrow^{1r8, 1r9, 1r14}$$

$$q_i a_j a_s \mathbf{a_s} a'_j a'_s e_j J_0 J_0 \boxed{\mathbf{q_1 d_j} J_1 J_2 J_3 \boxed{b_j b_s c_k \#}} \Rightarrow^{1r1, 1r13}$$

$$\mathbf{q_1} q_i a_j a_s a'_j a'_s \mathbf{d_j} J_0 J_0 \boxed{\mathbf{a_s} e_j J_1 J_2 J_3 \boxed{b_j \mathbf{b_s} \#}} \Rightarrow^{1r14, 2r1}$$

$$q_i a_j \mathbf{a_s} a'_j \mathbf{a'_s} e_j J_0 J_0 \boxed{\mathbf{b_s q_1} J_1 \mathbf{J_2} J_3 \boxed{\mathbf{a_s} d_j \#}} \Rightarrow^{1r1, 1r3, 2r2}$$

$$q_l q_i a_j a'_j \mathbf{b_s} e_j \mathbf{J_0 J_0} \boxed{\mathbf{a_s a_s a'_s J_1} J_2 J_3 \boxed{d_j \#}}$$

So, scenario 3 takes place and the computation never halts.

## 3. END

$$R_{1,f} = \{\mathtt{1f1} : (T_1, out; F_1, in)\} \cup \{\mathtt{1f2} : (F_i, out; F_{i+1}, in) \mid 1 \le i \le 8\}$$
$$\cup \{\mathtt{1f3} : (T_2, out; P_1), \ \mathtt{1f4} : (P_2, out), \ \mathtt{1f5} : (F_0, out; P_2, in)\}.$$

$$R_{2,f} = \{\mathtt{2f1} : (T_1, out; F_0, in), \quad \mathtt{2f2} : (F_0, out), \quad \mathtt{2f3} : (T_2, out; F_0, in)\}$$
$$\cup \{\mathtt{2f4} : (P_1, in), \quad \mathtt{2f5} : (P_1, out; J_1, in), \quad \mathtt{2f6} : (P_1, out; J_2, in)\}$$
$$\cup \{\mathtt{2f7} : (P_1, out; J_3, in), \quad \mathtt{2f8} : (J_3, out; L, in), \quad \mathtt{2f9} : (P_2, out; F_9, in)\}.$$

Once the counter automaton reaches the final state, $q_f$ is in region 1 and it exchanges with object $a_n$ (rule 1r2) and object $F_0$ will be moved to region 1 in several steps (rules 1r15). Further symbol $F_0$ takes $T_1$ and $T_2$ to region 1, in either order. The duty of $T_2$ is to bring $P_1$ from the environment to region 2, where $P_1$ pumps objects $J_1, J_2, J_3$ from region 1 to region 2. If on the previous steps of simulation of counter automaton $M$ object $J_3$ was moved to the environment (by rules 1r11, 1r12), scenario 0 takes place and the computation never halts, as there is only one possibility to stop an infinite loop with object $L$, i.e. to move it to region 2 by rule 2f8.

$T_1$ starts a chain of exchanges of objects $F_i$, as a result object $F_9$ will be moved to region 1 and then object $P_2$ will be moved to the environment, where it pumps object $F_0$ to the environment. So, at the end of the computation there are only objects $c_k, k \in D$ in region 1. The entire simulation shows $N(\Pi_1) \supseteq N(M)$.

The converse inclusion holds because the system may only halt if it has correctly simulated a computation of the counter automaton (according to the design of the system) from state $q_0$ to state $q_f$, while if behavior of $M$ is not simulated correctly, then the computation never halts and hence does not contribute to $N(\Pi_1)$. This shows that P systems with two membranes and symport/antiport rules of size one with the output in the skin membrane generate all recursively enumerable sets of natural numbers. Since the power of such systems cannot exceed that of Turing machines, the statement of the theorem is an equality. □

**Theorem 2.** $\mathbb{N}^{skin}OP_2(sym_2) = \mathbb{N}RE$.

*Proof.* As in the proof of Theorem 1 we simulate a counter automaton $M = (d, Q, q_0, q_f, P)$ that starts with empty counters. We suppose that all instructions from $P$ are bijectively labelled with elements of $\{1, \ldots, n\} = I$, $n$ is a label of the *halt* instruction, $I' = I \setminus \{n\}$, and $I = I_+ \cup I_- \cup I_{=0}$, where we denote by $I_+, I_-,$ and $I_{=0}$ the set of labels for the "increment" -, "decrement" -, and "test for zero" -instructions, respectively. We use also the next notations: $C = \{c_i \mid 1 \le i \le d\}$, $F = \{0, \ldots f\}$ and $F' = F \setminus \{f\}$. We construct the P system $\Pi_2$ as follows:

$$\Pi_2 = (O, [_1 \ [_2 \ ]_2 \ ]_1, w_1, w_2, E, R_1, R_2, 1),$$
$$O = E \cup Q \cup \{b_j, d_j, d'_j \mid j \in I'\} \cup \{a_1, \#_1, \#_2, L, t_1, t_2, t_4, t_5, t_6, t_8\}$$
$$E = C \cup \{a_j \mid j \in I' \setminus \{1\}\} \cup \{d''_j \mid j \in I'\} \cup \{t_0, t_3, t_7\}$$
$$w_1 = q_0 a_1 \#_1 L t_1 t_6 \prod_{j \in I'} b_j,$$
$$w_2 = \#_2 t_2 t_4^n t_5 t_8 \prod_{i \in F \setminus \{0\}} q_i \prod_{j \in I'} d_j \prod_{j \in I'} d'_j,$$
$$R_i = R_{i,s} \cup R_{i,r} \cup R_{i,f}, i \in \{1, 2\}.$$

We code the counter automaton as follows: the environment will hold the current state of the automaton, represented by a symbol $q_i \in Q$; region 1 will hold the value of all counters, represented by the number of occurrences of symbols $c_k \in C$, $k \in D$, where $D = \{1, \ldots, d\}$.

As in Theorem 1 we split our proof into several parts that depend on the logical separation of the behavior of the system and use the same agreements. The rules $R_i$ are given by three phases:

1. START: preparation of the system for the computation.
2. RUN: simulation of instructions of the counter automaton.
3. END: terminating the computation.

## 1. START

As in Theorem 1 we use the following idea: in our system we have a symbol $L$ which moves from region 1 to the environment and back in an infinite loop. This loop may be stopped only if all stages completed correctly.

$$R_{1,s} = \{\mathbf{1s1} : (L, out), \quad \mathbf{1s2} : (L, in)\},$$
$$R_{2,s} = \emptyset.$$

## 2. RUN

$$
\begin{aligned}
R_{1,r} = {} & \{\mathbf{1r1} : (q_i a_j, in) \mid (j : q_i \to q_l, c_k \gamma) \in P, \gamma \in \{+, -, = 0\}\} \\
& \cup \{\mathbf{1r2} : (a_j d_j, out) \mid j \in I'\} \\
& \cup \{\mathbf{1r3} : (d_j c_k, in) \mid (j : q_i \to q_l, k+) \in P, k \in D\} \\
& \cup \{\mathbf{1r4} : (d_j, in) \mid j \in I_-\} \\
& \cup \{\mathbf{1r5} : (d_j d_j'', in) \mid j \in I_{=0}\} \\
& \cup \{\mathbf{1r6} : (d_j', out) \mid j \in I_+\} \\
& \cup \{\mathbf{1r7} : (d_j' c_k, out) \mid (j : q_i \to q_l, k\gamma) \in P, k \in D, \gamma \in \{-, = 0\}\} \\
& \cup \{\mathbf{1r8} : (d_j' d_j'', in) \mid j \in I_+ \cup I_-\} \\
& \cup \{\mathbf{1r9} : (d_j'' q_l, out) \mid (j : q_i \to q_l, k\gamma) \in P, k \in D, \gamma \in \{-, +, = 0\}\} \\
& \cup \{\underline{\mathbf{1r10}} : (\#_2, in), \ \underline{\mathbf{1r11}} : (\#_2, out)\}, \\
R_{2,r} = {} & \{\mathbf{2r1} : (a_j b_j, in) \mid j \in I'\} \cup \{\mathbf{2r2} : (q_i, in) \mid i \in F'\} \\
& \cup \{\mathbf{2r3} : (a_j d_j, out), \ \mathbf{2r4} : (b_j d_j', out), \ \mathbf{2r5} : (d_j, in) \mid j \in I'\} \\
& \cup \{\mathbf{2r6} : (d_j' d_j'', in) \mid j \in I'\} \\
& \cup \{\mathbf{2r7} : (d_j'' q_l, out) \mid (j : q_i \to q_l, k\gamma) \in P, k \in D, \gamma \in \{-, +, = 0\}\} \\
& \cup \{\underline{\mathbf{2r8}} : (b_j \#_2, out), \ \underline{\mathbf{2r9}} : (d_j' \#_1, in) \mid j \in I'\} \\
& \cup \{\underline{\mathbf{2r10}} : (d_j' \#_1, in) \mid j \in I_-\} \\
& \cup \{\underline{\mathbf{2r11}} : (\#_1 \#_2, out)\}.
\end{aligned}
$$

Now we explain the behavior of simulating the instruction $(j : q_i \to q_l, c_k \gamma)$. Index $s$ stands for any possible instruction associated to state $q_l$.

**"Increment"**-instruction:

$$q_i a_j a_s d_j'' c_k \boxed{b_j \#_1 \boxed{q_l d_j d_j' \#_2}} \Rightarrow^{1r1} d_j'' a_s c_k \boxed{\mathbf{q_i a_j b_j} \#_1 \boxed{q_l d_j d_j' \#_2}} \Rightarrow^{2r1,2r2}$$

$$d_j'' a_s c_k \boxed{\#_1 \boxed{q_i q_l \mathbf{a_j d_j d_j' b_j} \#_2}} \Rightarrow^{2r3,2r4} d_j'' a_s c_k \boxed{\mathbf{a_j d_j d_j' b_j} \#_1 \boxed{q_i q_l \#_2}} \qquad (B)$$

Now there are two variants of computations (depending on the application of rule **1r2** or rule **2r5**). It is easy to see that the application of rule **2r5** leads to an infinite computation. In this case rule **2r1** will be applied, symbol $b_j$ again appears in region 2, but symbol $d_j'$ is absent in this region at that moment (it is situated in the environment) and we cannot apply rule **2r4**, so symbol $\#_2$ will be moved to region 1 eventually by rule <u>2r8</u>, that leads to an infinite computation (rules <u>1r10</u>, <u>1r11</u>).

So, consider applying rule **1r2**:

$$d_j'' a_s c_k \boxed{\mathbf{a_j d_j d_j' b_j} \#_1 \boxed{q_i q_l \#_2}} \Rightarrow^{1r2,1r6} \mathbf{d_j' d_j'' d_j} c_k a_j a_s \boxed{b_j \#_1 \boxed{q_i q_l \#_2}} \Rightarrow^{1r3,1r8}$$

$$a_j a_s \boxed{\mathbf{d_j' d_j'' d_j} c_k b_j \#_1 \boxed{q_i q_l \#_2}}$$

Now there are two variants of computations (depending on the application of rule **1r6** or rule **2r6**). It is easy to see that the application of rule **1r6** leads to an infinite computation. In this case rule <u>2r9</u> will be applied and symbol $\#_1$ will be moved to region 2 and after that symbol $\#_2$ will appear in region 1 (rule <u>2r11</u>) that leads to an infinite computation (rules <u>1r10</u>, <u>1r11</u>).
So, consider applying rule **2r6**:

$$a_j a_s \boxed{\mathbf{d_j' d_j'' d_j} c_k b_j \#_1 \boxed{q_i q_l \#_2}} \Rightarrow^{2r5,2r6} a_j a_s \boxed{c_k b_j \#_1 \boxed{q_i \mathbf{q_l} d_j'' d_j d_j' \#_2}} \Rightarrow^{2r7}$$

$$a_j a_s \boxed{\mathbf{q_l} d_j'' c_k b_j \#_1 \boxed{q_i d_j d_j' \#_2}} \Rightarrow^{1r9} a_j \mathbf{q_l a_s} d_j'' \boxed{c_k b_j \#_1 \boxed{q_i d_j d_j' \#_2}}$$

In that way, $q_i$ is replaced by $q_l$ and $c_k$ is moved from the environment into region 1. Notice that symbols $a_j, b_j, d_j, d_j', d_j''$ have returned to their original positions.

**"Decrement"** -instruction:
(i) *There is some $c_k$ in region 1.*
We consider configuration (B) above with symbol $c_k$ in region 1.

$$d_j'' a_s \boxed{\mathbf{a_j d_j d_j' c_k} b_j \#_1 \boxed{q_i q_l \#_2}} \Rightarrow^{1r2,1r7} \mathbf{d_j' d_j'' d_j} a_j a_s c_k \boxed{b_j \#_1 \boxed{q_i q_l \#_2}} \Rightarrow^{1r4,1r8}$$

$$a_j a_s c_k \boxed{\mathbf{d_j' d_j'' d_j} b_j \#_1 \boxed{q_i q_l \#_2}}$$

Now there are two variants of computations (depending on the application of rule **1r7** or rule **2r6**). It is easy to see that the application of rule **1r7** leads to an infinite computation. In this case rule **2r9** will be applied and symbol $\#_1$ will be moved to region 2 and after that symbol $\#_2$ will appear in region 1 (rule **2r11**) that leads to an infinite computation (rules **1r10**, **1r11**). So, consider applying rule **2r6**:

$$a_j a_s c_k \;\Big|\, \mathbf{d'_j d''_j d_j b_j} \#_1 \boxed{q_i q_l \#_2}\,\Big| \Rightarrow^{2r5,2r6} a_j a_s c_k \;\Big|\, b_j \#_1 \boxed{q_i \mathbf{q_l d''_j} d_j d'_j \#_2}\,\Big| \Rightarrow^{2r7}$$

$$a_j a_s c_k \;\Big|\, \mathbf{q_l d''_j} b_j \#_1 \boxed{q_i d_j d'_j \#_2}\,\Big| \Rightarrow^{1r9} a_j \mathbf{q_l a_s} d''_j c_k \;\Big|\, b_j \#_1 \boxed{q_i d_j d'_j \#_2}\,\Big|$$

In the way described above, $q_i$ is replaced by $q_l$ and $c_k$ is removed from region 1 to the environment. Notice that symbols $a_j, b_j, d_j, d'_j, d''_j$ have returned to their original positions.

(ii) *There is no $c_k$ in region 1.*
Again we start with configuration (B).

$$d''_j a_s \;\Big|\, \mathbf{a_j d_j d'_j} b_j \#_1 \boxed{q_i q_l \#_2}\,\Big|$$

In this case rule **2r10** will be applied eventually that leads to an infinite computation.

**"Test for zero"** -instruction:
Symbol $q_i$ is replaced by symbol $q_l$ if there is no $c_k$ in region 1, otherwise $d'_j$ in region 1 will moved in the environment and the computation will never stop.

(i) *There is no $c_k$ in region 1.*
Again we start with configuration (B).

$$d''_j a_s \;\Big|\, \mathbf{a_j d_j d'_j} b_j \#_1 \boxed{q_i q_l \#_2}\,\Big| \Rightarrow^{1r2} \mathbf{d''_j d_j} a_j a_s \;\Big|\, d'_j b_j \#_1 \boxed{q_i q_l \#_2}\,\Big| \Rightarrow^{1r5}$$

$$a_j a_s \;\Big|\, \mathbf{d''_j d'_j d_j} b_j \#_1 \boxed{q_i q_l \#_2}\,\Big| \Rightarrow^{2r5,2r6} a_j a_s \;\Big|\, b_j \#_1 \boxed{q_i \mathbf{q_l d''_j} d'_j d_j \#_2}\,\Big| \Rightarrow^{2r7}$$

$$a_j a_s \;\Big|\, \mathbf{q_l d''_j} b_j \#_1 \boxed{q_i d'_j d_j \#_2}\,\Big| \Rightarrow^{1r9} \mathbf{q_l a_s} a_j d''_j \;\Big|\, b_j \#_1 \boxed{q_i d'_j d_j \#_2}\,\Big|$$

In that way, $q_i$ is replaced by $q_l$. Notice that symbols $a_j, b_j, d_j, d'_j, d''_j$ have returned to their original positions.

(i) *There is some $c_k$ in region 1.*
We consider configuration (B) above with symbol $c_k$ in region 1.

$$d''_j a_s \;\Big|\, \mathbf{a_j d_j d'_j c_k} b_j \#_1 \boxed{q_i q_l \#_2}\,\Big| \Rightarrow^{1r2,1r7} d'_j \mathbf{d''_j d_j} a_j a_s c_k \;\Big|\, b_j \#_1 \boxed{q_i q_l \#_2}\,\Big| \Rightarrow^{1r5}$$

$$d'_j a_j a_s c_k \;\Big|\, \mathbf{d_j} b_j \mathbf{d''_j} \#_1 \boxed{q_i q_l \#_2}\,\Big|$$

Now rule **2r9** will be applied eventually that leads to an infinite computation.

## 3. END

$$R_{1,f} = \{\mathtt{1f1} : (q_f t_0, in), \ \mathtt{1f2} : (t_0 t_1, out), \ \mathtt{1f3} : (t_1 t_3, in) \ \mathtt{1f4} : (t_4 \#_1, out)\}$$
$$\cup \ \{ \ \mathtt{1f5} : (t_4 b_j, out) \mid j \in I' \}$$
$$\cup \ \{ \ \mathtt{1f6} : (t_5 t_6, out), \ \mathtt{1f7} : (t_6 t_7, in), \ \mathtt{1f8} : (t_8, out)\}.$$
$$R_{2,f} = \{\mathtt{2f1} : (q_f t_0, in), \ \mathtt{2f2} : (t_0 t_2, out), \ \mathtt{2f3} : (Lt_2, in), \ \mathtt{2f4} : (t_3, in),$$
$$\mathtt{2f5} : (t_3 t_4, out), \ \mathtt{2f6} : (t_3 t_5, out), \ \mathtt{2f7} : (t_7 t_1, in),$$
$$\mathtt{2f8} : (t_6 t_7, in), \ \mathtt{2f9} : (t_7 t_8, in)\}.$$

Once the counter automaton reaches the final state, $q_f$ is in the environment and brings in region 1 symbol $t_0$ (rule $\mathtt{1f1}$). Now there are two variants of computations (depending on the application of rule $\mathtt{1f2}$ or rule $\mathtt{2f1}$). It is easy to see that the application of rule $\mathtt{1f2}$ leads to an infinite computation as there is no chance for symbol $t_0$ to appear in region 2 and takes symbol $L$ in region 2, thus to stop an infinite loop (rules $\mathtt{2f2}, \mathtt{2f3}$).

So, consider the applying rule $\mathtt{2f1}$. In this case symbol $q_f$ brings symbol $t_0$ to region 2 and symbol $t_2$ stops an infinite loop (rules $\mathtt{2f2}, \mathtt{2f3}$). Now rule $\mathtt{1f2}$ will be applied and symbol $t_1$ will be moved to the environment where it takes symbol $t_3$ and they appear in region 1 (rule $\mathtt{1f3}$). Further symbol $t_3$ pumps symbols $t_4$ to region 1 from region 2 (rules $\mathtt{2f4}, \mathtt{2f5}$), there they take symbols $b_j$ and symbol $\#_1$ and bring them to the environment (rules $\mathtt{1f4}, \mathtt{1f5}$). Symbol $t_3$ also pumps symbol $t_5$ to region 1 from region 2 (rule $\mathtt{2f6}$) and after that symbol $t_7$ appears in region 1 (rules $\mathtt{1f6}, \mathtt{1f7}$). Now with help of symbol $t_7$ two symbols $t_6$ and $t_1$ will be moved to region 2 (rules $\mathtt{2f7}, \mathtt{2f8}, \mathtt{2f9}$). Finally, symbol $t_8$, that appears in region 1 by rule $\mathtt{2f9}$, will be moved to the environment (rule $\mathtt{1f8}$).

So, at the end of the computation there are only objects $c_k, k \in D$ in region 1. The entire simulation shows the inclusion $N(\Pi_2) \supseteq N(M)$.

The converse inclusion also holds because the system may only halt if it has correctly simulated a computation of the counter automaton (according to the design of the system) from state $q_0$ to state $q_f$, while if behavior of $M$ is not simulated correctly, then the computation never halts and hence does not contribute to $N(\Pi_2)$. This shows that P systems with two membranes and symport rules of size two with the output in the skin membrane generate all recursively enumerable sets of natural numbers. Since the power of such systems cannot exceed that of Turing machines, the statement of the theorem is an equality.                                                                                                □

**Program Check.** P systems in both theorems were checked for errors with the help of a program that simulates P systems, originally developed by the first author and modified by Galina Magariu and Tatiana Verlan with assistance of Vladimir Rogojin.

## 4  Conclusions

In this paper we prove the new results that any recursively enumerable set of natural numbers is generated by symport/antiport (and purely symport) P systems with two membranes and minimal cooperation where the output membrane is the skin membrane. It contrasts with the previous result where an elementary membrane is used as the output membrane, where at least one superfluous object is necessary in the output membrane in order to get universality. Thus we answered the question of Francesco Bernardini about computational power of symport/antiport P systems with two membranes and minimal cooperation where the output membrane is the skin membrane.

## Acknowledgements

The authors acknowledge the project 06.411.03.04P from the Supreme Council for Science and Technological Development of the Academy of Sciences of Moldova. The first author gratefully acknowledges the support by Academy of Finland, project 203667 and the second author gratefully acknowledges the support of European Commission, project MolCIP, MIF1-CT-2006-021666. The authors acknowledge Galina Magariu and Tatiana Verlan for suggestions, most of them incorporated in the present version of the paper.

## References

1. Alhazov, A., Freund, R., Rogozhin, Y.: Computational Power of Symport/Antiport: History, Advances, and Open Problems. In: Freund, R., Păun, G., Rozenberg, G., Salomaa, A. (eds.) WMC 2005. LNCS, vol. 3850, pp. 1–30. Springer, Heidelberg (2006)
2. Alhazov, A., Freund, R., Rogozhin, Y.: Some Optimal Results on Communicative P Systems with Minimal Cooperation. In: [17], pp. 23–36
3. Alhazov, A., Margenstern, M., Rogozhin, V., Rogozhin, Y., Verlan, S.: Communicative P Systems with Minimal Cooperation. In: Mauri, G., Păun, G., Pérez-Jiménez, M.J., Rozenberg, G., Salomaa, A. (eds.) WMC 2004. LNCS, vol. 3365, pp. 161–177. Springer, Heidelberg (2005)
4. Alhazov, A., Rogozhin, Y.: Minimal Cooperation in Symport/Antiport P Systems with One Membrane. In: [18], pp. 29–34
5. Alhazov, A., Rogozhin, Y.: Towards a Characterization of P Systems with Minimal Symport/Antiport and Two Membranes. In: Hoogeboom, H.J., Păun, G., Rozenberg, G., Salomaa, A. (eds.) WMC 2006. LNCS, vol. 4361, pp. 135–153. Springer, Heidelberg (2006)
6. Alhazov, A., Rogozhin, Y.: Generating Languages by P Systems with Minimal Symport/Antiport. Computer Science Journal of Moldova 14, 3(42), 299–323 (2006)
7. Alhazov, A., Rogozhin, Y., Verlan, S.: Symport/Antiport Tissue P Systems with Minimal Cooperation. In: [17], pp. 37–52
8. Alhazov, A., Rogozhin, Y., Verlan, S.: Minimal Cooperation in Symport/Antiport Tissue P Systems. International Journal of Foundation of Computer Science 18(1), 163–179 (2007)
9. Bernardini, F., Gheorghe, M.: On the Power of Minimal Symport/Antiport. In: Martín-Vide, C., Mauri, G., Păun, G., Rozenberg, G., Salomaa, A. (eds.) Membrane Computing. LNCS, vol. 2933, pp. 72–83. Springer, Heidelberg (2004)

10. Bernardini, F., Păun, A.: Universality of Minimal Symport/Antiport: Five Membranes Suffice. In: Martín-Vide, C., Mauri, G., Păun, G., Rozenberg, G., Salomaa, A. (eds.) Membrane Computing. LNCS, vol. 2933, pp. 43–45. Springer, Heidelberg (2004)
11. Freund, R., Oswald, M.: GP Systems with Forbidding Context. Fundamenta Informaticae 49, 1–3, 81–102 (2002)
12. Freund, R., Oswald, M.: P Systems with Activated/Prohibited Membrane Channels. In: [29], pp. 261–268
13. Freund, R., Păun, A.: Membrane Systems with Symport/Antiport: Universality Results. In: [29], pp. 270–287
14. Frisco, P.: About P Systems with Symport/Antiport. In: Păun, G., Riscos-Núñez, A., Romero-Jiménez, A., Sancho-Caparrini, F. (eds.) Second Brainstorming Week on Membrane Computing. TR *01/2004*, Research Group on Natural Computing, University of Seville, pp. 224–236 (2004)
15. Frisco, P., Hoogeboom, H.J.: Simulating Counter Automata by P Systems with Symport/Antiport. In: [29], pp. 288–301
16. Frisco, P., Hoogeboom, H.J.: P Systems with Symport/Antiport Simulating Counter Automata. Acta Informatica 41, 2–3, 145–170 (2004)
17. Gutiérrez-Naranjo, M.A., Păun, G., Pérez-Jiménez, M.J. (eds.): Cellular Computing (Complexity Aspects). ESF PESC Exploratory Workshop. Fénix Editora, Sevilla (2005)
18. Gutiérrez-Naranjo, M.A., Riscos-Núñez, A., Romero-Campero, F.J., Sburlan, D. (eds.): Third Brainstorming Week on Membrane Computing. RGNC TR *01/2005*, University of Seville. Fénix Editora, Sevilla (2005)
19. Kari, L., Martín-Vide, C., Păun, A.: On the Universality of P Systems with Minimal Symport/Antiport Rules. In: Jonoska, N., Păun, G., Rozenberg, G. (eds.) Aspects of Molecular Computing. LNCS, vol. 2950, pp. 254–265. Springer, Heidelberg (2003)
20. Margenstern, M., Rogozhin, V., Rogozhin, Y., Verlan, S.: About P Systems with Minimal Symport/Antiport Rules and Four Membranes. In: [22], pp. 283–294
21. Martín-Vide, C., Păun, A., Păun, G.: On the Power of P Systems with Symport Rules. Journal of Universal Computer Science 8(2), 317–331 (2002)
22. Mauri, G., Păun, G., Pérez-Jiménez, M.J., Rozenberg, G., Salomaa, A. (eds.): WMC 2004. LNCS, vol. 3365. Springer, Heidelberg (2005)
23. Minsky, M.L.: Finite and Infinite Machines. Prentice Hall, Englewood Cliffs, New Jersey (1967)
24. Păun, A., Păun, G.: The Power of Communication: P Systems with Symport/Antiport. New Generation Computing 20, 295–305 (2002)
25. Păun, G.: Computing with Membranes. Journal of Computer and Systems Science 61, 108–143 (2000)
26. Păun, G.: Membrane Computing. An Introduction. Springer, Heidelberg (2002)
27. Păun, G.: Further Twenty Six Open Problems in Membrane Computing. In: [18], pp. 249–262 (2005)
28. Păun, G.: 2006 Research Topics in Membrane Computing. In: Gutiérrez-Naranjo, M.A., Păun, G., Riscos-Núñez, A., Romero-Campero, F.J. (eds.) Fourth Brainstorming Week on Membrane Computing, vol. 1, pp. 235–251. Fénix Edit., Sevilla (2006)
29. Păun, G., Rozenberg, G., Salomaa, A., Zandron, C. (eds.): Membrane Computing. LNCS, vol. 2597. Springer, Heidelberg (2003)
30. Rozenberg, G., Salomaa, A. (eds.): Handbook of Formal Languages, vol. 3. Springer, Berlin (1997)

31. Vaszil, G.: On the Size of P Systems with Minimal Symport/Antiport. In: [22], pp. 422–431
32. Verlan, S.: Optimal Results on Tissue P Systems with Minimal Symport/ Antiport. In: EMCC meeting, Lorentz Center, Leiden (2004)
33. Verlan, S.: Tissue P Systems with Minimal Symport/Antiport. In: Calude, C.S., Calude, E., Dinneen, M.J. (eds.) DLT 2004. LNCS, vol. 3340, pp. 418–430. Springer, Heidelberg (2004)
34. P Systems Webpage, http://psystems.disco.unimib.it

# On the Reachability Problem in P Systems with Mobile Membranes

Bogdan Aman[2] and Gabriel Ciobanu[1,2]

[1] "A.I.Cuza" University of Iaşi, Faculty of Computer Science
Blvd. Carol I no.11, 700506 Iaşi, Romania
[2] Romanian Academy, Institute of Computer Science
Blvd. Carol I no.8, 700505 Iaşi, Romania
baman@iit.tuiasi.ro, gabriel@info.uaic.ro

**Abstract.** We investigate the problem of reaching a configuration from another configuration in mobile membranes, and prove that the reachability can be decided by reducing it to the reachability problem of a version of pure and public ambient calculus without the capability open.

## 1  Introduction

Membrane systems (called also P systems) are introduced by Gh. Păun in [9] as a class of parallel computing devices inspired by biology. The definition of this computing model starts from the observation that any biological system is a complex hierarchical structure, with a flow of materials and information which underlies their functioning. The membrane computing deals with the evolution of systems composed by objects, rules and membranes nested in other membranes. The P systems with mobile membranes [6] is a model which expresses mobility by the movement of membranes in such a system. The movement is given mainly by two operations: exocytosis and endocytosis.

Ambient calculus is a formalism introduced in [3] to describe concurrent and mobile computation. In contrast with other formalisms for mobile processes such as the $\pi$-calculus [8] based on the notion of communication, the ambient calculus is based on the notion of movement. An ambient is a named location, and represents a unit of movement. Ambients mobility is controlled by the capabilities in, out, and open; the mobile ambients describe the migration of processes between certain boundaries.

The membrane systems and mobile ambients have similar structures and common concepts. Both have a hierarchical structure, work with an explicit notion of location, and are used to model various aspects on the distributed systems. The distributed features of mobile ambients are described in [3], and distributed algorithms for membrane systems are presented in [4].

In this paper we investigate the problem of reaching a certain configuration in mobile membranes starting from a given configuration. We prove that reachability in mobile membranes can be decided by reducing it to the reachability problem of a version of pure and public ambient calculus from which the open

G. Eleftherakis et al. (Eds.): WMC8 2007, LNCS 4860, pp. 113–123, 2007.

capability has been removed. It is proven in [1] that the reachability for this fragment of ambient calculus is decidable by reducing it to marking reachability for Petri nets, which is proven to be decidable in [7]. Problems like reachability and boundedness are investigated in [5] for other classes of P systems, namely for extensions of PB systems with volatile membranes.

The structure of the paper is as follows. In Section 2 we present the mobile membrane systems, whereas in Section 3 we present a version of pure and public mobile ambients without the capability open. The core of the paper is represented by Section 4, where we investigate the reachability problem for mobile membranes. Conclusions and references end the paper.

## 2   Mobile Membranes Systems

**Definition 1.** *A* mobile membrane system *is a construct*
$$\prod = (V \cup \overline{V}, H \cup \overline{H}, \mu, w_1, \dots, w_n, R), \quad where:$$

1. *$n \geq 1$ is the* degree *of the system, given by the initial total number of membranes;*
2. *$V \cup \overline{V}$ is an alphabet (its elements are called* objects*), where $V \cap \overline{V} = \emptyset$;*
3. *$H \cup \overline{H}$ is a finite set of* labels *for membranes, where $H \cap \overline{H} = \emptyset$;*
4. *$\mu$ is a membrane structure consisting of $n$ membranes labeled (not necessarily in a one-to-one manner) with elements of $H$;*
5. *$w_1, w_2, \dots, w_n$ are* multisets of objects *placed in the membranes of the system;*
6. *$R$ is a finite set of* developmental rules*, of the following forms:*

   (a) $\overline{a}\!\downarrow \rightarrow \overline{a}\!\downarrow a\!\downarrow$, *for $a\!\downarrow \in V$, $\overline{a}\!\downarrow \in \overline{V}$;*      replication rule
       *The objects $\overline{a}\!\downarrow$ are used to create new objects $a\!\downarrow$ without being consumed.*

   (b) $\overline{a}\!\uparrow \rightarrow \overline{a}\!\uparrow a\!\uparrow$, *for $a\!\uparrow \in V$, $\overline{a}\!\uparrow \in \overline{V}$;*      replication rule
       *The objects $\overline{a}\!\uparrow$ are used to create new objects $a\!\uparrow$ without being consumed.*

   (c) $[\, a\!\downarrow \,]_h [\,]_a \rightarrow [\, [\,]_h \,]_a$, *for $a, h \in H$, $a\!\downarrow \in V$;*      endocytosis
       *An elementary membrane labeled $h$ enters the adjacent membrane labeled $a$ under the control of object $a\!\downarrow$. The labels $h$ and $a$ remain unchanged during this process; however object $a\!\downarrow$ is consumed during the operation. Membrane $a$ is not necessarily elementary.*

   (d) $[\, [\, a\!\uparrow \,]_h \,]_a \rightarrow [\,]_h [\,]_a$, *for $a, h \in H$, $a\!\uparrow \in V$;*      exocytosis
       *An elementary membrane labeled $h$ is sent out of a membrane labeled $a$, under the control of object $a\!\uparrow$. The labels of the two membranes remain unchanged, and object $a\!\uparrow$ of membrane $h$ is consumed during this operation. Membrane $a$ is not necessarily elementary.*

   (e) $[\,]_{\overline{h}} \rightarrow [\,]_{\overline{h}}[\,]_h$ *for $h \in H$, $\overline{h} \in \overline{H}$*      division rules
       *An elementary membrane labeled $\overline{h}$ is divided into two membranes labeled by $\overline{h}$, and $h$ and having the same objects.*

The condition $H \cap \overline{H} = \emptyset$ above states that the membranes having labels from the set $\overline{H}$ can participate only in rules of type $(e)$. Similarly, the condition $V \cap \overline{V} = \emptyset$ states that the objects from $\overline{V}$ can participate only in rules of type $(a)$ and $(b)$.

A configuration in mobile membranes describes a distribution of objects from $\Gamma$ into the membranes of $\prod$.

The rules are applied using the following principles:

1. In biological systems, the molecules are divided into classes of different types. Consequently, we divide the objects into four classes: $a \downarrow$ - objects which control the *endocytosis*, $a \uparrow$ - objects which control the *exocytosis*, and $\overline{a} \downarrow$, $\overline{a} \uparrow$ - objects which produce new objects from the first two classes without being consumed.

2. All the rules of type $(c), (d)$ are applied in parallel, non-deterministically choosing the rules, the membranes and the objects, in such a way that the parallelism is maximal; this means that in each step we apply a set of rules such that no further rule of type $(c), (d)$ can be added to the set, and no further membrane or object can evolve at the same time.

3. Membrane $a$ from each rule of type $(c), (d)$ is said to be passive, while membrane $h$ is said to be active. In any step of a computation, any object and any active membrane can be involved in at most one rule; however the passive membranes are not considered involved in the use of rules, and so they can be used by several rules at the same time (as passive membranes).

4. When a membrane is moved across another membrane (by endocytosis or exocytosis), its whole content is moved.

5. If a membrane is divided, then its content is replicated into the two new copies.

6. The skin membrane can never be divided.

7. Not all the rules of type $(a), (b), (e)$ are applied whenever it is possible; we choose non-deterministically whether the rules of these types are applied.

According to these rules, we get transitions among the configurations of the system. For two configurations $M$ and $N$ we say that $M$ reduces to $N$ if there is a sequence of rules applicable to the configuration $M$ in order to obtain the configuration $N$.

## 3   Mobile Ambients

We describe a variant of pure and public mobile ambients (mobile ambients in which communication and name restriction are omitted); more details can be found in [1]. Given an infinite set of names $\mathcal{N}$ (ranged over by $m, n, \dots$), we define the set $\mathcal{A}$ of mobile ambients (denoted by $A, A', B, \dots$) together with their capabilities (denoted by $C, C', \dots$) as follows:

| $C$ | $::=$ | $in\ n$ | $\mid$ | $out\ n$ | | | **Capabilities** |
|---|---|---|---|---|---|---|---|
| $A$ | $::=$ | $C.\,A$ | $\mid$ | $n[\,A\,]$ | $\mid$ | $A\mid B$ | $\mid$ | $!A$ | **Processes** |

A movement $C.\,A$ is provided by a capability $C$, followed by the execution of process $A$. An entry capability *in n* instructs the surrounding ambient to enter a sibling ambient labeled by $n$, while an exit capability *out n* instructs the surrounding ambient to exit its parent ambient labeled by $n$. An ambient

$n[A]$ represents a bounded place labeled by $n$ in which a process $A$ is executed. $A \mid B$ is a parallel composition of processes $A$ and $B$. $!A$ denotes an unbounded replication of process $A$.

Processes of this calculus are grouped into equivalence classes, up to trivial syntactic restructuring, by the **structural congruence** relation $\equiv$ which is the least congruence satisfying the following requirements:

$$A \mid B \equiv B \mid A \qquad\qquad A \equiv B \text{ implies } A \mid A' \equiv B \mid A'$$
$$(A \mid B) \mid A' \equiv A \mid (B \mid A') \qquad A \equiv B \text{ implies } !A \equiv !B$$
$$A \equiv A \qquad\qquad A \equiv B \text{ implies } n[A] \equiv n[B]$$
$$A \equiv B \text{ implies } B \equiv A \qquad A \equiv B \text{ implies } C.A \equiv C.B$$
$$A \equiv B, \ B \equiv A' \text{ implies } A \equiv A'$$

The *operational semantics* of the mobile ambients is defined in terms of a reduction relation $\Rightarrow$ by the following axioms and rules:

**Axioms:**

(**In**)  $\quad n[\, in \ m.\, A \mid A'\,] \mid m[\, B\,] \ \Rightarrow \ m[\, n[\, A \mid A'\,] \mid B\,] \ ;$

(**Out**)  $\quad m[\, n[\, out \ m.\, A \mid A'\,] \mid B\,] \ \Rightarrow \ n[\, A \mid A'\,] \mid m[\, B\,] \ ;$

(**Repl**)  $\quad !A \Rightarrow A \mid !A \ .$

**Rules:**

(**Comp**)  $\quad \dfrac{A \ \Rightarrow \ A'}{A \mid B \ \Rightarrow \ A' \mid B} \qquad$ (**Amb**)  $\dfrac{A \ \Rightarrow \ A'}{n[\, A\,] \ \Rightarrow \ n[\, A'\,]}$

$\qquad\qquad$ (**Struc**)  $\dfrac{A \equiv A', \ A' \ \Rightarrow \ B', \ B' \equiv B}{A \ \Rightarrow \ B} \ .$

The axioms represent the one-step reductions for *in* and *out*, and the unfolding of replication. The rules propagate reduction across ambient nesting, parallel composition and allow the use of structural congruence during reduction. According to (**Comp**), the axioms are applied in an interleaving manner.

We denote by $\Rightarrow^*$ the reflexive and transitive closure of the binary relation $\Rightarrow$.

## 4    Reachability Problem

In this section we prove that the problem of reaching a configuration starting from a certain configuration is decidable for the special class of mobile membranes systems introduced in Section 2.

**Theorem 1.** *For two arbitrary configurations $M_1$ and $M_2$ in a mobile membrane system, it is decidable whether $M_1$ reduces to $M_2$.*

The main steps of the proof are as follows:

1. mobile membranes systems are reduced to pure and public mobile ambients without the capability *open*;
2. the reachability problem for two arbitrary configurations can be expressed as the reachability problem for the corresponding mobile ambients.

3. the reachability problem is decidable for a fragment of pure and public mobile ambients without the capability *open*.

The following subsections are devoted to the proof of Theorem 1.

### 4.1  From Mobile Membranes to Mobile Ambients

We use the following translation steps:

1. any object $a\downarrow$ is translated into a capability *in a*;
2. any object $a\uparrow$ is translated into a capability *out a*;
3. any object $\bar{a}\downarrow$ is translated into a replication $!in$ $a$
4. any object $\bar{a}\uparrow$ is translated into a replication $!out$ $a$
5. a membrane $h$ is translated into an ambient $h$
6. an elementary membrane $\bar{h}$ is translated into a replication $!h[\ ]$ where all the objects inside membrane $h$ are translated into capabilities in ambient $h$ using the above steps.

A correspondence exists between the rules of the mobile membrane systems and the reduction rules of the mobile ambients as follows:

- rule $(c)$ corresponds to rule **(In)**;
- rule $(d)$ corresponds to rule **(Out)**;
- rules $(a), (b), (e)$ correspond to instances of rule **(Repl)**.

If we start with a configuration $M$ of a mobile membrane system, we denote by $\mathcal{T}(M)$ the mobile ambient obtained using the above translation steps. For example, starting from the configuration $M = [m\downarrow\ m\uparrow]_n[\ ]_m$ we obtain $\mathcal{T}(M) = n[in\ m\ |\ out\ m]\ |\ m[\ ]$.

**Proposition 1.** *For configuration $M$ and $N$, $M$ reduces to $N$ by applying one rule if and only if $\mathcal{T}(M)$ reduces to $\mathcal{T}(N)$ by applying only one reduction rule.*

*Proof (Sketch).* Since $M$ reduces $N$ by applying one rule, then one of the rules of type $(a), \ldots, (e)$ is applied. We treat only the case when a rule of type $(a)$ is applied, the others being treated in a similar manner.

If a rule $\bar{a}\downarrow \rightarrow \bar{a}\downarrow\ a\downarrow$ is applied, only one object from the configuration $M$ is used (namely $\bar{a}\downarrow$) to create a new object $a\downarrow$, thus obtaining the configuration $N$. By translating the configuration $M$ into $\mathcal{T}(M)$, we have that $\bar{a}\downarrow$ is translated in $!in\ a$. By applying the reduction rule corresponding to $(a)$ (namely the rule **(Repl)**) to $!in\ m$, then we have that $!in\ a \Rightarrow in\ a\ |\ !in\ a$, and so a new capability $in\ a$ is created. We note that $\mathcal{T}(\bar{a}\downarrow\ a\downarrow) = !in\ a\ |\ in\ a$, which means that the obtained mobile ambient is $\mathcal{T}(N)$ (in fact it is structural congruent to $\mathcal{T}(N)$).

According to Proposition 1 the reachability problem for mobile membranes can be reduced to a similar problem for mobile ambients.

## 4.2   From Mobile Ambients to Petri Nets

After translating the mobile membranes into a fragment of mobile ambients, we present the algorithm used in [1] to translate this fragment of mobile ambients into a fragment of Petri nets which is known to be decidable from [7]. The fragment of mobile ambients used here is a subset of the fragment of mobile ambients used in [1] and the difference is provided by the extra-rule $!A \Rightarrow !A \mid !A$ used in [1].

We note that applying a reduction rule over a process either increases the number of ambients or leaves it unchanged. The only reduction rule which increases the number of ambients when applied is the rule (**Repl**), while the other reduction rules leave the number of ambients unchanged. If we reach process $B$ starting from process $A$, then the number of ambients of process $B$ is known. Therefore, we can use this information to know how many times the reduction rule (**Repl**) is applied to replicate ambients. A similar argument does not hold for capabilities as they can be consumed by the reduction rules (**In**) and (**Out**).

An ambient context $C$ is a process in which may occur some holes (denoted by □). Using the ambient contexts, we split a process into two parts: one is a context containing ambients, whereas the other is a process without ambients. In order to uniquely identify all the occurrences of replication, ambient, capability or hole □ within an ambient context or a process, we introduce a labeling system. Using a countable set of labels, we say that a process $A$ or an ambient context $C$ is well-labeled if any label occurs at most once in $A$ or $C$. We denote by $Amb(C)$ the multiset of ambients occurring in an ambient context $C$. We say that two processes are label-free-equivalent if after removing all the labels from the two processes, they are structurally congruent.

**I) Labeled Transition System.** For the reachability problem $A \Rightarrow^* B$, we denote by $C_A$ a well-labeled ambient context, and by $\theta_A$ a mapping from the set of holes in $C_A$ to some labeled processes without replicable ambients such that $\theta_A(C_A)$ is well-labeled, and $\theta_A(C_A) = A$ where labels are ignored.

A labeled transition system $L_{A,B}$ describes all possible reductions for a context $C_A$: this includes reductions of replications and capabilities contained in $C_A$ and in the processes associated with the holes of the context. The states of the labeled transition system $L_{A,B}$ are associative-commutative equivalent classes of ambient contexts, and for simplicity, we often identify a state as one of the representatives of its class.

We define a mapping $\theta_{L_{A,B}}$ which extends the mapping $\theta_A$. Initially, $L_{A,B}$ contains (the equivalence class of) $C_A$ as a unique state, and we have $\theta_{L_{A,B}} = \theta_A$. We present in what follows the construction steps of $\theta_{L_{A,B}}$, where cap stands for in or out:

1. For any ambient context $C$ from $L_{A,B}$ and for any labeled capability $cap^w n$ in $C$, if this capability can be executed using one of the rules (**In**) or (**Out**) leading to some ambient context $C'$, then a state $C'$ and a transition from $C$ to $C'$ labeled by $cap^w n$ are added to $L_{A,B}$.

2. For any ambient context $C$ from $L_{A,B}$ and for any labeled replication $!^w$ in $C$ such that the reduction rule **(Repl)** is applied, we define the ambient context $C'$ as follows: $C'$ is identical to $C$ except that the subcontext $!^w C_a$ in $C$ is replaced by $!^w C_a \mid \gamma(C_a)$ in $C'$; the mapping $\gamma$ relabels $C_a$ with fresh labels, such that $C'$ is well-labeled. If $Amb(C') \subseteq Amb(B)$, then state $C'$ and a transition from $C$ to $C'$ labeled by $!^w$ is added to $L_{A,B}$. Additionally, we define $\theta'_{L_{A,B}}$ as an extension of $\theta_{L_{A,B}}$ such that for all $\square^{w'}$ in $C_a$ we have:

   (i) $\theta'_{L_{A,B}}(\gamma(\square^{w'}))$ and $\theta_{L_{A,B}}(\square^{w'})$ are label-free-equivalent,

   (ii) labels in $\theta'_{L_{A,B}}(\gamma(\square^{w'}))$ are fresh in the currently built transition system $L_{A,B}$,

   (iii) $\theta'_{L_{A,B}}(\gamma(\square^{w'}))$ is well-labeled.

   Finally, we set $\theta_{L_{A,B}}$ to be $\theta'_{L_{A,B}}$.

3. For any ambient context $C$ from $L_{A,B}$, for any labeled hole $\square^w$ in $C$ and for any capability $cap^w n$ in the process $\theta_{L_{A,B}}(\square^w)$, we consider the ambient context $C_m$ identical to $C$ except that $\square^w$ in $C$ has been replaced by $\square^w \mid cap^w n$ in $C_m$. If the capability $cap^w n$ can be consumed in $C_m$ using one of the rules **(In)** or **(Out)** leading to an ambient context $C'$, then state $C'$ and a transition from $C$ to $C'$ labeled by $cap^w n$ are added to transition system $L_{A,B}$.

4. For any ambient context $C$ from $L_{A,B}$ and for any labeled hole $\square^w$ in $C$ associated by $\theta_{L_{A,B}}$ with a process of the form $!^{w'} A'$, if a replication $!^{w'}$ can be reduced in process $\theta_{L_{A,B}}(C)$ using rule **(Repl)**, then a transition from $C$ to itself labeled by $!^{w''}$ is added to $L_{A,B}$ for any replication $!^{w''}$ in $\theta_{L_{A,B}}(\square^w)$.

In the second step, the reduction of a replication contained in the ambient context by means of the rule **(Repl)** is done only when the number of ambients in the resulting process is smaller than the number of ambients in the target process $B$, namely $Amb(C') \subseteq Amb(B)$. This requirement is crucial as it implies that the transition system $L_{A,B}$ has only finitely many states.

As an example, we give in Figure 1 the labeled transition system associated with the process $n[!^1 in\ m.!^2 out\ m] \mid m[\ ]$ (we omit in this process unnecessary labels). We use the labeled replications $!^1$ and $!^2$ to distinguish between different replication operators which appear in this process.

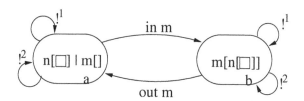

**Fig. 1.** A labeled transition system for the process $n[!^1 in\ m.!^2 out\ m] \mid m[\ ]$

It is worth to note that the labeled transitions in $L_{A,B}$ for replications and capabilities of an ambient context correspond to the reductions performed over processes. As shown in steps 3 and 4, the transitions applied for any capabilities or replications associated with the holes are independently of the fact that they are effectively available to perform a transition (at this point).

## II) From Processes Without Ambients to Petri Nets.
In what follows we show how to build a Petri net from a labeled process without ambients. We denote by $\mathcal{E}(E)$ the set of all multisets which can be built with elements from the set $E$.

We recall that a Petri net is given by a 5-tuple $(\mathcal{P}, \mathcal{P}_i, \mathcal{T}, Pre, Post)$, where

- $\mathcal{P}$ is a finite set of places;
- $\mathcal{P} \subseteq \mathcal{P}_i$ is a set of initial places;
- $\mathcal{T}$ is a finite set of transitions;
- $Pre, Post : \mathcal{T} \to \mathcal{E}(\mathcal{P})$ are mappings from transitions to multisets of places.

We say that an ambient-free process is rooted if it is of the form $cap^w n.A'$ or of the form $!^w A'$. We define the Petri net $PN_{A'}$ associated with a rooted process $A'$ as follows: the places of $PN_{A'}$ are precisely the rooted subprocesses of $A'$, and $A'$ itself is the unique initial place; the transitions are defined as the set of all capabilities $in^w n$, $out^{w'} n$ and replications $!^w$ occurring in $A'$. Finally, $Pre$ and $Post$ are defined for all transitions as follows:

- $Pre(cap^w n) = \{cap^w n\}$ and $Post(cap^w n) = \emptyset$ if $cap^w n$ is a place in $PN_{A'}$.
- $Pre(cap^w n) = \{cap^w n.(A_1 \mid \ldots \mid A_k)\}$ and $Post(cap^w n) = \{A_1 \mid \ldots \mid A_k\}$ if $cap^w n.(A_1 \mid \ldots \mid A_k)$ is a place in $PN_{A'}$ ($A_1 \mid \ldots \mid A_k$ being rooted processes).
- $Pre(!^w) = Pre(!^w) = \{!^w A'\}$, $Post(!^w) = \{!^w A', A'\}$ and $Post(!^w) = \{!^w A'\}$ if $!^w A'$ is a place in $PN_{A'}$.

For $!^1 in\ m.!^2 out\ m$, we obtain the Petri net given in Figure 2:

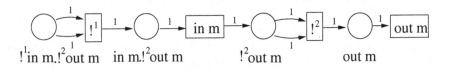

**Fig. 2.** A Petri net for the process $!^1 in\ m.!^2 out\ m$

We denote by $PN_{\square^w}$ the Petri net $PN(\theta_{L_{A,B}}(\square^w))$, that is, the Petri net corresponding to the rooted ambient-free process associated with $\square^w$ by $\theta_{L_{A,B}}$. In what follows we show how to combine the transition system $L_{A,B}$ and the Petri nets $PN_{\square^w}$ into one single Petri net.

## III) Combining the Transition Systems and Petri Nets.
We first turn the labeled transition system $L_{A,B}$ into a Petri net $PN_L = (\mathcal{P}_L, \mathcal{P}_L^i, \mathcal{T}_L, Pre_L, Post_L)$ where

- $\mathcal{P}_L$ is a set of states of $L_{A,B}$;
- $\mathcal{P}_L^i$ is a singleton set containing the state corresponding to the ambient context $\mathcal{C}_A$ of $A$;
- $\mathcal{T}_L$ is the set of transitions of the form $(s, l, s')$, with
  - $s$ and $s'$ states from $L_{A,B}$,
  - a transition $l$ from $s$ to $s'$ in $L_{A,B}$;
- $Pre(t) = s$ and $Post(t) = \{s'\}$ for all transitions $t = (s, l, s')$.

We define a Petri net $PN_{A,B} = (\mathcal{P}_{A,B}, \mathcal{P}_{A,B}^i, \mathcal{T}_{A,B}, Pre_{A,B}, Post_{A,B})$ by

- places (initial places) of $PN_{A,B}$ are the union of places (initial places) of $PN_L$ and of each of the Petri nets $PN_{\square^w}$ (for $\square^w$ occurring in one of the states of $L_{A,B}$);
- transitions of $PN_{A,B}$ are precisely the transitions of $PN_L$;
- the mappings $Pre_{A,B}$ and $Post_{A,B}$ are defined for all transitions $t = (a, f, b)$ as:

  (i) $Pre_{A,B}(t) = \{a\}$ and $Post_{A,B}(t) = \{b\}$ if $f$ does not occur as a transition in any $PN_{\square^w}$ (for $\square^w$ occurring in one of the states of $L_{A,B}$),
  (ii) if $f$ is a transition of $PN_{\square^w}$, then $Pre_{A,B}(t) = \{a\} \cup Pre_{\square^w}(f)$ and $Post_{A,B}(t) = \{b\} \cup Post_{\square^w}(f)$, where $Pre_{\square^w}$ and $Post_{\square^w}$ are the mappings $Pre$ and $Post$ of $PN_{\square^w}$), respectively.

### 4.3  Deciding Reachability

We recall that for a Petri net $PN = (\mathcal{P}, \mathcal{P}^i, \mathcal{T}, Pre, Post)$, a marking $m$ is a multiset from $\mathcal{E}(P)$. A transition $t$ is enabled by a marking $m$ if $Pre(t) \subseteq m$. Executing an enabled transition $t$ for a marking $m$ gives a marking $m'$ defined as $m' = (m \setminus Pre(t)) \cup Post(t)$ (where $\setminus$ stands for the multiset difference). A marking $m'$ is reachable from $m$ if there exists a sequence $m_0, \ldots, m_k$ of markings such that $m_0 = m$, $m_k = m'$ and for each $m_i, m_{i+1}$, there exists an enabled transition for $m_i$ whose execution gives $m_{i+1}$.

**Theorem 2 ([7]).** *For all Petri nets $P$, for all markings $m, m'$ of $P$, one can decide whether $m'$ is reachable from $m$.*

For the reachability problem $A \Rightarrow^* B$ over ambients, we consider the Petri net $PN_{A,B}$ and the initial marking $m_A$ defined as $m_A = \mathcal{P}_{A,B}^i$. In Figure 3 is depicted the initial marking for process $n[!^1 in\ m.!^2 out\ m] \mid m[\ ]$ as a combination of the labeled transition system of Figure 1 and the Petri net of Figure 2.

It should be noticed that for any marking $m$ reachable from $m_A$, $m$ contains exactly one occurrence of a place from $\mathcal{P}_L$. Roughly speaking, to any reachable marking corresponds exactly one ambient context. Moreover, the execution of one transition in the Petri net $PN_{A,B}$ simulates a reduction from $\Rightarrow$.

We define now the set $\mathcal{M}_B$ of markings of $PN_{A,B}$ corresponding to $B$. Intuitively, a marking $m$ belongs to $\mathcal{M}_B$ if $m$ contains exactly one occurrence $\mathcal{C}$ of a place from $\mathcal{P}_L$ (that is, representing some ambient context) and in the context $\mathcal{C}$,

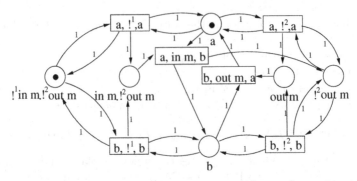

**Fig. 3.** The Petri net for the labeled process $n[!^1 in\ m.!^2 out\ m]\ |\ m[\ ]$

the holes can be replaced with processes without ambients to obtain $B$. Each of the processes without replication must correspond to a marking of the sub-Petri net associated with the hole it fills up. $\mathcal{M}_B$ is defined as the set of markings $m$ for $PN_{A,B}$ satisfying:

(i) there exists exactly one ambient context $\mathcal{C}_m$ in $m$;
(ii) $\sigma_m(\mathcal{C}_m)$ and $B$ are label-free-equivalent, for any substitution $\sigma_m$ from holes $\square^w$ occurring in $\mathcal{C}_m$ to processes without ambients defined as $\sigma_m(\square_m) = P_1\ |\ldots|\ P_k$ for $\{P_1,\ldots,P_k\}$ the multiset corresponding to the restriction of $m$ to the places of $PN_{\square^w}$
(iii) for all holes $\square^w$ occurring in a state of the transition system $L_{A,B}$ but not in $\mathcal{C}_m$, the restriction of $m$ to places of $PN_{\square^w}$ is precisely the set of initial places of $PN_{\square^w}$.

We adapt the results presented in [1] to our restricted fragment of mobile ambients.

**Proposition 2.** *For a Petri net $PN_{A,B}$, there are only finitely many markings corresponding to a process $B$, and the set $\mathcal{M}_B$ can be computed.*

The translation correctness is ensured by the following result.

**Proposition 3.** *For all processes $A, B$ we have that $A \Rightarrow B$ if and only if there exists a marking from $\mathcal{M}_B$ such that $m_B$ is reachable from $m_A$ in $PN_{A,B}$.*

Using Proposition 3 and Theorem 2, we can decide whether an ambient $A$ can be reduced to an ambient $B$.

**Theorem 3.** *For two arbitrary ambients $A$ and $B$ from our restricted fragment, it is decidable whether $A$ reduces to $B$.*

## 5    Conclusion

In this paper we have investigated the problem of reaching a certain configuration of a system of mobile membranes starting from another configuration. In order to do this we use the result of [1] where the reachability problem for the pure and public ambient calculus without the capability open is proven to be decidable. The same problem is tackled in [2], where the authors do not take into account the replication of ambients which is used in our case to simulate the division rules in mobile membranes. We proved that the reachability can be decided by reducing this problem to the reachability problem for a fragment of ambient calculus.

## Acknowledgements

Research partially supported by CEEX Grant 47/2005.

## References

1. Boneva, I., Talbot, J.-M.: When Ambients Cannot be Opened. In: Gordon, A.D. (ed.) ETAPS 2003 and FOSSACS 2003. LNCS, vol. 2620, pp. 169–184. Springer, Heidelberg (2003)
2. Busi, N., Zavattaro, G.: Deciding Reachability in Mobile Ambients. In: Sagiv, M. (ed.) ESOP 2005. LNCS, vol. 3444, pp. 248–262. Springer, Heidelberg (2005)
3. Cardelli, L., Gordon, A.: Mobile Ambients. In: Nivat, M. (ed.) ETAPS 1998 and FOSSACS 1998. LNCS, vol. 1378, pp. 140–155. Springer, Heidelberg (1998)
4. Ciobanu, G.: Distributed Algorithms over Communicating Membrane Systems. BioSystems 70, 123–133 (2003)
5. Delzanno, G., Van Begin, L.: On the Dynamics of PB Systems with Volatile Membranes. In: Eleftherakis, G., Kefalas, P., Păun, G., Rozenberg, G., Salomaa, A. (eds.) WMC 2007, vol. 4860, pp. 240–256. Springer, Heidelberg (2007)
6. Krishna, S.N., Păun, G.: P Systems with Mobile Membranes. Natural Computing 4(3), 255–274 (2005)
7. Mayr, E.W.: An Algorithm for the General Petri Net Reachability Problem. SIAM Journal of Computing 13(3), 441–460 (1984)
8. Milner, R.: Communicating and Mobile Systems: The $\pi$-Calculus. Cambridge University Press, Cambridge (1999)
9. Păun, G.: Computing with membranes. Journal of Computer and System Sciences 61(1), 108–143 (2000)
10. Păun, G.: Membrane Computing. An Introduction. Springer, Heidelberg (2002)

# Modeling Symport/Antiport P Systems with a Class of Hierarchical Petri Nets

Luca Bernardinello, Nicola Bonzanni, Marco Mascheroni, and Lucia Pomello

Dipartimento di Informatica, Sistemistica e Comunicazione
Universit degli Studi di Milano–Bicocca
via Bicocca degli Arcimboldi 8, I-20126 Milano, Italy
bernardinello@disco.unimib.it

**Abstract.** A model of P systems with symport/antiport rules is given in terms of hypernets, a generalization of a class of hierarchical Petri nets introduced for modeling mobility inside the nets-within-nets paradigm. The hierarchical structure of a P system is reflected by the associated hypernet, where molecules are modeled by unstructured agents (simple tokens) and membranes by agents. Each agent is modeled by a net which may contain in its places unstructured agents or other agents. Agents can exchange tokens with their sub- or super-agents and thus the hierarchy may change. The main result of the paper shows a correspondence between reachable configurations of the P system and reachable hyper-markings of the related hypernet, in such a way that if the P system can evolve from one configuration to another one then in the hypernet there exists a corresponding transformation of hypermarkings.

## 1  Introduction

In recent years the notion of *system of mobile agents* has gained importance in computer science and engineering. These systems are formed by *agents* which move around a space, interacting with each other. Often, these agents are pieces of software traveling across a network of hosts, where they can be executed in a local environment. Such a development has led to envisage formal models in which one can represent mobile agents, their environment, and their interactions. Since agents move and run in parallel with others, concurrency theory is a natural framework in which to look for adequate models.

In 1986, Valk proposed a kind of Petri nets in which tokens can be nets, which can be moved across the places of a hosting net, possibly interacting with it (see [15]). Building on this idea, *hypernets* were defined in [1]. A hypernet is formed by agents, each modeled by a Petri net. In a given configuration, each agent, except one, is also a token residing in a place of another agent (the exception consists in the highest level agent, which acts as an environment for all others). The relation of containment can dynamically change as an effect of firing transitions; agents can exchange their sub-agents by forming so called *consortia*.

The hierarchy of agents in a hypernet resembles the hierarchy of membranes in a P system, and the mechanism of consortia can be seen as a way to exchange

G. Eleftherakis et al. (Eds.): WMC8 2007, LNCS 4860, pp. 124–137, 2007.

molecules across a membrane. This idea is the subject of the present paper, where we define a translation from P systems with symport/antiport rules to a class of hypernets. Such class is a generalization of the class defined in [1]. The main idea of this translation is quite simple: each membrane and each individual molecule in the P system is represented by an agent in the hypernet. Molecule agents are unstructured, that is, they are simple tokens, like in usual nets, and can only be passively moved by the active components. Membrane agents, viceversa, are nets, with places that can contain molecule agents, and places that can contain other membrane agents. Consortia correspond to rules of the P system, whereby molecules can be exchanged across a membrane.

It should be noted that hypernets would allow, in themselves, movement of membrane agents, so that the hierarchical structure of membranes could change. This capability is not exploited here, since we deal with P systems where only molecules move around, but might be useful in modeling more general kinds of systems.

In this paper, we are not interested in the computational aspects of the theory of P systems, but rather focus on modeling aspects. Consequently, we compare the two models on the basis of their reachable configurations.

After recalling the basic definitions related to the class of P systems with symport/antiport rules (Section 2), we define hypernets in Section 3. Section 4 shows how to build a hypernet from a P system, and states in which sense the two models can be considered as equivalent. Finally, in Section 5, we draw some considerations, and suggest possible developments.

## 2    P Systems with Symport/Antiport Rules

Many kinds of membrane systems have been investigated during the last years. One of the most studied variant of the general model of P systems was introduced in [10] under the name of systems with symport/antiport rules. Those terms came from two membrane transport mechanisms. Whereas the term symport stands for the biological process by which two molecules pass together across a membrane, when the two molecules pass simultaneously, but in the opposite direction, the process is called antiport.

The class of membrane systems with symport/antiport rules is a class of purely communicating P systems, where the objects involved in the computation only pass through membranes. This means that the objects involved never change and a sort of conservation law for objects is observed during the entire evolution of the system.

Many results on this kind of P systems, especially about their computational power, can be found in [11], [7], [8], [4]. Here we provide a simplified version of the definition of P system with symport/antiport rules, [12].

### 2.1    Formal Definition

Formally, we define a P system with symport/antiport rules (of degree $m$), as a construct of the form

$$\Pi = (O, \mu, w_1, w_2, \ldots, w_m, R_1, R_2, \ldots, R_m),$$

where:

- $O$ is the (finite and non empty) alphabet of objects.
- The membrane structure $\mu = (N, E, i)$ is a rooted tree underlying $\Pi$, where $N = \{1, 2, \ldots, m\}$ is the set of nodes and each node in $N$ defines a membrane of $\Pi$. The set $E \subseteq N \times N$ defines the edges. For each node $j \in N$, the membrane associated to the node $j$ contains all the membranes associated to the children of $j$. $i$ is the root of the tree and hence the skin membrane (the outermost membrane of the system).
- $w_1, w_2, \ldots, w_m$ are multisets over $O$ representing the objects present in the regions $1, 2, \ldots, m$ of the membrane structure $\mu$ in the initial configuration of the system (in the following, multisets will be described either by strings, with exponents denoting the multiplicity of elements, or by the usual characteristic function of multisets).
- $R_1, R_2, \ldots, R_m$ are finite sets of evolution rules associated with the membranes of $\mu$. Moreover we impose $R_i = \emptyset$, where $i$ is the skin of the membrane structure. This clause ensures that the external membrane is impermeable and hence the total number of objects involved in the computation is finite (and constant); this is required if we want to build hypernets with a finite number of agents.

In the following we often use the term *molecule* when referring to an object in a membrane of the P system.

As said above, each rule governs the communication through a specific membrane and can be of two kinds, symport rule or antiport rule. A symport rule is of the form $(u, in)$ or $(u, out)$, where $u$ is a multiset over $O$, stating that all the objects of $u$ pass together through a membrane, entering in the former case and exiting in the latter. For example, in a membrane $i$, after the application of the symport rule $(u, in)$, the multiset associated to this membrane will contain all the objects previously present, plus the objects present in $u$. The multiset associated to the membrane that contains $i$, will contain all the objects previously present, minus those in $u$. Similarly, an antiport rule is of the form $(u, out; v, in)$, where $u$ and $v$ are multisets over $O$, stating that when $u$ exits, at the same time, a multiset $v$ must enter the membrane.

The P system described above evolves from configuration to configuration by the application of a multiset of rules in each membrane. Formally, a configuration is a tuple $C = (v_1, v_2, \ldots, v_m)$ and $C \overset{\hat{R}}{\Rightarrow} C'$ denotes that $C$ evolves into $C'$ due to the application of $\hat{R}$, where $\hat{R} = (\bar{R}_1, \bar{R}_2, \ldots, \bar{R}_m)$ is a multi-rules vector applicable to $C$ and $\bar{R}_j$ is a multiset over $R_j$.

The evolution of the system is non-deterministic and maximally parallel: at each step, the configuration changes by applying a maximal multiset of rules, chosen in a non deterministic way; the rules must be all applicable without mutual interferences in the current configuration.

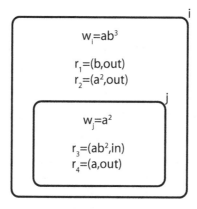

**Fig. 1.** Fragment of a symport/antiport P system

## 2.2 Example

Fig. 1 shows a fragment of a P system with symport rules. The system depicted here consists of two nested membranes: $j$, the inner membrane, and $i$, the outer one, which we assume to be a membrane contained in a larger membrane structure. The set of rules of $i$ is $R_i = \{(b, out), (a^2, out)\}$, and the set of rules of $j$ is $R_j = \{(ab^2, in), (a, out)\}$. In the same way we define the initial multisets of objects $w_i = ab^3$ and $w_j = a^2$.

In this configuration the rules $r_1, r_3, r_4$ are enabled and a multi-rules vector can be built with this rules in a maximally parallel manner, i.e., the multi-rules vector $\hat{R} = (\{r_1\}, \{r_3, r_4, r_4\})$ is applicable to the initial configuration. Note that other multi-rules vectors can be applied to the same configuration. The application of $\hat{R}$ leads to a new state where the objects in the membrane $i$ are $a^2$ and the objects in the membrane $j$ are $ab^2$.

# 3  Hypernets

In this section we introduce a generalization of Petri hypernets [1] that for simplicity we call also here *hypernets*. A hypernet is defined by a fixed set of agents, each agent is modeled by a net and can manipulate other agents as tokens, while being manipulated as token by another agent at the same time. This yields a hierarchy of agents. The highest level agent acts as an environment for all other agents, these latter are located each one in some place of another agent. Agents can exchange tokens with their sub- or super-agents and thus the hierarchy may change.

In what follows we first define the structure of hypernets giving the definition of agent and of hypernet, then we define the behaviour of hypernets, and at the end of the section we illustrate hypernets on an example modeling the P system given in Subsection 2.2 as it will be discussed in Section 4.

## 3.1   Structure of Hypernets

An agent is modeled by a Petri net, a bipartite oriented graph, whose nodes are of two types: places and transitions. Places are partitioned into two disjoint sets: the set of local places, which are locations in which other agents can stay, and the set of virtual places, which represent communication channels along which agents exchange tokens each others. Places and transitions are interconnected by weighted oriented arcs, which define how many tokens are taken away from an input place and how many are put into an output place, when a transition fires. For each transition the sum of the weights of the input arcs must be equal to the sum of the weights of the output arcs. In this way the amount of tokens will not change while transitions fire. Moreover, to each triple of interconnected elements place-transition-place it is assigned, by a function $\phi_A$, a value which defines, in a way compatible with the arc weights, the number of tokens which flow along the path identified by the triple. In other words, $\phi_A$ defines how the tokens taken away from an input place of a transition will be distributed into the output places, when the transition will fire; and this distribution will be the same for each occurrence of the same transition. For basic definitions and notions on Petri nets, see, for example, [14].

**Definition 1.** *An agent is a tuple* $A = (P_A \cup V_A, T_A, F_A, \phi_A)$, *where* $(P_A \cup V_A, T_A, F_A)$ *is a, possibly empty, finite Petri net in which:*

- $P_A$ *is the set of* local *places and* $V_A$ *is the set of* virtual *places, (or communication places), with* $P_A \cap V_A = \emptyset$;
- $T_A$ *is the set of* transitions;
- *the function* $F_A : ((V_A \cup P_A) \times T_A) \cup (T_A \times (V_A \cup P_A)) \longrightarrow \mathbb{N}$ *defines the* flow, *assigning a weight to each arc identified by the pair of elements* $x$, $y$ *such that* $F_A(x,y) > 0$, *in such a way that*
  $\forall t \in T_A, \sum_{p \in \bullet t} F_A(p,t) = \sum_{p \in t\bullet} F_A(t,p)$, *where* $p \in \bullet t$ *iff* $F_A(p,t) > 0$ *and* $p \in t^\bullet$ *iff* $F_A(t,p) > 0$;

*and the function* $\phi_A : (V_A \cup P_A) \times T_A \times (V_A \cup P_A) \longrightarrow \mathbb{N}$ *defines the* paths, *i.e., the triples* $(p,t,q)$ *such that* $\phi_A(p,t,q) > 0$, *by assigning a weight to them in such a way that:*

$\forall p \in {}^\bullet t, \ F_A(p,t) = \sum_{q \in t\bullet} \phi_A(p,t,q)$
$\forall q \in t^\bullet, \ F_A(t,q) = \sum_{p \in \bullet t} \phi_A(p,t,q)$

In the following $(p,t,q) \in \phi_A$ iff $\phi_A(p,t,q) > 0$; moreover, given a subset of agents $X \subseteq \mathcal{N}$, we use the following notation: $P_X = \bigcup_{A \in X} P_A$, $V_X = \bigcup_{A \in X} V_A$, $T_X = \bigcup_{A \in X} T_A$, $\phi_X = \bigcup_{A \in X} \phi_A$.

A hypernet is defined by a set of agents and by a relation $\Delta$. Agents have disjoint sets of places. A transition may belong to different agents, modeling synchronous interaction among them. Transitions connected to virtual places model interchanges of tokens among sub-/super-agents. Said *output paths* the path ending with a virtual place and *input paths* the ones starting with a virtual place, the relation $\Delta$ identifies communication channels by defining, for a given transition belonging to different agents, a correspondence (output path - input path) in a way compatible with path weights.

**Definition 2.** Let $\mathcal{N} = \{A_1, A_2, \ldots, A_n\}$ be a family of agents, and let $S^o = \{(p, t, v) \in \phi_A | A \in \mathcal{N} \text{ and } v \in V_A\}$ and $S^i = \{(v, t, q) \in \phi_A | A \in \mathcal{N} \text{ and } v \in V_A\}$ be the sets of output paths and input paths, respectively. (Note that a path can be both an output and an input path.)

A hypernet is a pair $H = (\mathcal{N}, \Delta)$, where

- The agents in $\mathcal{N}$ have disjoint sets of places:

$$\forall A_i, A_j \in \mathcal{N}, \ (P_{A_i} \cup V_{A_i}) \cap (P_{A_j} \cup V_{A_j}) = \emptyset;$$

- and $\Delta \subseteq S^o \times S^i$ is a relation which associates, for a given transition, output paths to input paths with the same weight and belonging to different agents, i.e.,

$\forall t \in T_{\mathcal{N}}, \ \forall (p, t, q) \in \phi_{A_i} \text{ and } \forall (p', t, q') \in \phi_{A_j} \text{ such that } A_i, A_j \in \mathcal{N}$:
$((p, t, q), (p', t, q')) \in \Delta \Rightarrow A_i \neq A_j \text{ and } \phi_{A_i}(p, t, q) = \phi_{A_j}(p', t, q').$

**Definition 3.** Let $\mathcal{N} = \{A_1, A_2, \ldots, A_n\}$ be a family of agents. A map $\mathcal{M} : \{A_2, \ldots, A_n\} \longrightarrow P_{\mathcal{N}}$, assigning to each agent different from $A_1$ the local place in which is located, is a hypermarking of $\mathcal{N}$ iff, considering the relation $\uparrow_{\mathcal{M}} \subseteq \mathcal{N} \times \mathcal{N}$ defined by: $A_i \uparrow_{\mathcal{M}} A_j \Leftrightarrow \mathcal{M}(A_i) \in P_{A_j}$, then the graph $\langle \mathcal{N}, \uparrow_{\mathcal{M}} \rangle$ is a tree with root $A_1$.

**Definition 4.** A marked hypernet is a pair $(H, \mathcal{M})$ where $H$ is a hypernet and $\mathcal{M}$ is a hypermarking defining the initial configuration.

In a configuration the system results hierarchically structured. The highest level agent $A_1$, the root of the tree describing the hierarchy, plays the role of the environment containing all the other agents. The relation of containment between agents, and then the hierarchical structure, can change as an effect of firing transitions as formalized in the following subsection.

## 3.2 Behaviour of Hypernets

Let $H = (\mathcal{N}, \Delta)$, with $\mathcal{N} = \{A_1, A_2, \ldots, A_n\}$, be a hypernet.

A consortium is a set of interconnected active agents, cooperating in performing a transition $t$, moving other passive agents along the paths containing $t$.

**Definition 5.** A consortium is a tuple $\Gamma = (t, \tau, \delta, \gamma)$ where:

- $t \in T_{\mathcal{N}}$ is the name of the consortium,
- $\tau \subseteq \{A \in \mathcal{N} | t \in T_A\}$, $\tau \neq \emptyset$, is the non empty set of active agents. To this set we can associate $\phi_{\tau_t} = \{(p, t, q) \in \phi_\tau \mid p, q \in P_\tau \cup V_\tau\}$, the set of paths of the agents $\tau$ containing the transition $t$.

- $\delta$ defines a bijective correspondence between output paths containing $t$ and input paths containing $t$ of active agents, without contradicting the relation $\Delta$. Let $\phi_{o,\tau_t} = \phi_{\tau_t} \cap S^o$ and $\phi_{i,\tau_t} = \phi_{\tau_t} \cap S^i$. If $\phi_{o,\tau_t} \neq \emptyset$, $\delta : \phi_{o,\tau_t} \longrightarrow \phi_{i,\tau_t}$ is a bijection such that $\forall s \in \phi_{o,\tau_t}$, $\delta(s) = s' \Rightarrow (s, s') \in \Delta$, while if $\phi_{o,\tau_t} = \emptyset$, then $\delta$ is the empty map. Note that $\delta$ relates paths belonging to different agents.
- The passive agents which are moved when the consortium occurs are selected through the map $\gamma$. Let $C \subseteq \mathcal{N} \backslash A_1$ be a chosen set of passive agents, then $\gamma : C \longrightarrow \phi_{\tau_t} \backslash S^i$ is surjective and associates as many passive agents to each path containing $t$ and belonging to an active agent as the weight of the path itself, i.e., $\forall s \in \phi_{\tau_t} \backslash S^i$, $|\gamma^{-1}(s)| = \phi_{\mathcal{N}}(s)$. Note that an agent can be active and passive at the same time.

Moreover the following conditions must be satisfied:

- the set of active agents $\tau$ is a minimal one, in the sense that the agents in $\tau$ must be each other interconnected through the interaction $t$, i.e., the undirected graph $G_1 = (\tau, E_1)$ is connected, where $E_1 = \{(A_i, A_j) \mid A_i, A_j \in \tau \text{ and } \exists s_i \in \phi_{A_i}, \exists s_j \in \phi_{A_j} : \delta(s_i) = s_j\}$ and
- the undirected graph $G_2 = (\tau \cup C, E_2)$ is acyclic, where $E_2$ connects $A_i$ to $A_j$ if $A_i$ will be put inside $A_j$ through $t$, i.e., considered the recursively defined map $\delta^* : \phi_{\tau_t} \longrightarrow \phi_{\tau_t}$ such that

$$\delta^*(s) = \begin{cases} s & \text{if } s \notin \phi_{o,\tau_t} \\ \delta^*(\delta(s)) & \text{otherwise} \end{cases}$$

$E_2 = \{(A_i, A_j) | \delta^*(\gamma(A_i)) \in \phi_{A_j}, A_i \in C, A_j \in \tau\}$.

The intuition behind the last condition of the previous definition is the following. By subsequent applications of the map $\delta$ it is possible to construct chains of paths interrelated through paths with only virtual places. However, the meaningful chains are the one which starts with a path with a real input place, the one from which an agent will be taken out, and ends with a path with a real output place, the one in which the agent will be put into. The last condition requires that these chains are not closed.

In [2] it is proven that chains containing a real place can be prolonged to finite chains containing at most two real places, one in an input path and one in an output path.

**Definition 6.** Let $H = (\mathcal{N}, \Delta)$ be a generalized hypernet and $\mathcal{M}$ be a hypermarking.

A consortium $\Gamma = (t, \tau, \delta, \gamma)$ is enabled in $\mathcal{M}$, denoted $\mathcal{M}[\Gamma\rangle$, iff the following two conditions hold

- $\forall A \in C, \gamma(A) = (p, t, q) \Rightarrow \mathcal{M}(A) = p$
- $\forall A_i, A_j \in \tau, \forall s \in S^o \cap \phi_{A_i}, \delta(s) \in \phi_{A_j} \Rightarrow A_i \uparrow_{\mathcal{M}} A_j \lor A_j \uparrow_{\mathcal{M}} A_i$

*If $\mathcal{M}[\Gamma\rangle$, then the* occurrence *of $\Gamma$ leads to the new hypermarking $\mathcal{M}'$, denoted $\mathcal{M}[\Gamma\rangle\mathcal{M}'$, such that $\forall A \in \mathcal{N}$:*

$$\mathcal{M}'(A) = \begin{cases} \mathcal{M}(A) & \text{if } A \notin C; \\ q & \text{if } A \in C \text{ and } \delta^*(\gamma(A)) = (p, t, q). \end{cases}$$

It is possible to prove [2] that $\mathcal{M}'$ is a hypermarking, i.e., that the class of hypermarkings of a hypernet is closed under the occurrence of a consortium.

Two consortia $\Gamma_1 = (t_1, \tau_1, \delta_1, \gamma_1)$ and $\Gamma_2 = (t_2, \tau_2, \delta_2, \gamma_2)$ are *independent* iff the maps $\gamma_1$ and $\gamma_2$ select two different sets of passive agents, i.e., iff $C_1 \cap C_2 = \emptyset$

If two independent consortia are both enabled in a hypermarking $\mathcal{M}$ then they can *concurrently* occur in $\mathcal{M}$.

Let $\Gamma_H$ be the set of possible consortia in $H$. A set of consortia $U \subseteq \Gamma_H$ is a *step* enabled in a hypermarking $\mathcal{M}$, denoted $\mathcal{M}[U\rangle$, iff

- $\forall \Gamma_i, \Gamma_j \in U$, $\Gamma_i$ and $\Gamma_j$ are independent,
- $\forall \Gamma_i \in U$, $\mathcal{M}[\Gamma_i\rangle$

If $\mathcal{M}[U\rangle$, then the occurrence of the step $U$ leads to the new hypermarking $\mathcal{M}'$, denoted $\mathcal{M}[U\rangle\mathcal{M}'$, such that $\forall A \in \mathcal{N}$:

$$\mathcal{M}'(A) = \begin{cases} \mathcal{M}(A) & \text{if } \forall \Gamma_i \in U, A \notin C_i; \\ q & \text{if } \exists \Gamma_i \in U : (A \in C_i \text{ and } \delta_i^*(\gamma_i(A)) = (p, t_i, q)). \end{cases}$$

$U$ is a *maximal step* enabled in $\mathcal{M}$, and its occurrence yields $\mathcal{M}'$, iff $\mathcal{M}[U\rangle\mathcal{M}'$ and $\forall U' \supset U : not(\mathcal{M}[U'\rangle)$.

In [2] it is shown how it is possible to associate to each hypernet a 1-safe net in such a way that there is a strict correspondence between their behaviors, i.e., in terms of Petri net theory, in such a way that the case graph of the 1-safe net is isomorphic to the transition system generated by the reachable hypermarkings of the hypernet.

Since 1-safe nets are a basic class model in Petri net theory, this translation shows that hypernets are well rooted inside the theory of Petri nets.

## 3.3  Example

The Fig. 2 shows the structure of two hypernet's agents. The unfilled circles are local places while the filled ones are virtual places. The agent $A_j$ is nested in the agent $A_i$, in fact $\mathcal{M}(A_j) = a_j^i$, so $A_j \uparrow_\mathcal{M} A_i$. Moreover we assume $u_1, u_2, u_3, u_4$ to be unstructured agents such that $\mathcal{M}(u_1) = \mathcal{M}(u_2) = \mathcal{M}(u_3) = b^i$ and $\mathcal{M}(u_4) = a^i$. Now consider the consortium $\Gamma = (r_3, \tau, \delta, \gamma)$ where

- the set of active agents is $\tau = \{A_i, A_j\}$,
- the bijection $\delta$ builds two communication channels between $A_i$ and $A_j$ gluing two pair of paths: $\delta(a^i, r_3, \bar{a}^i) = (\bar{a}^j, r_3, a^j)$ and $\delta(b^i, r_3, \bar{b}^i) = (\bar{b}^j, r_3, b^j)$,
- the set of passive agents is $C = \{u_1, u_2, u_4\}$ and $\gamma(u_1) = \gamma(u_2) = (b^i, r_3, \bar{b}^i)$ and $\gamma(u_4) = (a^i, r_3, \bar{a}^i)$.

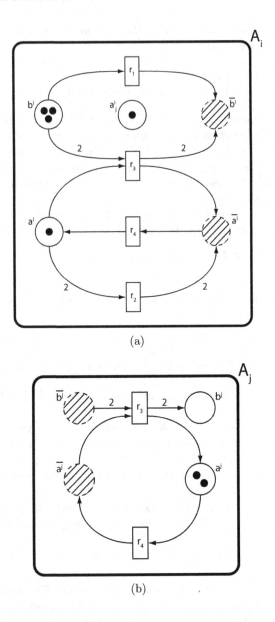

**Fig. 2.** Fragment of a hypernet

The consortium $\Gamma$ is valid and enabled in the initial hypermarking. When $\Gamma$ occurs the system reaches a new hypermarking $\mathcal{M}'$ where $\mathcal{M}'(u_1) = \mathcal{M}'(u_2) = b^j$ and $\mathcal{M}'(u_4) = a^j$. Note that the agents $u_1, u_2, u_4$ pass through the communication channels established by $\delta$ from the agent $A_i$ to the agent $A_j$.

# 4    Membrane Systems as Hypernets

Our goal in this section is to show how a P system with symport/antiport rules and with an impermeable external membrane can be modeled as a hypernet.

In the following, we write $i \lhd j$ to mean that membrane $i$ is directly contained in membrane $j$.

Let $\Pi = (O, \mu, w_1, \ldots, w_m, R_1, \ldots, R_m)$ be a P system of degree $m$, with symport/antiport rules. We assume that 1 is the outer membrane, with no rules, so that $R_1 = \emptyset$.

The hypernet associated to $\Pi$ will be denoted by $H = (\mathcal{N}, \Delta)$. The hypernet $H$ contains one agent for each membrane, and one agent for each individual molecule in $\Pi$. Notice that, in the P systems we handle, molecules are neither created nor deleted.

Let $W = \sum_{i=1}^{m} w_i$. $W$ is a multiset giving the total number of objects for each type in the system. Define

$$\text{MOL} = \{(x, i) | x \in O \wedge 1 \leq i \leq W(x)\}.$$

For each $(x, i)$ in MOL, we define an unstructured agent in the hypernet $H$.

$$\mathcal{N} = \{A_1, A_2, \ldots, A_m\} \cup \text{MOL}.$$

Agent $A_i$ corresponds to membrane $i$ of the P system. It has one place for each membrane directly contained in $i$, and one for each type of molecule; moreover, it has one virtual place for each type of molecule, to be used in exchanging tokens.

$$P_i = \{a_j^i | j \lhd i\} \cup \{x^i | x \in O\},$$
$$V_i = \{\bar{x}^i | x \in O\}.$$

The set of transitions of agent $A_i$ has one transition for each rule in membrane $i$, and one for each rule in membranes directly contained in $i$.

$$T_i = \{r | r \in R_i\} \cup \{r | r \in R_j \wedge j \lhd i\}.$$

We now turn to define the flow function and the paths for agent $A_i$.

- For each rule $r = (u, in) \in R_i$, and for each rule $r = (u, in; v, out) \in R_i$:

$$F(\bar{x}^i, r) = F(r, x^i) = \phi((\bar{x}^i, r, x^i)) = u(x).$$

- For each rule $r = (v, out) \in R_i$, and for each rule $r = (u, in; v, out) \in R_i$:

$$F(x^i, r) = F(r, \bar{x}^i) = \phi_i((x^i, r, \bar{x}^i)) = v(x).$$

Let $j \lhd i$. Then,

- For each rule $r = (u, in) \in R_j$, and for each rule $r = (u, in; v, out) \in R_j$:

$$F(x^i, r) = F(r, \bar{x}^i) = \phi_i((x_i, r, \bar{x}^i)) = u(x).$$

- For each rule $r = (v, out) \in R_j$, and for each rule $r = (u, in; v, out) \in R_j$:

$$F(\bar{x}^i, r) = F(r, x_i) = \phi_i((\bar{x}^i, r, x_i)) = v(x).$$

Define now the $\Delta$ relation. For all $i, j$ such that $i \triangleleft j$:

$$\forall r = (u, in) \in R_i, \forall x \in O : u(x) > 0, (x^j, r, \bar{x}^j) \Delta (\bar{x}^i, r, x^i),$$
$$\forall r = (u, out) \in R_i, \forall x \in O : u(x) > 0, (x^i, r, \bar{x}^i) \Delta (\bar{x}^j, r, x^j),$$
$$\forall r = (u, in; v, out) \in R_i, \forall x \in O : u(x) > 0, (x^j, r, \bar{x}^j) \Delta (\bar{x}^i, r, x^i),$$
$$\forall r = (u, in; v, out) \in R_i, \forall x \in O : v(x) > 0, (x^j, r, \bar{x}^j) \Delta (\bar{x}^i, r, x^i).$$

The initial hypermarking $\mathcal{M}$ reflects the initial configuration of $\Pi$. Membrane agents are placed according to the hierarchical structure of $\Pi$:

$$\forall i \in \{2, \ldots, m\} : \mathcal{M}(A_i) = a_i^j \text{ iff } i \triangleleft j.$$

All agents $(x, k)$ corresponding to molecules are initially distributed in the corresponding places $x^i$ in membrane agents so that a place $x^i$ contains $w_i(x)$ unstructured agents of type $(x, k)$.

In order to state the exact relation between the dynamics of a P system $\Pi$ and the dynamics of the corresponding hypernet $H$, we need to define two relations. The first defines a correspondence between configurations of $\Pi$ and hypermarkings of $H$. The other defines a correspondence between steps of $\Pi$ and maximal steps of $H$. Define

$$\mathbf{Conf} = \{(v_1, \ldots, v_m) | \sum_1^m v_i = \sum_1^m w_i\}.$$

as the set of all potential configurations of $\Pi$ with the same number and type of molecules as the initial configuration. Define **HM** as the set of all hypermarkings of $H$.

Let $\mathcal{M} : \{A_1, \ldots, A_m\} \cup \mathrm{MOL} \to P$ be an element of **HM**, where $P$ is the set of all local places of $H$, and $C = (v_1, \ldots v_m) \in \mathbf{Conf}$, where $v_i : O \to \mathbb{N}$. We also need some auxiliary definition. By $I(x, i, \mathcal{M})$ we denote the set of agents representing molecules of type $x$ hosted in the corresponding place of agent $A_i$ in $\mathcal{M}$.

$$I(x, i, \mathcal{M}) = \{(x, i) | (x, i) \in \mathrm{MOL} \wedge \mathcal{M}((x, i)) = x^i\}$$

**Definition 7.** *The hypermarking $\mathcal{M}$ simulates configuration $C$ (denoted by $\mathcal{M} \sim C$) iff*

1. $\mathcal{M}(A_i) \in P_j$ *iff* $i \triangleleft j$, *for* $i \in \{2, \ldots, m\}$,
2. $|I(x, i, \mathcal{M})| = v_i(x)$.

Notice that $\sim$ is a partial surjective function: each configuration of $\Pi$ has at least one corresponding hypermarking. The hypermarkings corresponding to one given configuration differ only for the distribution of molecules of the same kind in membrane agents. These molecules are identical in the P system, while their corresponding agents are distinguished.

We now define a correspondence between maximal steps in the P system and maximal steps of consortia in the hypernet. This correspondence is based on another one, associating single rules and consortia.

Let $r$ be a rule of membrane $i$ in $\Pi$. By construction, the associated hypernet has two transitions labeled by $r$, one in the agent corresponding to $i$, and one in the agent corresponding to the membrane containing $i$; assume it is $j$. A consortium simulating the execution of $r$ involves $i$ and $j$ as active agents, and a number of passive agents taken from MOL.

We consider here an antiport rule $r = (u, in; v, out)$, where $u$ and $v$ are multiset on $O$. Symport rules can be seen as special cases where either $u$ or $v$ is the empty multiset.

**Definition 8.** *Let $\Gamma = (r, \tau, \delta, \gamma)$ be a consortium. Then $\Gamma \sim r$ iff the following conditions hold.*

1. $\tau = \{A_i, A_j\}$
2. *The output paths involved in $\Gamma$ are either of the form $(y^i, r, \bar{y}^i)$ if $v(y) > 0$, or of the form $(x^j, r, \bar{x}^j)$ if $u(x) > 0$.*
3. *The function $\delta$ is defined by*

$$\delta((y^i, r, \bar{y}^i)) = (\bar{y}^j, r, y^j),$$
$$\delta((x^j, r, \bar{x}^j)) = (\bar{x}^i, r, x^i).$$

4. *Let $Z = \{z \in \text{MOL} | z = (x, k) \wedge \gamma(z) = (x^j, r, \bar{x}^j)\}$; then $|Z| = u(x)$.*
5. *Let $Z = \{z \in \text{MOL} | z = (y, k) \wedge \gamma(z) = (y^i, r, \bar{y}^i)\}$; then $|Z| = v(y)$.*

A transition in a P system is a multiset of independently executable rules. Let $R = \bigcup_{i=1}^{m} R_i$ be the set of all rules of $\Pi$, and $\rho : R \to \mathbb{N}$ be a multiset of rules. A set $U$ of consortia in $H$ simulates $\rho$ (denoted by $U \sim \rho$) if, for each $r \in R$, $U$ contains $\rho(r)$ consortia which simulate $r$, and the consortia in $U$ are pairwise independent.

We are now ready to state the main result of this section. The following lemma states that any change of configuration in $\Pi$ can be simulated by a set of mutually independent consortia in $H$. Let $\Pi$ be a P-system with symport and antiport rules, such that 1 is the outer membrane with $R_1 = \emptyset$, and $H = (\mathcal{N}, \Delta)$ be the associated hypernet, with initial hypermarking $\mathcal{M}$.

**Lemma 1.** *Let $C$ be a configuration of $\Pi$, and $\rho$ be a multiset of rules, enabled at $C$, with $C \overset{\rho}{\Rightarrow} C'$. Then, for all $\mathcal{M} \in \mathbf{HM}$,*

$$\mathcal{M} \sim C \Rightarrow \exists U \subseteq \Gamma_H : U \sim \rho, \; \mathcal{M}[U\rangle\mathcal{M}', \; \mathcal{M}' \sim C'.$$

Notice that the consortia forming $U$ can always be chosen to be pairwise independent. From this lemma, one can prove, by induction from the initial configuration, that the evolution of the P system can be simulated by the hypernet.

### 4.1  Example

Fig. 2, already discussed above (Section 3) as a generic hypernet, shows the fragment of the hypernet corresponding to the P system of Fig. 1. The two membranes $i$ and $j$ are modeled by the agents $A_i$ (Fig. 2(a)) and $A_j$ (Fig. 2(b)). The local place $a_j^i \in P_i$, which contains (as token) the agent $A_j$, reflects the fact that the membrane $j$ is nested inside $i$, while the local places $a^i, b^i$ represent the presence of molecules $a$ and $b$ respectively, inside the agent $A_i$ (this is also true for $a^j, b^j$ and the membrane $j$). Then $\{r_1, r_2, r_3, r_4\} \subseteq T_N$ are transitions built from the evolution rules of the membrane system. The initial hypermarking matches the initial configuration of the P system.

## 5  Conclusions

In this paper we have considered P systems with symport/antiport rules and we have shown how they can be modeled by a class of hierarchical Petri net systems, a generalization of hypernets [1].

The hierarchical structure of the P system is reflected by the agents's hierarchy of the hypernet, where molecules are modeled by unstructured agents (hence empty nets or simple tokens) and membranes by agents, nets which may contain in their places unstructured agents or other agents.

The exchange of molecules through a membrane of a P system, as defined by a symport or an antiport rule, corresponds to a consortium involving two active agents, that represent the two nested membranes which exchange each other passive unstructured tokens (molecules).

The main result, as given in Section 4, states a correspondence between reachable configurations of the P system and reachable hypermarkings of the related hypernet. If the P system can evolves from a configuration to another one as the result of the application of a multi-rules vector, then in the hypernet exists an associated set of consortia transforming a hypermarking, corresponding to the first configuration, and another corresponding to the second one.

A translation, that takes a hypernet and returns a 1-safe Petri net (one of the basic models in Petri net theory) such that the case graphs of the latter is isomorphic to the transition system generated by the execution of consortia of the former, has been shown in [2]. This transformation proves that hypernets are well rooted in net theory. In [9] a definition of non sequential processes for hypernets was given. This can be used to derive an alternative semantics for P systems based on a purely causal dependency notion.

In the literature other works have investigated the relation between P system and Petri nets [6], [13], [5]. It is a matter of future work a deeper comparison with these approaches and with other computational models, inspired by biological membranes and derived from calculi of concurrency and mobility, as for example those proposed by Cardelli [3].

Hypernets allow movement of structured agents from one level to another one, so that the hierarchy of agents may change. In terms of P systems, this means to consider movements of membrane agents. This capability is not exploited here, however it would be interesting in future to study the modeling of P systems with active membranes [12].

**Acknowledgements.** Work partially supported by MIUR and CNR - IPI PAN. The authors wish to thank Claudio Zandron and Alberto Leporati for their helpful suggestions.

# References

1. Bednarczyk, M.A., Bernardinello, L., Pawłowski, W., Pomello, L.: Modelling mobility with Petri hypernets. In: Fiadeiro, J.L., Mosses, P.D., Orejas, F. (eds.) WADT 2004. LNCS, vol. 3423, pp. 28–44. Springer, Heidelberg (2005)
2. Bonzanni, N.: P systems e reti di Petri ad alto livello. Universit degli studi di Milano-Bicocca. Dipartimento di Informatica Sistemistica e Comunicazione. Graduation thesis (March 2007)
3. Cardelli, L.: Brane calculi. In: Danos, V., Schachter, V. (eds.) CMSB 2004. LNCS (LNBI), vol. 3082, pp. 257–278. Springer, Heidelberg (2005)
4. Frisco, P.: About P systems with symport/antiport. Soft Computing 9(9), 664–672 (2005)
5. Frisco, P.: P systems, Petri nets, and program machines. In: Freund, R., Păun, G., Rozenberg, G., Salomaa, A. (eds.) WMC 2005. LNCS, vol. 3850, pp. 209–223. Springer, Heidelberg (2006)
6. Kleijn, J.H.C.M., Koutny, M., Rozenberg, G.: Towards a Petri net semantics for membrane systems. In: Freund, R., Păun, G., Rozenberg, G., Salomaa, A. (eds.) WMC 2005. LNCS, vol. 3850, pp. 292–309. Springer, Heidelberg (2006)
7. Martín-Vide, C., Păun, A., Păun, G.: On the power of P systems with symport rules. Journal of Universal Computer Science 8(2), 317–331 (2002)
8. Martín-Vide, C., Păun, A., Păun, G., Rozenberg, G.: Membrane systems with coupled transport: Universality and normal forms. Fundamenta Informaticae 49(1-3), 1–15 (2002)
9. Mascheroni, M.: Hypernet e processi non sequenziali. Universit degli studi di Milano-Bicocca. Dipartimento di Informatica Sistemistica e Comunicazione. Graduation thesis (April 2007)
10. Păun, A., Păun, G.: The power of communication: P systems with symport/antiport. New Generation Computing 20(3), 295–305 (2002)
11. Păun, A., Păun, G., Rozenberg, G.: Computing by communication in networks of membranes. International Journal of Foundations of Computer Science 13(6), 779–798 (2002)
12. Păun, G.: Introduction to membrane computing. In: First brainstorming Workshop on Uncertainty in Membrane Computing, Palma de Mallorca, Spain (2004)
13. Qi, Z., You, J., Mao, H.: P systems and Petri nets. In: Martín-Vide, C., Mauri, G., Păun, G., Rozenberg, G., Salomaa, A. (eds.) Membrane Computing. LNCS, vol. 2933, pp. 286–303. Springer, Heidelberg (2004)
14. Reisig, W., Rozenberg, G. (eds.): Lectures on Petri Nets I: Basic Models. LNCS, vol. 1491. Springer, Heidelberg (1998)
15. Valk, R.: Nets in computer organisation. In: Brauer, W., Reisig, W., Rozenberg, G. (eds.) Advances in Petri Nets 1986. LNCS, vol. 255, pp. 218–233. Springer, Heidelberg (1987)

# A Hybrid Approach to Modeling Biological Systems

Francesco Bernardini[1], Marian Gheorghe[2],
Francisco José Romero-Campero[3], and Neil Walkinshaw[2]

[1] Leiden Institute of Advanced Computer Science, University of Leiden
Niels Bohrweg 1, 2333 CA Leiden, The Netherlands
bernardi@liacs.nl
[2] Department of Computer Science, University of Sheffield
Regent Court, Portobello Street, Sheffield S1 4DP, UK
M.Gheorghe@dcs.shef.ac.uk
[3] Research Group on Natural Computing
Department of Computer Science and
Artificial Intelligence, University of Sevilla
Avda. Reina Mercedes s/n, 41012, Sevilla, Spain
fran@us.es

**Abstract.** This paper investigates a hybrid approach to modeling molecular interactions in biology. P systems, $\pi$-calculus, and Petri nets models, and two tools, Daikon, used in software reverse-engineering, and PRISM, a probabilistic model checker, are investigated for their expressiveness and complementary roles in describing and analyzing biological systems. A simple case study illustrates this approach.

## 1 Introduction

In the last decade there has been a great interest in using theoretical computer science models in biology, based on different paradigms (process algebras, cellular automata, Lindenmayer systems, Petri nets, Boolean functions, P systems, etc.) with the aim of providing an understandable, extensible and computable modeling framework while keeping the needed formalization to perform mathematical analysis. Every such model covers certain aspects of a system and combining two or more leads to obtaining a better and more comprehensive modeling approach. In order to include quantitative and qualitative aspects, there have been suggested various variants of certain models with new features like: Petri nets [10,22], stochastic $\pi$-calculus [28], and stochastic P systems [19].

In this paper we investigate the concerted use of different methods that will reveal a new vision on modeling biological systems by combining different complementary approaches. This is quite different from the hybrid approach discussed by [1] where it is shown how to switch between deterministic and stochastic behavior.

Section 2 introduces the three modeling approaches used in the paper: P systems, *pi*-calculus and Petri nets, as well as Daikon tool and a simple example involving a regulatory network that will be modeled within each approach.

G. Eleftherakis et al. (Eds.): WMC8 2007, LNCS 4860, pp. 138–159, 2007.

Section 3 presents Daikon's findings and the analysis of the invariants provided. The following two sections show how PRISM and a Petri net tool, PIPE, are used in order to confirm some of the properties suggested by Daikon. The final section summarizes our findings.

## 2 Modeling Paradigms

In this section we present three modeling approaches, namely P systems, $\pi$-calculus and Petri nets and a simple case study to illustrate the approach. This example is written directly into these three modeling paradigms and will be executed with a P systems simulator. A system of differential equations is associated to this example and the results obtained are compared to the stochastic behavior exhibited by the P systems simulator and PRISM. In the next three sections Daikon is used to reveal certain properties of our models as they appear through data sets generated by simulators, and PIPE, a Petri nets tool, to ascertain some invariants of the system. and two tools, namely PRISM and PIPE, that are used to analyze and verify properties identified by Daikon.

The aim of this investigation is not to study the relationships between the results produced by using differential equations and those generated by P systems and/or PRISM. The relationships between a special class of P systems working in a deterministic manner according to a metabolic algorithm and differential equations models has been already considered [7]. In this study we are using differential equations only as a substitute for real data in order to illustrate our approach that allows us to "guess" certain properties of the model and then to verify whether they hold or not as general properties or just only happens to be true for the instances generated by simulation.

Nowadays ordinary differential equations (ODE) constitute the most widely used approach in modeling molecular interaction networks in cell systems. They have been used successfully to model kinetics of conventional macroscopic chemical reactions. Nevertheless the realization of a reaction network as a system of ODEs is based on two assumptions. First, cells are assumed to be well stirred and homogeneous volumes so that concentrations do not change with respect to space. Whether or not this is a good approximation depends on the time and space scales involved. In bacteria it has been shown that molecular diffusion is sufficiently fast to mix proteins. This is not the case in eukaryotic cells where the volume is considerably bigger and it is structured in different compartments like nucleus, mitochondria, Golgi body, etc. The second basic assumption is that chemical concentrations vary continuously over time. This assumption is valid if the number of molecules of each species in the reaction volume (the cell or the subcellular compartment) are sufficiently large and the reactions are fast. A sufficiently large number of molecules is considered to be at least thousands of molecules; for hundreds or fewer molecules the continuous approach is questionable.

Writing and solving numerically a system of ODE describing a chemical reaction network can be largely automated. Each species is assigned a single variable $X(t)$ which represents the concentration of the species at time $t$. Then, for each molecular species, a differential equation is written to describe its concentration change over time due to interactions with other species in the system. The rate of each reaction is represented using a kinetic rate law, which commonly depends on one or more rate constants. Exponential decay law, mass action law and Michaelis-Menten dynamic are the most widely used kinetic mechanisms. Finally in order to solve the system of ODEs we must impose a set of initial condition representing the initial concentration of each species involved.

Due to the limitations of ODEs to handle cellular systems with low number of molecules and spatial heterogeneity, some computational approaches have been recently proposed. In what follows we discuss three different approaches, P systems, $\pi$-calculus, and Petri nets.

## 2.1  P Systems

Membrane computing is an emergent branch of natural computing introduced by Gh. Păun in [18]. The models defined in this context are called P systems. In the sequel we will use membrane computing and P systems with the same meaning. Roughly speaking, a P system consists of a cell-like membrane structure, in the compartments of which one places multisets of objects and strings which evolve according to given rules. Recently P systems have been used to model biological phenomena within the framework of computational systems biology presenting models of oscillatory systems [6], signal transduction [19], gene regulation control [20], quorum sensing [27] and metapopulations [21]. In this respect, P systems present a formal framework for the specification and simulation of cellular systems which integrates structural and dynamic aspects in a comprehensive and relevant way while providing the required formalization to perform mathematical and computational analysis.

In the original approach of P systems the rules are applied in a maximally parallel way. This produces two inaccuracies: the reactions represented by the rules do not take place at the correct rate, and all time steps are equal and do not represent the time evolution of the real system. In order to solve these two problems stochastic P systems have been introduced in [19].

**Definition 1.** *A stochastic P system is a construct*

$$\Pi = (O, L, \mu, M_1, M_2, \ldots, M_n, R_1, \ldots, R_n),$$

*where:*

- *$O$ is a finite alphabet of symbols representing objects;*
- *$L$ is a finite alphabet of symbols representing labels for compartments;*
- *$\mu$ is a membrane structure containing $n \geq 1$ membranes labeled with elements from $L$;*
- *$M_i = (l_i, w_i, s_i)$, for each $1 \leq i \leq n$, is the initial configuration of membrane $i$ with $l_i \in L$, the label of this membrane, $w_i \in O^*$ a finite multiset of objects and $s_i$ a finite set of strings over $O$;*

- $R_i$, for each $1 \leq i \leq n$, is a finite set of rewriting rules associated with membrane $i$, of one of the following two forms:

• *Multiset rewriting rules:*

$$obj_1 \, [\, obj_2 \,]_l \xrightarrow{k} obj_1' \, [\, obj_2' \,]_l \qquad (1)$$

with $obj_1, obj_2, obj_1', obj_2' \in O^*$ some finite multisets of objects and $l$ a label from $L$. A multiset of objects, $obj$ is represented as $obj = o_1 + o_2 + \ldots + o_m$ with $o_1, \ldots, o_m \in O$.

These rules are multiset rewriting rules that operate on both sides of membranes, that is, a multiset $obj_1$ placed outside a membrane labeled by $l$ and a multiset $obj_2$ placed inside the same membrane can be simultaneously replaced with a multiset $obj_1'$ and a multiset $obj_2'$, respectively.

• *String rewriting rules:*

$$[\, obj_1 + str_1; \ldots; obj_p + str_p \,]_l \xrightarrow{k} \qquad (2)$$
$$[\, obj_1' + str_{1,1}' + \ldots str_{1,i_1}'; \ldots; obj_p' + str_{p,1}' + \ldots str_{p,i_p}' \,]_l$$

A string $str$ is represented as follows $str = \langle s_1.s_2.\cdots.s_i \rangle$ where $s_1, \ldots, s_i \in O$. In this case each multiset of objects $obj_j$ and string $str_j$, $1 \leq j \leq p$, are replaced by a multiset of objects $obj_j'$ and strings $str_{j,1}' \ldots str_{j,i_j}'$.

The stochastic constant $k$ is used to compute the propensity of the rule by multiplying it by the number of distinct possible combinations of the objects and substrings present on the left-side of the rule with respect to the current contents of membranes involved in the rule. The propensity associated with each rule is used to compute the probability and time needed to apply it.

Cellular systems consisting of molecular interactions taking place in different locations of living cells are specified using stochastic P systems as follows. Different regions and compartments are specified using membranes. Each molecular species is represented by an object in the multiset associated with the region or compartment where the molecule is located. The multiplicity of each object represents the number of molecules of the molecular species represented by the object. Strings are used to specify the genetic information encoded in DNA and RNA. Molecular interactions, compartment translocation and gene expression are specified using rewriting rules on multisets of objects and strings - see Table 1.

**Table 1.** Modeling principles in P systems

| Biochemistry | P System |
|---|---|
| Compartment | Region defined by a membrane |
| Molecule | Object |
| Molecular Population | Multiset of objects |
| Biochemical Transformation | Rewriting rule |
| Compartment Translocation | Boundary rule |

In stochastic P systems [19] constants are associated with rules in order to compute their probabilities and time needed to be applied according to Gillespie algorithm. This approach is based on a Monte Carlo algorithm for stochastic simulation of molecular interactions taking place inside a single volume [8]. In contrast to this, in P systems we have a membrane structure delimiting different compartments (volumes), each one with its own set of rules (molecular interactions) and multiset of objects and strings (molecules). In this respect, a scheduling algorithm called the Multicompartmental Gillespie algorithm [19] is used so that each compartment evolves according to a different Gillespie algorithm. In this point our approach differs from other computational approaches which run a single Gillespie algorithm across the whole system without taking into account the compartmentalized cellular structure [10,28].

We illustrate our approach with a biomolecular system consisting in positive, negative and constitutive expression of a gene. Our model includes the specification of a gene, its transcribed RNA, the corresponding translated protein and activator and repressor molecules which bind to the gene producing an increase in

$$\Pi = (\{gene, rna, protein, act, rep, act\text{-}gene, rep\text{-}gene\}, \{b\}, [\ ]_b, (b, M_i, \emptyset),$$
$$\{r_1, \ldots, r_9\})$$

Initial multisets: $M_{0,1} = gene$; $M_{0,2} = gene + act... + act$ and
$M_{0,3} = gene + rep... + rep$ where $act$ and $rep$ occur 10 times each.
Rules:

$r_1 : \ [\ gene\ ]_b \xrightarrow{c_1} [\ gene + rna\ ]_b \quad c_1 = 0.347\ min^{-1}$

$r_2 : \ [\ rna\ ]_b \xrightarrow{c_2} [\ rna + protein\ ]_b \quad c_2 = 0.174\ min^{-1}$

$r_3 : \ [\ rna\ ]_b \xrightarrow{c_3} [\ ]_b \quad c_3 = 0.347\ min^{-1}$

$r_4 : \ [\ protein\ ]_b \xrightarrow{c_4} [\ ]_b \quad c_4 = 0.0116\ min^{-1}$

$r_5 : \ [\ act + gene\ ]_b \xrightarrow{c_5} [\ act\text{-}gene\ ]_b \quad c_5 = 6.6412087\ molec^{-1}min^{-1}$

$r_6 : \ [\ act\text{-}gene\ ]_b \xrightarrow{c_6} [\ act + gene\ ]_b \quad c_6 = 0.6\ s^{-1}$

$r_7 : \ [\ act\text{-}gene\ ]_b \xrightarrow{c_7} [\ act\text{-}gene + rna\ ]_b \quad c_7 = 3.47\ min^{-1}$

$r_8 : \ [\ rep + gene\ ]_b \xrightarrow{c_8} [\ rep\text{-}gene\ ]_b \quad c_8 = 6.6412087\ molec^{-1}min^{-1}$

$r_9 : \ [\ rep\text{-}gene\ ]_b \xrightarrow{c_9} [\ rep + gene\ ]_b \quad c_9 = 0.6\ min^{-1}$

**Fig. 1.** P system model of gene expression

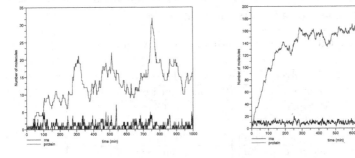

**Fig. 2.** Constitutive expression and positive regulation

transcription rate or prevent the gene from being transcribed, respectively. The bacterium where the system is located is represented using a membrane. The stochastic constants used in our model are taken from the gene control system in the lac operon in E. coli [2,13,14]. In this case transcription and translation have been represented using rewriting rules on multisets of objects, a more detailed description of the concurrent processes of transcription and translation using rewriting rules on strings is presented in [20]. The P systems model is formally defined in Figure 1. It consists of one single compartment labeled $b$, with no strings, and consequently using only multiset rewriting rules. The model refers to three distinct initial conditions, denoted by multisets $M_{0,i}$, and corresponding to constitutive expression, positive and negative regulations, respectively. Simulations of constitutive expression and positive regulation case studies are presented in Figure 2 using a tool available at [30]. A set of ordinary differential equations and their associated graphs, modeling the same examples, are provided in Figure 3. The ODE model is not used here to show its relationship to the previous P systems approach, but to provide a set of data that normally is taken through biological experiments. This will only be used to provide data measurements that will help identifying and validating properties of the P systems model.

$$\frac{dr}{dt} = c_1 - c_3 r + c_7 \frac{act}{act + K} \quad \text{where } K \text{ is the Michaelis-Menten constant}$$
$$\frac{dp}{dt} = c_2 r - c_4 p$$

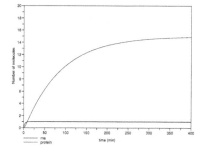 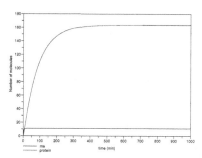

**Fig. 3.** Constitutive and positive expression using ODE model

## 2.2   $\pi$-Calculus

The $\pi$-calculus approach was introduced as a formal language to describe mobile concurrent processes [17]. It is now a widely accepted model for interacting systems with dynamically evolving communication topology. The $\pi$-calculus has a simple semantics and a tractable algebraic theory. Starting with atomic actions and simpler processes, complex processes can be then constructed. The process expressions are defined by guarded processes, parallel composition $P|Q$, nondeterministic choice $P + Q$, replication $!P$, and a restriction operator $(\nu x)P$ creating a local fresh channel $x$ for a process $P$.

Different variants have been used to model molecular interactions [28]. A $\pi$-calculus specification of our system is provided by Figure 4. As usual for this type of modeling approach, each chemical element will be represented as a process and its definition will refer to all possible interactions of it. The initial process may be any of $S_{0,i}$. The process called *gene* defines all possible interactions of a constitutive reaction, producing messenger RNA, a positive regulation, leading to a complex denoted by *act-gene*, or negative regulation, that gets the complex *rep-gene*. This process definition corresponds in a P systems model to rules $r_1$, $r_5$ and $r_8$. In this way we can see, at least syntactically, similarities and differences between the two modeling approaches for expressing chemical interactions. More about the use of both P systems and $\pi$-calculus to model chemical interactions is provided by [26].

**Table 2.** Modeling Principles in $\pi$-calculus

| Biochemistry | $\pi$-calculus |
|---|---|
| Compartment | Private communication channel |
| Molecule | Process |
| Molecular Population | Systems of communicating processes |
| Biochemical Transformation | Communication channel |
| Compartment Translocation | Extrusion of a private channel's scope |

Initial processes: $S_{0,1} = gene$; $S_{0,2} = gene \mid act \mid \ldots \mid act$ and
$S_{0,3} = gene \mid rep \mid \ldots \mid rep$
Processes:
$gene := \tau_{c_1}.(\,gene \mid rna\,) + a_{c_5}?.act\text{-}gene + r_{c_8}?.rep\text{-}gene$
$rna := \tau_{c_2}.(\,rna \mid protein\,) + \tau_{c_3}.0$
$protein := \tau_{c_4}.0$
$act := a_{c_5}!.0$
$act\text{-}gene := \tau_{c_6}.(\,act \mid gene\,) + \tau_{c_7}.(\,act\text{-}gene \mid rna\,)$
$rep := r_{c_8}!.0$
$rep\text{-}gene := \tau_{c_9}.(\,rep \mid gene\,)$

**Fig. 4.** $\pi$-calculus model of gene expression

## 2.3   Petri Nets

Petri nets are a mathematical and computational tool for modeling and analysis of discrete event systems typically with a concurrent behavior. Petri nets offer a formal way to represent the structure of a discrete event system, simulate its behavior, and prove certain properties of the system. Petri nets have applications in many fields of system engineering and computer science. Here we only recall some basic concepts of Petri nets and refer to the current literature [9,23,24,25] for details regarding the theory and applications of Petri nets. In particular, we focus only on a specific class of Petri nets called place-transition nets or PT-nets, for short.

Informally, a PT-net is a directed graph formed by two kinds of nodes called places and transitions respectively. Directed edges, called arcs, connect places to transitions, and transitions to places; each arc has associated a weight. Thus, for each transition, one identifies a set of input places, the places which have at least one arc directed to that transition, and a set of output places, the places which the outgoing arcs of that transitions are directed to. Then, a non-negative integer number of tokens is assigned to each place; these numbers of tokens define the state of the PT-net also called the marking of the PT-net. In a PT-net, a transition is enabled when the number of tokens in each input place is greater than or equal to the weight of the arc connecting that place to the transition. An enabled transition can fire by consuming tokens from its input places and producing tokens in its output places; the number of tokens produced and consumed are determined by the weights of the arcs involved. The firing of a transition can be understood as the movement of tokens from some input places to some output places.

More precisely, we give the following definition.

**Definition 2.** *A* PT-net *is a construct* $N = (P, T, W, M_0)$ *where:* $P$ *is a finite set of* places, $T$ *is a finite set of* transitions, *with* $P \cap T = \emptyset$, $W : (P \times T) \cup (T \times P) \to \mathbf{N}$ *is the* weight function, $M_0$ *is a multiset over* $P$ *called the* initial marking, *and* $L$ *is a* location mapping.

PT-nets are usually represented by diagrams where places are drawn as circles, transitions are drawn as squares, and an arc $(x, y)$ is added between $x$ and $y$ if $W(x, y) \geq 1$. These arcs are then annotated with their weight if this is 2 or more.

Given a PT-net $N$, the *pre-* and *post-multiset* of a transition $t$ are respectively the multiset $pre_N(t)$ and the multiset $post_N(t)$ such that, for all $p \in P$, $|p|_{pre_N(t)} = W(p, t)$ and $|p|_{post_N(t)} = W(t, p)$. A configuration of $N$, which is called a *marking*, is any multiset over $P$; in particular, for every $p \in P$, $|p|_M$ represents the number of *tokens* present inside place $p$. A transition $t$ is *enabled* at a marking $M$ if the multiset $pre_N(t)$ is contained in the multiset $M$. An enabled transition $t$ at marking $M$ can *fire* and produce a new marking $M'$ such that $M' = M - pre_N(t) + post_N(t)$ (i.e., for every place $p \in P$, the firing transition $t$ consumes $|p|_{pre_N(t)}$ tokens and produces $|p|_{post_N(t)}$ tokens).

In order to reason about some basic properties, it is convenient to introduce a matrix-based representation for PT-nets. Specifically, let $N = (P, T, W, M_0)$ be a PT-net and let $\pi : P \to |P|$ and $\tau : T \to |T|$ be two bijective functions. We call place $j$ the place $p$ with $\pi(p) = j$, and we call transition $i$ the transition $t$ with $\tau(t) = i$. Then, a marking $M$ is represented as a $|P|$ vector which contains in each position $j$ the number of tokens currently present inside place $j$. The *incidence matrix* of $N$ is the $|T| \times |P|$ matrix $A$ such that, for every element $a_{ij}$ of $A$, $a_{ij} = |\pi^{-1}(j)|_{post_N(\tau^{-1}(i))}| - |\pi^{-1}(j)|_{pre_N(\tau^{-1}(i))}|$ (i.e., $a_{ij}$ denotes the change in the number of tokens in place $j$ due to the firing of transition $i$). A *control vector* $u$ is a $|T|$ vector containing 1 in position $i$ to denote the firing of transition $i$, 0 otherwise. Thus, if a particular marking $M_n$ is reached from the initial marking $M_0$ through a firing sequence $u_1, u_2, \ldots, u_n$ of enabled transitions, we obtain

$$M_n = M_0 + A^T \cdot \sum_{k=1}^{n} u_k$$

which represents the *reachable-marking equation*.

The aforementioned representation of a PT-net $N$ allows us to introduce the notions of *P-invariants* and *T-invariants*. P-invariants are the positive solutions of the equation $A \cdot y = 0$; the non-zero entries of a solution $y$ represents the set of places whose total number of tokens does not change with any firing sequence from $M_0$. T-invariants instead are the positive solutions of the equation $A^T \cdot x = 0$; a solution vector $x$ represents the set of transitions which have to fire from some marking $M$ to return to the same marking $M$. Then, a PT-net is said to be *bounded* if there exists a $|P|$ vector $B$ such that, for all marking $M$ reachable from $M_0$, we have $M \leq B$; a PT-net is said to be *alive* if, for all marking $M$ reachable from $M_0$, there exists at least one transition enabled at marking $M$.

As pointed out in [10,22], a PT-net model for a system of molecular interactions can be obtained by representing each molecular species as a different place and each biochemical transformation as a different transition. Tokens inside a place can then be used to indicate the presence of a molecule in certain proportions. This modeling approach is summarised in Table 3. Thus, a biochemical system is represented as a discrete event system whose structural properties are useful for drawing conclusions about the behavior and structure of the original biochemical system [22]. For instance, P-invariants determine the set of molecules whose total net concentrations remain unchanged during the application of certain biochemical transformations; T-invariants instead indicate the presence of cyclic reactions which lead to a condition where some reactions are in a state of continuous operation. Also, the property of liveness is useful to determine the absence of metabolic blocks which may hinder the progress of the biochemical system.

**Table 3.** Modeling principles in PT-nets

| Biochemistry | PT-net |
|---|---|
| Molecule | Place |
| Molecular Population | Marking |
| Biochemical Transformation | Transition |
| Reactant | Input Place |
| Product | Output Place |

Finally, we recall that it was shown in [4,15,16] how to transform a P system into a corresponding PT-net. This is done by considering a transition for each rule in the P system that has the left-hand side of the rule as pre-multiset and the right-hand side of the rule as post-multiset. In particular, in order to model the localization of rules and objects inside the membranes, one considers in the corresponding PT-net a distinct place for each object possibly present inside a membrane. Thus, the transformation of objects inside the membranes and the

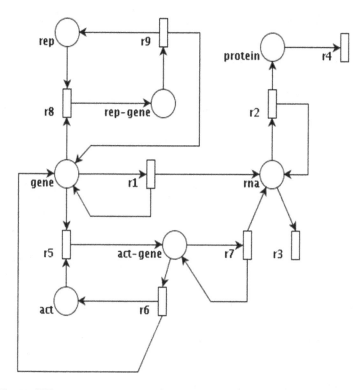

**Fig. 5.** PT-net representation of gene positive and negative regulation

communication of objects between membranes is mapped into the movement of tokens between places of a PT-net. This translation is briefly illustrated in Table 4. Thus, we have a direct way for obtaining a PT-net representation of a given P system model that offers us the possibility of analyzing the P system model in terms of PT-net properties. We illustrate this approach by showing in Figure 5 the PT-net translation of the P system model of Figure 1.

Clearly, we have that transition $r_i$ corresponds to rule $ri$, $1 \leq i \leq 9$. For the PT-net of Figure 5, if we set $M_0$ as initial marking, where $M_0$ contains one token in the place *gene*, then we have only constitutive expression; if we set $M_0$ as having one token in *gene* and $n$ in *act* with $n \geq 1$, then we have positive regulation; if we

**Table 4.** Translation of a P system into a PT-net

| P System | PT-net |
|---|---|
| Object $a$ inside membrane $i$ | Place $a_i$ |
| Multiplicity of an object | Number of tokens inside a place |
| Rule | Transition |
| Left-hand side of a rule | Input places |
| Right-hand side of a rule | Output places |

set $M_0$ with one token in *gene* and $m$ in *rep* with $m \geq 1$, then we have negative regulation. The incidence matrix for PT-net of Figure 5 is reported in Appendix 1 together with its P-invariants and T-invariants. The relevance of these invariants with respect to this specific case study is discussed in Section 5. These are obtained by using PIPE [29], a freely available Petri net tool. As well as this, PIPE allows us to check for the properties of boundedness and liveness (i.e., absence of deadlock). This type of investigation reveals qualitative aspects of the problem modeled as it relies on qualitative behavior expressed by the Petri nets tool.

### 2.4 Daikon Tool

Daikon [5] is a tool that was initially developed to reverse-engineer specifications from software systems. The specifications are in terms of invariants, which are rules that must hold true at particular points as the program executes. To detect invariants the program is executed multiple times and the values of the variables are recorded at specific points (e.g., the start and end of a program function). Daikon infers the invariants by attempting to fit sets of predefined rules to the values of program variables at every recorded program point. Usually the most valuable invariants are preconditions, postconditions and system invariants. These specify the conditions that must hold between variables before a function is executed, after a function has finished executing, and throughout the program execution respectively. As a trivial example, a precondition for the function $div(a, b)$ that divides $a$ by $b$ would be $b \geq 0$. Daikon provides about 70 predefined invariants [5], such as $x > y$, $a < x < b$, $y = ax + b$, and can also be extended to check for new user-supplied invariants.

The idea of using a set of executions to infer rules that govern system behavior, as espoused by Daikon, is particularly useful in the context of biological models. The ability to automatically infer invariants from model simulations is useful for the following reasons: (1) obvious invariants will confirm that the model is behaving as it should, (2) anomalous invariants can indicate a fault in the model and its parameter values or (3) could even suggest novel, latent relationships between model variables.

## 3   Finding Functional Relationships in Raw (Wet Real) Data: Data Analysis Using Daikon

This section demonstrates the use of Daikon to discover relationships between variables in the output from P system simulations of the gene regulation model. The aim is to identify invariants that govern model behavior for negative, constitutive and positive gene regulation. Here we select a sample of the generated invariants and show how they relate to the high-level functionality of the system, and how they can be of use for further model analysis. Invariants are only useful if they are representative of a broad range of model behavior. A single simulation can usually not be considered to be representative, especially if the model is non-trivial and contains stochastic behavior. For this analysis the model was simulated 30 times, ten times for negative, constitutive and positive regulation respectively.

Using Daikon to generate invariants from simulation output is relatively straightforward. It takes as input two files, one of which declares the types of invariants that are of interest, along with the set of relevant variables for each type of invariant. The other file contains the variable values from model simulations, and lists them under their respective declared invariants. In our case the output is in the form of a linear time series, where variable values are provided for $t = 0...n$ time points in the simulation. To analyze this with Daikon, the declaration file contains the three invariant types described above (preconditions, postconditions and system invariants), along with the key model variables under each type ($gene, act$-$gene, rep$-$gene, rna, prot, rep, act$). The data trace file maps any variable values at $t = 0$ to the preconditions, the relationship between every pair of variables $t$ and $t - 1$ to the postconditions, and the variable values for every value of $t$ to the system invariants. To guarantee accurate invariants, the data trace has to be constructed from a set of simulations that can be deemed to be sufficiently representative of the system's behavior.

Figure 6 contains a sample of the invariants that were discovered. These provide a number of insights into the behavior of the system that would be difficult to ascertain from passively observing the simulations. Here we provide an overview of some of these results.

The preconditions show precisely which model parameters are altered for the positive, negative and constitutive sets of simulations. For all simulations, $gene$ starts off as active, and all other variables are zero, apart from the activators variable $act$ for positive and the repressor variable $rep$ for negative regulation.

The postconditions provide a number of insights into the dynamics of the model because they summarize the rules that govern the change in variable values for every single time-step. For positive regulation we learn that the number of proteins will never decrease back to zero throughout the simulation (protein can only be zero if it was already zero at the previous value of $t$), and that the gene must become active to produce $rna$, which can only happen when $act$-$gene$ becomes 1. For negative regulation it shows that the amount of activators remain constantly zero. For constitutive regulation, similarly to positive regulation, the number of proteins can never decrease to zero.

The invariants are rules that hold throughout the entire simulation. These usually cover the range of values a variable can hold, e.g., $gene$ can either be on or off for positive or negative regulation, but is constantly on for constitutive regulation, or the number of proteins is always between zero and 205 for positive regulation. It also points out rules that can sometimes be fundamental to the behavior of the model. For example, in positive regulation $act$-$gene$ and $gene$ can never be on at the same time, which makes sense because $act$-$gene$ is responsible for activating the $gene$ when it is not active. The same holds for $rep$-$gene$ and $gene$ in negative regulation. In positive regulation it also points out that, without any $rna$, there can be no proteins.

These rules provide a number of useful insights into the behavior of the model, many of which are expected, but some of which may either be anomalous or might

| | Positive | Negative | Constitutive |
|---|---|---|---|
| Pre-conditions | $gene = 1$<br>$rna = prot$<br>$= rep\text{-}gene$<br>$= rep$<br>$= act\text{-}gene$<br>$= 0$<br>$act = 10$ | $gene = 1$<br>$rna = prot$<br>$= rep\text{-}gene$<br>$= act$<br>$= act\text{-}gene$<br>$= 0$<br>$rep = 10$ | $gene = 1$<br>$rna = prot$<br>$= rep\text{-}gene$<br>$= rep$<br>$= act\text{-}gene$<br>$= act$<br>$= 0$ |
| Post-condtions | $(prot = 0) \rightarrow$<br>$(orig(prot) = 0)$<br>$(rna = 0) \rightarrow (gene = 0)$<br>$(orig(rna) = 0) \rightarrow$<br>$(gene = 0)$<br>$gene \leq rna$<br>$(prot = 0) \rightarrow (gene = 0)$<br>$(rna = 0) \rightarrow (orig(act\text{-}$<br>$gene = 0)$ | $act = orig(act)$<br>$= orig(act\text{-}gene)$<br>$rna < orig(rep)$<br>$rep > orig(prot)$ | $(prot = 0) \rightarrow$<br>$(orig(prot) = 0)$<br>$gene = orig(gene)$<br>$rep = orig(rep)$ |
| Invariants | $gene = one\,of\,\{0,1\}$<br>$rep = rep\text{-}gene$<br>$= 0$<br>$0 \leq rna \leq 24$<br>$0 \leq prot \leq 205$<br>$act = one\,of\,\{9,10\}$<br>$act\text{-}gene = one\,of\,\{0,1\}$<br>$(gene \wedge act\text{-}gene) = 0$<br>$(rna = 0) \rightarrow (prot = 0)$ | $gene = one\,of\,\{0,1\}$<br>$act = act\text{-}gene$<br>$= 0$<br>$rna = one\,of\,\{0,1\}$<br>$rep = one\,of\,\{9,10\}$<br>$rep\text{-}gene =$<br>$one\,of\,\{0,1\}$<br>$prot =$<br>$one\,of\,\{0,1,2,3\}$<br>$(gene \wedge rep\text{-}gene) = 0$<br>$rna < rep$<br>$rep > prot$ | $rep = rep\text{-}gene$<br>$= act$<br>$= act\text{-}gene$<br>$= 0$<br>$gene = 1$<br>$0 \leq rna \leq 7$<br>$0 \leq prot \leq 32$<br>$rna \geq rep$ |

**Fig. 6.** Invariants discovered by Daikon

identify relationships that had not been previously considered. As an example, the preconditions, which simply summarize the input parameters, are obviously expected, but in practice identified that a small number of our experimental simulations had been mistakenly executed with the wrong parameter values (the precondition for positive regulations stated $act = one\,of\,\{9,10,19,20\}$ instead of just 9 and 10). Rules such as $rna \geq rep$ in constitutive regulation and $rep > prot$ in positive regulation are obviously statistically justified by the simulations, but had not been considered explicitly. New rules like these are useful seeds for further experimentation and analysis, and the following section will show how we have investigated these novel properties with the PRISM model checker.

## 4   PRISM Analysis of the System

Most research in systems biology focuses on the development of models of different biological systems in order to be able to simulate them, accurately enough such as to be able to reveal new properties that can be difficult or impossible to discover through direct experiments. One key question is what one can do with a model, other than just simulate trajectories. This question has been considered in detail for deterministic models, but less for stochastic models. Stochastic systems defy conventional intuition and consequently are harder to conceive.

The field is widely open for theoretical advances that help us to reason about systems in greater detail and with finer precision. An attempt in this direction consists of using model checking tools to analyze in an automatic way various properties of the model. Probabilistic model checking is a formal verification technique. It is based on the construction of a precise mathematical model of a system which is to be analyzed. Properties of this system are then expressed formally using temporal logic and analyzed against the constructed model by a probabilistic model checker. Our current attempt uses a probabilistic symbolic model checking approach based on PRISM (Probabilistic and Symbolic Model Checker) [11,12]. PRISM supports three different types of probabilistic models, discrete time Markov chains (DTMC), Markov decision processes (MDP) and continuous time Markov chains (CTMC). PRISM supports systems specifications through two temporal logics, PCTL (probabilistic computation tree logic) for DTMC and MDP and CSL (continuous stochastic logic) for CTMC.

In order to construct and analyze a model with PRISM, it must be specified in the PRISM language, a simple, high level, state-based language. The fundamental components of the PRISM language are modules, variables and commands. A model is composed of a number of modules which can interact with each other. A module contains a number of local variables and commands.

The values of these variables at any given time constitute the states of the module. The space of reachable states is computed using the range of each variable and its initial value. The global state of the whole model is determined by the local state of all modules.

The behavior of each module is described by a set of commands. A command takes the form:

$$[ \text{ action } ] \; g \to \lambda_1 : u_1 + \cdots + \lambda_n : u_n;$$

The guard $g$ is a predicate over all the variables of the model. Each update $u_i$ describes the new values of the variables in the module specifying a transition of the module. The expressions $\lambda_i$ are used to assign probabilistic information, rates, to transitions.

The label action placed inside the square brackets are used to synchronize the application of different commands in different modules. This forces two or more modules to make transitions simultaneously. The rate of this transition is equal to the product of the individual rates, since the processes are assumed to be independent.

The main components of a P system are a membrane structure consisting of a number of membranes that can interact with each other, an alphabet of objects and a set of rules associated to each membrane. These components can easily be mapped into the components of the PRISM language using modules to represent membranes, variables to describe the alphabet and commands to specify the rules.

A PRISM specification of our system is provided in Appendix 2. Appendix 3 shows the probability that some molecules concentrations will reach certain values at steady state. The ranges of values provided by Daikon represent an indication of possible levels for various molecular concentrations, but in order to know the likely values around steady states, PRISM provides a set of properties that help in this respect. For example, for positive regulation, Daikon provides the range 0 to 24, for *rna* molecules, but PRISM shows that values between 0 and 15 are more likely to be obtained than values greater than 15, and values over 20 are very unlikely to be reached. These values are also confirmed by the graphs provided by differential equations and P system simulator.

Other properties, suggested by Daikon analysis, like *rna* < *rep*, *prot* < *rep*, are also validated by PRISM by showing they take place with a higher probability for values of *rep* less than 5 - see Appendix 4. The average or expected behavior of the stochastic system is also provided and this is very close to ODE behavior.

## 5  Petri Net Analysis of the System

In this section we will show how different invariants will emerge from the analysis of the Petri net and how Daikon hypotheses are formally verified or new problems are formulated.

T-invariants in Appendix 1 show that:

- If we fire transition $r1$ and then transition $r3$, the current marking of the network remains unchanged because we first produce a molecule of *rna* and then we consume it; the same happens if we first fire transition $r7$ and than transition $r3$, or if we first fire transition $r2$ and then transition $r4$ (i.e., we first produce a molecule of *protein* and then we consume it).
- The operation of binding the activator to the gene and its de-binding are one the reverse of the other, hence firing transition $r5$ followed by transition $r6$ (or vice versa) has no effect on the current marking; these two transitions constitute a continuous loop.
- The operation of binding the repressor to the gene and its de-binding are one the reverse of the other, hence firing transition $r8$ followed by transition $r9$ (or vice versa) has no effect on the current marking; these two transitions constitute a continuous loop.

P-invariants computed by PIPE, in Appendix 1, for some initial marking with one element in *gene*, $n$ in *act* and $m$ in *rep*, where $n, m \geq 0$ show that:

- The gene is always present and it can assume three different states: *gene*, *act-gene*, and *rep-gene*; these three states are mutually exclusive; in the case of constitutive expression (i.e, $n = m = 0$), we have $M(gene) = 1$ indicating that the gene is always present - this confirms Daikon invariant $gene = 1$; in the case of positive regulation (i.e., $n \geq 1$ and $m = 0$), we have $M(gene) + M(act\text{-}gene) = 1$ indicating two mutually exclusive states - this confirms Daikon invariant $gene \wedge act\text{-}gene = 0$; in the case of negative regulation (i.e., $m \geq 1$ and $n = 0$, we have $M(gene) + M(rep\text{-}gene) = 1$ indicating two mutually exclusive states - this confirms Daikon invariant $gene \wedge rep\text{-}gene = 0$.

- For positive regulation, the number of activator molecules cannot be increased but can be decreased only by 1 - similar invariant is found by Daikon.
- For negative regulation, the number of repressor molecules cannot be increased but can be decreased only by 1 - similar invariant is found by Daikon.

PIPE also shows that the network is not bounded but it is alive. In fact, *gene* is always present and we can keep firing transition $r1$ to increase indefinitely the amount of *rna*. The liveness here comes from the above invariant and shows that the system will be working forever. The boundedness instead produces a result that apparently contradicts PRISM findings, where the probability that the number of *rna*'s is greater than 7 is almost 0! This comes from the fact that PIPE uses a non-deterministic system instead of a probabilistic one considered by PRISM and P system simulator. It will be interesting to check this property with a probabilistic Petri net tool.

# 6    Conclusions

In this paper we have investigated the concerted use of different methods, and shown how these can provide complementary insights into different facets of biological system behavior. Individual modeling techniques have their own respective benefits and usually excel at reasoning about a system from a particular perspective. This paper shows how these benefits can be leveraged by using different modeling techniques in concert.

As a case study, we have constructed a P system model of a small gene expression system and produced equivalent specifications using Petri net and π-calculus approaches. Simulations of the P system model were analyzed by Daikon (to identify potential rules that govern model output), and some of the most interesting suggested rules were checked using the PRISM probabilistic model checker. The Petri net model was analyzed with PIPE, a general Petri net analysis tool. The results show how analysis results from different models of the same system are useful for the purposes of both validating and improving each other.

The gene expression model was chosen because it is manageable, and thus forms a useful basis for a case study to compare different modeling techniques. Our future work will apply the techniques shown in this paper to a larger and more realistic case study. This should provide further insights into the benefits that arise in modeling increasingly complex systems when the modeler is increasingly reliant upon the use of various automated tools to study the model behavior.

# Acknowledgements

The research of Francesco Bernardini is supported by NWO, Organization for Scientific Research of The Netherlands, project 635.100.006 "VIEWS". The authors would like to thank the anonymous reviewers for their comments and suggestions that have helped to improve the paper.

# References

1. Bianco, L., Fontana, F.: Towards a Hybrid Metabolic Algorithm. In: Hoogeboom, H.J., Păun, G., Rozenberg, G., Salomaa, A. (eds.) WMC 2006. LNCS, vol. 4361, pp. 183–196. Springer, Heidelberg (2006)
2. Blundell, M., Kennell, D.: Evidence for Endonucleolytic Attack in Decay of Lac Messenger RNA in Escherichia Coli. J. Mol. Biol. 83, 143–161 (1974)
3. Calder, M., Vyshemirsky, V., Gilbert, D., Orton, R.: Analysis of Signalling Pathways Using Continuous Time Markov Chains. In: Priami, C., Plotkin, G. (eds.) Transactions on Computational Systems Biology VI. LNCS (LNBI), vol. 4220, pp. 44–67. Springer, Heidelberg (2006)
4. Dal Zilio, S., Formenti, E.: On the Dynamics of PB Systems: A Petri Net View. In: Martín-Vide, C., Mauri, G., Păun, G., Rozenberg, G., Salomaa, A. (eds.) Membrane Computing. LNCS, vol. 2933, pp. 153–167. Springer, Heidelberg (2004)
5. Ernst, M., Cockrell, J., Griswold, W., Notkin, D.: Dynamically Discovering Likely Program Invariants to Support Program Evolution. IEEE Transactions on Software Engineering 27(2), 99–123 (2001)
6. Fontana, F., Bianco, L., Manca, V.: P Systems and the modeling of Biochemical Oscillations. In: Freund, R., Păun, G., Rozenberg, G., Salomaa, A. (eds.) WMC 2005. LNCS, vol. 3850, pp. 199–208. Springer, Heidelberg (2006)
7. Fontana, F., Manca, V.: Discrete Solutions of Differential Equations by Metabolic P Systems. Theoretical Computer Science 372(2-3), 165–182 (2007)
8. Gillespie, D.T.: Exact Stochastic Simulation of Coupled Chemical Reactions. The Journal of Physical Chemistry 81(25), 2340–2361 (1977)
9. Girault, C., Valk, R.: Petri Nets for Systems Engineering. Springer, Heidelberg (2003)
10. Goss, P.J.E., Peccoud, J.: Quantitative Modeling of Stochastic Systems in Molecular Biology using Stochastic Petri Nets. Proc. Natl. Acad. Sci. USA 95, 6750–6755 (1999)
11. Heath, J., Kwiatkowska, M., Norman, G., Parker, D., Tymchyshyn: Probabilistic Model Checking of Complex Biological Pathways. In: Priami, C. (ed.) CMSB 2006. LNCS (LNBI), vol. 4210, pp. 32–47. Springer, Heidelberg (2006)
12. Hinton, A., Kwiatkowska, M., Norman, G., Parker, D.: PRISM: A Tool for Automatic Verification of Probabilistic Systems. In: Hermanns, H., Palsberg, J. (eds.) TACAS 2006 and ETAPS 2006. LNCS, vol. 3920, pp. 441–444. Springer, Heidelberg (2006)
13. Hlavacek, S., Savageau, M.A.: Subunit Structure of Regulator Proteins Influences the Design of Gene Circuitry Analysis of Perfectly Coupled and Uncoupled Circuits. J. Mol. Biol. 248, 739–755 (1995)
14. Kennell, D., Riezman, H.: Transcription and Translation Initiation Frequencies of the Escherichia Coli Lac Operon. J. Mol. Biol. 114, 1–21 (1977)
15. Kleijn, K., Koutny, K.: Synchrony and Asynchrony in Membrane Systems. In: Hoogeboom, H.J., Păun, G., Rozenberg, G., Salomaa, A. (eds.) WMC 2006. LNCS, vol. 4361, pp. 66–85. Springer, Heidelberg (2006)
16. Kleijn, K., Koutny, M., Rozenberg, G.: Towards a Petri Net Semantics for Membrane Systems. In: Freund, R., Păun, G., Rozenberg, G., Salomaa, A. (eds.) WMC 2005. LNCS, vol. 3850, pp. 292–309. Springer, Heidelberg (2006)
17. Milner, R.: Communication and Mobile Systems: The $\pi$-calculus. Cambridge University Press, Cambridge (1999)

18. Păun, G.: Computing with Membranes. Journal of Computer and System Sciences 61(1), 108–143 (2000)
19. Pérez-Jiménez, M.J., Romero-Campero, F.J.: P Systems, a New Computationl modeling Tool for Systems Biology. In: Priami, C., Plotkin, G. (eds.) Transactions on Computational Systems Biology VI. LNCS (LNBI), vol. 4220, pp. 176–197. Springer, Heidelberg (2006)
20. Pérez-Jiménez, M.J., Romero-Campero, F.J.: Modeling Gene Expression Control Using P Systems: The Lac Operon, A Case Study (submitted, 2007)
21. Pescini, D., Besozzi, D., Mauri, C., Zandron, C.: Dynamical Probabilistic P systems. International Journal of Foundations of Computer Science 17(1), 183–204 (2007)
22. Reddy, V.N., Liebman, M.N., Mavrouniotis, M.L.: Qualitative Analysis of Biochemical Reaction Systems. Computers in Biology & Medicine 26(1), 9–24 (1996)
23. Reisig, W.: Elements of Distributed Algorithms, modeling and Analysis with Petri Nets. Springer, Heidelberg (1998)
24. Reisig, W., Rozenberg, G. (eds.): Lectures on Petri Nets I: Basic Models. LNCS, vol. 1491. Springer, Heidelberg (1998)
25. Reisig, W., Rozenberg, G. (eds.): Lectures on Petri Nets II: Applications. LNCS, vol. 1492. Springer, Heidelberg (1998)
26. Romero-Campero, F.J., Gheorghe, M., Ciobanu, G., Auld, J.M., Pérez-Jiménez, M.J.: Cellular modeling Using P Systems and Process Algebra. Progress in Natural Science 17(4), 375–383 (2007)
27. Romero-Campero, F.J., Pérez-Jiménez, M.J.: A Model of the Quorum Sensing System in Vibrio Fischeri Using P Systems (submitted, 2007)
28. Regev, A., Shapiro, E.: The π-calculus as an Abstraction for Biomolecular Systems. In: Ciobanu, G., Rozenberg, G. (eds.) Modeling in Molecular Biology, Springer, Heidelberg (2007)
29. Platform Independent Petri Net Editor: http://pipe2.sourceforge.net
30. P System Simulator: http://www.dcs.shef.ac.uk/~marian/PSimulatorWeb/PSystemMF.htm

## APPENDIX 1

Transpose of the incidence matrix for PT-net of Figure 5:

```
          r1 r2 r4 r3 r5 r7 r6 r8 r9
gene       0  0  0  0 -1  0  1 -1  1
rna        1  0  0 -1  0  1  0  0  0
protein    0  1 -1  0  0  0  0  0  0
act        0  0  0  0 -1  0  1  0  0
act-gene   0  0  0  0  1  0 -1  0  0
rep        0  0  0  0  0  0  0 -1  1
rep-gene   0  0  0  0  0  0  0  1 -1
```

which shows the variations on the number of tokens determined by each transition.

T-invariants obtained in PIPE:

| r1 | 1 | 0 | 0 | 0 | 0 |
|----|---|---|---|---|---|
| r2 | 0 | 1 | 0 | 0 | 0 |
| r4 | 0 | 1 | 0 | 0 | 0 |
| r3 | 1 | 0 | 0 | 1 | 0 |
| r5 | 0 | 0 | 1 | 0 | 0 |
| r7 | 0 | 0 | 0 | 1 | 0 |
| r6 | 0 | 0 | 1 | 0 | 0 |
| r8 | 0 | 0 | 0 | 0 | 1 |
| r9 | 0 | 0 | 0 | 0 | 1 |

P-invariants computed by PIPE for some initial marking contains one token in *gene* $n$ in *act* and $m$ in *rep*, with $n, m \geq 0$:

| gene | 1 | 0 | 0 | |
|------|---|---|---|---|
| rna | 0 | 0 | 0 | $M(gene) + M(act\text{-}gene) + M(rep\text{-}gene) = 1$ |
| protein | 0 | 0 | 0 | $M(act) + M(act\text{-}gene) = n$ |
| act | 0 | 1 | 0 | $M(rep) + M(rep\text{-}gene) = m$ |
| act-gene | 1 | 1 | 0 | |
| rep | 0 | 0 | 1 | |
| rep-gene | 1 | 0 | 1 | |

## APPENDIX 2

```
// Gene expression control
// Model is stochastic
stochastic
// Bounds to the number of molecules
const int rna_bound;
const int protein_bound;
const int number_activators;
const int number_repressors;
const int initact;
const int initrep;
// Stochastics constants associated with each
command/rule/molecular interaction
const double c1 = 0.347;      // [ gene ]_b -c1-> [ gene + rna ]_b
const double c2 = 0.174;      // [ rna ]_b -c2-> [ rna + protein ]_b \\
const double c3 = 0.347;      // [ rna ]_b -c3-> [    ]_b
const double c4 = 0.0116;     // [ protein ]_b -c4-> [    ]_b
const double c5 = 6.6412087;  // [ act + gene ]_b -c5-> [ actgene ]_b
const double c6 = 0.6;        // [ actgene ]_b -c6-> [ act + gene ]_b
const double c7 = 3.47;       // [ actgene ]_b -c7-> [ actgene + rna ]_b
const double c8 = 6.6412087;  // [ rep + gene ]_b -c8-> [ repgene ]_b
const double c9 = 0.6;        // [ repgene ]\_b -c9-> [ rep + gene ]_b
// Module representing a bacterium
module bacterium
    gene : [ 0 .. 1 ] init 1;
    actgene : [ 0 .. 1 ] init 0;
    repgene : [ 0 .. 1 ] init 0;
    act : [ 0 .. 1 ] init initact;
    rep : [ 0 .. 1 ] init initrep;
    rna : [ 0 .. rna_bound ] init 0;
    protein : [ 0 .. protein_bound ] init 0;
    // [ gene ]_b -c1-> [ gene + rna ]_b
    [ ] gene = 1 & rna < rna_bound -> c1 : (rna' = rna + 1);
    // [ rna ]_b -c2-> [ rna + protein ]_b
    [ ] rna > 0 & protein < protein_bound -> c2*rna :
(protein' = protein + 1);
    // [ rna ]_b -c3-> [ ]_b
    [ ] rna > 0 -> c3*rna : (rna' = rna - 1);
    // [ protein ]_b -c4-> [    ]_b
    [ ] protein > 0 -> c4*protein : (protein' = protein - 1);
    // [ act + gene ]_b -c5-> [ actgene ]_b
    [ ] act = 1 & gene = 1 -> c5*number_activators : (gene'
= 0) & (act' = 0) & (actgene' = 1);
    // [ actgene ]_b -c6-> [ act + gene ]_b
    [ ] actgene = 1 & act = 0 -> c6 : (actgene' = 0) &
(act' = 1) & (gene' = 1);
```

```
    // [ actgene ]_b -c7-> [ actgene + rna ]_b
    [ ] actgene = 1 & rna < rna_bound -> c7 : (rna' = rna +
1);
    // [ rep + gene ]_b -c8-> [ repgene ]_b
    [ ] rep = 1 & gene = 1 -> c8*number_repressors : (gene'
= 0) & (rep' = 0) & (repgene' = 1);
    // [ repgene ]_b -c9-> [ rep + gene ]_b
    [ ] repgene = 1 & rep = 0 -> c9 : (repgene' = 0) &
(rep' = 1) & (gene' = 1);
endmodule
```

## APPENDIX 3

Ranges of molecules
```
P = ? [ true U <= T rna > bound ]
P = ? [ true U <= T protein > bound ]
```
Constitutive regulation
```
rna <= 7
rna >= 0
prot <= 32
prot >= 0
```

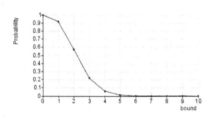
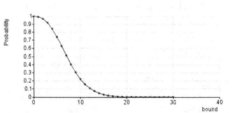

Positive regulation
```
rna <= 24
rna >= 0
prot <= 205
prot >= 0
```

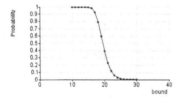
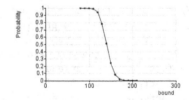

Negative regulation
```
rna one of { 0, 1 }
prot one of { 0, 1, 2, 3 }
```

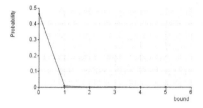

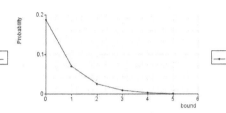

## APPENDIX 4

**Relationship between the number of repressors
and rna and protein molecules.**
```
rna < rep
rep > prot
P = ? [ true U<=T rna > rep ]
P = ? [ true U<=T protein > rep ]
```

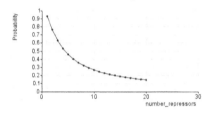

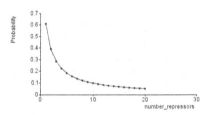

**Expected number of molecules
R = ? [ I = T ]**

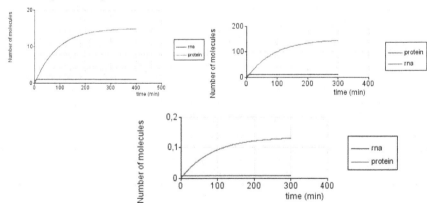

**Other invariants**
```
P = ? [ true U gene = actgene ] ⇒ Result: 0.0
P = ? [ true U gene = repgene ] ⇒ Result: 0.0
```

# Causality in Membrane Systems

Nadia Busi

**Abstract.** P systems are a biologically inspired model introduced by Gheorghe Păun with the aim of representing the structure and the functioning of the cell. P systems are usually equipped with the maximal parallelism semantics; however, since their introduction, some alternative semantics have been proposed and investigated.

We propose a semantics that describes the causal dependencies occurring between the reactions of a P system. We investigate the basic properties that are satisfied by such a semantics. The notion of causality turns out to be quite relevant for biological systems, as it permits to point out which events occurring in a biological pathway are necessary for another event to happen.

## 1 Introduction

Membrane computing is a branch of natural computing, initiated by Gheorghe Păun with the definition of P systems in [22,23]. The aim is to provide a formal modeling of the structure and the functioning of the cell, making use especially of automata, languages, and complexity theoretic tools.

Membrane systems are based upon the notion of *membrane structure,* which is a structure composed by several cell-membranes, hierarchically embedded in a main membrane called the *skin membrane.* A plane representation of a membrane structure can be given by means of a Venn diagram, without intersected sets and with a unique superset. The membranes delimit *regions* and we associate with each region a set of *objects,* described by some symbols over an alphabet, and a set of *evolution rules.*

In the basic variant, the objects evolve according to the evolution rules, which can modify the objects to obtain new objects and send them outside the membrane or to an inner membrane. The evolution rules are applied in a maximally parallel manner: at each step, all the objects which can evolve should evolve.

A computation device is obtained: we start from an initial configuration, with a certain number of objects in certain membranes, and we let the system evolve. If a computation *halts,* that is no further evolution rule can be applied, the result of the computation is defined to be the number of objects in a specified membrane (or expelled through the skin membrane). If a computation never halts (i.e., one or more object can be rewritten forever), then it provides no output.

Since their introduction, plenty of variants of P systems have been proposed, and a lot of research effort has been carried out, especially concerned with the study of the expressivity and the universality of the proposed models and with the ability to solve NP-complete problems in polynomial time.

G. Eleftherakis et al. (Eds.): WMC8 2007, LNCS 4860, pp. 160–171, 2007.

The aim of this work is to start an investigation of the causal dependencies arising among reactions occurring in P systems. The main motivation for this work comes from system biology, as the understanding of the causal relations occurring between the events of a complex biological pathway could be of precious help, e.g., for limiting the search space in the case some unpredicted event occurs.

In this paper we concentrate on P systems with cooperative rules, namely systems whose evolution rules are of the form $u \to v$, representing the fact that the objects in $u$ are consumed and the objects in $v$ are produced.

The study of causal semantics in concurrency theory is quite old. For example, the study of a causal semantics for process algebras dates back to the early nineties for CCS [20] (see, e.g., [13, 11, 18]), and to the mid nineties for the $\pi$-calculus [21] (see, e.g., [3, 5, 14, 15]).

To the best of our knowledge, the only other works that deal with causality in bio-inspired calculi with membranes and compartments are the following. In [7] a causal semantics for the Mate/Bud/Drip Brane Calculus [9] is proposed. In [17] a causal semantics for Beta Binders [26, 27] – based on the $\pi$-calculus semantics and on the enhanced operational semantics approach of [15] – is defined. One of the main differences between Beta Binders on one side, and Brane Calculi and P systems on the other side, is that the membrane structure in Beta Binders is flat, whereas in Brane Calculi and in P systems the membranes are nested to form a hierarchical structure.

The paper is organized as follows. After providing some basic definitions in Section 2, in Section 3 we define (cooperative) P systems. Section 4 recalls a detailed definition of maximal parallelism semantics that will be used in the following to provide a comparison between the causal and the maximal parallelism semantics. Section 5 is devoted to the definition of the causal semantics; after an informal introduction, a formal definition is provided, and finally some result on the properties enjoyed by the causal semantics are given. Section 6 reports some conclusive remarks.

## 2   Basic Definitions

In this section we provide some definitions that will be used throughout the paper.

**Definition 1.** *Given a set $S$, a finite multiset over $S$ is a function $m : S \to \mathbb{N}$ such that the set $dom(m) = \{s \in S \mid m(s) \neq 0\}$ is finite. The multiplicity of an element $s$ in $m$ is given by the natural number $m(s)$. The set of all finite multisets over $S$, denoted by $\mathcal{M}_{fin}(S)$, is ranged over by $m$. A multiset $m$ such that $dom(m) = \emptyset$ is called empty. The empty multiset is denoted by $\emptyset$.*

*Given the multisets $m$ and $m'$, we write $m \subseteq m'$ if $m(s) \leq m'(s)$ for all $s \in S$ while $\oplus$ denotes their multiset union, i.e., $m \oplus m'(s) = m(s) + m'(s)$. The operator $\setminus$ denotes multiset difference: $(m \setminus m')(s) = $ if $m(s) \geq m'(s)$ then $m(s) - m'(s)$ else 0. The scalar product, $j \cdot m$, of a number $j$ with $m$ is $(j \cdot m)(s) =$*

$j \cdot (m(s))$. *The cardinality of a multiset is the number of occurrences of elements contained in the multiset:* $|m| = \sum_{s \in S} m(s)$.

The powerset of a set $S$ is defined as $\mathcal{P}(S) = \{X \mid X \subseteq S\}$.

**Definition 2.** *Let $m$ be a finite multiset over $S$ and $X \subseteq S$. The multiset $m|_X$ is defined as follows: for all $s \in S$, $m|_X(s) = m(s)$ if $s \in X$, and $m|_X(s) = 0$ otherwise.*

**Definition 3.** *A string over $S$ is a finite (possibly empty) sequence of elements in $S$. Given a string $u = x_1 \ldots x_n$, the length of $u$ is the number of occurrences of elements contained in $u$ and is defined by $|u| = n$.*

*With $S^*$ we denote the set of strings over $S$, and $u, v, w, \ldots$ range over $S$. Given $n \geq 0$, with $S^n$ we denote the set of strings of length $n$ over $S$.*

*Given a string $u = x_1 \ldots x_n$ and $i$ such that $1 \leq i \leq n$, with $(u)_i$ we denote the $i$-th element of $u$, namely, $(u)_i = x_i$.*

*Given a string $u = x_1 \ldots x_n$, the multiset corresponding to $u$ is defined as follows: for all $s \in S$, $m_u(s) = |\{i \mid x_i = s \wedge 1 \leq i \leq n\}|$. With abuse of notation, we use $u$ to denote also $m_u$.*

**Definition 4.** *With $S \times T$ we denote the Cartesian product of sets $S$ and $T$, with $\times_n S$, $n \geq 1$, we denote the Cartesian product of $n$ copies of set $S$ and with $\times_{i=1}^n S_i$ we denote the Cartesian product of sets $S_1, \ldots, S_n$, i.e., $S_1 \times \ldots \times S_n$. The $i$th projection of $(x_1, \ldots, x_n) \in \times_{i=1}^n S_i$ is defined as $\pi_i(x) = x_i$, and lifted to subsets $X \subseteq \times_{i=1}^n S_i$ as follows: $\pi_i(X) = \{\pi_i(x) \mid x \in X\}$.*

Given a binary relation $R$ over a set $S$, with $R^n$ we denote the composition of $n$ instances or $R$, with $R^+$ we denote the transitive closure of $R$, and with $R^*$ we denote the reflexive and transitive closure of $R$.

## 3   P Systems

We recall the definition of catalytic P systems without priorities on rules. For a thorough description of the model, motivation, and examples see, e.g., [8, 12, 22, 23, 24]. To this aim, we start with the definition of *membrane structure*:

**Definition 5.** *Given the alphabet $\{[,]\}$, the set $MS$ is the least set inductively defined by the following rules:*

- $[\,] \in MS$,
- *if $\mu_1, \mu_2, \ldots, \mu_n \in MS$, $n \geq 1$, then $[\mu_1 \ldots \mu_n] \in MS$.*

*We define the following relation over $MS$: $x \sim y$ iff the two strings can be written as $x = [_1 \ldots [_2 \ldots]_2 \ldots [_3 \ldots]_3 \ldots]_1$ and $y = [_1 \ldots [_3 \ldots]_3 \ldots [_2 \ldots]_2 \ldots]_1$ (i.e., if two pairs of parentheses that are neighbors can be swapped together with their contents).*

*The set $\overline{MS}$ of membrane structures is defined as the set of equivalence classes w.r.t. the relation $\sim^*$.*

We call a *membrane* each matching pair of parentheses appearing in the membrane structure. A membrane structure $\mu$ can be represented as a Venn diagram, in which any closed space (delimited by a membrane and by the membranes immediately inside) is called a *region* of $\mu$.

**Definition 6.** *A* P system *(of degree d, with $d \geq 1$) is a construct*

$$\Pi = (V, \mu, w_1^0, \ldots, w_d^0, R_1, \ldots, R_d, i_0), \quad \text{where:}$$

1. *V is a finite alphabet whose elements are called* objects;
2. *$\mu$ is a* membrane structure *consisting of d membranes (usually labeled with i and represented by corresponding brackets $[_i$ and $]_i$, with $1 \leq i \leq d$);*
3. *$w_i^0$, $1 \leq i \leq d$, are strings over V associated with the regions $1, 2, \ldots, d$ of $\mu$; they represent multisets of objects present in the regions of $\mu$ at the beginning of computation (the multiplicity of a symbol in a region is given by the number of occurrences of this symbol in the string corresponding to that region);*
4. *$R_i$, $1 \leq i \leq d$, are finite sets of* evolution rules *over V associated with the regions $1, 2, \ldots, d$ of $\mu$; these evolution rules are of the form $u \to v$, where u and v are strings from $(V \times \{here, out, in\})^*$;*
5. *$i_0 \in \{1, \ldots, d\}$ specifies the* output membrane *of $\Pi$.*

The membrane structure and the multisets represented by $w_i^0$, $1 \leq i \leq d$, in $\Pi$ constitute the *initial state*[1] of the system. A transition between states is governed by an application of the evolution rules which is done in parallel; all objects, from all membranes, which *can be* the subject of local evolution rules *have to* evolve simultaneously.

The application of a rule $u \to v$ in a region containing a multiset $m$ results in subtracting from $m$ the multiset identified by $u$, and then in adding the multiset defined by $v$. The objects can eventually be transported through membranes due to the targets *in* and *out* (we usually omit the target *here*).

The system continues parallel steps until there remain no applicable rules in any region of $\Pi$; then the system halts. We consider the number of objects from $V$ contained in the output membrane $i_0$ when the system halts as the *result* of the underlying computation of $\Pi$.

We introduce a couple of functions on membrane structures that will be useful in the following:

**Definition 7.** *Let $\mu$ be a membrane structure consisting of d membranes, labeled with $\{1, \ldots, d\}$.*

*Given two membranes i and j in $\mu$, we say that i is contained in j if the surface delimited by the perimeter of i in the Venn diagram representation of $\mu$ is contained inside the perimeter of j.*

---

[1] Here we use the term *state* instead of the classical term *configuration* because we will define a (essentially equivalent but syntactically) different notion of configuration in Section 5.

*We say that $i$ is the* father *of $j$ (and $j$ is a* child *of $i$) if the membrane $j$ is contained in $i$, and no membrane exists that contains $j$ and is contained in $i$.*

*The partial function* $father : \{1, \ldots, d\} \rightarrow \{1, \ldots, d\}$ *returns the father of a membrane $i$, or is undefined if $i$ is the external membrane.*

*The function* $children : \{1, \ldots, d\} \rightarrow \mathcal{P}(\{1, \ldots, d\})$ *returns the set of children of a membrane.*

For example, take $\mu = [_1[_2[_3 \ ]_3]_2 \ [_4 \ ]_4]_1$; then, $father(2) = father(4) = 1$, $father(3) = 2$ and $father(1)$ is undefined; moreover, $children(4) = \emptyset$ and $children(1) = \{2, 4\}$.

## 4    Maximal Parallelism Semantics for P Systems

In order to compare the classical maximal parallelism semantics with the causal semantics, in this section we recall a detailed definition of the computation of a P system, proposed in [4], where a maximal parallelism evolution step is represented as a (maximal) sequence of simple evolution steps, which are obtained by the application of a single evolution rule.

Throughout this section, we let $\Pi = (V, \mu, w_1^0, \ldots, w_d^0, R_1, \ldots, R_d, i_0)$ be a P system.

To represent the states of the system reached after the execution of a non maximal sequence of simple evolution rules, we introduce the notion of *partial configuration* of a system. In a partial configuration, the contents of each region is represented by two multisets:

- The multiset of *active objects* contains the objects that were in the region at the beginning of the current maximal parallelism evolution step. These objects can be used by the next simple evolution step.
- The multiset of *frozen objects* contains the objects that have been produced in the region during the current maximal parallelism evolution step. These objects will be available for consumption in the next maximal parallelism evolution step.

**Definition 8.** *A partial configuration of $\Pi$ is a tuple* $((w_1, \bar{w}_1), \ldots, (w_d, \bar{w}_d)) \in \times_d(V^* \times V^*)$.

*We use* $\times_{i=1}^{d}(w_i, \bar{w}_i)$ *to denote the partial configuration above.*

*The set of partial configurations of $\Pi$ is denoted by $Conf_\Pi$. We use $\gamma, \gamma', \gamma_1, \ldots$ to range over $Conf_\Pi$.*

In the above definition, $w_1, \ldots, w_d$ represent the active multisets, whereas $\bar{w}_1, \ldots, \bar{w}_d$ represent the frozen multisets.

A *configuration* is a partial configuration containing no frozen objects; configurations represent the states reached after the execution of a maximal parallelism computation step.

**Definition 9.** *A configuration of $\Pi$ is a partial configuration* $\times_{i=1}^{d}(w_i, \bar{w}_i)$ *satisfying the following:* $\bar{w}_i = \emptyset$ *for $i = 1, \ldots, d$.*

*The initial configuration of $\Pi$ is the configuration* $\times_{i=1}^{d}(w_i^0, \emptyset)$.

The size of a partial configuration is the number of active objects contained in the configuration.

**Definition 10.** *Let* $\gamma = \times_{i=1}^{d}(w_i, \bar{w}_i)$ *be a partial configuration. The* size *of* $\gamma$ *is* $\#(\gamma) = \sum_{i=1}^{d} |w_i|$.

The execution of a simple evolution rule is formalized by the notion of reaction relation, defined as follows:

**Definition 11.** *The* reaction relation $\mapsto$ *over* $Conf_\Pi \times Conf_\Pi$ *is defined as follows:*

$\times_{i=1}^{d}(w_i, \bar{w}_i) \mapsto \times_{i=1}^{d}(w_i', \bar{w}_i')$ *iff there exist* $k$, *with* $1 \leq k \leq d$, *an evolution rule* $u \to v \in R_k$ *and a migration string* $\rho \in \{1, \ldots, d\}^{|v|}$ *such that*

- $u \subseteq w_{k,,}$
- $w_k' = w_k \setminus u$
- $\forall i : 1 \leq i \leq d$ *and* $i \neq k$ *implies* $w_i' = w_i$,
- $\forall j : 1 \leq j \leq |v|$ *the following holds:*
    - *if* $\pi_2((v)_j) = here$ *then* $(\rho)_j = k$,
    - *if* $\pi_2((v)_j) = out$ *then* $(\rho)_j = father(k)$,[2]
    - *if* $\pi_2((v)_j) = in$ *then* $(\rho)_j \in children(k)$,[3]
- $\forall i, 1 \leq i \leq d : \bar{w}_k' = \bar{w}_k \oplus \bigoplus_{1 \leq j \leq |v|, (\rho)_j = k} (v)_j$.

Note that the size of a configuration represents an upper bound to the length of the sequences of reactions starting from that configuration. Hence, infinite sequences of reactions are not possible.

**Proposition 1.** *Let* $\gamma$ *be a configuration. If* $\gamma \mapsto^n \gamma'$ *then* $n \leq \#(\gamma)$.

The *heating function* heated transforms the frozen objects of a configuration in active objects, and will be used in the definition of the maximal parallelism computation step.

**Definition 12.** *Let* $\times_{i=1}^{d}(w_i, \bar{w}_i)$ *be a partial configuration of* $\Pi$.
*The* heating function heated : $Conf_\Pi \to Conf_\Pi$ *is defined as follows:*

$$heated(\times_{i=1}^{d}(w_i, \bar{w}_i)) = \times_{i=1}^{d}(w_i \oplus \bar{w}_i, \emptyset).$$

Now we are ready to define the maximal parallelism computational step $\Mapsto$:

**Definition 13.** *The* maximal parallelism computational step $\Mapsto$ *over (nonpartial) configurations of* $\Pi$ *is defined as follows:* $\gamma_1 \Mapsto \gamma_2$ *iff there exists a partial configuration* $\gamma'$ *such that* $\gamma_1 \mapsto^+ \gamma'$, $\gamma' \not\mapsto$ *and* $\gamma_2 = heated(\gamma')$.

An operational semantics for P systems with maximal parallelism semantics has been defined for P systems in [1, 2, 10]. The main difference w.r.t. our approach is concerned with the fact that, while in this section a maximal parallelism computational step is defined as a maximal sequence of reactions, in [1, 2, 10] no notion of reaction is provided, and the notion equivalent to the maximal parallelism computational step is defined directly by SOS rules [25]. A detailed comparison of the two approaches is beyond the scope of the present paper and deserves further investigation.

---

[2] As $\rho \in \{1, \ldots, d\}^{|v|}$, this implies that $father(k)$ is defined.
[3] This implies that $children(k)$ is not empty.

# 5    A Causal Semantics for P Systems

In this section we provide a causal semantics for cooperative P systems. To define a causal semantics, we follow the approach used in [18] for CCS, and in [3] for the $\pi$-calculus.

## 5.1    An Informal Explanation

The idea consists in decorating the reaction relation with two pieces of information:

- a fresh name $k$, that is associated to the reaction and it is taken from the set of causes $\mathcal{K}$;
- a set $H \subseteq \mathcal{K}$, containing all the names associated to the already occurred reactions, that represent a cause for the current reaction.

To keep track of the names of the already occurred reactions that may represent a cause for the reactions that may happen in the future, we introduce a notion of causal configuration that associates to each object an information on its causal dependencies. As in [3], for the sake of clarity we only keep track of the so called immediate causes, as the set of general causes can be reconstructed by transitive closure of the immediate causal relation. We will provide more explanation on this point with an example in the following part of the paper.

Now we start with an informal introduction of causality in P systems. Consider the following system with a unique membrane:

$$\Pi_1 = (\{a, b, c, d, e, f\}, [_1 \ ]_1, ae, \{a \to bc, c \to d, e \to f\}, 1).$$

If we consider the reaction relation $\mapsto$ defined in the previous section, we have that the system $\Pi_1$ can perform either a reaction obtained by the application of the rule $a \to bc$ followed by a reaction obtained by the application of rule $e \to f$, or a sequence of two reactions where the application the rule $e \to f$ is followed by the application of $a \to bc$. The applications of the two rules are independent, in the sense that all the objects consumed by both the rules are already present in the initial configuration. Hence, the two rules can be applied in the same maximal parallelism step, and no one of the rules is causally dependent on the other one.

Consider now the system

$$\Pi_2 = (\{a, b, c, d, e, f\}, [_1 \ ]_1, a, \{a \to bc, c \to d, e \to f\}, 1),$$

obtained from $\Pi_1$ by removing object $e$ from the initial state. In this case, only rule $a \to bc$ can be applied. After the application of such a rule, an instance of object $c$ is created by the application of rule $a \to bc$. Now, a further reduction step can be performed, consisting in applying rule $c \to d$. However, the applications of the two rules $a \to bc$ and $c \to d$ cannot be swapped, and the two rules cannot be applied in the same maximal parallelism computational step. This is because the object $c$ consumed by rule $c \to d$ has been produced by rule $a \to bc$. In this case, we say that the reduction step consisting in the application of rule $c \to d$ *causally depends* on the reduction step consisting in the application of rule $a \to bc$.

If we consider again system $\Pi_1$, we have that, after the application of the two rules $a \to bc$ and $e \to f$, the rule $c \to e$ can be applied, and it is caused by the application of rule $a \to bc$.

We would like to note that the causal semantics is in some sense "finer" than the maximal parallelism step semantics, as it permits to identify exactly which rule(s) represent a cause for the execution of another rule. Consider, e.g., the system

$$\Pi_3 = (\{a, b, c, d, e, f\}, [_1 \ ]_1, ae, \{a \to bc, cf \to d, e \to f\}, 1).$$

According to the maximal parallelism semantics, the two systems cannot be distinguished, as both can perform a maximal parallelism step containing two rules (i.e., $\{a \to bc, e \to f\}$), followed by a maximal parallelism step containing a single rule (resp. $\{c \to d\}$ for $\Pi_1$ and $\{cf \to d\}$ for $\Pi_3$). On the other hand, if we consider the causal semantics, we have that the application of rule $c \to d$ in $\Pi_1$ causally depends only on one of the two rules applied in the previous maximal parallelism step, i.e., $a \to bc$, whereas the application fo the rule $cf \to d$ in $\Pi_3$ causally depends on both the rules applied in the previous maximal parallelism step.

## 5.2    The Formal Definition of Causal Semantics

In this section we provide a formal definition of the notions introduced in the previous section.

Let $\mathcal{K}$ be a denumerable set of cause names, disjoint from the set $V$ of objects.

Throughout this section, we let $\Pi = (V, \mu, w_1^0, \ldots, w_d^0, R_1, \ldots, R_d, i_0)$ be a P system.

To be able to define the set of causes of a reaction, we proceed in the following way: we associate a fresh (i.e., never used before) cause name to each reaction performed in the system. Then, each instance of object in a configuration of the system is decorated with the causal name of the reaction that produced it, or with $\emptyset$ if the object is already present in the initial configuration.[4] To keep track of such causal information, we introduce the notion of *causal configuration* fo a system.

**Definition 14.** *A causal configuration of $\Pi$ is a tuple $z_1, \ldots, z_d$, where $z_i \in (V \times \mathcal{P}(\mathcal{K}))^*$ for $i = 1, \ldots, d$.*
*We use $\times_{i=1}^d z_i$ to denote the causal configuration above.*
*The set of causal configurations of $\Pi$ is denoted by $CConf_\Pi$.*

---

[4] For homogeneity with other classes of P systems, actually we decorate each object with a – possibly empty – set of cause names, even if, in the class of P systems considered in this paper, a single cause name is sufficient.

*We use $\gamma, \gamma', \gamma_1, \ldots$ to range over $CConf_\Pi$.[5]*

*Let $w_i^0 = o_{i,1} o_{i,2} \ldots o_{i,n_i}$ for $i = 1, \ldots, d$. The* initial causal configuration *of $\Pi$ is the configuration $\times_{i=1}^d (o_{i,1}, \emptyset)(o_{i,2}, \emptyset) \ldots (o_{i,n_i}, \emptyset)$ .*

For example, $((a, \emptyset)(e, \emptyset))$ represents the initial causal configuration of the P system $\Pi_1$ in the previous subsection, and $((b, k_1)(c, k_1)(e, \emptyset))$ represents another configuration of $\Pi_1$, reached after the firing of rule $a \to bc$ (for the sake of clarity we omit the surrounding braces if the set of causes is a singleton).

Now we are ready to define the causal semantics for P systems. We write $\gamma \xrightarrow{h;H} \gamma'$ to denote the fact that system $\Pi$ in configuration $\gamma$ performs an action – to which we associate the cause name $h$ – that is caused by the (previously occurred) actions whose action names form the set $H$. The cause name $h$ is a fresh name: this means that it has not been used yet in the current computation.

The execution of an evolution rule is formalized by the notion of causal reaction relation.

Before providing the definition of causal reaction relation, we need some auxiliary definitions.

**Definition 15.** *The function $drop : (V \times \mathcal{P}(\mathcal{K}))^* \to V^*$ removes the causality information:*

$$drop(\varepsilon) = \varepsilon,$$
$$drop((o, H)w) = o \, drop(w).$$

*The function $drop$ is extended to configurations in the obvious way:*

$$drop(\times_{i=1}^d z_i) = \times_{i=1}^d drop(z_i).$$

*The function $causes : V \times \mathcal{P}(\mathcal{K}))^* \to \mathcal{P}(\mathcal{K})$ produces the set of causal labels in a string:*

$$causes(\varepsilon) = \emptyset,$$
$$causes((o, H)w) = H \cup causes(w).$$

*The function $deco : V^* \to V \times \mathcal{P}(\mathcal{K}))^*$ decorates each object in a string with a given set of causal labels:*

$$deco(\varepsilon, H) = \emptyset,$$
$$deco(ow, H) = (o, H)deco(w, H).$$

**Definition 16.** *The* causal reaction relation *$\xrightarrow{h;H}$ over $CConf_\Pi \times CConf_\Pi$ is defined as follows:*

*$\times_{i=1}^d z_i \xrightarrow{h;H} \times_{i=1}^d z_i'$ iff there exist $k$, with $1 \leq k \leq d$, a string $w \in (V \times \mathcal{P}(\mathcal{K}))^*$, an evolution rule $u \to v \in R_k$ and a migration string $\rho \in \{1, \ldots, d\}^{|v|}$ such that*

- *$u = drop(w)$,*
- *$H = causes(w)$,*

---

[5] With abuse of notation, we use $\gamma, \gamma', \gamma_1, \ldots$ to denote both partial configurations and causal configurations. It will be clear from the context to which kind of configuration we are referring to.

- $w \subseteq z_k$,
- $z'_k = z_k \setminus w \oplus deco(v, \{h\})$,
- $\forall j : 1 \leq j \leq |v|$ the following holds:
  - if $\pi_2((v)_j) = here$ then $(\rho)_j = k$,
  - if $\pi_2((v)_j) = out$ then $(\rho)_j = father(k)^6$,
  - if $\pi_2((v)_j) = in$ then $(\rho)_j \in children(k)$,[7]
- $\forall i, 1 \leq i \leq d$ and $i \neq k$: $z'_i = z_i \oplus \bigoplus_{1 \leq j \leq |v|, (\rho)_j = i}((v)_j, h)$,

## 5.3   Properties of the Causal Semantics

The causal semantics for the class of P systems considered in this paper enjoys some nice properties.

The first property is the *retrievability* of the maximal parallelism step semantics from the causal semantics. According to such a property, there is no loss of information when moving from the maximal parallelism to the causal semantics, as we can reconstruct the maximal parallelism semantics of a system by looking at its causal execution:

**Theorem 1.** $\times_{i=1}^d (w_i, \emptyset) \mapsto \times_{i=1}^d (w'_i, \emptyset)$ *is a maximal parallelism computational step if and only if there exist* $\gamma, \gamma' \in CConf(\Pi)$, $h_1, \ldots, h_n$, $H_1, \ldots H_n$ *such that*

- $drop(\gamma) = \times_{i=1}^d (w_i, \emptyset)$,
- $drop(\gamma') = \times_{i=1}^d (w'_i, \emptyset)$,
- $\gamma \xrightarrow{h_1; H_1} \cdots \xrightarrow{h_n; H_n} \gamma'$
- $h_i \notin H_j$ *for all* $i, j: 1 \leq i, j \leq n$,
- *if there exist* $h, H$ *such that* $\gamma' \xrightarrow{h; H}$ *then there exists* $i$ *such that* $1 \leq i \leq n$ *and* $h_i \in H$.

The other property is the so-called diamond property, stating that if two non-causally related actions can happen one after the other, then they can happen also in the other order, and at the end they reach the same system.

**Theorem 2.** *If* $\gamma \xrightarrow{h_1; H_1}_{r_1} \gamma' \xrightarrow{h_2; H_2}_{r_2} \gamma''$ *and* $h_1 \notin H_2$, *then there exists a causal configuration* $\gamma'''$ *such that* $\gamma \xrightarrow{h_2; H_2}_{r_2} \gamma''' \xrightarrow{h_1; H_1}_{r_1} \gamma''$.

# 6   Conclusion

In this paper we tackled the problem of defining a causal semantics for a basic class of P systems. We think that the study of the causal dependencies that arise between the actions performed by a system is of primary importance for models inspired by the biology, because of its possible application to the analysis of complex biological pathways.

---

[6] As $\rho \in \{1, \ldots, d\}^{|v|}$, this implies that $father(k)$ is defined.
[7] This implies that $children(k)$ is not empty.

This paper represents a first step in this direction, but a lot of work remains to be done. For example, if we move to other classes of membrane systems, such as, e.g., P systems with promoters and inhibitors, we have to deal with more involved causal relations among reactions, and it could happen that some of the properties enjoyed by the causal semantics for basic P systems presented in this work no longer hold. Another interesting research topic is the investigation of the causal semantics for classes of P systems whose membrane structure is dynamically evolving (e.g., we can consider dissolution rules, duplication, gemmation or either brane-like operations). Once we have completed the definition of a causal semantics for systems with an evolving structure, we will start investigating the causal dependencies arising in biological pathways involving membranes, such as, e.g., the LDL Cholesterol Degradation Pathway [19], that was modeled in P systems in [6].

# References

1. Andrei, O., Ciobanu, G., Lucanu, D.: Operational semantics and rewriting logic in membrane computing. In: Proceedings SOS Workshop, ENTCS (2005)
2. Andrei, O., Ciobanu, G., Lucanu, D.: A rewriting logic framework for operational semantics of membrane systems. Theoretical Computer Science 373(3), 163–181 (2007)
3. Boreale, M., Sangiorgi, D.: A fully abstract semantics for causality in the $\pi$-calculus. Acta Inf., 35(5): 353–400, 1998. An extended abstract appeared in Mayr, E.W., Puech, C. (eds.) STACS 1995. LNCS, vol. 900, pp. 243–254. Springer, Heidelberg (1995)
4. Busi, N.: Using well-structured transition systems to decide divergence for catalytic P systems. Theoretical Computer Science 372(2-3), 125–135 (2007)
5. Busi, N., Gorrieri, R.: A Petri net semantics for $\pi$- calculus. In: Lee, I., Smolka, S.A. (eds.) CONCUR 1995. LNCS, vol. 962, pp. 145–159. Springer, Heidelberg (1995)
6. Busi, N., Zandron, C.: Modeling and analysis of biological processes by mem(brane) calculi and systems. In: WSC 2006. Proceedings of the Winter Simulation Conference, ACM, New York (2006)
7. Busi, N.: Towards a causal semantics for brane calculi. In: Proc. Fifth Brainstorming Week on Membrane Computing, Sevilla, pp. 97–112 (2007)
8. Calude, C.S., Păun, G.: Computing with Cells and Atoms. Taylor & Francis, London (2001)
9. Cardelli, L.: Brane calculi - Interactions of biological membranes. In: Danos, V., Schachter, V. (eds.) CMSB 2004. LNCS (LNBI), vol. 3082, Springer, Heidelberg (2005)
10. Ciobanu, G., Andrei, O., Lucanu, D.: Structural operational semantics of P systems. In: Freund, R., Păun, G., Rozenberg, G., Salomaa, A. (eds.) WMC 2005. LNCS, vol. 3850, Springer, Heidelberg (2006)
11. Darondeau, P., Degano, P.: Causal trees. In: Ronchi Della Rocca, S., Ausiello, G., Dezani-Ciancaglini, M. (eds.) ICALP 1989. LNCS, vol. 372, pp. 234–248. Springer, Heidelberg (1989)
12. Dassow, J., Păun, G.: On the power of membrane computing. J. Univ. Comput. Sci. 5(2) (1999)

13. Degano, P., De Nicola, R., Montanari, U.: Partial ordering descriptions and observations of nondeterministic concurrent processes. In: de Bakker, J.W., de Roever, W.-P., Rozenberg, G. (eds.) Linear Time, Branching Time and Partial Order in Logics and Models for Concurrency. LNCS, vol. 354, pp. 438–466. Springer, Heidelberg (1989)

14. Degano, P., Priami, C.: Causality for mobile processes. In: Fülöp, Z., Gecseg, F. (eds.) Automata, Languages and Programming. LNCS, vol. 944, pp. 660–671. Springer, Heidelberg (1995)

15. Degano, P., Priami, C.: Non interleaving semantics for mobile processes. Theoretical Computer Science 216(1-2), 237–270 (1999)

16. Danos, V., Pradalier, S.: Projective brane calculus. In: Danos, V., Schachter, V. (eds.) CMSB 2004. LNCS (LNBI), vol. 3082, Springer, Heidelberg (2005)

17. Guerriero, M.L., Priami, C.: Causality and concurrency in beta-binders. TR-01-2006 The Microsoft Research - University of Trento Centre for Computational and Systems Biology (2006)

18. Kiehn, A.: Proof Systems for cause based equivalences. In: Borzyszkowski, A.M., Sokolowski, S. (eds.) MFCS 1993. LNCS, vol. 711, Springer, Heidelberg (1993)

19. Lodish, H., Berk, A., Matsudaira, P., Kaiser, C.A., Krieger, M., Scott, M.P., Zipursky, S.L., Darnell, J.: Molecular Cell Biology, 4th edn. W.H. Freeman and Company (1999)

20. Milner, R.: Communication and Concurrency. Prentice-Hall, Englewood Cliffs (1989)

21. Milner, R., Parrow, J., Walker, D.: A calculus of mobile processes. Information and Computation 100, 1–77 (1992)

22. Păun, G.: Computing with membranes: an Introduction. Bull. EATCS 67 (1999)

23. Păun, G.: Computing with membranes. Journal of Computer and System Sciences 61(1), 108–143 (2000)

24. Păun, G.: Membrane Computing. An Introduction. Springer, Heidelberg (2002)

25. Plotkin, G.D.: Structural operational semantics. Journal of Logic and Algebraic Programming 60, 17–139 (2004)

26. Priami, C., Quaglia, P.: Beta binders for biological interactions. In: Danos, V., Schachter, V. (eds.) CMSB 2004. LNCS (LNBI), vol. 3082, pp. 20–33. Springer, Heidelberg (2005)

27. Priami, C., Quaglia, P.: Operational patterns in beta-binders. T. Comp. Sys. Biology 1, 50–65 (2005)

# Simulating the Bitonic Sort Using P Systems

Rodica Ceterchi[1], Mario J. Pérez-Jiménez[2], and Alexandru Ioan Tomescu[1]

[1] Faculty of Mathematics and Computer Science, University of Bucharest
Str. Academiei 14, RO-010014, Bucharest, Romania
[2] Research Group on Natural Computing
Department of Computer Science and Artificial Intelligence
University of Sevilla
Avda. Reina Mercedes s/n, 41012 Sevilla, Spain
rceterchi@gmail.com, mario.perez@cs.us.es, alexandru.tomescu@gmail.com

**Abstract.** This paper gives a version of the parallel bitonic sorting algorithm of Batcher, which can sort $N$ elements in time $O(\log^2 N)$. We apply it to the 2D mesh architecture, using the shuffled row-major indexing function. A correctness proof of the proposed algorithm is given. Two simulations with P systems are introduced and discussed. The first one uses dynamic communication graphs and follows the guidelines of the mesh version of the algorithm. The second simulation requires only symbol rewriting rules in one membrane.

## 1 Introduction

P systems, introduced in [19], are powerful computational models, with non-deterministic as well as parallel features. Deterministic P systems can also be considered, and the power of their parallel features compared against the power of other computational models which enjoy parallelism. Along this line we refer to previous work, which relates P systems with parallel networks of processors, functioning according to the SIMD paradigm (Single Instruction Multiple Data machines), in [8], [9], for shuffle-exchange networks, and in [6] for 2D mesh networks. The comparison was approached by designing P systems which simulate the functioning of a specific architecture, when solving a specific problem. In [7] the general features of this type of approach were abstracted, giving a "blueprint" for the design of a class of deterministic P systems, with *dynamic communication graphs*, which simulate a given parallel architecture, functioning to implement a given algorithm.

Among the choices to be made for the problem to solve, the *static sorting* imposes itself, being a central theme in computer science. Although it is well known that comparison-based sorting (sequential) algorithms require at least $N \log N$ comparisons to sort $N$ items, performing many comparisons in parallel can reduce the sorting time. This paper analyzes the bitonic sorting algorithm, one of the fastest parallel sorting algorithms where the sequence of comparisons is not data-dependent. The bitonic sorting network was discovered by Batcher [3], who also discovered the network for odd-even sort. These were the first

G. Eleftherakis et al. (Eds.): WMC8 2007, LNCS 4860, pp. 172–192, 2007.
© Springer-Verlag Berlin Heidelberg 2007

networks capable of sorting $N$ elements in time $O(\log^2 N)$. Stone [24] maps the bitonic sort onto a perfect-shuffle interconnection network, sorting $N$ elements by using $N$ processors in time $O(\log^2 N)$. Siegel [23] shows that bitonic sort can also be performed on the hypercube in time $O(\log^2 N)$. The shuffled row-major indexing formulation of bitonic sort on a mesh-connected computer is presented by Thompson and Kung [25]. They also show how the odd-even merge sort can be used with snakelike row-major indexing. Nassimi and Sahni [16] present a row-major indexed bitonic sort formulation for a mesh with the same performance as shuffled row-major indexing.

Static sorting algorithms have been developed and proposed also in the P systems area. Among the first approaches, made independently, we mention [2] and [4], [5]. The problem of sorting with P systems occupies Chapter 8, [1], of [10].

We analyze in this paper a version of the bitonic sorting algorithm of Batcher, and its implementation on the 2D mesh architecture. Section 2 introduces the mesh topology and the model of computation. In Subsection 2.2 we present the algorithm, and the main result, Theorem 1, whose Corollary is the correctness proof of the algorithm. Other results in this subsection, like Lemma 2, and the Remarks, are subsequently used to prove assertions about the algorithm, and, in Section 3, about the simulations with P systems.

Section 3 is devoted to the presentation of two different simulations of the algorithm with P systems. The first simulation uses dynamic communication graphs, as in [7]. A generative approach to the sequence of graphs used to communicate values between the membranes is a novel feature. We outline the main features of a second simulation, which uses only one membrane, and symbol rewriting rules.

## 2    Preliminaries: The Bitonic Sort on the 2D-Mesh

### 2.1    Model of Computation and Indexing Function

The presentation of the bitonic sort on the 2D-mesh architecture is made here based mainly on the paper [25]. It is the same algorithm as in [21], but with more emphasis on the routings necessary to compare elements situated at greater distances on the mesh. Also, some restrictions imposed in [25], will be eliminated, or re-examined, since they were dictated by their explicit connection to the ILLIAC IV-type parallel computer. In general, our references to parallel machines/architectures will be at the level of generalization to be found for instance in [21].

Let us assume, as in [25], that we have a parallel computer with $N = n \times n$ identical processors, disposed in a 2D-mesh structure. A processor is connected to all of its four vertical or horizontal neighbors, except for the processors situated on the perimeter, which have at most two or three neighbors, as no "wrap-around connections" are permitted.

Another assumption is that it is a SIMD (Single Instruction Multiple Data) machine. During each time unit, a single instruction is executed by a set of processors. In what follows, only two processor registers and two instructions are needed. For inter-processor data moves, we will use a routing instruction which copies the value of a register to a register of a neighbor processor. The second instruction is the internal comparison between the values of the two registers of a processor.

We define $t_R$ the time for one-unit distance routing step, and $t_C$ the time required for one comparison step. Concurrent data movement is allowed, as long as it is in the same direction. Moreover, any number of parallel comparisons can be made simultaneously.

In order not to make the notation cumbersome, we let the same letter, say $i$, stand for an integer in $\{0, 1, \ldots, n - 1\}$, and for its binary representation as a string. For $n = 2^k$, as the case will be, $i$ will be a binary string of length $k$. Whenever necessary, we complete with zeros (obviously, to the left) to obtain strings of the same length. When we refer to bits of such a string, we count from 1 to $n$, starting from right to left, such that the "first" bit will be that of the least significant digit, and so forth. However, when we write such a string, we will write it with bits numbered from right to left.

We will consider the *Shuffled row-major indexing* function on the processors, illustrated in Figure 1.

**Definition 1.** *The shuffled row-major indexing function sRM is defined as*

$$sRM_k : \{0, 1, \ldots, 2^k - 1\} \times \{0, 1, \ldots, 2^k - 1\} \longrightarrow \{0, 1, \ldots, 2^{2k} - 1\},$$

$$sRM_k(i_1 i_2 \cdots i_k, j_1 j_2 \cdots j_k) = i_1 j_1 i_2 j_2 \cdots i_k j_k.$$

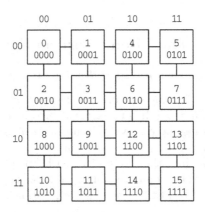

**Fig. 1.** Shuffled row-major indexing scheme for a $4 \times 4$ mesh

Assuming that initially $N$ integers are loaded in the $N$ processors, the sorting problem is defined as moving the $j$th smallest element to the processor indexed by $j$, for all $j \in \{0, 1, \ldots N - 1\}$.

Let $sRM^{-1}$ be the inverse of $sRM$, and let us denote as below the projection of the first and second argument of $sRM^{-1}$.

$$sRM_{row}^{-1}, sRM_{col}^{-1} : \{0, 1, \ldots, 2^{2k} - 1\} \rightarrow \{0, 1, \ldots 2^k - 1\},$$
$$sRM_{row}^{-1}(i_1 j_1 i_2 j_2 \cdots i_k j_k) = i_1 i_2 \cdots i_k,$$
$$sRM_{col}^{-1}(i_1 j_1 i_2 j_2 \cdots i_k j_k) = j_1 j_2 \cdots j_k.$$

## 2.2 The Bitonic Sorting Algorithm

A bitonic sequence is a concatenation of two monotonic sequences, one ascending and the other one descending, or a sequence such that a cyclic shift of its elements would put them in such a form.

In Batcher's bitonic sorting network [3] of order $n$, the input is a bitonic sequence $a$ of $n/2$ increasing elements followed by $n/2$ decreasing elements. These two sequences are merged by first applying $n/2$ comparators to $a_0$ and $a_{n/2}$, $a_1$ and $a_{(n/2)+1}$, ..., $a_{n/2}$ and $a_{n-1}$. This first-phase partitions the elements into two bitonic sequences of $n/2$ smaller elements and of $n/2$ larger elements. These two bitonic sequences are further sorted by applying two bitonic merging networks of size $n/2$ to each sequence. A bitonic sorting network for 16 elements appears in Figure 2.

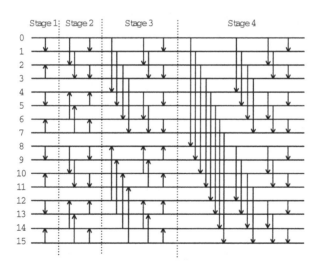

**Fig. 2.** A bitonic sorting network of size 16

The following lemma is due to K. Batcher [3].

**Lemma 1.** *Given a bitonic sequence* $\langle a_0, a_1, \ldots, a_{2n-1} \rangle$ *the following hold:*

1. $d = \langle \min\{a_0, a_n\}, \min\{a_1, a_{n+1}\}, \ldots, \min\{a_{n-1}, a_{2n-1}\} \rangle$ *is bitonic.*
2. $e = \langle \max\{a_0, a_n\}, \max\{a_1, a_{n+1}\}, \ldots, \max\{a_{n-1}, a_{2n-1}\} \rangle$ *is bitonic.*

*3.* $\max(d) < \min(e)$ *(i.e., every element of the sequence d is smaller that any element of sequence e).*

By an abuse of notations, we will refer to a sequence of processors as the sequence of integers stored in one designated register $A$ of the processors at a certain moment. Similarly, we will use $\min / \max\{P_i, P_j\}$ meaning $\min / \max\{P_i[A], P_j[A]\}$ and refer to such operations as a comparison and interchange of values between processors $P_i$ and $P_j$.

We will give a generic algorithm for Batcher's bitonic sorter on an array $\langle P_0, \ldots, P_{2^{2k}-1} \rangle$ of processors, independent of the indexing function used. The algorithm (as illustrated in Figure 2 for $k = 2$) will consist of $2k$ stages, numbered from 1 to $2k$. After each Stage $i$, $(1 \leq i \leq 2k)$, the sequence $\langle P_{2^i j}, \ldots, P_{2^i j + 2^i - 1} \rangle$ with $0 \leq j \leq 2^{2k-i} - 1$ will be an ascending sequence for all $j$ even, and a descending sequence, for all $j$ odd.

> **Input.** an array $\langle P_0, \ldots, P_{2^{2k}-1} \rangle$ of processors
> **Output.** the sequence $\langle P_0, \ldots, P_{2^{2k}-1} \rangle$ is ascending
>
> **Stage($i$)**
> |   **for** $t \leftarrow i$ **downto** 1 **do**
> |       // compare processors with indices differing on bit $t$
> |       **forall** $j \leftarrow 0$ **to** $2^{2k-t} - 1$ **in parallel do**
> |         **if** $2^t j$ div $2^i$ is even **then** order = ascending
> |         **else** order = descending
> |         **Merge**$(2^t j, 2^t j + 2^t - 1, order)$
> **end**
>
> **Bitonic-Sort**
> |   **for** $i \leftarrow 1$ **to** $2k$ **do**
> |      **Stage**($i$)
> **end**

**Algorithm 1.** Bitonic sort on an array of $2^{2k}$ processors

Given a bitonic sequence of processors $\langle P_0, P_1, \ldots, P_{2n-1} \rangle$, by **Merge**$(0, 2n - 1, ascending)$ we mean an operation which yields the sequence:

$$\langle \min\{P_0, P_n\}, \min\{P_1, P_{n+1}\}, \ldots, \min\{P_{n-1}, P_{2n-1}\},$$
$$\max\{P_0, P_n\}, \max\{P_1, P_{n+1}\}, \ldots, \max\{P_{n-1}, P_{2n-1}\} \rangle.$$

Analogously, a call to **Merge**$(0, 2n - 1, descending)$ produces

$$\langle \max\{P_0, P_n\}, \max\{P_1, P_{n+1}\}, \ldots, \max\{P_{n-1}, P_{2n-1}\},$$
$$\min\{P_0, P_n\}, \min\{P_1, P_{n+1}\}, \ldots, \min\{P_{n-1}, P_{2n-1}\} \rangle.$$

**Theorem 1.** *After each Stage $i$, the sequence $\langle P_{2^i j}, \ldots, P_{2^i j + 2^i - 1} \rangle, 0 \le j \le 2^{k-i} - 1$ will be an ascending sequence for all $j$ even, and a descending sequence, for all $j$ odd.*

*Proof.* We shall reason by induction on $i$. For the base case $i = 1$ it is immediate that the statement holds. Now let the statement be true for $i$ and show that is it also true for $i + 1$.

First, $t = i + 1$ and $0 \le j \le 2^{k-i-1} - 1$. The sequence $S$ for the $i + 1$ case can be written as

$$S = \langle P_{2^{i+1} j}, \ldots, P_{2^{i+1} j + 2^i - 1} \rangle =$$

$$\langle P_{2^i 2j}, \ldots, P_{2^i 2j + 2^i - 1}, P_{2^i (2j+1)}, \ldots, P_{2^i (2j+1) + 2^i - 1} \rangle.$$

From the induction hypothesis, we have that the sub-sequence

$$S_1 = \langle P_{2^i 2j}, \ldots, P_{2^i 2j + 2^i - 1} \rangle$$

is ascending as $2j$ is even for any $j$, and that

$$S_2 = \langle P_{2^i (2j+1)}, \ldots, P_{2^i (2j+1) + 2^i - 1} \rangle$$

is descending as $2j + 1$ is odd for any $j$. Therefore, the whole sequence $S$ is bitonic.

At this point we apply the **Merge** operation on $S$, and get $S' = S_1' S_2'$. By Lemma 1 we have that $S_1'$ and $S_2'$ are both bitonic. Moreover, when doing an ascending merge, $\max(S_1') < \min(S_2')$ and when doing a descending merge, $\min(S_1') > \max(S_2')$. This ensures that the two sequences are relatively ordered and can be sorted independently in parallel.

For $1 \le t < i + 1$ the **Merge** operations are the same as in a merging network. We note that for all $2^{i+1} j \le l < 2^{i+1}(j + 1)$, $l$ div $2^{i+1} = j$ and therefore all subsequent **Merge** operations for $t < i + 1$ on these processors will have the same *order* as when $t = i + 1$.                                       □

**Corollary 1.** *Given a sequence $\langle P_0, \ldots, P_{2^{2k} - 1} \rangle$ of processors, Algorithm 1 is correct.*

**Lemma 2.** *Given a $2^k \times 2^k$ 2D-mesh indexed with the function sRM and using Algorithm 1, for any two processors $x = 2^t j + l$ and $y = 2^t j + l + 2^{t-1}$, with $0 \le l \le 2^{t-1} - 1$, $1 \le t \le i$, and $0 \le j \le 2^{k-t} - 1$, which compare and interchange values inside a call of the form **Merge** $(2^t j, 2^t j + 2^t - 1, order)$, the following hold:*

    *(i) the binary representations of $x$ and $y$ differ only on bit $t$;*

    *(ii) if $t$ is even then $x$ and $y$ reside on the same vertical line of the mesh; if $t$ is odd they are on the same horizontal line;*

   *(iii) the distance on the mesh between $x$ and $y$ is $2^{\lceil t/2 \rceil - 1}$;*

   *(iv) all processors situated on the same line between $x$ and $y$ are involved in the same **Merge** operation (i.e., have indices between $2^t j$ and $2^t j + 2^t - 1$).*

*Proof.* (*i*) Since $x = 2^t j + l$ and $l \leq 2^{t-1} - 1$, we have that $l$ contributes to bits 1 to $t - 1$ and that $2^t j$ contributes to bits $t + 1$ to $2k$. Therefore bit $t$ of $x$ is 0. Similarly, since $y = x + 2^{t-1}$, bit $t$ of $y$ is 1, and all other bits are the same as those of $x$.

(*ii*) Let $i_1$, and $j_1$ be the row and column indices of $x$, $sRM_{row}^{-1}(x)$ and $sRM_{col}^{-1}(x)$, respectively. Analogously, let $i_2$ and $j_2$ be the row and column indices of $y$. From *i*) we have that $x$ and $y$ differ on bit $t$, and hence the following two cases hold: $t$ is even and $i_1 \neq i_2$, $j_1 = j_2$, or $t$ is odd and $i_1 = i_2$, $j_1 \neq j_2$. In the first case $x$ and $y$ are on the same column, and in the latter, they are on the same row.

(*iii*) Using the notations above, let us assume that $t$ is even and $i_1 \neq i_2$, $j_1 = j_2$. If $x$ and $y$ differ on bit $t$, then $i_1$ and $i_2$ will differ on bit $t/2$, and therefore $|i_1 - i_2| = 2^{t/2-1}$. From *ii*) $x$ and $y$ are on the same line of the mesh and the distance between them is $|i_1 - i_2| = 2^{t/2-1}$. Similarly, when $t$ is odd and $i_1 = i_2$, $j_1 \neq j_2$, we have that $j_1$ and $j_2$ differ on bit $\lceil t/2 \rceil$. As before, the distance between $x$ and $y$ is $2^{\lceil t/2 \rceil - 1}$.

(*iv*) Consider again the case $t$ even and $i_1 \neq i_2$, $j_1 = j_2$. We have to show that for all numbers $i$ with $i_1 \leq i \leq i_2$, we have $2^t j \leq sRM(i, j_1) \leq 2^t j + 2^t - 1$. But since $i_1 \leq i \leq i_2$, form the definition of $sRM$, we have that $x \leq sRM(i, j_1) \leq y$, which concludes our proof as $2^t j \leq x$ and $y \leq 2^t j + 2^t - 1$. Analogously for $t$ odd. $\square$

## 2.3    Applying the Bitonic Sorting Algorithm to the 2D-Mesh

Thompson and Kung [25], and Orcutt [18] showed that Batcher's bitonic sorting algorithm can be applied to sorting on a mesh-connected parallel computer, once the indexing function is chosen. In [25] it is noted that a necessary condition for optimality is that a comparison-interchange on the $j$th bit be no more expensive than the $(j+1)$th bit, for all $j$. From (*iii*) of Lemma 2 we have that the "shuffled row-major" indexing scheme satisfies such condition, and leads to a complexity of $(14(n - 1) - 8 \log n)t_R + (2 \log^2 n + \log n)t_C$ according to [25]. The algorithm for a $4 \times 4$ mesh with the shuffle row-major indexing is illustrated below and in Figure 3, where by "well ordered" we refer to the corresponding comparison directions from Figure 2.

**Stage 1.** Bitonic sort on pairs of adjacent $1 \times 1$ matrices by the comparison interchange indicated, result: "well ordered" $1 \times 2$ matrices. Time: $2t_R + t_C$.
**Stage 2.** Bitonic sort on $1 \times 2$ matrices, result: $2 \times 2$ matrices. Time: $4t_R + 2t_C$.
**Stage 3.** Bitonic sort on $2 \times 2$ matrices, result: $2 \times 4$ matrices. Time: $8t_R + 3t_C$.
**Stage 4.** Bitonic sort on the two $2 \times 4$ matrices. Time: $12t_R + 4t_C$.

At each stage of Algorithm 1, we have a comparison and interchange of values between two processors. We have seen in Lemma 2 that using the $sRM$ indexing function, these two processors will sit on the same vertical or horizontal line of the mesh. When they are not directly connected, they have to route their values through neighbor processors, residing on the shortest path between them

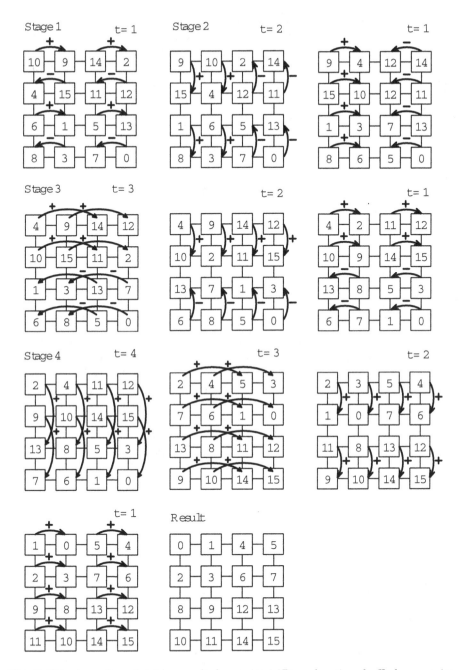

**Fig. 3.** Bitonic sorting algorithm applied on a $4 \times 4$ 2D-mesh, using shuffled row-major indexing

(i.e., the line of the mesh on which they are placed). At each Stage $i$, we have a comparison and interchange between processors whose indices differ only on bit $t$, with $1 \leq t \leq i$. Keeping in mind the parallel structure of our machine, the merging operation becomes a merging of square or rectangular portions of the mesh. Therefore, using the $sRM$ indexing, the **Merge** operation defined previously becomes a **Merge** operation on sub-arrays of processors situated on the same line of the mesh. We denote such operation **compare-interchange**.

For a better understanding of the way a call **Merge** $(2^t j, 2^t j + 2^t - 1, order)$ $(1 \leq i \leq 2k, 1 \leq t \leq i, 0 \leq j \leq 2^{2k-t} - 1)$ is translated to the $2^k \times 2^k$ mesh topology, we make the following observations:

*Remark 1.* The portion of the mesh has dimensions $2^{t-\lceil t/2 \rceil} \times 2^{\lceil t/2 \rceil}$ (i.e., $2^{t-\lceil t/2 \rceil}$ rows and $2^{\lceil t/2 \rceil}$ columns). This is true since $2^t$ processors are involved in the **Merge** and from Lemma 2 the maximal length of the sub-arrays involved in the **Merge** situated on the same line is $2 \cdot 2^{\lceil t/2 \rceil - 1}$.

*Remark 2.* For $t$ even, we have a merging of square portions of the mesh of size $2^{t/2} \times 2^{t/2}$ and the compare interchange operations are done between processors residing on the same column of the mesh, For $t$ odd we have a merging of rectangular portions of the mesh of size $2^{\lceil t/2 \rceil - 1} \times 2^{\lceil t/2 \rceil}$, and the compare interchange operation are done between processors residing on the same row of the mesh.

Let us see what are the necessary routings for the case for $t = 1$ (the processors are directly connected since $2^{\lceil t/2 \rceil - 1} = 1$). Consider a call of the form **Merge**$(x, x + 1, order)$. Let processors $P_x$ and $P_{x+1}$ have the two registers denoted by $A$ and $B$. Then the first instruction performed is a routing from $P_x[A]$ to $P_{x+1}[B]$. Next, perform a comparison operation in processor $P_{x+1}$, and store the minimum/maximum in register $B$. Finally, route back to $P_x$ the value of the $B$ register of $P_{x+1}$, with a total time is $2t_R + t_C$. The pseudo-code is written below, where by **compare**$(P_{x+1}, ascending/descending)$ we understand an internal comparison in processor $P_{x+1}$, which places the minimal/maximal value in register $B$.

**Input.** index $x$ and sorting order *order*
**Output.** the sequence $\langle P_x, P_{x+1} \rangle$ is ordered w.r.t. *order*

**route**$(P_x[A], P_{x+1}[B])$
**compare**$(P_{x+1}, order)$
**route**$(P_{x+1}[B], P_x[A])$

**Algorithm 2.** Compare-interchange operation for adjacent processors

The case when we have to merge an array $a$ of $2^i$ processors situated on the same line of the mesh, indexed from 0 to $2^i - 1$, and such that $P_{a[j]}$ is neighbor with $P_{a[j+1]}$ for all $0 \leq j < 2^i - 1$ is specified in Algorithm 3. The basic idea is that we have to shift the values of the first half of the array in the $B$ registers of the second half, perform a comparison operation in parallel in these processors, and then shift back the minimal/maximal values. Hence a total time of $2^i t_R + t_C$.

**Input.** array $a$ of indices, integer $i$, and sorting order *order*
**Output.** the sequence $\langle P_{a[0]}, P_{a[1]}, \ldots, P_{a[2^i-1]} \rangle$ is ordered w.r.t. *order*

compare-interchange$(a, i, order)$
    forall $j \leftarrow 0$ to $2^{i-1} - 1$ in parallel do
        route$(P_{a[j]}[A], P_{a[j+1]}[B])$ // route right one unit
    for $k \leftarrow 1$ to $2^{i-1} - 1$ do
        forall $j \leftarrow 0$ to $2^{i-1} - 1$ in parallel do
            route$(P_{a[j+k]}[B], P_{a[j+1+k]}[B])$ // shift the values to the second
            half of the array

    forall $j \leftarrow 2^{i-1}$ to $2^i - 1$ in parallel do
        compare$(P_{a[j]}, order)$ // compare internally

    for $k \leftarrow 2^{i-1} - 1$ downto 1 do
        forall $j \leftarrow 0$ to $2^{i-1} - 1$ in parallel do
            route$(P_{a[j+k+1]}[B], P_{a[j+k]}[B])$ // shift back the results

    forall $j \leftarrow 0$ to $2^{i-1} - 1$ in parallel do
        route$(P_{a[j+1]}[B], P_{a[j]}[A])$ // final routing back in the $A$ registers
end

**Algorithm 3.** Compare-interchange operation for an array of neighbor processors situated on the same line of the mesh

In order to give a version of the bitonic sorting algorithm on a 2D-mesh of $2^{2k}$ processors, we make the following conventions and remarks.

The right and down neighbors of $r$ (defined whenever possible) are:

$$right(r) = sRM(sRM_{row}^{-1}(r), sRM_{col}^{-1}(r) + 1)$$
$$down(r) = sRM(sRM_{row}^{-1}(r) + 1, sRM_{col}^{-1}(r))$$

For a specification independent of the parity of bit $t$, denote (whenever possible):

$$next_t(r) = \begin{cases} right(r), & \text{if } t \text{ odd;} \\ down(r), & \text{if } t \text{ even.} \end{cases}$$

Let $next_t^{(c)}(r)$ be the $c$-th iterate of the function $next_t$, i.e., $(next_t \circ \cdots \circ next_t)(r)$ $c$ times.

Algorithm 4 represents the bitonic sort on a 2D-mesh of processors. The line **Merge**$(2^t j, 2^t j + 2^t - 1, order)$ of Algorithm 1 has been replaced by its corresponding operations on the 2D-mesh. We need a **compare-interchange** subroutine for every line of the sub-matrix of the mesh involved in the **Merge**. From Remarks 1-2 we know that the dimensions of the sub-matrix are $2^{\lceil t/2 \rceil} \times 2^{t-\lceil t/2 \rceil}$. Hence the maximal length of a sequence of processors situated on the same line which compare and interchange values in a **Merge** operation is $2^{\lceil t/2 \rceil}$. Moreover, the indices of the first processors on every line $l$ (i.e., the smallest indices on every line $l$) in a $\mathbf{Merge}(2^t j, 2^t j + 2^t - 1, order)$ are $sRM(sRM_{row}^{-1}(2^t j) + l, sRM_{col}^{-1}(2^t j))$, with $0 \leq l \leq 2^{t - \lceil t/2 \rceil} - 1$.

**Input.** an array $\langle P_0, \ldots, P_{2^{2k} - 1} \rangle$ of processors
**Output.** the sequence $\langle P_0, \ldots, P_{2^{2k} - 1} \rangle$ is ascending

**Stage**$(i)$

    **for** $t \leftarrow i$ **downto** 1 **do**
        // compare processors with indices differing on bit $t$
        **forall** $j \leftarrow 0$ **to** $2^{2k - t} - 1$ **in parallel do**
            **if** $2^t j$ **div** $2^i$ **is even then** order $=$ ascending
            **else** order $=$ descending
            // for every line in the Merge operation
            **forall** $l \leftarrow 0$ **to** $2^{t - \lceil t/2 \rceil} - 1$ **in parallel do**
                // build array $a$ containing processor indices on line $l$
                **for** $c \leftarrow 0$ **to** $2^{\lceil t/2 \rceil} - 1$ **do**
                    $a[c] \leftarrow next_t^{(c)}(sRM(sRM_{row}^{-1}(2^t j) + l, sRM_{col}^{-1}(2^t j)))$
                **compare-interchange**$(a, \lceil t/2 \rceil, order)$

**end**

**Bitonic-Sort**
    **for** $i \leftarrow 1$ **to** $2k$ **do**
        **Stage**$(i)$
**end**

**Algorithm 4.** Bitonic sort on a 2D-mesh of $2^{2k}$ processors

# 3   Modeling with Membranes

Given the embedded parallel structure of a P system, modeling a 2D-mesh is a natural and straightforward approach. In what follows, we present two such systems.

## 3.1   A P System with Dynamic Communication of 2D-Mesh Type

The first P system we introduce is along the same general lines as the model proposed in [6]. For each of the processors $P_i$, $i \in \{0, 1, \ldots, 2^{2k} - 1\}$ we have an associated membrane, which we label $i$. The two registers $A$ and $B$ of each processor are coded by two different symbols, say $a$ for the $A$ register and $b$ for the $B$ register. The number of occurrences of $a$ represents the value of the $A$ register, and analogously for $b$. Similarly to tissue-like P systems, we will have a collection of elementary membranes, connected by certain graphs, at certain moments of their evolution in time. The graphs we consider will be sub-graphs of the total graph of the 2D-mesh network, also sub-graphs of the identity graph of the 2D-mesh network.

In a slightly different manner from [8] or [9], we refer to the communication graph associated to a given architecture with the following conventions: the vertices of the graph are the processors, and the edges are the network connections characteristic of the architecture.

In the case of the $2^k \times 2^k$ 2D-mesh with the $sRM$ indexing function, let $G_{total}$ be the underlying communication graph consisting of all edges necessary to the architecture. We introduce the following notation for the set of vertices of $G_{total}$:

$$V(G_{total}) = \{0, 1, \ldots, 2^{2k} - 1\}.$$

Hence, the set of edges is

$$E(G_{total}) = \{(sRM(i,j), sRM(i,j+1),) \mid 0 \leq i \leq 2^k - 1, 0 \leq j \leq 2^k - 2\}$$
$$\cup \{(sRM(i,j), sRM(i+1,j)) \mid 0 \leq i \leq 2^k - 2, 0 \leq j \leq 2^k - 1\}.$$

Note that at a certain step of the sorting algorithm not all edges are involved in communication. Therefore we will call *active sub-graphs* of $G_{total}$ those graphs containing only such edges. We also introduce the *identity* graph, with

$$V(Id) = \{0, 1, \ldots, 2^{2k} - 1\},$$
$$E(Id) = \{(sRM(i,j), sRM(i,j)) \mid 0 \leq i \leq 2^k - 1, 0 \leq j \leq 2^k - 1\},$$

for modeling internal processing steps.

As in [6], the P system which we will consider in the sequel, departs from the classical P systems in two respects:

- The connections between individual membranes of a P system, $\mu$, which was a tree-like structure of membranes (see [19]), and which in tissue-like P systems becomes a graph structure, is now, a *sequence of graphs*.
- The rules of a P system, usually associated to *membranes*, will now be associated to *communication graphs* between membranes.

  (a) We simulate the internal computations performed by a subset of processors by the action of symbol or object rewriting rules, at work simultaneously inside the corresponding subset of membranes. We will associate such rules to the corresponding active subsets of $Id$.
  (b) We simulate the exchange of data performed by the processors with communication rules (symport/antiport rules) between membranes. The communication rules will be associated to the active sub-graphs of $G_{total}$. Auxiliary rewriting is necessary before and after the application of a communication rule.

In order to describe the evolution of a P system which simulates the behavior of the bitonic sorting algorithm in the 2D-mesh architecture, we use pairs of the type $[graph, rules]$. We have $graph$ a sub-graph of $G_{total}$ or $Id$ and $rules$ a mapping from the set of all edges of $graph$, $E(graph)$, to the set of all symbol/object rewriting rules for routing or comparison operations.

The general presentation of such a P system is of the form

$$\Pi = (V = \{a, b, a^*, b^*\}, \langle [a^{x_0}]_0, \ldots, [a^{x_{2^{2k}-1}}]_{2^{2k}-1} \rangle, R_\mu),$$

where the membrane indices are $\{0, 1, \ldots, 2^{2k} - 1\}$. Each integer $x_i$, with $i \in \{0, 1, \ldots, 2^{2k} - 1\}$, is codified as the number of occurrences of symbol $a$ inside membrane $i$. Finally, $R_\mu$ is the finite sequence of pairs $[graph, rules]$ which simulates Algorithm 4. We will see in the sequel that $R_\mu$ is generated algorithmically, by concatenating sequences of pairs $[graph, rules]$[1]. The additional symbols $a^*$ and $b^*$ are necessary for the auxiliary rewriting needed by the communication (routing) operations.

In order to give such a sequence, we have to closely follow Algorithm 4. The routing operations in the **compare-interchange** sub-routine will be replaced by their corresponding ones in terms of the P system formalism described above. Consider two adjacent processors $P_x$ and $P_y$ which interchange values. The three possible routing operations are: **route**$(P_x[A], P_y[B])$, **route**$(P_x[B], P_y[B])$, **route**$(P_x[B], P_y[A])$. The implementation with rewriting and communication rules of the first operation follows the lines: rewrite $a \to a^*$ into membrane $x$, apply the communication rule $(a^*, out)$ along the edge $(x, y)$, transporting all the $a^*$ symbols from membrane $x$ into $y$, and then, in membrane $y$, rewrite $a^*$ back to the desired symbol, in this case $a^* \to b$. Below, we give a specification of a sequence $[graph, rules]$ accomplishing this routing operation.

$$[Id_1, rules_1] \cdot [G, rules] \cdot [Id_2, rules_2], \text{ such that} \qquad (\text{rAB})$$
$$Id_1 \subseteq Id, (x, x) \in E(Id_1), rules_1((x, x)) = \{a \to a^*\},$$
$$G \subseteq G_{total}, (x, y) \in E(G), rules((x, y)) = \{(a^*, out)\},$$
$$Id_2 \subseteq Id, (y, y) \in E(Id_2), rules_2((y, y)) = \{a^* \to b\}.$$

Similarly, an operation **route**$(P_x[B], P_y[A])$ is specified as:

$$[Id_1, rules_1] \cdot [G, rules] \cdot [Id_2, rules_2], \text{ such that} \qquad (\text{rBA})$$
$$Id_1 \subseteq Id, (x, x) \in E(Id_1), rules_1((x, x)) = \{b \to b^*\},$$
$$G \subseteq G_{total}, (x, y) \in E(G), rules((x, y)) = \{(b^*, out)\},$$
$$Id_2 \subseteq Id, (y, y) \in E(Id_2), rules_2((y, y)) = \{b^* \to a\}.$$

In the case of a **route**$(P_x[B], P_y[B])$, only one communication graph is needed. The reason for not having supplementary rewriting is that such routings are done in parallel. The value from $P_x[B]$ is routed to $P_y[B]$ in parallel with the routing of $P_y[B]$ to a $B$ register of a neighbor processors. Hence the number of symbols $b$ in membrane $y$ is the desired one $P_x[B]$.

$$[G, rules], \text{ such that} \qquad (\text{rBB})$$
$$G \subseteq G_{total}, (x, y) \in E(G), rules((x, y)) = \{(b, out)\}.$$

---

[1] We denote the empty sequence by $\lambda$, and the concatenation of two sequences by ".".

**Input.** array $a$ of membrane indices, integer $i$, and sorting order *order*
**Output.** the sequence of integers stored in the sequence of membranes
with indices $\langle a[0], a[1], \ldots, a[2^i - 1] \rangle$ is ordered w.r.t. *order*

**compare-interchange-membr**$(a, i, order)$

    **forall** $j \leftarrow 0$ **to** $2^{i-1} - 1$ **in parallel do**
        // route right one unit in the $B$ registers - (rAB) rule
        $E(Id_1^t) \leftarrow E(Id_1^t) \cup \{(j, j)\}$; $rules_{0,1}^t((j, j)) \leftarrow \{a \rightarrow a^*\}$
        $E(G_0^t) \leftarrow E(G_0^t) \cup \{(j, j+1)\}$; $rules_0^t((j, j+1)) \leftarrow \{(a^*, out)\}$
        $E(Id_2^t) \leftarrow E(Id_2^t) \cup \{(j+1, j+1)\}$;
        $rules_{0,2}^t((j+1, j+1)) \leftarrow \{a^* \rightarrow b\}$

    **for** $k \leftarrow 1$ **to** $2^{i-1} - 1$ **do**
        // shift the values to the second half of the array - (rBB) rule
        **forall** $j \leftarrow 0$ **to** $2^{i-1} - 1$ **in parallel do**
            $E(G_k^t) \leftarrow E(G_k^t) \cup \{(j+k, j+1+k)\}$
            $rules_k^t((j+k, j+1+k)) \leftarrow \{(b, out)\}$

    **forall** $j \leftarrow 2^{i-1}$ **to** $2^i - 1$ **in parallel do**
        // compare internally - (C) rule
        $E(Id_C^t) \leftarrow E(Id_C^t) \cup \{(j, j)\}$
        **if** *order is ascending* **then**
            $rules_C^t((j, j)) \leftarrow \{ab \rightarrow ab, a \rightarrow b, b \rightarrow b\}$
        **else**
            $rules_C^t((j, j)) \leftarrow \{ab \rightarrow ab, a \rightarrow a, b \rightarrow a\}$

    **for** $k \leftarrow 2^{i-1} - 1$ **downto** $1$ **do**
        // shift back the results - (rBB) rule
        **forall** $j \leftarrow 0$ **to** $2^{i-1} - 1$ **in parallel do**
            $E(G_{2^i-k}^t) \leftarrow E(G_{2^i-k}^t) \cup \{(j+k+1, j+k)\}$
            $rules_{2^i-k}^t((j+1+k, j+k)) \leftarrow \{(b, out)\}$

    **forall** $j \leftarrow 0$ **to** $2^{i-1} - 1$ **in parallel do**
        // final routing back in the $A$ registers - (rBA) rule
        $E(Id_{2^i,1}^t) \leftarrow E(Id_{2^i,1}^t) \cup \{(j+1, j+1)\}$;
        $rules_{2^i,1}^t((j+1, j+1)) \leftarrow \{b \rightarrow b^*\}$
        $E(G_{2^i}^t) \leftarrow E(G_{2^i}^t) \cup \{(j+1, j)\}$; $rules_{2^i}^t((j+1, j)) \leftarrow \{(b^*, out)\}$
        $E(Id_{2^i,2}^t) \leftarrow E(Id_{2^i,2}^t) \cup \{(j, j)\}$; $rules_{2^i,2}^t((j, j)) \leftarrow \{b^* \rightarrow a\}$

**end**

**Algorithm 5.** Compare-interchange operation for an array of neighbor membranes associated to processors situated on the same line of the mesh

**Input.** an array $\langle [a^{x_0}]_0, \ldots, [a^{x_{2^{2k}-1}}]_{2^{2k}-1}\rangle$ of membranes containing
integers $\{x_i \mid 0 \le i \le 2^{2k} - 1\}$ codified as appearances of letter $a$
**Output.** the sequences $R_\mu^i$ of pairs $[graph, rules]$ simulating Stage $i$, and
finally the sequence $R_\mu$ simulating bitonic sort

**Sim-Stage**($i$)
$\quad$ $R_\mu^i \leftarrow \lambda$
$\quad$ **for** $t \leftarrow i$ **downto** 1 **do**
$\quad\quad$ **forall** $s \in \{0, \ldots, 2^{\lceil t/2 \rceil} - 1\}$ **do**
$\quad\quad\quad$ $G_s^t \leftarrow \lambda;\ rules_s^t \leftarrow \emptyset$
$\quad\quad$ // compare processors with indices differing on bit $t$
$\quad\quad$ **forall** $j \leftarrow 0$ **to** $2^{2k-t} - 1$ **in parallel do**
$\quad\quad\quad$ **if** $2^t j$ div $2^i$ is even **then** order = ascending
$\quad\quad\quad$ **else** order = descending
$\quad\quad\quad$ // for every line in the Merge operation
$\quad\quad\quad$ **forall** $l \leftarrow 0$ **to** $2^{t-\lceil t/2 \rceil} - 1$ **in parallel do**
$\quad\quad\quad\quad$ // build array $a$ containing processor indices on line $l$
$\quad\quad\quad\quad$ **for** $c \leftarrow 0$ **to** $2^{\lceil t/2 \rceil} - 1$ **do**
$\quad\quad\quad\quad\quad$ $a[c] \leftarrow next_t^{(c)}(sRM(sRM_{row}^{-1}(2^t j) + l, sRM_{col}^{-1}(2^t j)))$
$\quad\quad\quad\quad$ **compare-interchange-membr**($a, \lceil t/2 \rceil, order$)
$\quad\quad$ // route to the second half
$\quad\quad$ $R_\mu^i \leftarrow R_\mu^i \cdot [Id_{0,1}^t, rules_{0,1}^t][G_0^t, rules_0^t][Id_{0,2}^t, rules_{0,2}^t]$
$\quad\quad$ **for** $s \leftarrow 1$ **to** $2^{\lceil t/2 \rceil - 1} - 1$ **do**
$\quad\quad\quad$ $R_\mu^i \leftarrow R_\mu^i \cdot [G_s^t, rules_s^t]$
$\quad\quad$ // compare internally
$\quad\quad$ $R_\mu^i \leftarrow R_\mu^i \cdot [Id_C^t, rules_C^t]$
$\quad\quad$ //route back to the first half
$\quad\quad$ **for** $s \leftarrow 2^{i-1} + 1$ **to** $2^i - 1$ **do**
$\quad\quad\quad$ $R_\mu^i \leftarrow R_\mu^i \cdot [G_s^t, rules_s^t]$
$\quad\quad$ $R_\mu^i \leftarrow R_\mu^i \cdot [Id_{2^i,1}^t, rules_{2^i,1}^t][G_{2^i}^t, rules_{2^i}^t][Id_{2^i,2}^t, rules_{2^i,2}^t]$
**end**
**Sim-Bitonic-Sort**
$\quad$ $R_\mu = \lambda$
$\quad$ **for** $i \leftarrow 1$ **to** $2k$ **do**
$\quad\quad$ **Sim-Stage**($i$)
$\quad\quad$ $R_\mu = R_\mu \cdot R_\mu^i$
**end**

**Algorithm 6.** Generating the sequence of pairs $R_\mu = R_\mu^1 \cdot R_\mu^2 \cdot \ldots \cdot R_\mu^{2k}$ simulating the bitonic sorting algorithm on the $2^k \times 2^k$ 2D mesh

Consider now an internal comparison operation in processor $P_x$, **compare**($P_x$, order) which places **max**($P_x[A], P_x[B]$) in register $B$ if the order is ascending, or in register $A$ if the order is descending. This can be formalized as:

$$[Id_C, rules], \quad \text{such that } Id_C \subseteq Id, \; (x,x) \in E(Id_C), \qquad (C)$$

$$rules_C((x,x)) = \begin{cases} \{ab \to ab, a \to b, b \to b\}, & \text{if order is ascending}, \\ \{ab \to ab, a \to a, b \to a\}, & \text{if order is descending}. \end{cases}$$

For every bit $t$, we have a sequence $G_s^t$ (sub-graphs of $G_{total}$) of communication graphs which simulate all parallel routing operations, illustrated by rules (rAB), (rBB), (rBA), (C), where $0 \le s \le 2^{\lceil t/2 \rceil - 1} - 1$. From the above considerations and the steps of Algorithm 3, we introduce the sub-routine **compare-interchange-membr**$(a, i, order)$ in Algorithm 5.

After the communication graphs and their corresponding rules have been generated, all there remains to be done is concatenate the pairs $[graph, rules]$. This is done in Algorithm 6.

**Theorem 2.** *The P system $\Pi$ with $R_\mu = R_\mu^1 \cdot R_\mu^2 \cdot \ldots \cdot R_\mu^{2k}$ generated by Algorithm 6 sorts in ascending order integers codified in membranes as numbers of occurrences of letter $a$.*

*Proof.* This is a consequence of Theorem 1 and its corollary and the way the simulation was conceived. □

## 3.2 Bitonic Sorting in One Membrane

We propose here a simulation of the bitonic sorting, which uses only one membrane. We will use (cooperative) symbol rewriting rules. The cooperation will be minimal, i.e., of degree two, since we follow closely the algorithm, and thus the whole process is based on comparators.

Consider an alphabet with $2^{2k}$ symbols, $V = \{v_0, v_1, \cdots v_{2^{2k}-1}\}$. We will call it the *primary alphabet*.

We will also consider *auxiliary* alphabets, which we will specify in the sequel, in order to achieve sorting by rewriting.

We want to sort in ascending order the sequence of *distinct* integers

$$\langle x_0, x_1, \cdots x_{2^{2k}-1} \rangle,$$

codified over $V$ as the multiset

$$w = v_0^{x_0} v_1^{x_1} \cdots v_{2^{2k}-1}^{x_{2^{2k}-1}}.$$

We want to design a P system which, by rewriting acting in a maximal parallel manner and competing for objects, produces, from the initial configuration $w$, the configuration

$$w_f = v_0^{\sigma(x_0)} v_1^{\sigma(x_1)} \cdots v_{2^{2k}-1}^{\sigma(x_{2^{2k}-1})},$$

where $\sigma$ is the permutation which yields the total order, i.e., such that $\sigma(x_0) < \sigma(x_1) < \cdots < \sigma(x_{2^{2k}-1})$.

Consider the alphabet $V$ as ordered by the natural order given by the indices, and let $v = v_0 v_1 \cdots v_{2^{2k}-1}$ be the *alphabet word* (see [1]), i.e., the word obtained by concatenating the letters of $V$ in their natural order. We call *extended alphabet words* over $V$, all words in $V^*$ in which all the letters appear in their natural order. Note that both $w$ and $w_f$, the initial and the final configuration of our P system, are extended alphabet words. Actually, all the intermediate configurations over $V$ will be of this type.

Let $M_j(u)$ denote the multiplicity of letter $v_j$ in a word $u \in V^*$. Then

$$w = v_0{}^{x_0} v_1{}^{x_1} \cdots v_{2^{2k}-1}{}^{x_{2^{2k}-1}} = v_0{}^{M_0(w)} \cdots v_{2^{2k}-1}{}^{M_{2^{2k}-1}(w)}.$$

Consider first the case $n = 2$ $(k = 0)$. We have 2 integers codified over $\{v_0, v_1\}$ as an extended alphabet word. Consider the auxiliary alphabets

- $\{a, b\}$, for writing sources of a comparator,
- $\{c^+, d^+\}$, for writing targets of a $\oplus$-comparator,
- $\{c^-, d^-\}$, for writing targets of a $\ominus$-comparator.

Consider the rules:

$$C_\oplus = \{v_0 \to a, v_1 \to b\} \cup \{ab \to c^+d^+, a \to d^+, b \to d^+\} \cup \{c^+ \to v_0, d^+ \to v_1\}.$$

The first group rewrites all $v_0$s to $a$s and $v_1$s to $b$s, the second group performs the comparison and produces the ascending order, and the last group rewrites back into the original alphabet. We have the sequence of configurations

$$v_0{}^{x_0} v_1{}^{x_1} \to a^{x_0} b^{x_1} \to c^{+min(x_0,x_1)} d^{+max(x_0,x_1)} \to v_0{}^{min(x_0,x_1)} v_1{}^{max(x_0,x_1)}.$$

Similarly, the rules:

$$C_\ominus = \{v_0 \to a, v_1 \to b\} \cup \{ab \to c^-d^-, a \to c^-, b \to c^-\} \cup \{c^- \to v_0, d^- \to v_1\},$$

achieve a descending comparator, generating the sequence of configurations

$$v_0{}^{x_0} v_1{}^{x_1} \to a^{x_0} b^{x_1} \to c^{-max(x_0,x_1)} d^{-min(x_0,x_1)} \to v_0{}^{max(x_0,x_1)} v_1{}^{min(x_0,x_1)}.$$

**Lemma 3.** *On a two-letter alphabet, starting from an initial configuration $w = v_0{}^{x_0} v_1{}^{x_1}$, by applying rules in $C_\oplus$ we obtain $w_f$ such that $(M_i(w_f))_i$ is ascending, and by applying rules in $C_\ominus$ we obtain $w_f$ such that $(M_i(w_f))_i$ is descending.*

Note that rules $C_\oplus$ simulate a **Merge**$(0, 1, +)$, and $C_\ominus$ a **Merge**$(0, 1, -)$.

We now want to simulate a whole family of merge operations done in parallel.

We take 2 auxiliary alphabets, $S^+$ and $S^-$ to codify sources of $+$ or $-$ comparators, and another pair, $T^+$ and $T^-$, to codify outputs (targets) of $+$ or $-$ comparators. We label them in a bijective correspondence with $V$.

$$S^+ = \{s_0{}^+, \cdots s_{2^{2k}-1}^+\},$$

$$T^+ = \{t_0{}^+, \cdots t_{2^{2k}-1}^+\},$$

and similarly for $-$. (For the time being, only 4 copies of the initial alphabet. We will probably need 4 different copies for every stage, in order to keep them independent.)

At Stage (1) we have to simulate $\mathbf{Merge}(2j, 2j + 1, order)$, for all $0 \leq j \leq 2^{2k-1} - 1$, where $order = +$ for all $j$ even, and $order = -$ for all $j$ odd.

This is equivalent to:

- Rewrite all symbols of $V$ into start symbols for appropriate comparators, using the sets of rules

$$\{v_{2j} \to s_{2j}{}^+, v_{2j+1} \to s_{2j+1}{}^+ \mid 0 \leq j \leq 2^{2k-1} - 1 \text{ , j even}\}$$
$$\cup \{v_{2j} \to s_{2j}{}^-, v_{2j+1} \to s_{2j+1}{}^- \mid 0 \leq j \leq 2^{2k-1} - 1 \text{ , j odd}\}.$$

- Apply in parallel the rewriting of symbols which correspond to the simulations of the comparators:

$$\{s_{2j}{}^+ s_{2j+1}{}^+ \to t_{2j}{}^+ t_{2j+1}{}^+, s_{2j}{}^+ \to t_{2j+1}{}^+, s_{2j+1}{}^+ \to t_{2j+1}{}^+ \mid$$
$$0 \leq j \leq 2^{2k-1} - 1 \text{ , j even}\}$$
$$\cup \{s_{2j}{}^- s_{2j+1}{}^- \to t_{2j}{}^- t_{2j+1}{}^-, s_{2j}{}^- \to t_{2j}{}^-, s_{2j+1}{}^- \to t_{2j}{}^- \mid$$
$$0 \leq j \leq 2^{2k-1} - 1 \text{ , j odd}\}.$$

- Rewrite back all symbols of $T$'s into $V$.

$$\{v_{2j} \leftarrow t_{2j}{}^+, v_{2j+1} \leftarrow t_{2j+1}{}^+ \mid 0 \leq j \leq 2^{2k-1} - 1 \text{ , j even}\}$$
$$\cup \{v_{2j} \leftarrow t_{2j}{}^-, v_{2j+1} \leftarrow t_{2j+1}{}^- \mid 0 \leq j \leq 2^{2k-1} - 1 \text{ , j odd}\}.$$

The general scheme is given in Algorithm 7.

The calls to $\mathbf{Merge}(2^t j, 2^t j + 2^t - 1, order)$ are equivalent to parallel calls to $\mathbf{Merge}(x, y, order)$, where $x$ and $y$ are like in Lemma 2. The same result ensures us that, both the rewriting which feeds the comparators, and the rewriting which implements the comparators can be done in parallel. For $\mathbf{Merge}(x, y, order = -)$, we use

$$\{s_x{}^- s_y{}^- \to t_x{}^- t_y{}^-, s_x{}^- \to t_x{}^-, s_y{}^- \to t_x{}^-\}.$$

We propose the following sets of rules for simulating iteration $t$ at **Sim-Stage**$(i)$:

(WF) Rewriting to $S$'s, with $* = \begin{cases} +, & \text{if } 2^t j \text{ div } 2^i \text{ is even,} \\ -, & \text{if } 2^t j \text{ div } 2^i \text{ is odd,} \end{cases}$

$$\{v_x \to s_x{}^* \in S_t{}^* \mid x \in [2^t j, 2^t j + 2^{t-1}), 0 \leq j \leq 2^{2k-t+1} - 1\}.$$

(C) Rewriting which simulates the comparators, for appropriate pairs of indices:

$$\{s_x{}^+ s_y{}^+ \to t_x{}^+ t_y{}^+, s_x{}^+ \to t_y{}^+, s_y{}^+ \to t_y{}^+ \mid$$
$$x \in [2^t j, 2^t j + 2^{t-1}), \ y = x + 2^{t-1}, \ 0 \leq j \leq 2^{2k-t} - 1\},$$
$$\{s_x{}^- s_y{}^- \to t_y{}^- t_x{}^-, s_x{}^- \to t_x{}^-, s_y{}^- \to t_x{}^- \mid$$
$$x \in [2^t j, 2^t j + 2^{t-1}), \ y = x + 2^{t-1}, \ 0 \leq j \leq 2^{2k-t} - 1\}.$$

(WB) Rewriting from $T$'s:

$$\{v_x \leftarrow t_x{}^* \in T_t{}^* \mid x \in [2^t j, 2^t j + 2^{t-1}), 0 \le j \le 2^{2k-t+1} - 1\}.$$

**Input.** an extended alphabet word $w$ over $V$
**Output.** the extended alphabet word $w_f$ over $V$, such that $\langle M_i(w_f) \rangle_i$ is ascending

**Sim-Stage**($i$)
  for $t \leftarrow i$ **downto** 1 **do**
    Take 4 extra copies of the start and the terminal alphabets, $S_t^+$, $S_t^-$, $T_t^+$, $T_t^+$, different for each value of $t$. For $t$'s smaller than $i$ we can re-use the alphabets of previous stages.
    **forall** $j \leftarrow 0$ **to** $2^{2k-t} - 1$ **in parallel do**
      **if** $2^t j$ **div** $2^i$ **is even then** order = ascending
      **else** order = descending
      // Simulate the calls **Merge**($2^t j, 2^t j + 2^t - 1, order$)
      (WF) Rewrite all symbols in $V$ with the appropriate symbol in $S_t^+ \cup S_t^-$.
      (C) Apply the rewriting which simulates the appropriate comparators.
      (WB) Rewrite back all symbols in $T_t^+ \cup T_t^-$ to symbols of $V$.
**end**

**Sim-Bitonic-Sort**
  for $i \leftarrow 1$ **to** $2k$ **do**
    **Sim-Stage**($i$)
**end**

**Algorithm 7.** Simulating bitonic sort on an alphabet of $2^{2k}$ letters $V$

## 4   Conclusions and Open Problems

We have presented a bitonic sorting algorithm which can be implemented on a 2D mesh of processors. We have not yet found in the literature a formal proof of the correctness of bitonic sorting, an equivalent, or an analogue of our Theorem 1.

Next, we have proposed two simulations with P systems. The first simulation, derived in a "straightforward" manner from the functioning of the algorithm on the mesh, is inspired from work in [6], [8], [9], and [7], where the general framework was abstracted. It introduces a generative approach to the sequence of communication graphs, a novel feature to be explored in subsequent work. Note that, although in our case we have simulated a deterministic algorithm with a deterministic P system, dynamic communication graphs can be used with non-deterministic systems as well.

The second simulation is at the opposite pole: it requires no routings of values at all, just an appropriate codification of the symbols. It is in this area that other versions of the algorithm could be implemented, independent of the topology of a given structure, and the parallel features of the P systems can be compared against those of other computational devices.

# References

1. Alhazov, A., Sburlan, D.: Static Sorting P Systems, ch. 8 in [10]
2. Arulanandham, J.J.: Implementing Bead – Sort with P Systems. In: Calude, C.S., Dinneen, M.J., Peper, F. (eds.) UMC 2002. LNCS, vol. 2509, pp. 115–125. Springer, Heidelberg (2002)
3. Batcher, K.: Sorting Networks and Their Applications. In: Proc. of the AFIPS Spring Joint Computing Conf. 32(2968), pp. 307–314
4. Ceterchi, R., Martín-Vide, C.: Dynamic P Systems. In: Păun, G., Rozenberg, G., Salomaa, A., Zandron, C. (eds.) Membrane Computing. LNCS, vol. 2597, pp. 146–186. Springer, Heidelberg (2003)
5. Ceterchi, R., Martín-Vide, C.: P Systems with Communication for Static Sorting. In: Cavaliere, M., Martín-Vide, C., Păun, G. (eds.) GRLMC Report 26, Rovira i Virgili University, Tarragona (2003)
6. Ceterchi, R., Pérez Jiménez, M.J.: On Two-Dimensional Mesh Networks and Their Simulation with P Systems. In: Mauri, G., Păun, G., Pérez-Jiménez, M.J., Rozenberg, G., Salomaa, A. (eds.) WMC 2004. LNCS, vol. 3365, pp. 259–277. Springer, Heidelberg (2005)
7. Ceterchi, R., Pérez Jiménez, M.J.: On Simulating a Class of Parallel Architectures. International Journal of Foundations of Computer Science 17, 91–110 (2006)
8. Ceterchi, R., Pérez Jiménez, M.J.: Simulating Shuffle – Exchange Networks with P Systems. In: Păun, G., Riscos, A., Sancho, F., Romero, A. (eds.) Proceedings of the Second Brainstorming Week on Membrane Computing. Report RGNC 01/04, pp. 117–129 (2004)
9. Ceterchi, R., Pérez Jiménez, M.J.: A Perfect Shuffle Algorithm for Reduction Processes and its Simulation with P Systems. In: Dzitac, I., Maghiar, T., Popescu, C. (eds.) Proceedings of the International Conference on Computers and Communications ICCC 2004, Editura Univ. Oradea, pp. 92–97 (2004)
10. Ciobanu, G., Păun, G., Pérez Jiménez, M.J. (eds.): Applications of Membrane Computing. Springer, Heidelberg (2006)
11. Corbett, P.F., Scherson, I.D.: Sorting in Mesh Connected Multiprocessors. IEEE Transactions on Parallel and Distributed Systems 3, 626–632 (1992)
12. Dowd, M., Perl, Y., Rudolph, L., Saks, M.: The Periodic Balanced Sorting Network. Journal ACM 36, 738–757 (1989)
13. Han, Y., Igarashi, Y., Truszczynski, M.: Indexing Functions and Time Lower Bounds for Sorting on a Mesh-connected Computer. Discrete Applied Mathematics 36, 141–152 (1992)
14. Knuth, D.E.: The Art of Computer Programming. Sorting and Searching, vol. 3. Addison-Wesley, Reading, Mass (1973)
15. Layer, C., Pfleiderer, H.-J.: A Reconfigurable Recurrent Bitonic Sorting Network for Concurrently Accessible Data. In: Becker, J., Platzner, M., Vernalde, S. (eds.) FPL 2004. LNCS, vol. 3203, pp. 648–657. Springer, Heidelberg (2004)

16. Nassimi, D., Sahni, S.: Bitonic Sort on a Mesh Connected Parallel Computer. IEEE Transactions on Computers C28(1), 2–7 (1979)
17. Nassimi, D., Sahni, S.: An Optimal Routing Algorithm for Mesh-Connected Parallel Computers. Journal ACM 27, 6–29 (1980)
18. Orcutt, S.E.: Computer Organization and Algorithms for Very High Speed Computations. Ph.D. Th., Stanford U., Stanford, Calif., ch. 2, pp. 20–23 (1974)
19. Păun, G.: Computing with Membranes. Journal of Computer and System Sciences 61, 108–143 (2000)
20. Păun, G.: Membrane Computing. An Introduction. Springer, Heidelberg (2002)
21. Quinn, M.J.: Parallel Computing. Theory and Practice. McGraw – Hill Series in Computer Science (1994)
22. Sado, K., Igarashi, Y.: Some Parallel Sorts on a Mesh-connected Processor Array and Their Time Efficiency. Journal of Parallel and Distributed Computing 3, 398–410 (1986)
23. Siegel, H.J.: The Universality of Various Types of SIMD Machine Interconnection Networks. In: Proceedings of the 4th Annual Symposium on Computer Architecture, pp. 23–25 (1977)
24. Stone, H.S.: Parallel Processing with the Perfect Shuffle. IEEE Transactions on Computers C-20, 153–161 (1971)
25. Thompson, C.D., Kung, H.T.: Sorting on a Mesh-connected Parallel Computer. Communications of the ACM 20, 263–271 (1977)
26. The membrane computing web page: http://psystems.disco.unimib.it

# On the Number of Agents in P Colonies

Luděk Cienciala[1], Lucie Ciencialová[1], and Alica Kelemenová[1,2]

[1] Institute of Computer Science, Silesian University in Opava, Czech Republic
[2] Department of Computer Science, Catholic University Ružomberok, Slovakia
{ludek.cienciala,lucie.ciencialova,alica.kelemenova}@fpf.slu.cz

**Abstract.** We continue the investigation of P colonies introduced in [8], a class of abstract computing devices composed of independent membrane agents, acting and evolving in a shared environment.

We decrease the number of agents sufficient to guarantee computational completeness of P colonies with one and with two objects inside each agent, respectively, owing some special restrictions to the type of programs. We characterize the generative power of the partially blind machine by the generative power of special P colonies.

## 1 Introduction

P colonies were introduced in [8] as formal models of a computing device inspired by membrane systems ([10]) and by grammar systems called colonies ([6]). This model intends to structure and functioning of a community of living organisms in a shared environment.

The independent organisms living in a P colony are called agents. Each agent is given by a collection of objects embedded in a membrane. The number of objects inside the agent is the same for each one of them. The environment contains several copies of a basic environmental object denoted by $e$. The number of the copies of $e$ is unlimited.

A set of programs is associated with each agent. The program determines the activity of the agent by rules. In every moment of computation all the objects inside of the agent are being either evolved (by an evolution rule) or transported (by a communication rule). Two such rules can also be combined into checking rule which specifies two possible actions: if the first rule is not applicable then the second one should be applied. So it sets the priority between two rules.

The computation starts in the initial configuration. Using their programs the agents can change their objects and possibly objects in the environment. This gives possibility to affect the behavior of the other agents in next steps of computation. In each step of the computation, each agent with at least one applicable program nondeterministically chooses one of them and executes it. The computation halts when no agent can apply any of its programs. The result of the computation is given by the number of some specific objects present at the environment at the end of the computation.

There are several different ways used how to define the beginning of the computation. (1) At the beginning of computation the environment and all agents

G. Eleftherakis et al. (Eds.): WMC8 2007, LNCS 4860, pp. 193–208, 2007.
© Springer-Verlag Berlin Heidelberg 2007

contain only copies of object $e$. (2) All the agents can contain various objects at the beginning of computation - the agents are in different initial states. The environment contains only copies of object $e$. (3) The initial state of the environment is nonempty (there are some object different from the object $e$) - the environment contains initial "parameters" for future computation, while the agents start with $e$-s.

In [4,7,8] the authors study P colonies with two objects inside the agents. In this case programs consist of two rules, one for each object. If the former of these rules is an evolution and the latter is a communication or checking, we speak about restricted P colonies. If also another combination of the types of the rules is used, we obtain non-restricted P colonies. The restricted P colonies with checking rules are computationally complete [3,4].

In the present paper we study properties of restricted P colonies without checking rules and computational power of P colonies with one object and the minimal number of agents.

We start with definitions in Section 2.

In Section 3 we will deal with P colonies with one object inside each agent. In [1] there was shown that at most seven programs for each agent as well as five agents guarantee the computational completeness of these P colonies. In the preset paper we look for the generative power of P colonies with less than five agents. Two results are achieved in this direction. First, we show, that four agents are enough for computational completeness of P colonies. The second result gives a lower bound for the generative power the P colonies with two agents. Even a restricted variant of these P colonies is at least as powerful as the partially blind register machines.

Restricted P colonies are studied in Section 4. It is known that one agent is sufficient to obtain computational completeness of restricted P systems with checking rules ([4]). For the restricted P colonies that do not use checking rules we will prove that two agents are sufficient to obtain the universal computational power.

## 2    Definitions

Throughout the paper we assume the reader to be familiar with the basics of the formal language theory. For more information on membrane computing, we recommend [11]. We briefly summarize the notation used in the present paper.

We use $NRE$ to denote the family of the recursively enumerable sets of non-negative integers and $N$ to denote the set of non-negative integers.

Let $\Sigma$ be an alphabet. Let $\Sigma^*$ be the set of all words over $\Sigma$ (including the empty word $\varepsilon$). We denote the length of the word $w \in \Sigma^*$ by $|w|$ and the number of occurrences of the symbol $a \in \Sigma$ in $w$ by $|w|_a$.

A multiset of objects $M$ is a pair $M = (V, f)$, where $V$ is an arbitrary (not necessarily finite) set of objects and $f$ is a mapping $f : V \to N$; $f$ assigns to each object in $V$ its multiplicity in $M$. The set of all finite multisets over the finite set $V$ is denoted by $V^\circ$. The support of $M$ is the set $supp(M) = \{a \in V \mid f_M(a) \neq 0\}$.

The cardinality of $M$, denoted by $|M|$, is defined by $|M| = \sum_{a \in V} f_M(a)$. Any finite multiset $M$ over $V$ can be represented as a string $w$ over alphabet $V$ with $|w|_a = f_M(a)$ for all $a \in V$. Obviously, all words obtained from $w$ by permuting the letters can also represent the same $M$, and $\varepsilon$ represents the empty multiset. For multiset $M$ represented by word $w$ we use the notation $_\star w$.

## 2.1  P Colonies

We briefly recall the notion of P colonies introduced in [8]. A P colony consists of agents and environment. Both the agents and the environment contain objects. With every agent a set of programs is associated. There are two types of rules in the programs. The first type, called evolution rules, are of the form $a \rightarrow b$. It means that object $a$ inside of the agent is rewritten (evolved) to the object $b$. The second type of rules, called communication rules, are of the form $c \leftrightarrow d$. When this rule is performed, the object $c$ inside the agent and the object $d$ outside of the agent change their positions, so, after execution of the rule object $d$ appears inside the agent and $c$ is placed outside in the environment.

In [7] the ability of agents was extended by checking rule. Such a rule gives to the agents an opportunity to choose between two possibilities. It has the form $r_1/r_2$. If the checking rule is performed, the rule $r_1$ has higher priority to be executed than the rule $r_2$. It means that the agent checks the possibility to use rule $r_1$. If it can be executed, the agent has to use it. If the rule $r_1$ cannot be applied, the agent uses the rule $r_2$.

**Definition 1.** *A P colony of the capacity $c$ is a construct*
$$\Pi = (A, e, f, _\star v_E, B_1, \ldots, B_n), \text{ where:}$$

- *$A$ is an alphabet whose elements are called objects,*
- *$e$ is the basic object of the colony, $e \in A$,*
- *$f$ is the final object of the colony, $f \in A$,*
- *$_\star v_E$ is a multiset over $A - \{e\}$,*
- *$B_i$, $1 \leq i \leq n$, are agents; each agent is a construct $B_i = (_\star o_i, P_i)$, where*
  - *$_\star o_i$ is a multiset over $A$ which determines the initial state (content) of agent $B_i$ and $|_\star o_i| = c$,*
  - *$P_i = \{p_{i,1}, \ldots, p_{i,k_i}\}$ is a finite set of programs, where each program contains exactly $c$ rules, which are in one of the following forms each:*
    - *$a \rightarrow b$, called an evolution rule,*
    - *$c \leftrightarrow d$, called a communication rule,*
    - *$r_1/r_2$, called a checking rule; $r_1, r_2$ are evolution or communication rules.*

The initial configuration of the P colony is the $(n + 1)$-tuple of multisets of objects present in the P colony at the beginning of the computation, i.e., $(_\star o_1, \ldots, _\star o_n, _\star v_E)$. Formally, a configuration of P colony $\Pi$ is given by $(_\star w_1, \ldots, _\star w_n, _\star w_E)$, where $|_\star w_i| = c$, $1 \leq i \leq n$, $_\star w_i$ represents all the objects placed inside the $i$-th agent and $_\star w_E \in (A - \{e\})^\circ$ represents all the objects in the environment different from the object $e$.

In this paper, the parallel model of P colonies will be studied. At each step of a parallel computation, each agent which can use one of its programs should use one. If the number of applicable programs is higher than one, the agent nondeterministically chooses one of them.

Let the programs of each $P_i$ be labeled in a one-to-one manner by labels in a set $lab\,(P_i)$ and $lab\,(P_i) \cap lab\,(P_j) = \emptyset$ for $i \neq j$, $1 \leq i,j \leq n$.

To express derivation step formally we introduce the following four functions. For a rule $r$ being $a \to b$, $c \leftrightarrow d$, and $c \leftrightarrow d/c' \leftrightarrow d'$, respectively, and for multiset $_\star w \in A^\circ$ we define:

$$left\,(a \to b, _\star w) = _\star a \qquad\qquad left\,(c \leftrightarrow d, _\star w) = _\star \varepsilon$$
$$right\,(a \to b, _\star w) = _\star b \qquad\qquad right\,(c \leftrightarrow d, _\star w) = _\star \varepsilon$$
$$export\,(a \to b, _\star w) = _\star \varepsilon \qquad\qquad export\,(c \leftrightarrow d, _\star w) = _\star c$$
$$import\,(a \to b, _\star w) = _\star \varepsilon \qquad\qquad import\,(c \leftrightarrow d, _\star w) = _\star d$$

$$left\,(c \leftrightarrow d/c' \leftrightarrow d', _\star w) = _\star \varepsilon$$
$$right\,(c \leftrightarrow d/c' \leftrightarrow d', _\star w) = _\star \varepsilon$$
$$\left.\begin{array}{l} export\,(c \leftrightarrow d/c' \leftrightarrow d', _\star w) = _\star c \\ import\,(c \leftrightarrow d/c' \leftrightarrow d', _\star w) = _\star d \end{array}\right\} \text{ for } |_\star w|_d \geq 1$$
$$\left.\begin{array}{l} export\,(c \leftrightarrow d/c' \leftrightarrow d', _\star w) = _\star c' \\ import\,(c \leftrightarrow d/c' \leftrightarrow d', _\star w) = _\star d' \end{array}\right\} \text{for } |_\star w|_d = 0 \text{ and } |_\star w|_{d'} \geq 1$$

For a program $p$ and any $\alpha \in \{left, right, export, import\}$, let
$$\alpha\,(p, _\star w) = \cup_{r \in p} \alpha\,(r, _\star w).$$
A transition from a configuration to another one is denoted as
$$(_\star w_1, \ldots, _\star w_n, _\star w_E) \Rightarrow (_\star w_1', \ldots, _\star w_n', _\star w_E'),$$ where the following
conditions are satisfied:

- There is a set of program labels $P$ with $|P| \leq n$ such that
  - $p, p' \in P$, $p \neq p'$, $p \in lab\,(P_j)$, $p' \in lab\,(P_i)$, $i \neq j$,
  - for each $p \in P$, $p \in lab\,(P_j)$, $left\,(p, _\star w_E) \cup export\,(p, _\star w_E) = _\star w_j$, and $\bigcup_{p \in P} import\,(p, _\star w_E) \subseteq _\star w_E$.
- Furthermore, the chosen set $P$ is maximal, that is, if any other program $r \in \bigcup_{1 \leq i \leq n} lab\,(P_i)$, $r \notin P$, is added to $P$, then the conditions above are not satisfied.

Finally, for each $j$, $1 \leq j \leq n$, for which there exists a $p \in P$ with $p \in lab\,(P_j)$, let $w_j' = right\,(p, _\star w_E) \cup import\,(p, _\star w_E)$. If there is no $p \in P$ with $p \in lab\,(P_j)$ for some $j$, $1 \leq j \leq n$, then let $_\star w_j' = _\star w_j$ and moreover, let
$$_\star w_E' = _\star w_E - \bigcup_{p \in P} import\,(p, _\star w_E) \cup \bigcup_{p \in P} export\,(p, _\star w_E).$$
Union and "$-$" here are multiset operations.

A configuration is halting if the set of program labels $P$ satisfying the conditions above cannot be chosen to be other than the empty set. A set of all possible halting configurations is denoted by $H$. With a halting computation we can associate a result of the computation, given by the number of copies of the special symbol $f$ present in the environment. The set of numbers computed by a P colony $\Pi$ is defined as

$$N\left(\Pi\right) = \left\{ |_{*}w_{E}|_{f} \mid (_{*}o_{1}, \ldots, {}_{*}o_{n}, {}_{*}v_{E}) \Rightarrow^{*} (_{*}w_{1}, \ldots, {}_{*}w_{n}, {}_{*}w_{E}) \in H \right\},$$

where $(_{*}o_{1}, \ldots, {}_{*}o_{n}, {}_{*}v_{E})$ is the initial configuration, $(_{*}w_{1}, \ldots, {}_{*}w_{n}, {}_{*}w_{E})$ is a halting configuration, and $\Rightarrow^{*}$ denotes the reflexive and transitive closure of $\Rightarrow$.

Given a P colony $\Pi = (A, e, f, {}_{*}v_{E}, B_{1}, \ldots, B_{n})$ the maximal number of programs associated with the agents in P colony $\Pi$ is called the *height* of P colony $\Pi$. The *degree* of P colony $\Pi$ is the number of agents in P colony $\Pi$. The third parameter characterizing a P colony is the *capacity* of P colony $\Pi$, describing the number of the objects inside each of the agents.

Let us use the following notations: $NPCOL_{par}(c, n, h)$ is the family of all sets of numbers computed by P colonies working in parallel, using no checking rules, and with the capacity at most $c$, the degree at most $n$, and the height at most $h$. If the checking rules are allowed, the family of all sets of numbers computed by P colonies is denoted by $NPCOL_{par}K$. If the P colonies are restricted, we use notation $NPCOL_{par}R$ and $NPCOL_{par}KR$, respectively.

## 2.2   Register Machines

In this paper we characterize the size of the families $NPCOL_{par}(c, n, h)$ comparing them with the recursively enumerable sets of numbers. To achieve this aim we use the notion of a register machine.

**Definition 2.** [9] *A register machine is a construct $M = (m, H, l_{0}, l_{h}, P)$ where $m$ is the number of registers, $H$ is the set of instruction labels, $l_{0}$ is the start label, $l_{h}$ is the final label, $P$ is a finite set of instructions injectively labeled with the elements from the set $H$.*

The instructions of the register machine are of the following forms:

$l_{1} : (ADD(r), l_{2}, l_{3})$   Add 1 to the content of the register $r$ and proceed to the instruction (labeled with) $l_{2}$ or $l_{3}$.

$l_{1} : (SUB(r), l_{2}, l_{3})$   If the register $r$ stores a value different from zero, then subtract 1 from its content and go to instruction $l_{2}$, otherwise proceed to instruction $l_{3}$.

$l_{h} : HALT$   Halt the machine. The final label $l_{h}$ is only assigned to this instruction.

Without loss of generality, one can assume that in each $ADD$-instruction $l_{1} : (ADD(r), l_{2}, l_{3})$ and in each $SUB$-instruction $l_{1} : (SUB(r), l_{2}, l_{3})$ the labels $l_{1}, l_{2}, l_{3}$ are mutually distinct.

The register machine $M$ computes a set $N(M)$ of numbers in the following way: it starts with all registers empty (hence storing the number zero) with the instruction labeled $l_{0}$ and it proceeds to apply the instructions as indicated by the labels (and made possible by the contents of registers). If it reaches the halt instruction, then the number stored at that time in the register 1 is said to be computed by $M$ and hence it is introduced in $N(M)$. (Because of the nondeterminism in choosing the continuation of the computation in the case of

$ADD$-instructions, $N(M)$ can be an infinite set.) It is known (see, e.g., [9]) that in this way we compute all Turing computable sets.

Moreover, we call a register machine partially blind [5], if we interpret a subtract instruction $l_1 : (SUB(r); l_2; l_3)$ in the following way: if the value of register $r$ is different from zero, then subtract one from its contents and go to instruction $l_2$ or to instruction $l_3$; if in register $r$ is stored zero, then the program ends without yielding a result.

When the partially blind register machine reaches the final state, the result obtained in the first register is taken into account if the remaining registers store value zero. The family of sets of non-negative integers generated by partially blind register machines is denoted by $NRM_{pb}$. The partially blind register machines accept a proper subset of $NRE$.

## 3    P Colonies with One Object Inside the Agent

In this section we analyze the behavior of P colonies with only one object inside each agent. Each program in this case is formed by only one rule, either an evolution or a communication.

If all the agents have their programs with evolution rules, the agents "live only for themselves" and do not communicate with the environment.

In [1] the following results were proved:

$NPCOL_{par}K(1, *, 7) = NRE,$
$NPCOL_{par}K(1, 5, *) = NRE.$

The number of agents in the second result can be decreased. This is demonstrated by the following theorem.

**Theorem 1.** $NPCOL_{par}K(1, 4, *) = NRE.$

*Proof.* We construct a P colony simulating the computation of a register machine. Because there are only copies of $e$ in the environment and inside the agents in the initial configuration, we will initialize a computation by generating the initial label $l_0$. After generating the symbol $l_0$ this agent stops and it can start its activity only by using a program with a communicating rule.

Two agents will cooperate in order to simulate the ADD and SUB instructions.

Let us consider a register machine $M = (m, H, l_0, l_h, P)$. We can represent the content $m_i$ of the register $i$ by $m_i$ copies of the specific object $a_i$ in the environment. We construct the P colony $\Pi = (A, e, f, {}_*\varepsilon, B_1, \ldots, B_4)$ with:

- alphabet $A = \{l, l'|l \in H\}$
  $\cup \{E_i, E_i', F_i, F_i', F_i'' \mid \text{ for each } l_i \in H\}$
  $\cup \{a_i \mid 1 \leq i \leq m\} \cup \{e, d, m, C\},$
- $f = a_1,$
- $B_i = ({}_*e, P_i),$ $1 \leq i \leq 4,$ where $P_i$ will be specified in the next steps of the proof.

The programs in $P_1$ serve for the initialization of the computation and in the simulation of $SUB$ instructions, programs in $P_2$ have an auxiliary character. The programs in $P_3$ and in $P_4$ realize $ADD$ and $SUB$ instructions.

(1) To initialize the simulation of a computation of $M$ we take an agent $B_1 = (_\star e, P_1)$ with the set of programs:

$P_1$ :

$1 : \langle e \rightarrow l_0 \rangle , 2 : \langle l_0 \leftrightarrow d \rangle$ ;

(2) We need one more agent to generate a special object $d$. While object $C$ is not in the environment the agent $B_2$ places a further copy of $d$ to the environment.

$P_2$ :

$3 : \langle e \rightarrow d \rangle , 4 : \langle d \leftrightarrow C/d \leftrightarrow e \rangle$ ;

The P colony $\Pi$ starts its computation in the initial configuration $(_\star e, _\star e, _\star e, _\star e, _\star \varepsilon)$. In the first subsequence of steps of P colony $\Pi$ only agents $B_1$ and $B_2$ can apply their programs.

| | configuration of $\Pi$ | | | | | labels of applicable programs | | | |
| step | $B_1$ | $B_2$ | $B_3$ | $B_4$ | $Env$ | $P_1$ | $P_2$ | $P_3$ | $P_4$ |
| --- | --- | --- | --- | --- | --- | --- | --- | --- | --- |
| 1. | $_\star e$ | $_\star e$ | $_\star e$ | $_\star e$ | | 1 | 3 | | |
| 2. | $_\star l_0$ | $_\star d$ | $_\star e$ | $_\star e$ | | | 4 | | |
| 3. | $_\star l_0$ | $_\star e$ | $_\star e$ | $_\star e$ | $_\star d$ | 2 | 3 | | |
| 4. | $_\star d$ | $_\star d$ | $_\star e$ | $_\star e$ | $_\star l_0$ | | | | |

(3) To simulate the ADD-instruction $l_1 : (ADD(r), l_2, l_3)$ two agents $B_3$ and $B_4$ are used in $\Pi$. These agents help each other to add one copy of object $a_r$ and object $l_2$ or $l_3$ to the environment using the following programs:

| $P_3$ | $P_3$ | $P_4$ | $P_4$ |
| --- | --- | --- | --- |
| $5 : \langle e \leftrightarrow l_1 \rangle ,$ | $11 : \langle E'_1 \rightarrow l'_2 \rangle ,$ | $15 : \langle e \leftrightarrow E_1 \rangle ,$ | $21 : \langle e \leftrightarrow l'_2 \rangle ,$ |
| $6 : \langle l_1 \rightarrow E_1 \rangle ,$ | $12 : \langle E'_1 \rightarrow l'_3 \rangle ,$ | $16 : \langle E_1 \rightarrow E'_1 \rangle ,$ | $22 : \langle e \leftrightarrow l'_3 \rangle ,$ |
| $7 : \langle E_1 \leftrightarrow d \rangle ,$ | $13 : \langle l'_2 \leftrightarrow e \rangle ,$ | $17 : \langle E'_1 \leftrightarrow e \rangle ,$ | $23 : \langle l'_2 \rightarrow l_2 \rangle ,$ |
| $8 : \langle d \rightarrow L_1 \rangle ,$ | $14 : \langle l'_3 \leftrightarrow e \rangle ,$ | $18 : \langle e \leftrightarrow L_1 \rangle ,$ | $24 : \langle l'_3 \rightarrow l_3 \rangle ,$ |
| $9 : \langle L_1 \leftrightarrow E'_1/L_1 \rightarrow m \rangle ,$ | | $19 : \langle L_1 \rightarrow a_r \rangle ,$ | $25 : \langle l_2 \leftrightarrow e \rangle ,$ |
| $10 : \langle m \rightarrow d \rangle ,$ | | $20 : \langle a_r \leftrightarrow e \rangle ,$ | $26 : \langle l_3 \leftrightarrow e \rangle .$ |

The agent $B_3$ consumes the object $l_1$, changes it to $E_1$ and places it to the environment. The agent $B_4$ borrows $E_1$ from the environment and returns $E'_1$. $B_3$ rewrites the object $d$ to some $L_i$. If this $L_i$ has the same index as $E'_i$ placed in the environment, the computation can go to the next phase. If indices of $L_i$ and $E_i$ are different, the agent $B_3$ tries to generate another $L_i$. If the computation gets over this checking step, agent $B_4$ generates one copy of object $a_r$ and places it into the environment (adding 1 to the content of register $r$). Then agent $B_3$ generates the helpful object $l'_2$ or $l'_3$ and places it into the environment. The agent $B_4$ exchanges it for the "valid label" $l_2$ or $l_3$.

An instruction $l_i : (ADD(r), l_j, l_k)$ is simulated by the following sequence of steps. Let the content of the agent $B_2$ be $d$.

| | configuration of $\Pi$ | | | | | labels of applicable programs | | | |
|---|---|---|---|---|---|---|---|---|---|
| step | $B_1$ | $B_2$ | $B_3$ | $B_4$ | $Env$ | $P_1$ | $P_2$ | $P_3$ | $P_4$ |
| 1. | $_\star d$ | $_\star d$ | $_\star e$ | $_\star e$ | $_\star l_i a_r^u d^v$ | | 4 | 5 | |
| 2. | $_\star d$ | $_\star e$ | $_\star l_i$ | $_\star e$ | $_\star a_r^u d^{v+1}$ | | 3 | 6 | |
| 3. | $_\star d$ | $_\star d$ | $_\star E_i$ | $_\star e$ | $_\star a_r^u d^{v+1} d$ | | 4 | 7 | |
| 4. | $_\star d$ | $_\star e$ | $_\star d$ | $_\star e$ | $_\star E_i a_r^u d^{v+1}$ | | 3 | 8 | 15 |
| 5. | $_\star d$ | $_\star d$ | $_\star L_i$ | $_\star E_i$ | $_\star a_r^u d^{v+1}$ | | 4 | | 16 |
| 6. | $_\star d$ | $_\star e$ | $_\star L_i$ | $_\star E_i'$ | $_\star a_r^u d^{v+2}$ | | 3 | | 17 |
| 7. | $_\star d$ | $_\star d$ | $_\star L_i$ | $_\star e$ | $_\star E_i' a_r^u d^{v+2}$ | | 4 | 9 | |
| 8. | $_\star d$ | $_\star e$ | $_\star E_i'$ | $_\star e$ | $_\star L_i a_r^u d^{v+3}$ | | 3 | **11 or 12** | 18 |
| 9. | $_\star d$ | $_\star d$ | $_\star l_j'$ | $_\star L_i$ | $_\star a_r^u d^{v+3}$ | | 4 | 13 | 19 |
| 10. | $_\star d$ | $_\star e$ | $_\star e$ | $_\star a_r$ | $_\star l_j' a_r^u d^{v+4}$ | | 3 | | 20 |
| 11. | $_\star d$ | $_\star d$ | $_\star e$ | $_\star e$ | $_\star l_j' a_r^{u+1} d^{v+4}$ | | 4 | | 21 |
| 12. | $_\star d$ | $_\star e$ | $_\star e$ | $_\star l_j'$ | $_\star a_r^{u+1} d^{v+5}$ | | 3 | | 23 |
| 13. | $_\star d$ | $_\star d$ | $_\star e$ | $_\star l_j$ | $_\star a_r^{u+1} d^{v+5}$ | | 4 | | 25 |
| 14. | $_\star d$ | $_\star e$ | $_\star e$ | $_\star e$ | $_\star l_j a_r^{u+1} d^{v+6}$ | | | | |

(4) For each SUB-instruction $l_1 : (SUB(r), l_2, l_3)$, the next programs are introduced in the sets $P_1$, $P_3$, and in the set $P_4$:

| $P_3$ | $P_3$ | $P_1$ | $P_4$ |
|---|---|---|---|
| $27 : \langle e \leftrightarrow l_1 \rangle,$ | $33 : \langle F_1'' \to l_3' \rangle,$ | $36 : \langle d \leftrightarrow F_1 \rangle,$ | $41 : \langle e \leftrightarrow l_2' \rangle,$ |
| $28 : \langle l_1 \to F_1 \rangle,$ | $34 : \langle l_2' \leftrightarrow e \rangle,$ | $37 : \langle F_1 \to F_1' \rangle,$ | $42 : \langle e \leftrightarrow l_3' \rangle,$ |
| $29 : \langle F_1 \leftrightarrow d \rangle,$ | $35 : \langle l_3' \leftrightarrow e \rangle;$ | $38 : \langle F_1' \leftrightarrow a_r / F_1' \to F_1'' \rangle,$ | $43 : \langle l_2' \to l_2 \rangle,$ |
| $30 : \langle d \leftrightarrow F_1' \rangle,$ | | $39 : \langle a_r \to d \rangle,$ | $44 : \langle l_3' \to l_3 \rangle,$ |
| $31 : \langle F_1' \to l_2' \rangle,$ | | $40 : \langle F_1'' \leftrightarrow d \rangle.$ | $45 : \langle l_2 \leftrightarrow e \rangle,$ |
| $32 : \langle d \leftrightarrow F_1'' \rangle,$ | | | $46 : \langle l_3 \leftrightarrow e \rangle.$ |

Agent $B_3$ starts the simulation of executing SUB-instruction $l_1$, the agent $B_1$ checks whether there is a copy of the object $a_r$ in the environment or not and gives this information ($F_1'$ – if there is some $a_r$; $F_1''$ – if there is no object $a_r$ in the environment) to the environment.

An instruction $l_i : (SUB(r), l_j, l_k)$ is simulated by the following sequence of steps. The computation for 0 in the register $r$ is given below.

| | configuration of $\Pi$ | | | | | labels of applicable programs | | | |
|---|---|---|---|---|---|---|---|---|---|
| step | $B_1$ | $B_2$ | $B_3$ | $B_4$ | $Env$ | $P_1$ | $P_2$ | $P_3$ | $P_4$ |
| 1. | $_\star d$ | $_\star d$ | $_\star e$ | $_\star e$ | $_\star l_i d^v$ | | 4 | 27 | |
| 2. | $_\star d$ | $_\star e$ | $_\star l_i$ | $_\star e$ | $_\star d^{v+1}$ | | 3 | 28 | |
| 3. | $_\star d$ | $_\star d$ | $_\star F_i$ | $_\star e$ | $_\star d^{v+1} d$ | | 4 | 29 | |
| 4. | $_\star d$ | $_\star e$ | $_\star d$ | $_\star e$ | $_\star F_i d^{v+1}$ | 36 | 3 | | |
| 5. | $_\star F_i$ | $_\star d$ | $_\star d$ | $_\star e$ | $_\star d^{v+2}$ | 37 | 4 | | |
| 6. | $_\star F_i'$ | $_\star e$ | $_\star d$ | $_\star e$ | $_\star d^{v+3}$ | 38 | 3 | | |

| step | configuration of $\Pi$ | | | | | labels of applicable programs | | | |
|---|---|---|---|---|---|---|---|---|---|
| | $B_1$ | $B_2$ | $B_3$ | $B_4$ | $Env$ | $P_1$ | $P_2$ | $P_3$ | $P_4$ |
| 7. | $_\star F_i''$ | $_\star d$ | $_\star d$ | $_\star e$ | $_\star d^{v+3}$ | 40 | 4 | | |
| 8. | $_\star d$ | $_\star e$ | $_\star d$ | $_\star e$ | $_\star F_i'' d^{v+3}$ | | 3 | 32 | |
| 9. | $_\star d$ | $_\star d$ | $_\star F_i''$ | $_\star e$ | $_\star d^{v+4}$ | | 4 | 33 | |
| 10. | $_\star d$ | $_\star e$ | $_\star l_k'$ | $_\star e$ | $_\star d^{v+5}$ | | 3 | 35 | |
| 11. | $_\star d$ | $_\star d$ | $_\star e$ | $_\star e$ | $_\star l_k' d^{v+5}$ | | 4 | | 42 |
| 12. | $_\star d$ | $_\star e$ | $_\star e$ | $_\star l_k'$ | $_\star d^{v+6}$ | | 3 | | 44 |
| 13. | $_\star d$ | $_\star d$ | $_\star e$ | $_\star l_k$ | $_\star d^{v+6}$ | | 4 | | 46 |
| 14. | $_\star d$ | $_\star e$ | $_\star e$ | $_\star e$ | $_\star l_k d^{v+7}$ | | | | |

The computation for a value different from 0 in the register $r$:

| step | configuration of $\Pi$ | | | | | labels of applicable programs | | | |
|---|---|---|---|---|---|---|---|---|---|
| | $B_1$ | $B_2$ | $B_3$ | $B_4$ | $Env$ | $P_1$ | $P_2$ | $P_3$ | $P_4$ |
| 1. | $_\star d$ | $_\star d$ | $_\star e$ | $_\star e$ | $_\star l_i a_r^u d^v$ | | 4 | 27 | |
| 2. | $_\star d$ | $_\star e$ | $_\star l_i$ | $_\star e$ | $_\star a_r^u d^{v+1}$ | | 3 | 28 | |
| 3. | $_\star d$ | $_\star d$ | $_\star F_i$ | $_\star e$ | $_\star a_r^u d^{v+1} d$ | | 4 | 29 | |
| 4. | $_\star d$ | $_\star e$ | $_\star d$ | $_\star e$ | $_\star F_i a_r^u d^{v+1}$ | 36 | 3 | | |
| 5. | $_\star F_i$ | $_\star d$ | $_\star d$ | $_\star e$ | $_\star a_r^u d^{v+2}$ | 37 | 4 | | |
| 6. | $_\star F_i'$ | $_\star e$ | $_\star d$ | $_\star e$ | $_\star a_r^u d^{v+3}$ | 38 | 3 | | |
| 7. | $_\star a_r$ | $_\star d$ | $_\star d$ | $_\star e$ | $_\star F_i a_r^{u-1} d^{v+3}$ | 39 | 4 | 30 | |
| 8. | $_\star d$ | $_\star e$ | $_\star F_i'$ | $_\star e$ | $_\star a_r^{u-1} d^{v+5}$ | | 3 | 31 | |
| 9. | $_\star d$ | $_\star d$ | $_\star l_j'$ | $_\star e$ | $_\star a_r^{u-1} d^{v+5}$ | | 4 | 34 | |
| 10. | $_\star d$ | $_\star e$ | $_\star e$ | $_\star e$ | $_\star l_j' a_r^{u-1} d^{v+6}$ | | 3 | | 41 |
| 11. | $_\star d$ | $_\star d$ | $_\star e$ | $_\star l_j'$ | $_\star a_r^{u-1} d^{v+6}$ | | 4 | | 43 |
| 12. | $_\star d$ | $_\star e$ | $_\star e$ | $_\star l_j$ | $_\star a_r^{u-1} d^{v+7}$ | | 3 | | 45 |
| 13. | $_\star d$ | $_\star d$ | $_\star e$ | $_\star e$ | $_\star l_j a_r^{u-1} d^{v+7}$ | | | | |

(5) The halting instruction $l_h$ is simulated by agent $B_3$ with subset of programs:

$$\frac{P_3}{47 : \langle e \leftrightarrow l_h \rangle, \ 48 : \langle l_h \rightarrow C \rangle, \ 49 : \langle C \leftrightarrow e \rangle.}$$

The agent consumes the object $l_h$ and in the environment there is no other object $l_m$. This agent places one copy of the object $C$ to the environment and stops working. In the next step the object $C$ is consumed by the agent $B_3$. No agent can start its work and the computation halts. The execution of the halting instruction $l_h$ stops all agents in colony $\Pi$:

| step | configuration of $\Pi$ | | | | | labels of applicable programs | | | |
|---|---|---|---|---|---|---|---|---|---|
| | $B_1$ | $B_2$ | $B_3$ | $B_4$ | $Env$ | $P_1$ | $P_2$ | $P_3$ | $P_4$ |
| 1. | $_\star d$ | $_\star d$ | $_\star e$ | $_\star e$ | $_\star l_h d^v$ | | 4 | 47 | |
| 2. | $_\star d$ | $_\star e$ | $_\star l_h$ | $_\star e$ | $_\star d^{v+1}$ | | 3 | 48 | |
| 3. | $_\star d$ | $_\star d$ | $_\star C$ | $_\star e$ | $_\star d^{v+1} d$ | | 4 | 49 | |
| 4. | $_\star d$ | $_\star e$ | $_\star e$ | $_\star e$ | $_\star C d^{v+1}$ | | 3 | | |
| 5. | $_\star d$ | $_\star d$ | $_\star e$ | $_\star e$ | $_\star C d^{v+2}$ | | 4 | | |
| 6. | $_\star d$ | $_\star C$ | $_\star e$ | $_\star e$ | $_\star d^{v+3}$ | - | - | - - - - - - | - - |

The P colony $\Pi$ correctly simulates the computation in the register machine $M$. The computation of $\Pi$ starts with no object $a_r$ placed in the environment in the same way as the computation in $M$ starts with zeros in all registers. The computation of $\Pi$ stops if the symbol $l_h$ and consequently object $C$ is placed inside the corresponding agent in the same way as $M$ stops by executing the halting instruction labeled $l_h$. Consequently, $N(M) = N(\Pi)$ and because the number of agents equals four, the proof is complete.     □

**Theorem 2.** $NRM_{pb} \subseteq NPCOL_{par}(1, 2, *)$.

*Proof.* Let us consider a partially blind register machine $M$ with $m$ registers. We construct a P colony $\Pi = (A, e, f, {}_*v_E, B_1, B_2)$ simulating a computation of the register machine $M$ with:

- $A = \{J, J', V, Q\} \cup \{l_i, l'_i, l''_i, L_i, L'_i, L''_i, E_i \mid l_i \in H\} \cup \{a_r \mid 1 \le r \le m\}$,
- $f = a_1$,
- $B_i = ({}_*e, P_i), i = 1, 2$.

The sets of programs are as follows:

(1) For initializing the simulation:

| $P_1$ : | $P_1$ : | $P_2$ : |
|---|---|---|
| 1: $\langle e \to J \rangle$, | 3: $\langle J \to l_0 \rangle$, | 5: $\langle e \leftrightarrow J \rangle$, |
| 2: $\langle J \leftrightarrow e \rangle$, | 4: $\langle Q \to Q \rangle$, | 6: $\langle J \to J' \rangle$, |
| | | 7: $\langle J' \leftrightarrow e \rangle$. |

At the beginning of the computation the first agent generates the object $l_0$ (the label of the starting instruction of $M$). It generates some copies of object $J$. The agent $B_2$ exchange them by $J'$.

| | configuration of $\Pi$ | | | labels of applicable programs | |
|---|---|---|---|---|---|
| | $B_1$ | $B_2$ | $Env$ | $P_1$ | $P_2$ |
| 1. | ${}_*e$ | ${}_*e$ | | 1 | − |
| 2. | ${}_*J$ | ${}_*e$ | | **2** *or* 3 | − |
| 3. | ${}_*e$ | ${}_*e$ | ${}_*J$ | 1 | 5 |
| 4. | ${}_*J$ | ${}_*J$ | | **2** *or* 3 | 6 |
| 5. | ${}_*l_0$ | ${}_*J'$ | | 8 *or* 24 *or* 34 | 7 |
| 6. | ? | ${}_*e$ | ${}_*J'$ | | |

(2) For every $ADD$-instruction $l_1 : (ADD(r), l_2, l_3)$, $P_1$ and $P_2$ contain:

| $P_1$ : | $P_1$ : | $P_2$ : |
|---|---|---|
| 8: $\langle l_1 \to l'_1 \rangle$, | 14: $\langle L_1 \leftrightarrow E_1 \rangle$, | 18: $\langle e \leftrightarrow l'_1 \rangle$, |
| 9: $\langle l'_1 \leftrightarrow J' \rangle$, | 15: $\langle L_1 \to Q \rangle$, | 19: $\langle l'_1 \to E_1 \rangle$, |
| 10: $\langle l'_1 \to Q \rangle$, | 16: $\langle E_1 \to l_2 \rangle$, | 20: $\langle E_1 \leftrightarrow e \rangle$, |
| 11: $\langle J' \to L''_1 \rangle$, | 17: $\langle E_1 \to l_3 \rangle$, | 21: $\langle e \leftrightarrow L_1 \rangle$, |
| 12: $\langle L''_1 \to L'_1 \rangle$, | | 22: $\langle L_1 \to a_r \rangle$, |
| 13: $\langle L'_1 \to L_1 \rangle$, | | 23: $\langle a_r \leftrightarrow e \rangle$. |

When there is an object $l_1$ inside agent $B_1$, the agent rewrites it to a copy of $l_1'$ and the agent sends it to the environment. The agent $B_2$ borrows $E_1$ from the environment and returns $E_1'$ back.

The agent $B_1$ rewrites the object $J'$ to some $L_i$. The first agent has to generate it in three steps to wait until the second agent generates the symbol $E_i'$ and places it into the environment. If this $L_i$ has the same index as $E_i'$ placed in the environment, the computation can go to the next phase. If the indices of $L_i$ and $E_i$ are different, the agent $B_1$ generates $Q$ and the computation never stops. If the computation gets over this checking step, $B_1$ generates object $l_2$ or object $l_3$.

| | configuration of $\varPi$ | | | labels of applicable programs | |
|---|---|---|---|---|---|
| | $B_1$ | $B_2$ | $Env$ | $P_1$ | $P_2$ |
| 1. | $_\star l_1$ | $_\star e$ | $_\star J'$ | 8 | – |
| 2. | $_\star l_1'$ | $_\star e$ | $_\star J'$ | 9 *or* 10 | – |
| 3. | $_\star J'$ | $_\star e$ | $_\star l_1'$ | 11 | 18 |
| 4. | $_\star L_1''$ | $_\star l_1'$ | | 12 | 19 |
| 5. | $_\star L_1'$ | $_\star E_1$ | | 13 | 20 |
| 6. | $_\star L_1$ | $_\star e$ | $_\star E_1$ | 14 *or* 15 | – |
| 7. | $_\star E_1$ | $_\star e$ | $_\star L_1$ | 16 *or* 17 | 21 |
| 8. | $_\star l_2$ | $_\star L_1$ | | 8 *or* 24 *or* 34 | 22 |
| 9. | ? | $_\star a_r$ | | 9 *or* 25 *or* 35 | 23 |
| 10. | ? | $_\star e$ | $_\star a_r$ | | |

(3) For every $SUB$-instruction $l_1 : (SUB(r), l_2, l_3)$ the following subsets of programs are in $P_1$ and $P_2$:

$P_1$ :
$24 : \langle l_1 \to l_1'' \rangle$,
$25 : \langle l_1'' \leftrightarrow a_r \rangle$,
$26 : \langle l_1'' \to Q \rangle$,
$27 : \langle a_r \to V \rangle$.

$P_1$ :
$28 : \langle V \leftrightarrow l_1''' \rangle$,
$29 : \langle l_1''' \to l_2 \rangle$,
$30 : \langle l_1''' \to l_3 \rangle$

$P_2$ :
$31 : \langle e \leftrightarrow l_1'' \rangle$,
$32 : \langle l_1'' \to l_1''' \rangle$,
$33 : \langle l_1''' \leftrightarrow e \rangle$,

In the first step the agent checks if there is any copy of $a_r$ in the environment (for zero in register $r$). Because of the nondeterminism of the computation in

| | configuration of $\varPi$ | | | $P_1$ | $P_2$ |
|---|---|---|---|---|---|
| | $B_1$ | $B_2$ | $Env$ | | |
| 1. | $_\star l_1$ | $_\star e$ | $_\star a_r$ | 24 | – |
| 2. | $_\star l_1''$ | $_\star e$ | $_\star a_r$ | 25 *or* 26 | – |
| 3. | $_\star a_r$ | $_\star e$ | $_\star l_1''$ | 27 | 31 |
| 4. | $_\star V$ | $_\star l_1''$ | | – | 32 |
| 5. | $_\star V$ | $_\star l_1'''$ | | – | 33 |
| 6. | $_\star V$ | $_\star e$ | $_\star l_1'''$ | 28 | – |
| 7. | $_\star l_1'''$ | $_\star e$ | | 29 *or* 30 | – |
| 8. | $_\star l_2$ | $_\star e$ | | | |

| | configuration of $\varPi$ | | | $P_1$ | $P_2$ |
|---|---|---|---|---|---|
| | $B_1$ | $B_2$ | $Env$ | | |
| 1. | $_\star l_1$ | $_\star e$ | | 24 | – |
| 2. | $_\star l_1''$ | $_\star e$ | | 26 | – |
| 3. | $_\star Q$ | $_\star e$ | | 4 | |
| 4. | $_\star Q$ | $_\star e$ | | | |

the positive case it can rewrite $a_r$ to $V$, in the other case $l_1''$ is rewritten to $Q$ and the computation will never halt. At the end of this simulation the agent $B_1$ generates one of the objects $l_2$, $l_3$.

(4) For the halting instruction $l_h$ the following programs are in sets $P_1$ and $P_2$:

| $P_1:$ | $P_2:$ | $P_2:$ |
|---|---|---|
| $34: \langle l_h \leftrightarrow J' \rangle,$ | $39: \langle e \leftrightarrow l_h \rangle,$ | $43: \langle L_h \leftrightarrow a_r \rangle, 1 < r \leq m$ |
| $35: \langle J' \rightarrow L_h \rangle,$ | $40: \langle l_h \rightarrow \overline{l_h} \rangle,$ | $44: \langle a_r \leftrightarrow e \rangle.$ |
| $36: \langle l_h \rightarrow Q \rangle,$ | $41: \langle \overline{l_h} \leftrightarrow e \rangle,$ | |
| $37: \langle L_h \rightarrow L_h \rangle,$ | $42: \langle e \leftrightarrow L_h \rangle,$ | |
| $38: \langle L_h \leftrightarrow \overline{l_h} \rangle,$ | | |

By using these programs, the P colony finishes the computation in the same way as the partially blind register machine halts its computation. Programs with labels 43 and 44 in $P_2$ check value zero stored in all except the first register. If there is some copy of object $a_r$, programs 43 and 44 are applied in a cycle and the computation never ends. Some copies of object $J'$ (for the the program with label 34) are present in the environment from the initialization of computation.

| all counters $r, 1 < r \leq m$ store zero | | | | |
|---|---|---|---|---|
| configuration of $\Pi$ | | | labels of applicable programs | |
| $B_1$ | $B_2$ | $Env$ | $P_1$ | $P_2$ |
| 1. $_*l_h$ | $_*e$ | $_*J'$ | **34** *or* 36 | − |
| 2. $_*J'$ | $_*e$ | $_*l_h$ | 35 | 39 |
| 3. $_*L_h$ | $_*l_h$ | | 37 | 40 |
| 4. $_*L_h$ | $_*\overline{l_h}$ | | 37 | 41 |
| 5. $_*L_H$ | $_*e$ | $_*\overline{l_h}$ | 38 | − |
| 6. $_*\overline{l_h}$ | $_*e$ | $_*L_h$ | − | 42 |
| 7. $_*\overline{l_h}$ | $_*L_h$ | | − | − |

| content of some counter $r, 1 < r \leq m$ is different from zero | | | | |
|---|---|---|---|---|
| configuration of $\Pi$ | | | labels of applicable programs | |
| $B_1$ | $B_2$ | $Env$ | $P_1$ | $P_2$ |
| 1. $_*l_h$ | $_*e$ | $_*J'a_r$ | **34** *or* 36 | − |
| 2. $_*J'$ | $_*e$ | $_*l_ha_r$ | 35 | 39 |
| 3. $_*L_h$ | $_*l_h$ | $_*a_r$ | 37 | 40 |
| 4. $_*L_h$ | $_*\overline{l_h}$ | $_*a_r$ | 37 | 41 |
| 5. $_*L_H$ | $_*e$ | $_*\overline{l_h}a_r$ | 38 | − |
| 6. $_*\overline{l_h}$ | $_*e$ | $_*L_ha_r$ | − | 42 |
| 7. $_*\overline{l_h}$ | $_*L_h$ | $_*a_r$ | − | 43 |
| 8. $_*\overline{l_h}$ | $_*a_r$ | $_*L_h$ | − | 44 |
| 9. $_*\overline{l_h}$ | $_*L_h$ | $_*a_r$ | − | 43 |

The P colony $\Pi$ correctly simulates any computation of the partially blind register machine $M$.    □

## 4    On the Computational Power of Restricted P Colonies Without Checking

For restricted P colonies the following results are known:

- $NPCOL_{par}KR(2,*,5) = NRE$ in [2,8],
- $NPCOL_{par}R(2,*,5) = NPCOL_{par}KR(2,1,*) = NRE$ in [4].

The next theorem determines the computational power of restricted P colonies working without checking rules.

**Theorem 3.** $NPCOL_{par}R(2,2,*) = NRE$.

*Proof.* Let us consider a register machine $M$ with $m$ registers. We construct a P colony $\Pi = (A, e, f, {}_*v_E, B_1, B_2)$ simulating the computations of register machine $M$ with:

- $A = \{G\} \cup \{l_i, l_i', l_i'', l_i''', l_i'''', \overline{l_i}, \overline{\overline{l_i}}, \underline{l_i}, \underline{\underline{l_i}}, L_i, L_i', L_i'', F_i \mid l_i \in H\} \cup$
  $\cup \{a_r \mid 1 \le r \le m\}$,
- $f = a_1$,
- $B_j = ({}_*ee, P_j), j = 1, 2$.

At the beginning of the computation the first agent generates the object $l_0$ (the label of starting instruction of $M$). Then it starts to simulate the instruction labeled $l_0$ and it generates the label of the next instruction. The set of programs is as follows:

(1) For initializing the simulation there is one program in $P_1$:

| $P_1$ |
| --- |
| $1 : \langle e \to l_0; e \leftrightarrow e \rangle$ |

The initial configuration of $\Pi$ is $({}_*ee, {}_*ee, {}_*\varepsilon)$. After the first step of the computation (only program 1 is applicable) the system enters configuration $({}_*l_0e, {}_*ee, {}_*\varepsilon)$.

(2) For every $ADD$-instruction $l_1 : (ADD(r), l_2, l_3)$ we add to $P_1$ the programs:

| $P_1$ | |
| --- | --- |
| $2 : \langle e \to a_r; l_1 \leftrightarrow e \rangle,$ | $3 : \langle e \to G; a_r \leftrightarrow l_1 \rangle,$ |
| $4 : \langle l_1 \to l_2; G \leftrightarrow e \rangle,$ | $5 : \langle l_1 \to l_3; G \leftrightarrow e \rangle.$ |

When there is an object $l_1$ inside the agent, it generates one copy of $a_r$, puts it into the environment and generates the label of the next instruction (it nondeterministically chooses one of the last two programs 4 and 5).

| | configuration of $\Pi$ | | | labels of applicable programs | |
| --- | --- | --- | --- | --- | --- |
| | $B_1$ | $B_2$ | $Env$ | $P_1$ | $P_2$ |
| 1. | ${}_*l_1e$ | ${}_*ee$ | ${}_*a_r^x$ | 2 | – |
| 2. | ${}_*a_re$ | ${}_*ee$ | ${}_*l_1a_r^x$ | 3 | – |
| 3. | ${}_*Gl_1$ | ${}_*ee$ | ${}_*a_r^{x+1}$ | 4 or 5 | – |
| 4. | ${}_*l_2e$ | ${}_*ee$ | ${}_*a_r^{x+1}G$ | | – |

(3) For every $SUB$-instruction $l_1 : (SUB(r), l_2, l_3)$, the next programs are added to sets $P_1$ and $P_2$:

| $P_1$ | $P_1$ | $P_2$ |
|---|---|---|
| $6 : \langle l_1 \rightarrow l_1'; e \leftrightarrow e \rangle,$ | $12 : \langle \overline{\overline{l_1}} \rightarrow \underline{l_2}; e \leftrightarrow L_1'' \rangle,$ | $18 : \langle e \rightarrow L_1; e \leftrightarrow l_1' \rangle,$ |
| $7 : \langle e \rightarrow l_1''; l_1' \leftrightarrow e \rangle,$ | $13 : \langle \overline{\overline{l_1}} \rightarrow \underline{l_3}; e \leftrightarrow L_1 \rangle,$ | $19 : \langle l_1' \rightarrow L_1'; L_1 \leftrightarrow l_1'' \rangle,$ |
| $8 : \langle e \rightarrow l_1'''; l_1'' \leftrightarrow e \rangle,$ | $14 : \langle L_1'' \rightarrow l_2; \underline{l_2} \leftrightarrow e \rangle,$ | $20 : \langle l_1'' \rightarrow L_1''; L_1' \leftrightarrow a_r \rangle,$ |
| $9 : \langle l_1''' \rightarrow l_1''''; e \leftrightarrow e \rangle,$ | $15 : \langle L_1 \rightarrow F_3; \underline{l_3} \leftrightarrow e \rangle,$ | $21 : \langle a_r \rightarrow e; L_1'' \leftrightarrow L_1 \rangle,$ |
| $10 : \langle l_1'''' \rightarrow \overline{l_1}; e \leftrightarrow e \rangle,$ | $16 : \langle e \rightarrow \underline{l_3}; F_3 \leftrightarrow \underline{l_3} \rangle,$ | $22 : \langle L_1 \rightarrow e; e \leftrightarrow e \rangle,$ |
| $11 : \langle \overline{l_1} \rightarrow \overline{\overline{l_1}}; e \leftrightarrow e \rangle,$ | $17 : \langle \underline{l_3} \rightarrow l_3; \underline{\underline{l_3}} \leftrightarrow e \rangle,$ | $23 : \langle l_1'' \rightarrow e; L_1' \leftrightarrow F_3 \rangle,$ |
| | | $24 : \langle F_3 \rightarrow e; e \leftrightarrow e \rangle.$ |

At the first phase of the simulation of the $SUB$ instruction the first agent generates object $l_1'$, which is consumed by the second agent. The agent $B_2$ generates symbol $L_1$ and tries to consume one copy of symbol $a_r$. If there is any $a_r$, the agent sends to the environment object $L_1''$ and consumes $L_1$. After this step the first agent consumes $L_1''$ or $L_1$ and rewrites it to $l_2$ or $l_3$. The objects $\underline{x}$, $\overline{x}$ and $\overline{\overline{x}}$ are used for a synchronization of the computation in both agents and for storing information about the state of the computation.

Instruction $l_1 : (SUB(r), l_2, l_3)$ is simulated by the following sequence of steps.

If the register $r$ stores a nonzero value:

| | configuration of $\Pi$ | | | labels of applicable programs | |
|---|---|---|---|---|---|
| | $B_1$ | $B_2$ | $Env$ | $P_1$ | $P_2$ |
| 1. | $_\star l_1 e$ | $_\star ee$ | $_\star a_r^x$ | 6 | – |
| 2. | $_\star l_1' e$ | $_\star ee$ | $_\star a_r^x$ | 7 | – |
| 3. | $_\star l_1'' e$ | $_\star ee$ | $_\star l_1' a_r^x$ | 8 | 18 |
| 4. | $_\star l_1''' e$ | $_\star L_1 l_1'$ | $_\star l_1'' a_r^x$ | 9 | 19 |
| 5. | $_\star l_1'''' e$ | $_\star L_1' l_1''$ | $_\star L_1 a_r^x$ | 10 | 20 |
| 6. | $_\star \overline{l_1} e$ | $_\star L_1'' a_r$ | $_\star L_1 L_1' a_r^{x-1}$ | 11 | 21 |
| 7. | $_\star \overline{\overline{l_1}} e$ | $_\star e L_1$ | $_\star L_1'' a_r^{x-1}$ | 12 | 22 |
| 8. | $_\star \underline{l_2} L_1''$ | $_\star ee$ | $_\star a_r^{x-1}$ | 14 | – |
| 9. | $_\star l_2 e$ | $_\star ee$ | $_\star a_r^{x-1} \underline{l_2}$ | | |

If the register $r$ stores value zero :

| | configuration of $\Pi$ | | | labels of applicable programs | |
|---|---|---|---|---|---|
| | $B_1$ | $B_2$ | $Env$ | $P_1$ | $P_2$ |
| 1. | $_\star l_1 e$ | $_\star ee$ | | 6 | – |
| 2. | $_\star l_1' e$ | $_\star ee$ | | 7 | – |
| 3. | $_\star l_1'' e$ | $_\star ee$ | $_\star l_1'$ | 8 | 18 |
| 4. | $_\star l_1''' e$ | $_\star L_1 l_1'$ | $_\star l_1''$ | 9 | 19 |
| 5. | $_\star l_1'''' e$ | $_\star L_1' l_1''$ | $_\star L_1$ | 10 | |

| | configuration of $\Pi$ | | | labels of applicable programs | |
|---|---|---|---|---|---|
| | $B_1$ | $B_2$ | $Env$ | $P_1$ | $P_2$ |
| 6. | $_\star \overline{l}_1 e$ | $_\star L_1' l_1''$ | $_\star L_1$ | 11 | |
| 7. | $_\star \overline{\overline{l}}_1 e$ | $_\star L_1' l_1''$ | $_\star L_1$ | 13 | |
| 8. | $_\star \underline{l}_3 L_1$ | $_\star L_1' l_1''$ | | 15 | – |
| 9. | $_\star F_3 e$ | $_\star L_1' l_1''$ | $_\star \underline{l}_3$ | 16 | – |
| 10. | $_\star \underline{l}_3 \underline{l}_3$ | $_\star L_1' l_1''$ | $_\star F_3$ | 17 | 23 |
| 11. | $_\star \underline{l}_3 e$ | $_\star F_3 e$ | $_\star \underline{\underline{l}}_3 L_1'$ | 2 or 6 or none | 24 |
| 12. | ? | $_\star ee$ | $_\star \underline{\underline{l}}_3 L_1'$ | | |

(4) For halting instruction $l_h$ no program is added to the sets $P_1$ and $P_2$.

The P colony $\Pi$ correctly simulates all computations of the register machine $M$ and the number contained in the first register of $M$ corresponds to the number of copies of the object $a_1$ present in the environment of $\Pi$. □

## 5 Conclusions

We have shown that the P colonies with capacity $c = 2$ and without checking programs, with height at most 2, are computationally complete. In Section 3 we have shown that the P colonies with capacity $c = 1$ and with checking/evolution programs and 4 agents are computationally complete.

We have verified also that partially blind register machines can be simulated by P colonies with capacity $c = 1$ without checking programs with two agents. The generative power of $NPCOL_{par}K(1, n, *)$ for $n = 2, 3$ remains open.

In Section 4 we have studied P colonies with capacity $c = 2$ without checking programs. Two agents guarantee the computational completeness in this case.

For more information on membrane computing, see [11], for more on computational machines and colonies in particular, see [9] and [6,7,8], respectively. Activities carried out in the field of membrane computing are currently numerous and they are available also at [12].

## Acknowledgements

This work has been supported by the Grant Agency of Czech Republic grants No. 201/06/0567 and by IGS SU 32/2007.

## References

1. Ciencialová, L., Cienciala, L.: Variations on the theme: P colonies. In: Kolář, D., Meduna, A. (eds.) Proceedings of the 1st International workshop WFM 2006, Ostrava, pp. 27–34 (2006)

2. Csuhaj-Varjú, E., Kelemen, J., Kelemenová, A., Păun, G., Vaszil, G.: Cells in environment: P colonies. Journal of Multiple-valued Logic and Soft Computing 12(3-4), 201–215 (2006)

3. Csuhaj-Varjú, E., Margenstern, M., Vaszil, G.: P colonies with a bounded number of cells and programs. In: Hoogeboom, H.J., Păun, G., Rozenberg, G. (eds.) Pre-Proceedings of the 7th Workshop on Membrane Computing, Leiden, the Netherlands, pp. 311–322 (2006)

4. Freund, R., Oswald, M.: P colonies working in the maximally parallel and in the sequential mode. In: Ciobanu, G., Păun, G. (eds.) Pre-Proceedings of the 1st International Workshop on Theory and Application of P Systems, Timisoara, Romania, pp. 49–56 (2005)

5. Greibach, S.A.: Remarks on blind and partially blind one-way multicounter machines. Theoretical Computer Science 7(1), 311–324 (1978)

6. Kelemen, J., Kelemenová, A.: A grammar-theoretic treatment of multi-agent systems. Cybernetics and Systems 23, 621–633 (1992)

7. Kelemen, J., Kelemenová, A.: On P colonies, a biochemically inspired model of computation. In: Proc. of the 6th International Symposium of Hungarian Researchers on Computational Intelligence, Budapest TECH, Hungary, pp. 40–56 (2005)

8. Kelemen, J., Kelemenová, A., Păun, G.: Preview of P colonies: A biochemically inspired computing model. In: Bedau, M., et al. (eds.) ALIFE IX. Workshop and Tutorial Proceedings, Ninth International Conference on the Simulation and Synthesis of Living Systems, Boston, Mass., pp. 82–86 (2004)

9. Minsky, M.L.: Computation: Finite and Infinite Machines. Prentice Hall, Englewood Cliffs, NJ (1967)

10. Păun, G.: Computing with membranes. Journal of Computer and System Sciences 61, 108–143 (2000)

11. Păun, G.: Membrane Computing: An Introduction. Springer, Berlin (2002)

12. P systems web page. http://psystems.disco.unimib.it

# Events, Causality, and Concurrency in Membrane Systems

Gabriel Ciobanu[1,2] and Dorel Lucanu[1]

[1] "A.I.Cuza" University of Iași, Faculty of Computer Science
[2] Romanian Academy, Institute of Computer Science, Iași
{gabriel,dlucanu}@info.uaic.ro

**Abstract.** This paper presents a modular approach to causality in membrane systems, using both string and multiset rewriting. In order to deal with membrane systems, the event structures are extended with notions like maximal concurrent transitions and saturated states with respect to concurrency. The event structure of a membrane system is defined in two steps: first the event structure of a maximal parallel step in membranes is defined, and then it is combined with a communication step. The main result of the paper proves that an event structure of a membrane corresponds to its operational semantics. Event structures for communicating membranes are also defined.

## 1 Introduction

The event structures represent a formal model for concurrent systems in which simultaneity among the events is fully considered. The event structures provide the causality of actions in a true concurrent system, and they were defined for various formalisms (Petri nets, CCS, $\pi$-calculus). The membrane systems describe a new model of computation inspired by biology. This model is given by a hierarchical structure, it is highly parallel, and the rules applied in parallel are chosen in a nondeterministic way. We study the event structure for membrane systems, defining both the causality and the conflict relations. It is worth to mention that the event structure of a membrane system was defined first time in [7]. A related work discussing the causality in membrane systems is [6]. The main motivation of defining the event structure of a membrane system comes from the fact that causality is an important aspect which can make the difference between the classic mathematical models and the new discrete models of the biological systems.

Membrane computing is a branch of natural computing, initiated by Păun [10] as a computing model inspired by biological systems which are complex hierarchical structures, with a flow of entities and information which underlies their functioning. Essentially, the membrane systems (called also P systems) are composed of various compartments with different tasks, all of them working simultaneously to accomplish a more general task. A membrane system consists of a hierarchy of nested membranes, placed inside a distinguishable membrane called *skin* surrounding all of them. A membrane contains multisets of

G. Eleftherakis et al. (Eds.): WMC8 2007, LNCS 4860, pp. 209–227, 2007.

*objects, evolution rules,* and possibly other *membranes.* The multisets of objects from a membrane correspond to the "chemicals swimming in the solution in the cell compartment", while the rules correspond to the "chemical reactions possible in the same compartment". The field of membrane computing is evolving quickly, and there are several applications in modeling biological systems [8].

Here we look at membrane systems as a non-standard computation mechanism based on parallelism and nondeterminism. Parallelism and nondeterminism represent in fact the essential features of the membrane computation which is strongly parallel and nondeterministic mainly due to its way of applying the rules of a membrane to its objects. The rules associated with a compartment are applied to the objects from a compartment in a parallel way, and the rules are chosen in a non-deterministic manner. Moreover, all compartments of the system evolve in parallel. Membrane systems are able to simulate Turing machines, hence they are computationally complete. There are mainly two types of results on membrane systems: computational universality and complexity. This paper is devoted to the causal dependencies, conflicts and concurrency in membrane systems. A membrane structure and the multisets of objects from its compartments identify a configuration of a P system. By a nondeterministic and parallel use of rules, the system can pass to another configuration; such a step is called a transition. A sequence of transitions constitutes a computation. A computation is successful if it halts. With a halting computation we can associate a result (in various ways). Because of the nondeterminism of the application of rules, starting from an initial configuration, we can get several successful computations, hence several results.

The models involving concurrency and nondeterminism are classified in [12] according to the fact they can faithfully take into account the difference between parallel and sequential computation (interleaving or non-interleaving model), or they can represent the branching structure of processes related to nondeterministic choices (linear time or branching time). Event structures represent the non-interleaving and branching-time models, and so they model the true concurrency and nondeterminism. We describe the nature of parallelism and nondeterminism of the membrane systems in terms of event structures [13], and describe formally the event structure of a membrane system.

## 2    Event Structures

In event-based models, a system is represented by a set of events (action occurrences) together with some structure on this set, determining the causality relations between the events. The causality between actions is expressed by a partial order, and the nondeterminism is expressed by a conflict relation on actions. For every two events $e$ and $e'$ it is specified either whether one of them is a prerequisite for the other ($e \leq e'$ or $e' \leq e$), whether they exclude each other ($e \# e'$), or whether they may happen in parallel ($e \ co \ e'$). The behavior of an event structure is formalized by associating to it a family of configurations representing sets of events which occur during (partial) runs of the system.

A concurrent step consists of simultaneously executing rules, each of them producing events which end up in the resulting event configuration. These steps are presumably cooperating to achieve a goal, and so they are not totally independent. They synchronize at certain points, and this is reflected in the events produced.

There are many levels of granularity at which one might describe the events that occur as a process executes. At the level of the components from which the system is composed, computation consists of events which reflect the rules. At a higher level, the system might be viewed in terms of parallel executions of its components. Thus, when we talk about the events which occur within a system, it is understood that we know the granularity of the representation, and that events are encoded to this granularity (degree of precision).

**Definition 1.** *An event structures is a triple* $(E, \leq, \#)$ *where*

- $E$ *is a set of events,*
- $\leq \subseteq E \times E$ *is a partial order, the* causality *relation,*
- $\# \subseteq E \times E$ *is an irreflexive and symmetric relation, the* conflict *relation,*

*satisfying the* principle of conflict heredity*:*

$$\forall e_1, e_2, e_3 \in E. \ e_1 \leq e_2 \wedge e_1 \# e_3 \Rightarrow e_2 \# e_3$$

*and the* finiteness assumption $\{e' \mid e' \leq e\}$ *is finite. Two events $e$ and $e'$ are* concurrent, *and write $e \, co \, e'$, if* $\neg(e \leq e' \vee e' \leq e \vee e \# e')$.

Note that the events of an event structure are corresponding in fact to event occurrences.

**Definition 2.** *Let* $(E, \leq, \#)$ *be an event structure. A* computation state (event configuration) *is a subset $x \subseteq E$ which is*

- conflict-free: $(\forall e, e' \in x) \neg(e \# e')$, *and*
- downwards-closed: $(\forall e, e')(e' \leq e \text{ and } e \in x) \implies e' \in x$.

**Definition 3.** *Let* $(E, \leq, \#)$ *be an event structure. Let $x$ and $x'$ be two finite computation states and let $e$ be an event. We write*

$$x \xrightarrow{e} x' \text{ if and only if } e \notin x \text{ and } x' = x \cup \{e\}.$$

**Proposition 1 (Diamond Property).** *Let* $(E, \leq, \#)$ *be an event structure. Two events $e_1$ and $e_2$ are concurrent, $e_1 \, co \, e_2$, if and only if there are configurations $x, x_1, x_2, x'$ such that:*

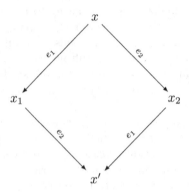

A more detailed presentation of the event structures can be found in [14].

## 2.1   Maximal Concurrent Transitions in an Event Structure

In this section we define a new transition system over the computation states of an event structure, where an action (a label of a transition) is given by the parallel occurrence of a maximal set of concurrent events.

Let $\max(x)$ denote the set of $\leq$-maximal events in the event set $x$, i.e., $e \in \max(x)$ if and only if $e \in x$ and there does not exist $e' \in x$ such that $e < e'$. By $\lceil e \rceil$ we denote the set $\{e' \mid e' \leq e\}$; if $x$ is a set of events, then $\lceil x \rceil = \{e' \mid (\exists e \in x)\, e' \leq e\}$.

**Definition 4.** *Let $(E, \leq, \#)$ be an event structure. A maximal concurrent action is a subset $a \subseteq E$ which*

- *consists only of concurrent events, i.e., $(\forall e, e' \in a)\ e\ co\ e'$, and*
- *is maximal with above property, i.e., $(\forall e \in E \setminus a)(\exists e' \in a)\ e \leq e' \vee e' \leq e \vee e \# e'$.*

We write $e \lessdot e'$ if and only if $e < e'$ and there does not exist $e''$ such that $e < e'' < e'$.

**Definition 5.** *Let $(E, \leq, \#)$ be an event structure. Let $x$ and $x'$ be two finite computation states and let $a$ be a maximal concurrent action. We write $x \xrightarrow{a} x'$ if and only if $x \cap a = \emptyset$ and $x' = x \cup a$.*

**Definition 6.** *Let $(E, \leq, \#)$ be an event structure. A computation state $x$ is saturated (with respect to concurrency) if for each event $e \in E$ and each computation state $x'$ such that $x \xrightarrow{e} x'$, then $\neg(\exists e' \in x)\ e\ co\ e'$.*

**Proposition 2.** *Let $(E, \leq, \#)$ be an event structure. Let $x$ and $x'$ be two finite computation states and a maximal concurrent action $a$ such that $x \xrightarrow{a} x'$. Then:*

1. *if $x = \emptyset$ then $x' = x \cup a = a = \lceil a \rceil$;*
2. *if $x$ is saturated, then $x'$ is saturated;*

3. *if $x$ is saturated and $x' = \lceil a \rceil$, then $x = \emptyset$;*
4. *if $x \neq \emptyset$ and $x$ is saturated, then $(\forall e' \in a)(\exists e \in \max(x))\ e < e'$.*

*Proof.* 1. We have $x' = a$, which is downwards-closed.

2. If $x' \xrightarrow{e} x''$ and $e'\ co\ e$, $e' \in x'$, then either $e' \in x$ or $e' \in a$. In both cases we get a contradiction.

3. We have $x'$ saturated, too. If $e \in x$, then we must have $e\ co\ e'$ for all $e' \in a$. Contradiction.

4. Let $e'$ be an arbitrary event in $a$. Since $x$ is downwards-closed and conflict-free, there exists an event $e$ in $x \cap \lceil a \rceil$. Since $x$ is finite, we can choose an $e$ such that $e < e'$.    □

## 2.2 Event Structure Associated to a Labeled Transition System

There are many cases when the operational semantics of a system is given by means of a labeled transition system (lts) describing all possible sequential computations. In order to study the concurrency properties, we must determine the event structure defined by such a labeled transition system.

Let $(S, \rightarrow, L, s_0)$ be a labeled transition system, where $S$ is a set of states, $\rightarrow$ is a transition relation consisting of triples $(s, \ell, s') \in S \times L \times S$, often written as $s \xrightarrow{\ell} s'$, $L$ is a set of labels (actions), and $s_0$ is an initial state. A (sequential) *computation* is a sequence $s_0 \xrightarrow{\ell_1} s_1 \ldots \xrightarrow{\ell_n} s_n$ such that $(s_{i-1}, \ell_i, s_i) \in \rightarrow$. Since all the computations start from $s_0$, each prefix of a computation is also a computation.

**Definition 7.** *Let $(S, \rightarrow, L, s_0)$ be a labeled transition system. Let $\sim$ be the smallest equivalence satisfying: if $(s, \ell_1, s_1)$, $(s, \ell_2, s_2)$, $(s_1, \ell_2, s_3)$, $(s_2, \ell_1, s_3) \in \rightarrow$ and $(\ell_1 \neq \ell_2$ or $s_1 \neq s_2)$, then $(s, \ell_1, s_1) \sim (s_2, \ell_1, s_3)$.*
*An* event $e$ *is a $\sim$-equivalence class written as $[s, \ell, s']$.*

Intuitively, two transitions are equivalent iff they are determined by the same action (label). This can be easier understood from the following picture:

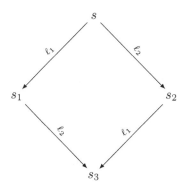

We have two events $e_1 = [s, \ell_1, s_1]$ and $e_2 = [s, \ell_2, s_2]$ corresponding to the transitions labeled by $\ell_1$ and $\ell_2$, respectively. The transitions labeled by $\ell_1$ and $\ell_2$ may occur in any order, i.e., the two events $e_1$ and $e_2$ are concurrent (see Proposition 1).

**Definition 8.** *Let* $(S, \rightarrow, L, s_0)$ *be an acyclic labeled transition system. An event* configuration *is a set of events* $e_i = [s_{i-1}, \ell_i, s_i]$ *corresponding to a computation* $s_0 \xrightarrow{\ell_1} s_1 \ldots \xrightarrow{\ell_n} s_n$.

**Theorem 1 (Event Structure of a lts).** *[9] Each lts* $(S, \rightarrow, L, s_0)$ *can be organized as an* event structure.

*Proof (Sketch).* The event structure $(E, \leq, \#)$ associated to $(S, \rightarrow, L, s_0)$ is defined by:

- $E$ is the set of events as defined above in Definition 7;
- $e_1 \leq e_2$ if every configuration which contains $e_2$ also contains $e_1$;
- $e_1 \# e_2$ if there is no configuration containing both $e_1$ and $e_2$.    □

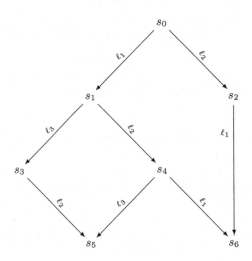

**Fig. 1.** An example of a labeled transition system

*Example 1.* The events defined by the lts in Figure 1 are:

$$e_1 = \{(s_0, \ell_1, s_1)\} \qquad e_4 = \{(s_1, \ell_2, s_4), (s_3, \ell_2, s_5)\}$$
$$e_2 = \{(s_0, \ell_2, s_2)\} \qquad e_5 = \{(s_2, \ell_1, s_6)\}$$
$$e_3 = \{(s_1, \ell_3, s_3), (s_4, \ell_3, s_5)\} \qquad e_6 = \{(s_4, \ell_1, s_6)\}$$

An event configuration describes a (partial) computation expressed in terms of events. This lts defines the following event configurations:

$$e_1 \qquad\qquad e_2 \; e_5$$
$$e_2 \qquad\qquad e_1 \; e_3 \; e_4$$
$$e_1 \; e_3 \qquad\qquad e_1 \; e_4 \; e_3$$
$$e_1 \; e_4 \qquad\qquad e_1 \; e_4 \; e_6$$

We have $e_1 < e_3$ because any event configuration containing $e_3$ also contains $e_1$. Since any occurrence of $e_3$ is always after an occurrence of $e_1$, it follows that there is causal relationship between the two events. We get $e_1 < e_4 < e_6$ and $e_2 < e_5$ in a similar way. Since there is no event configuration containing both $e_1$ and $e_2$, it follows that there is a conflict between the two events, i.e., $e_1 \# e_2$. We get $e_1 \# e_5$, $e_2 \# e_3$, $e_2 \# e_4$, $e_2 \# e_6$, $e_5 \# e_3$, $e_5 \# e_4$, and $e_5 \# e_6$ in a similar way.

## 3 What Is an Event for Membrane Systems?

In order to identify what is an event for a membrane system, we investigate the event structure of an evolution step.

*Example 2.* Let us consider a single membrane with the following three rules

$$\ell_1 : a \to b, \quad \ell_2 : b \to a, \quad \ell_3 : ab \to c$$

and having the content $aabc$. We investigate the space of all sequential rewriting corresponding to the application of rules in the evolution step, changing the content $aabc$ into $bbac$, in order to discover the events of this step. The exact definitions for an evolution step is given in the next subsections.

### 3.1 String Rewriting

We first assume that the sequential rewriting is executed over strings (non-commutative words). A *context* is a string of the form $w \bullet w'$, where $w, w'$ are strings of objects, and $\bullet$ is a special variable. Each rewriting step $wuw' \to wvw'$ is uniquely determined by the context $w \bullet w'$ and the rule $\ell : u \to v$. Thus the transition $(wuw', \ell, wvw') = wuw' \xrightarrow{\ell} wvw'$ is denoted by $(w \bullet w', \ell)$.

The maximal parallel rewriting over strings is defined as follows: $w \xRightarrow{mpr} w'$ if and only if there are $\ell_1, \ldots, \ell_n, \ell_i : u_i \to v_i$ $(i = \overline{1, n})$ such that $w = w_0 u_1 w_1 \ldots u_n w_n$, $w' = w_0 v_1 w_1 \ldots v_n w_n$ and $w_0 w_1 \ldots w_n$ irreducible (no rule can be applied). We may have $i \neq j$ and $l_i = l_j$ (a rule may be applied more than one time in a maximal parallel rewriting). Sometimes we write $w \Rightarrow_{(\ell_1, \ldots, \ell_n)} w'$ in order to emphasize the multiset of rules implied in the maximal parallel rewriting. Note that the pair $(w, w')$ does not uniquely identify the rules involved in $w \xRightarrow{mpr} w'$. For instance, if a membrane $M$ includes the evolution rules:

$$\ell_1 : a \to cd \qquad \ell_2 : a \to c \qquad \ell_3 : b \to f \qquad \ell_4 : b \to df$$

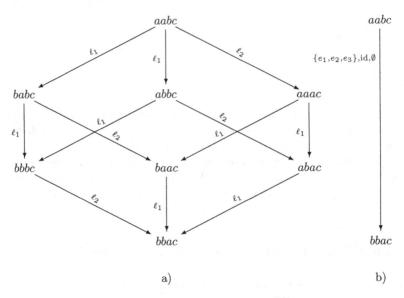

**Fig. 2.** A lts corresponding to $aabc \overset{mpr}{\Rightarrow} bbac$

then we have $ab \Rightarrow_{(\ell_1, \ell_3)} cdf$ and $ab \Rightarrow_{(\ell_2, \ell_4)} cdf$.

We consider first an example. The space of all sequential rewriting for the evolution step of Example 2 is represented in Figure 2.a.

By Definition 7, we have the following three concurrent events:

$$e_1 = \{(\bullet abc, \ell_1), (\bullet bbc, \ell_1), (\bullet bac, \ell_1), (\bullet aac, \ell_1)\}$$
$$e_2 = \{(a \bullet bc, \ell_1), (b \bullet bc, \ell_1), (b \bullet ac, \ell_1), (a \bullet ac, \ell_1)\}$$
$$e_3 = \{(aa \bullet c, \ell_2), (ab \bullet c, \ell_2), (bb \bullet c, \ell_2), (ba \bullet c, \ell_2)\}.$$

According to Definition 7, the event $e_1$ is also denoted by $[\bullet abc, \ell_1]$ or by $[\bullet bbc, \ell_1]$ and so on. Each event corresponds to the application of a certain evolution rule at a certain position in the string. The resulting event structure corresponds very well to the following description: distribute the object to the rules and then apply the evolutions rules in parallel. The parallel execution of all involved evolution rules is possible because the corresponding events are independent (no causalities, no conflicts). Figure 2.b represents the fact that $bbac$ is obtained from $aabc$ using the computation space described by the event structure $(\{e_1, e_2, e_3\}, \mathrm{id}, \emptyset)$.

*Remark 1.* During an evolution step, the content of a membrane is split in two sides: the objects produced in the current evolution step (these cannot contribute to apply new rules), and the "unused" objects which may contribute to apply new rules. In [3] there are used colors to distinguish the two sides. Here we omit to make explicit distinction because the objects consumed by a rule can be deduced from the definition of the events.

We give now the formal definition for the event structure associated to a mpr-step over strings.

**Definition 9.** *The labeled transition system associated to* $w \Rightarrow_{(\ell_1,\dots,\ell_n)} w'$ *is given by all sequential rewriting starting from* $w$ *and ending in* $w'$. *The event structure* $ES(w, w', \ell_1, \dots, \ell_n)$ *associated to* $w \Rightarrow_{(\ell_1,\dots,\ell_n)} w'$ *is the event structure associated to its labeled transition system.*

**Theorem 2.** *The event structure* $ES(w, w', \ell_1, \dots, \ell_n) = (E, \leq, \#)$ *associated to* $w \Rightarrow_{(\ell_1,\dots,\ell_n)} w'$ *consists only of concurrent events, i.e.,* $\leq = id$ *and* $\# = \emptyset$.

*Proof.* We have $w \Rightarrow_{(\ell_1,\dots,\ell_n)} w'$ iff $w = w_0 u_1 w_1 \dots u_n w_n$, $w' = w_0 v_1 w_1 \dots v_n w_n$, $\ell_i : u_i \to v_i$ is an evolution rule, for $i = 1, \dots, n$, and $w_0 w_1 \dots w_n$ is irreducible. The conclusion of the theorem follows from the fact that $\{[w_0 \dots \bullet w_i \dots w_n, \ell_i]\}$ is a configuration (any of events can occur first). $\qquad\square$

## 3.2   Multiset Rewriting

We assume now that the sequential rewriting is executed over multisets (commutative words).

We write $w =_c w'$ if and only if $w'$ is obtained from $w$ by a permutation of the objects, i.e., $w$ and $w'$ are equal modulo commutativity. Let $[w]$ denote the $=_c$-equivalence class of $w$, i.e., $[w] = \{w' \mid w =_c w'\}$.

A context is a multiset of the form $[\bullet w]$, where $w$ is a multiset of objects, and $\bullet$ is a special variable. It is easy to see now that the position of $\bullet$ in a context is not important, and therefore we write $\bullet$ at the beginning. Each rewriting step $[uw] \to [vw]$ is uniquely determined by the context $[\bullet w]$ and the rule $\ell : u \to v$. Therefore the transition $([uw], \ell, [vw]) = [uw] \xrightarrow{\ell} [vw]$ is denoted by $([\bullet w], \ell)$.

The maximal parallel rewriting over multisets is defined as follows: $[w] \overset{mpr}{\Rightarrow} [w']$ iff there are $\ell_1, \dots, \ell_n$, $\ell_i : u_i \to v_i$ ($i = \overline{1, n}$) such that $[w] = [u_1 \dots u_n r]$, $[w'] = [v_1 \dots v_n r]$ and $r$ is irreducible (no rule can be applied). Sometimes we write $[w] \Rightarrow_{[\ell_1,\dots,\ell_n]} [w']$ in order to emphasize the multiset of rules implied in the maximal parallel rewriting. It is worth to note that, in order to avoid possible infinite cycles and illegal transitions in a maximal parallel rewriting step, an object $v$ resulting from $[uw] \xrightarrow{\ell} [vw]$ is not used afterwards in that step.

The space of all rewriting for the evolution step in Example 2 is indicated in Figure 3.

We also have three events, but they are not totally causally independent:

$$e_1 = \{([\bullet abc], \ell_1), ([\bullet aac], \ell_1)\}, \quad e_1 < e_2$$
$$e_2 = \{([\bullet bbc], \ell_1), ([\bullet bac], \ell_1)\}$$
$$e_3 = \{([\bullet abc], \ell_2), ([\bullet bac], \ell_2), ([\bullet bbc]\ell_2)\}$$

According to Definition 7, the event $e_1$ is also denoted by $[[\bullet abc], \ell_1]$ or by $[[\bullet aac], \ell_1]$. The other events are similarly denoted. An event corresponds now to the application of an evolution rule at an arbitrary position. The position in strings cannot be used anymore to distinguish between events. Moreover, between the events $e_1$ and $e_2$ we have a causal dependency: $e_2$ may occur only after $e_1$. In fact, $e_1$ can be read as "the first application of the evolution rule $\ell_1$"

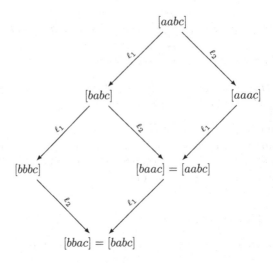

**Fig. 3.**

and $e_2$ as "the second application of the evolution rule $\ell_1$". We notice that the use of commutativity law changes dramatically the meaning of an event.

**Definition 10.** *The labeled transition system associated to* $[w] \Rightarrow_{[\ell_1,\ldots,\ell_n]} [w']$ *is given by all sequential rewriting starting from* $[w]$ *and ending in* $[w']$.

**Theorem 3.** *The event structure* $(E, \leq, \#)$ *associated to the labeled transition system defined by* $[w] \Rightarrow_{(\ell_1,\ldots,\ell_n)} [w']$ *has the following properties:*

- $[[\bullet w_1], \ell_1] < [[\bullet w_2], \ell_2]$ *if and only if* $\ell_1 = \ell_2$, $[[\bullet w_1], \ell_1]$ *corresponds to the i-th application of the rule* $\ell_1$, $[[\bullet w_2], \ell_1]$ *corresponds to the j-th application of the rule* $\ell_1$, *and* $i < j$;
- $\# = \emptyset$.

"Parallel" means "no causal dependency" between the events corresponding to the application of the rules. We may conclude either that working with multisets is not a good solution at this granularity, namely it is not possible to determine all the parallel rules applied in a mpr-step, or the procedure which determines the event structure from a lts finds "false" causalities for the particular case when the states are given by multisets. We believe that the later one is true; the causality relation given by the $i$-th application of a rule (when it is applied more than once) is artificial. Therefore we remove the false causal dependency in the definition of the event structure associated to a mpr-step.

**Definition 11.** *The event structure associated to* $[w] \Rightarrow_{(\ell_1,\ldots,\ell_n)} [w']$ *is* $(E, id, \emptyset)$, *where* $(E, \leq, \emptyset)$ *is the event structure associated to the labeled transition system defined by* $[w] \Rightarrow_{[\ell_1,\ldots,\ell_n]} [w']$. *We denote by* $ES([w], [w'], \ell_1, \ldots, \ell_n)$ *the event structure* $(E, id, \emptyset)$ *given above, and by* $E([w], [w'], \ell_1, \ldots, \ell_n)$ *its set of events* $E$.

## 3.3  Systems of Communicating Membranes

The constructions described in the previous subsections can be extended to communicating membranes. Communicating membranes (membrane systems with symport/antiport rules) are presented in detail in [11].

A *membrane system with symport/antiport rules* (of degree $n \geq 1$) is a construct of the form $\Pi = (O, \mu, w_1, \ldots, w_n, E, R_1, \ldots, R_n, i_o)$, where:

1. $O$ is the alphabet of objects,
2. $\mu$ is the membrane structure (of degree $n \geq 1$, with the membranes labeled in a one-to-one manner with $1, 2, \ldots, n$),
3. $w_1, \ldots, w_n$ are strings over $O$ representing the multisets of objects present in the $n$ compartments of $\mu$ in the initial configuration of the system,
4. $E \subseteq O$ is the set of objects appearing in the environment,
5. $R_1, \ldots, R_n$ are finite sets of rules associated with the $n$ membranes of $\mu$,
6. $i_o$ is the label of a membrane of $\mu$ indicating the *output* region.

The communication between membranes are expressed by rules of $R_i$; we can have *symport* rules of the forms $(x, in), (x, out)$, as well as *antiport* rules of the form $(u, out; v, in)$, where $x, u, v$ are strings over $O$. The length of $x$, respectively the maximum length of $u, v$, is called the *weight* of the corresponding (symport or antiport) rule. We refer mainly to symport rules.

Let us consider a P system with two membranes: $M$ consisting of the evolution rule $\ell : a \rightarrow b(c, in_{M'})$ and the content $[a]$, and $M'$ consisting of the evolution rule $\ell' : c \rightarrow d(a, out)$ and the content $[c]$. We assume that $M$ includes $M'$, i.e., the initial configuration of the system is $\langle M \mid [a] ; \langle M' \mid [c] \rangle \rangle$. We distinguish three cases.

*Simultaneous Evolution and Communication.* If we assume that the application of a communicating evolution rule, as $\ell$ or $\ell'$, is a single event, then the lts corresponding to all sequential computations is like in Figure 4. We have two concurrent events: one corresponding to the application of rule $\ell$, and the other one corresponding to the application of rule $\ell'$. Let $\Pi$ a system of communicating membranes and let $ES$ be the event structure associated to an evolution step of $\Pi$. If the contents of the membranes are represented as strings, then $ES$ consists of a set of concurrent events computed as in Section 3.1. If the contents of the membranes are represented by multisets, then $ES$ consists of a set of concurrent events computed as in Section 3.2, where the false causalities given by $i$-th application of a rule are removed.

*Separated Evolution and Communication.* If a rewrite semantics is used as in [3], the event is corresponding to an application of a rewrite rule. The application of the rule $\ell$ as a rewrite rule means replacing of the object $a$ with $b(c, in_{M'})$ in the content of $M$. The move of the *in*-message from $M$ to $M'$ is given with a new rewrite rule:

$$in : \langle M \mid [w(x, in_{M'})] ; \langle M' \mid [w'] \rangle \rangle \rightarrow \langle M \mid [w] ; \langle M' \mid [w'x] \rangle \rangle.$$

Similarly, the application of the rule $\ell'$ as a rewrite rule means replacing of the

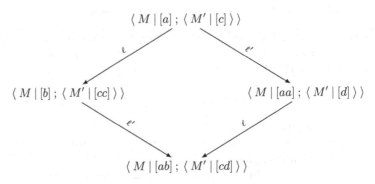

**Fig. 4.** Simultaneous evolution and communication

object $c$ with $d(a, out)$ in the content of $M'$. The move of the $out$-message from $M'$ to $M$ is given with the following rewrite rule:

$$out : \langle\, M \mid [w(x, in_{M'})]\,;\, \langle\, M' \mid [w']\,\rangle\,\rangle \to \langle\, M \mid [w]\,;\, \langle\, M' \mid [w'x]\,\rangle\,\rangle.$$

According to [2,3], an evolution step of the system consists of two main substeps: $mpr$-step consisting in the application of the evolution rules as rewrite rules, and $tar$-step consisting in the application of the rewrite rules moving the messages produced by the $mpr$-step to their target membrane. The $tar$-step is applied only after $mpr$-step is completely accomplished. The lts for our example is like in Figure 5. The notation for the system configurations was simplified in order to save space. We have four events: two corresponding to the evolution rules, and two corresponding to the communications (moving the messages to their targets). The later events are causally dependent of the former ones.

Let $\Pi$ a system of communicating membranes and let $ES = (E, \leq, \emptyset)$ be the event structure associated to an evolution step of $\Pi$. If $ES' = (E', \mathrm{id}, \emptyset)$ is the event structure associated to a $mpr$-step and $ES'' = (E'', \mathrm{id}, \emptyset)$ is the event structure associated to a $tar$-step, then $E = E' \cup E''$ (disjoint union) and $\leq\, =\, \mathrm{id} \cup \{e' < e'' \mid e' \in E', e'' \in E''\}$.

*Interleaving Evolution and Communication.* In [4] it is shown that the behavior of the P systems can be described as well by interleaving the applications of the evolutions rules with those of the communication rules.

Even if the lts corresponding to an evolution step is more complex (see Figure 6), we have the same number of events, namely four. The only causal dependencies are between the event corresponding to $in$-communication and the event corresponding to $\ell$, and between the event corresponding to $out$-communication and the event corresponding to $\ell'$.

For the general case, the event structure $ES = (E, \leq, \emptyset)$ associated with a system $\Pi$ is computed by extending the method described in Section 3.1 if the contents are represented by strings, or as in Section 3.2 if the contents are represented by multisets. We have $e \leq e'$ if and only if $e$ corresponds to an evolution rule having in the right hand side $in$- or $out$-messages and $e'$ is the event corresponding to a moving of such a message to its target.

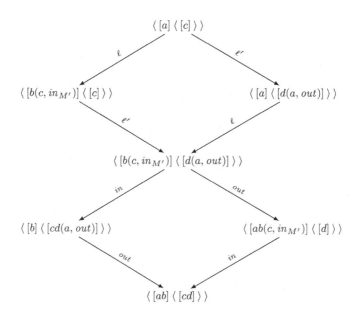

**Fig. 5.** Separated evolution and communication

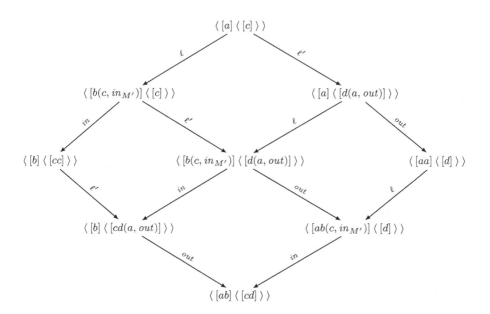

**Fig. 6.** Interleaving evolution and communication

*We have to answer two questions.*

Q1: What is the best representation for membrane contents: either strings or multisets?

Q2: What is the most suitable definition for the event structure of an evolution step including communication?

First we analyze Q1. The strings have the advantage that we can extract the event structure associated to an evolution step directly from the associated lts. However, as it is expected, the strings make too many distinctions and therefore do not supply the desired level of abstraction. If we permute the content of a membrane in such a way we may apply exactly the same rules but at different positions, then we get an event structure isomorphic to that corresponding to the initial content. The multisets supply the desired level of abstraction. The false causalities we get from the associated lts can be removed at a price which deserves to be paid.

Question Q2 has not yet a firm answer. The answer is depending on what biological system we want to model with these systems and on the granularity level we want to have in analyzing such systems. In [4] it is shown that the efficiency of a model checking algorithm is strong dependent on this granularity level. We will see later that the interleaving evolution and communication does not allow to retrieve the evolution steps from the event structures.

## 4    Event Structure of a Membrane

In this section we determine the event structure given by a membrane only considering the multiset approach. We first enumerate some problems which appear when we take into account the computation structure of a membrane.

*Fresh names.* The notation for events is no longer suitable for the case of membranes. We consider again the membrane of Example 2 with the content $aa$ this time, and the computation $[aa] \overset{mpr}{\Rightarrow} [bb] \overset{mpr}{\Rightarrow} [aa] \overset{mpr}{\Rightarrow} [bb]$. Since the events of the second step $[bb] \Rightarrow_{[\ell_2,\ell_2]} [aa]$ occur always before the events of the first step $[aa] \Rightarrow_{[\ell_1,\ell_1]} [bb]$, and the events of the third step $[aa] \Rightarrow_{[\ell_1,\ell_1]} [bb]$ occur before the events of the second step, we get $[a\bullet, \ell_1] < [b\bullet, \ell_2] < [a\bullet, \ell_1]$, i.e., the causality relation $<$ is cyclic. Therefore each event $[[\bullet u], \ell]$ in $E([w], [w'], \ell_1, \ldots, \ell_n)$ is denoted with a new fresh name $e$, and we define $action(e) = [\bullet u, \ell]$. We denote by $Fresh(x)$ the copy of $x$, where each event in $x$ is replaced by a fresh name, and by $action(x)$ the set $\{ action(e) \mid e \in x \}$. In this way, the event sets corresponding to different computation steps are disjoint.

*Causal dependency.* We consider the membrane including the following rules

$$\ell'_1 : a \to bc, \quad \ell'_2 : c \to d, \quad \ell'_3 : e \to f,$$

and the content $ae$. The lts given by all sequential rewriting is represented in Figure 7a) and it supplies the following events:

$$e_1' \text{ with } action(e_1') = [[\bullet e], \ell_1'] = \{([\bullet e], \ell_1'), ([\bullet f], \ell_1')\}$$
$$e_2' \text{ with } action(e_2') = [[\bullet a], \ell_3'] = \{([\bullet a], \ell_3'), ([\bullet bc], \ell_3')\}$$
$$e_3' \text{ with } action(e_3') = [[\bullet bf], \ell_2'] = \{([\bullet bf], \ell_2')\}$$

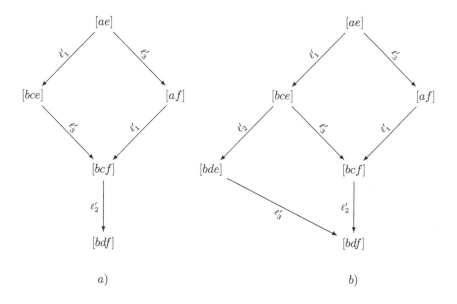

a)                                                b)

**Fig. 7.** Causal dependency and history causality

The only causalities we have are $e_1' \prec e_3'$ and $e_2' \prec e_3'$. Consequently, the only concurrency relation is $e_1' \, co \, e_2'$. This definition for causality is different from history causality introduced in [6], where $e \prec e'$ iff the rule involved in $e'$ uses objects produced by the rule involved in $e$. The history causality for our example produce the lts given in Figure 7b) and we have $e_1' \prec e_3'$, $e_1' \, co \, e_2'$ and $e_2' \, co \, e_3'$, where $action(e_2')$ is now $\{([\bullet a], \ell_3'), ([\bullet bc], \ell_3')([\bullet bd], \ell_3')\}$. As it is noted in [6], the history causal semantics does not faithfully reflect the maximal parallel rewriting; however, the maximal parallel rewriting can be retrieved by imposing some additional conditions. Another difference from [6] is that there each transition of the associated lts defines a new event; in particular, $([\bullet e], \ell_1')$ and $([\bullet f], \ell_1')$ defines two distinct events in the history causal semantics.

*Conflicts.* Let us consider the following membrane:

$$\ell_1'' : a \to c, \quad \ell_2'' : b \to d, \quad \ell_3'' : b \to e.$$

Two computations are possible for the content $ab$:
The events $e_2''$ and $e_3''$ with $action(e_2'') = [[\bullet a], \ell_2''] = \{([\bullet a], \ell_2''), ([\bullet c], \ell_2'')\}$ and $action(e_3'') = [[\bullet a], \ell_3''] = \{([\bullet a], \ell_3''), ([\bullet c], \ell_3'')\}$ are in conflict, i.e., $e_2'' \# e_3''$, because

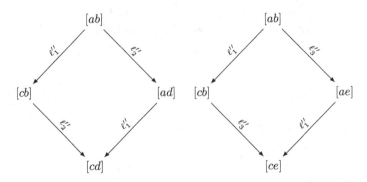

they compete on the same object (resource) $b$. The events $e_{11}''$ and $e_{12}''$ with $action(e_{11}'') = [[\bullet b], \ell_1''] = \{([\bullet b], \ell_1''), ([\bullet d], \ell_1'')\}$ and $action(e_{12}'') = [[\bullet b], \ell_1''] = \{([\bullet b], \ell_1''), ([\bullet e], \ell_1'')\}$ are also in conflict, i.e., $e_{11}'' \# e_{12}''$, because they correspond to different actions: the context $[\bullet e]$ is not possible in $e_{11}''$, and the context $[\bullet d]$ is not possible in $e_{12}''$. It is worth to note that $e_{11}'', e_2'' \in Fresh(E([ab], [cd], l_1'', l_2''))$ and $e_3'', e_{12}'' \in Fresh(E([ab], [cd], l_1'', l_3''))$. In a similar way, $e_{11}'' \# e_3''$ and $e_{12}'' \# e_2''$.

**Definition 12.** *In terms of multisets, a computation in $M$ is of the form $init \overset{mpr}{\Rightarrow} [w_1] \overset{mpr}{\Rightarrow} \cdots$, where init is the initial content of $M$. A multiset (content) $w$ is reachable in $M$ iff there is a computation from init to $[w]$.*

**Definition 13.** *Let $M$ be a membrane. An event structure $ES(M) = (E_M, \leq_M , \#_M)$ associated to a membrane $M$ is given as follows:*

1. $E_M$ *is the smallest set satisfying: for each reachable $[w]$, $[w']$ such that $[w] \Rightarrow_{(\ell_1,\ldots,\ell_n)} [w']$, $Fresh(E([w], [w'], \ell_1, \ldots, \ell_n)) \subseteq E_M$;*
2. $\leq_M$ *is the smallest partial order generated by:   if $[w] \Rightarrow_{[\ell_1,\ldots,\ell_m]} [w_1]$ and $[w_1] \Rightarrow_{[\ell_1',\ldots,\ell_n']} [w_2]$, then we have $e_1 \lessdot e_2 \in \leq_M$ for each $e_1$ in $Fresh(E([w], [w_1], \ell_1, \ldots, \ell_m))$ and $e_2$ in $Fresh(E([w_1], [w_2], \ell_1', \ldots, \ell_n'))$;*
3. $\#_M$ *is the smallest relation satisfying the principle of conflict heredity and including: if $([w] \Rightarrow_{[\ell_1,\ldots,\ell_m]} [w_1]) \neq ([w] \Rightarrow_{[\ell_1',\ldots,\ell_n']} [w_2])$, then we have $e_1 \#_M e_2$ for each $e_1$ in $Fresh(E([w], [w_1], \ell_1, \ldots, \ell_m))$ and $e_2$ in $Fresh(E([w], [w_2], \ell_1', \ldots, \ell_n'))$.*

We should note that each $Fresh(E([w], [w'], \ell_1', \ldots, \ell_n'))$ depends on the computation $init \overset{mpr}{\Rightarrow} \cdots \overset{mpr}{\Rightarrow} [w]$; otherwise we can have the following situation:

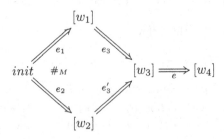

$e_1 \in Fresh(E(init, [w_1]), e_2 \in Fresh(E(init, [w_2]), e_3 \in Fresh(E([w_1], [w_3])$
and $e_3' \in Fresh(E([w_2], [w_3]), e \in Fresh(E([w_3], [w_4]);$ and we get

$$e_1 \#_M e_2, e_1 \leq e_3 \leq e, e_2 \leq e_3' \leq e$$

and thus $e \#_M e$.

Under this additional requirement to relate $Fresh(E([w], [w'], \ell_1', \ldots, \ell_n'))$ to
the computation $init \overset{mpr}{\Rightarrow} \cdots \overset{mpr}{\Rightarrow} [w]$, $ES(M)$ is indeed an event structure.

**Theorem 4.** *Given a membrane $M$, there is an event structure $ES(M) = (E_M, \leq_M, \#_M)$ such that if $[w], [w']$ are reachable in $M$ from init, and $[w] \Rightarrow_{(\ell_1, \ldots, \ell_n)} [w']$, then there exist two saturated computation steps $x, x'$ and a maximal concurrent action $a$ such that $x \overset{a}{\longrightarrow} x'$ is in $ES(M)$ and $action(a) = E([w], [w'], \ell_1, \ldots, \ell_n)$.*

*Proof.* We assume first that $[w] \Rightarrow_{(\ell_1, \ldots, \ell_n)} [w']$. Then $x$ is a computation space corresponding to a computation $init \overset{mpr}{\Rightarrow} \cdots \overset{mpr}{\Rightarrow} [w]$ in $M$, and $a$ is $Fresh(E([w], [w'], \ell_1, \ldots, \ell_n))$. If $[w] = init$, then $x = \emptyset$. It is easy to see that $x$ is conflict-free and downwards-closed, $a$ is a maximal set of concurrent events, and that $x \cap a = \emptyset$. $x' = x \cup a$ is also downwards-closed and conflict-free because it corresponds to the computation $init \overset{mpr}{\Rightarrow} \cdots \overset{mpr}{\Rightarrow} [w] \overset{mpr}{\Rightarrow} [w']$.

We assume now that $x \overset{a}{\longrightarrow} x'$ is in $ES(M)$. Let $depth(x')$ denote the length of the longest chain of causally dependent events in $x'$. We show by induction on $depth(x')$ that $x$ corresponds to a computation $init \overset{mpr}{\Rightarrow} \cdots \overset{mpr}{\Rightarrow} [w]$ and that $x'$ corresponds to $init \overset{mpr}{\Rightarrow} \cdots \overset{mpr}{\Rightarrow} [w] \overset{mpr}{\Rightarrow} [w']$. If $depth(x') = 0$, then $x = \emptyset$ and $\lceil a \rceil = a$, i.e., $a$ is a set of $\leq$-minimal events. The minimal events correspond to computations of the form $init \Rightarrow_{(\ell_1, \ldots, \ell_n)} [w]$. Two minimal events are concurrent if and only if both belong to the event structure corresponding to a computation $init \overset{mpr}{\Rightarrow} [w]$. Since $a$ is a maximal set of concurrent events, it follows that $a = Fresh(E(init, [w], \ell_1, \ldots, \ell_n))$. We assume now that $depth(x') > 0$. We have that $x$ corresponds to the computation $init \overset{mpr}{\Rightarrow} \cdots \overset{mpr}{\Rightarrow} [w]$ by inductive hypothesis. The $\leq$-maximal events in $x$ are those which contributed in the last evolution step of the derivation of $[w]$. Since each event in $a$ causally depends on a $\leq$-maximal event in $x$, it follows that $[w] \Rightarrow_{(\ell_1, \ldots, \ell_n)} [w']$ and $a = Fresh(E([w], [w'], \ell_1, \ldots, \ell_n))$ by the construction of $ES(M)$. $\qquad\square$

## 5   Event Structure for Communicating Membranes

For the case of systems of communicating membranes we have to distinguish between the three semantics of an evolution step including communication. Let $\Pi$ a system of communicating membranes.

*Simultaneous Evolution and Communication.* The algorithm computing the event structure $ES(\Pi)$ is similar to that from Definition 13. The only difference is that we have to consider configurations of $\Pi$ instead of membrane contents. The conclusion of Theorem 4 holds too, because the event structure of an evolution step does not include causalities.

*Separated Evolution and Communication.* The algorithm computing the event structure $ES(\Pi)$ is similar to that presented in the above paragraph, excepting the fact that $\leq$ must include now the causalities coming from the inside of evolution steps. The conclusion of Theorem 4 does not hold because the event structure of an evolution step includes causalities. However, because the communication is separated from the application of the evolution rules as rewrite rules, we can distinguish between a maximal concurrent actions corresponding to communications and a maximal concurrent actions corresponding to evolution rules. In this way, the conclusion of the theorem must be rephrased as follows: $[w] \Rightarrow [w']$ iff there is a maximal concurrent transitions $x \xrightarrow{a'} x' \xrightarrow{a''} x''$ in $ES(\Pi)$ such that $action(a') = E_{mpr}([w] \Rightarrow [w'])$ and $action(a') = E_{tar}([w] \Rightarrow [w'])$, where $E_{mpr}$ is the event structure corresponding to the *mpr*-step of $[w] \Rightarrow [w']$ and $E_{tar}$ is the event structure corresponding to the *mpr*-step of $[w] \Rightarrow [w']$.

*Interleaving Evolution and Communication.* The algorithm computing the event structure $ES(\Pi)$ is similar to that presented in the above paragraph. The conclusion of Theorem 4 does not hold from the same reasons as above. Moreover, we cannot retrieve the evolution steps $[w] \Rightarrow [w']$ from $ES(\Pi)$ because the events corresponding to the communication can be concurrent with those corresponding to evolution rules, and we have no criteria expressed in terms of causalities and concurrency to delimit the evolution steps.

## 6   Conclusion

Many of the formalisms using biology as inspiration can be naturally adapted to model and better understand the complex biomolecular systems. Membrane systems are used to model several biological phenomena [8]. For instance, a description of the sodium-potassium exchange pump is given in [5] by using membrane systems, and some dynamical aspects of the immune system are modeled by mobile membranes in [1]. However the new discrete time models are not so popular among the biologists as ordinary and partial differential equations are.

In this paper we emphasize a crucial feature in biology which is not described properly in the classical mathematical models. We refer to the causality of events in biological systems. For membrane systems, we overcome this drawback of the continuous time mathematical models in biology by defining and investigating a strong notion of causality provided by the event structures. We show that the notion of causality given by event structures is better for membrane systems than a weaker form of causality (history causality) presented in [6].

The event structure of a P system is constructed step-by-step, starting from the event structure of a maximally concurrent transition, followed by the event structures of the communicating membranes and by various combinations between the evaluation and communication steps. The interleaving between the evolution and the communication steps is also described in [4] in terms of operational semantics. An important result is given by Theorem 4, saying that causality of a membrane system corresponds properly to the operational semantics defined in [2,3].

**Acknowledgements.** This work has been supported by the research grant CEEX 47/2005.

# References

1. Aman, B., Ciobanu, G.: Describing the Immune System Using Enhanced Mobile Membranes. In: Proceedings FBTC (Satellite workshop of CONCUR 2007), pp. 1–14 (2007)
2. Andrei, O., Ciobanu, G., Lucanu, D.: A Structural Operational Semantics of the P Systems. In: Freund, R., Păun, G., Rozenberg, G., Salomaa, A. (eds.) WMC 2005. LNCS, vol. 3850, pp. 32–49. Springer, Heidelberg (2006)
3. Andrei, O., Ciobanu, G., Lucanu, D.: Operational Semantics and Rewriting Logic in Membrane Computing. Electronic Notes of Theoretical Computer Sci. 156, 57–78 (2006)
4. Andrei, O., Ciobanu, G., Lucanu, D.: A Rewriting Logic Framework for Operational Semantics of Membrane Systems. Theoretical Computer Sci. 373, 163–181 (2007)
5. Besozzi, D., Ciobanu, G.: A P System Description of the Sodium-Potassium Pump. In: Mauri, G., Păun, G., Pérez-Jiménez, M.J., Rozenberg, G., Salomaa, A. (eds.) WMC 2004. LNCS, vol. 3365, pp. 211–223. Springer, Heidelberg (2005)
6. Busi, N.: Causality in membrane systems. In: Eleftherakis, G., Kefalas, P., Păun, G., Rozenberg, G., Salomaa, A. (eds.) WMC 2007. LNCS, vol. 4860, pp. 160–171. Springer, Heidelberg (2007)
7. Ciobanu, G., Lucanu, D.: What is an event for membrane systems? In: Pre proceedings of Membrane Computing, International Workshop - WMC 8, Thessaloniki, Greece, pp. 255–266 (2007)
8. Ciobanu, G., Păun, G., Perez-Jimenez, M.J. (eds.): Applications of Membrane Computing. Natural Computing Series. Springer, Heidelberg (2006)
9. Nielsen, M., Rozenberg, G., Thiagarajan, P.S.: Transition Systems, Event Structures, and Unfoldings. Information and Computation 118, 191–207 (1995)
10. Păun, G.: Computing with Membranes. Journal of Computer and System Sciences 61, 108–143 (2000)
11. Păun, G.: Membrane Computing. An Introduction. Springer, Heidelberg (2002)
12. Sassone, V., Nielsen, M., Winskel, G.: Models for Concurrency: Towards a Classification. Theoretical Computer Sci. 170, 297–348 (1996)
13. Winskel, G.: Event structures. In: Brauer, W., Reisig, W., Rozenberg, G. (eds.) Advances in Petri Nets 1986. LNCS, vol. 255, pp. 325–392. Springer, Heidelberg (1987)
14. Winskel, G.: An introduction to event structures. In: de Bakker, J.W., de Roever, W.-P., Rozenberg, G. (eds.) Linear Time, Branching Time and Partial Order in Logics and Models for Concurrency. LNCS, vol. 354, pp. 364–397. Springer, Heidelberg (1989)
15. Winskel, G., Nielsen, M.: Models for Concurrency. In: Abramsky, S., Gabbay, D., Maibaum, T.S.E. (eds.) Handbook of Logic in Computer Science, Oxford University Press, Oxford (1995)

# P Systems with String Objects and with Communication by Request

Erzsébet Csuhaj-Varjú and György Vaszil

Computer and Automation Research Institute
Hungarian Academy of Sciences
Kende utca 13–17, H-1111 Budapest, Hungary
{csuhaj,vaszil}@sztaki.hu

**Abstract.** In this paper we study P systems using string-objects where the communication between the regions is indicated by the occurrence of so-called query symbols in the string. We define two variants of communication and prove that these systems with both types of communication are computationally complete, even having a number of membranes limited with relatively small constants.

## 1 Introduction

In this paper we continue our investigations on P systems with string objects and with communication by request. In [2], the authors studied tissue-like P systems over string objects where the evolution rules of the objects are represented by context-free rewriting rules which also describe the communication between the membranes by the help of communication symbols, called query symbols, one such symbol corresponding to each region of the system.

Membrane systems, or P systems, are distributed and parallel computing devices inspired by the functioning of the living cell [5]. A P system consists of a hierarchically embedded structure of membranes. Each membrane encloses a region that contains objects and might also contain other membranes. There are rules associated to the regions describing the evolution and the movement of the objects which together correspond to a computation.

For details on membrane systems, see the monograph [6] and the web-page http://psystems.disco.unimib.it.

While in the standard case a P system consists of a hierarchically embedded structure of membranes, tissue-like P systems are organized in another manner [4]. Instead of an individual cell, these correspond to groups of cells, like tissues or organs, interacting with each other either directly or with the use of the environment, but in any case, having the common property that the membrane structures are not necessarily described by a tree as those ones which correspond to individual cells.

The functioning of P systems with string objects consists of rewriting steps which rewrite the strings using context-free rewriting rules, and communication steps which exchange the strings between the regions.

G. Eleftherakis et al. (Eds.): WMC8 2007, LNCS 4860, pp. 228–239, 2007.

Communication in tissue-like P systems with string objects and with communication by request was defined as follows: When one or more query symbols are introduced in a string, then the rewriting of that string stops and the queries are satisfied by replacing the query symbols with strings which do not contain further query symbols from the region indicated by the query symbol, in all possible combinations. If no query symbol free string exists in the queried region, then the string containing the query disappears.

This model has some biological resemblance: if the strings are considered as descriptions of simple organisms, the query symbols as their "weak points", possibly infected or attacked by another organism, then the communication mimics some features of an infection or parasitism. Inspired by these resemblances, we call a communication of type $i$ (infection) if after communicating the copies of the strings, the strings themselves remain in the region, while the communication is called of type $p$ (parasitism), if after communication the communicated string itself disappears from its original region. The model is called an MPC system in short.

MPC systems can also be considered as modified variants of parallel communicating (PC) grammar systems defined over multisets of strings. PC grammar systems are networks of grammars organized in a communicating system to generate a single language. The reader interested in the theory of grammar systems is referred to [1,7].

In [2], the authors proved that MPC systems with 7 membranes and working with $i$-communication are able to describe all recursively enumerable languages. The computational completeness of these systems working with $p$-communication holds as well, even for a subclass consisting of systems having only 9 membranes.

In this paper we define the two types of communication for standard P systems and examine the computational power and the size complexity of these models. We call the new constructs RPC systems in short. In this case, the requested string can only be communicated either to the parent membrane or to one of the child membranes, depending on the issued query symbol. Thus, query symbols refer only to the neighboring regions. According to the above mentioned biological resemblance, both infection and parasitism are very local phenomena regarding their spread, i.e., in one step only the neighbors can be infected and parasitism can be developed only between two closely related, i.e., neighbor components.

As for MPC systems, the computational completeness can be proved for RPC systems with both types of communication: in the case of $i$-communication systems with 10 membranes and in the case of $p$-communication systems with 30 membranes are enough for demonstrating the power of the Turing machines. The reader can observe that both MPC systems and P systems are able to obtain the computational completeness even with relatively small number of membranes. Moreover, in the case of $i$-communication the difference between the two numbers is very small, i.e., the difference in the underlying structure of the membrane system has not too much influence on the computational power of the system.

## 2    Preliminaries and Definitions

We first recall the notions and the notations we use. The reader is assumed to be familiar with the basics of formal language theory, for details see [7]. Let $\Sigma$ be an alphabet and let $\Sigma^*$ be the set of all words over $\Sigma$, that is, the set of finite strings of symbols from $\Sigma$, and let $\Sigma^+ = \Sigma^* - \{\varepsilon\}$ where $\varepsilon$ denotes the empty word. For $w \in \Sigma^*$ and $S \subseteq \Sigma$, let $|w|_S$ denote the number of occurrences of symbols from $S$ in the string $w$ (if $S = \{a\}$ is a singleton set, we may write $|w|_a$ instead of $|w|_{\{a\}}$).

Let $V$ be a set of objects, and let $\mathbb{N}$ denote the set of non-negative integers. A multiset is a mapping $M : V \to \mathbb{N}$ which assigns to each object $a \in V$ its multiplicity $M(a)$ in $M$. The support of $M$ is the set $supp(M) = \{a \mid M(a) \geq 1\}$. If $supp(M)$ is a finite set, then $M$ is called a finite multiset. The set of all finite multisets over the set $V$ is denoted by $V^\circ$.

We say that $a \in M$ if $M(a) \geq 1$. For two multisets $M_1, M_2 : V \to \mathbb{N}$, $M_1 \subseteq M_2$ if for all $a \in V$, $M_1(a) \leq M_2(a)$. The union of $M_1$ and $M_2$ is defined as $(M_1 \cup M_2) : V \to \mathbb{N}$ with $(M_1 \cup M_2)(a) = M_1(a) + M_2(a)$ for all $a \in V$, the difference is defined for $M_2 \subseteq M_1$ as $(M_1 - M_2) : V \to \mathbb{N}$ with $(M_1 - M_2)(a) = M_1(a) - M_2(a)$ for all $a \in V$, and the intersection is $(M_1 \cap M_2) : V \to \mathbb{N}$ with $(M_1 \cap M_2)(a) = min(M_1(a), M_2(a))$ for $a \in V$, where $min(x, y)$ denotes the minimum of $x, y \in \mathbb{N}$. We say that $M$ is empty, denoted by $\epsilon$, if its support is empty, $supp(M) = \emptyset$.

In the following we sometimes list elements $a_1, \ldots, a_n$ of a multiset as $M = \{\{a_1, \ldots, a_n\}\}$, by using double brackets to distinguish from the usual set notation.

A P system is a structure of hierarchically embedded membranes, each having a label and enclosing a region containing a multiset of objects and possibly other membranes. The out-most membrane which is unique, is called the skin membrane. The membrane structure is denoted by a sequence of matching parentheses where the matching pairs have the same label as the membranes they represent. If membrane $l_i$ of a given membrane structure $\mu$ contains membrane $l_j$, and there is no other membrane, $l_k$, such that $l_k$ contains $l_j$ and $l_i$ contains $l_k$, then we say that membrane $l_i$ is the parent membrane of $l_j$, denoted as $parent_\mu(l_j) = l_i$, and $l_j$ is one of the child membranes of $l_i$, denoted as $l_j \in child_\mu(l_i)$. We also define for any region $l_i$ the set of regions $neighbor_\mu(l_i) = \{l_i\} \cup \{parent_\mu(l_i)\} \cup child_\mu(l_i)$.

The evolution of the contents of the regions of a P system is described by rules associated to the regions. Applying the rules synchronously in each region, the system performs a computation by passing from one configuration to another one. Several variants of the basic notion have been introduced and studied proving the power of the framework, see the monograph [6] for a summary of notions and results of the area.

In the following we focus on systems where the objects are represented with strings, object evolution is modeled by context-free string rewriting rules, and communication is performed by dynamically emerging requests with the use of query symbols appearing in the string objects.

**Definition 1.** A string rewriting P system with communication by request or an RPC system (of degree $m \geq 1$) is a construct

$$\Pi = (V, \mu, (M_1, R_1), \ldots, (M_m, R_m), i_o),$$

where:

- $V = N \cup T \cup K$ with $N, T, K$ being pairwise disjoint alphabets of nonterminals, terminals, and *query symbols*, respectively, and $K = \{Q_1, \ldots, Q_m\}$ (one query symbol is associated to each region of $\Pi$);
- $\mu$ is a membrane structure of $m$ membranes;
- $M_1, \ldots, M_m$ are finite multisets over $(N \cup T)^*$;
- $R_1, \ldots, R_m$ are finite sets of context-free rewriting rules of the form $A \rightarrow u$, with $A \in N$ and $u \in V^*$, satisfying the property that $A \rightarrow u \in R_i$ and $|u|_{\{Q_j\}} > 0$ implies $j \in neighbor_\mu(i)$;
- $i_o \in \{1, 2, \ldots, m\}$ is the index of the *output* membrane of $\Pi$.

The work of such a system starts from the initial configuration $(M_1, \ldots, M_m)$. It passes from a configuration $(M_1', \ldots, M_m')$, consisting of multisets of strings over $N \cup T \cup K$ placed in the $m$ regions of the system, to another configuration $(M_1'', \ldots, M_m'')$ in the following way. If no query symbol is present in the strings contained by the system, then each string from each multiset $M_i'$ is rewritten which can be rewritten according to the rules from $R_i, 1 \leq i \leq m$. This means the use of one rule from $R_i$, non-deterministically chosen, for each string. The strings which cannot be rewritten (no rule can be applied to them) remain unchanged. The resulting multisets of strings are $M_i'', 1 \leq i \leq m$. Note that the rewriting of strings is maximally parallel, in the sense that all strings which can be rewritten must be rewritten, and that the process is non-deterministic, the choice of rules and the places where the rules are applied can lead to several possible new multisets of strings.

If any query symbol is present in any of the strings contained by $M_i', 1 \leq i \leq n$, then a communication is performed: Each symbol $Q_j$ introduced in a string present in region $i$ (that is, in the multiset $M_i'$), where $j$ is the index of one of the neighboring regions, is replaced with all strings from this neighboring region $j$ which do not contain query symbols. If in region $j$ there are several strings without query symbols, then each of them is used, hence the string from region $i$ is replicated: a copy is created for each query symbol free string from region $j$ and the occurrences of $Q_j$ are replaced with different strings from region $j$ in each copy. If there are several query symbols in the same string from component $i$, then all of them are replaced (we also say that they are *satisfied*) at the same time, in all possible combinations. If a query symbol $Q_j$ cannot be satisfied (region $j$ contains no string without query symbols), then the string containing $Q_j$ is removed (it is like replacing it with the strings from an empty language). We call such a system *i*-communicating if copies of the requested strings are communicated to the requesting components, and *p*-communicating if after replacing the query symbols with the requested strings, these strings are removed from the multiset associated to the queried region.

In this way, all query symbols introduced by the rewriting rules disappear, they are either satisfied (replaced by strings without query symbols) or they disappear together with the string which contain them (in the case when they cannot be satisfied). The multisets obtained in this way in one communication step are $M_1'', \ldots, M_m''$, constituting the next configuration of the system.

We give now the formal definition of the transition.

**Definition 2.** Let $\Pi = (V, \mu, (M_1, R_1), \ldots, (M_m, R_m), i_o)$ be an RPC system as above, and let $(M_1', \ldots, M_m')$ and $(M_1'', \ldots, M_m'')$ be two configurations of $\Pi$. We say that $(M_1', \ldots, M_m')$ directly derives $(M_1'', \ldots, M_m'')$, if one of the following two cases holds.

1. There is no string containing query symbols, that is, $x \in (N \cup T)^*$ for all $x \in \bigcup_{i=1}^{m} M_i$. In this case, if $M_i' = \{\{x_{i,1}, \ldots, x_{i,t_i}\}\}$, then $M_i'' = \{\{y_{i,1}, \ldots, y_{i,t_i}\}\}$ where either $x_{i,j} \Rightarrow y_{i,j}$ according to a context-free rule of $R_i$, or $y_{i,j} = x_{i,j}$ if there is no rule in $R_i$ which can be applied to $x_{i,j}$, $1 \le j \le t_i$, $1 \le i \le m$.
2. There is at least one $x \in \bigcup_{i=1}^{m} M_i$ such that $|x|_K > 0$. In this case, rewriting is stopped and a communication step must be performed as follows. Let

$$
M_i^{req} = \begin{cases} \{\{x \in M_i' \mid |x|_K = 0\}\} & \text{if there is a } j \in neighbor_\mu(i), \text{ such} \\ & \text{that } y \in M_j' \text{ with } |y|_{Q_i} > 0, \\ \emptyset & \text{otherwise,} \end{cases}
$$

let

$$
M_i^{avail} = \{\{x \in M_i' \mid |x|_K = 0\}\},
$$

for all $i$, $1 \le i \le m$, and let for an $x = x_1 Q_{i_1} x_2 Q_{i_2} \ldots Q_t x_{t+1}$, $x_j \in (N \cup T)^*$, $Q_{i_j} \in K$, $1 \le j \le t+1$,

$$
Sat(x) = \{\{x_1 y_{i_1} x_2 y_{i_2} \ldots y_{i_t} x_{t+1} \mid y_{i_j} \in M_{i_j}^{avail}, 1 \le j \le t\}\}.
$$

Note that if $M_{i_j}^{avail} = \emptyset$ for some $i_j$, $1 \le j \le t$, then also $Sat(x) = \emptyset$. Now, for all $i$, $1 \le i \le m$,

$$
M_i'' = M_i' - M_i^{req} - \{\{x \in M_i' \mid |x|_K > 0\}\} + \bigcup_{x \in M_i', |x|_K > 0} Sat(x)
$$

in the $p$-communicating mode, and

$$
M_i'' = M_i' - \{\{x \in M_i' \mid |x|_K > 0\}\} + \bigcup_{x \in M_i', |x|_K > 0} Sat(x)
$$

in the $i$-communicating mode.

Let us denote the transitions from one configuration to another, $(M_1', \ldots, M_m')$ to $(M_1'', \ldots, M_m'')$, by $(M_1', \ldots, M_m') \Rightarrow_X (M_1'', \ldots, M_m'')$ with $X = i$ and $X = p$ for $i$-communicating and $p$-communicating systems, respectively.

The language generated by the RPC system consists of all terminal strings produced in region $i_o$ during any possible computation in $\Pi$.

$$L_X(\Pi) = \{x \in T^* \mid (M_1, \ldots, M_m) \Rightarrow_X^* (M'_1, \ldots, M'_m) \text{ and } x \in M'_{i_o}\}$$

for $X \in \{i, p\}$, where $\Rightarrow_X^*$ denote the reflexive and transitive closure of $\Rightarrow_X$.

The families of all languages generated in this way by RPC systems of degrees at most $m \geq 1$ with $i$-communication or $p$-communication, is denoted by $iRPC_mCF$ and $pRPC_mCF$, respectively. If we use systems of an arbitrary degree, then we replace the subscript $m$ with $*$. Let us also denote the class of recursively enumerable languages by $RE$.

Before presenting our results, we recall the notion of a two-counter machine from [3]. A *two-counter machine* $TCM = (T \cup \{Z, B\}, E, R)$ is a 3-tape Turing machine where $T$ is an *alphabet*, $E$ is a set of *internal states* with two distinct elements $q_0, q_F \in E$, and $R$ is a set of *transition rules*. The machine has a read-only input tape and two semi-infinite storage tapes (the counters). The alphabet of the storage tapes contains only two symbols, $Z$ and $B$ (blank), while the alphabet of the input tape is $T \cup \{B\}$. $R$ contains transition rules of the form $(q, x, c_1, c_2) \to (q', e_1, e_2)$ where $x \in T \cup \{\varepsilon\}$ corresponds to the symbol scanned on the input tape in state $q \in E$, and $c_1, c_2 \in \{Z, *\}$ correspond to the symbols scanned on the storage tapes. If $c_i = Z$, then the symbol scanned on the $i$th counter tape is $Z$, if $c_i = *$, then the symbol scanned is either $Z$ or $B$. By a rule of this form, $M$ enters state $q' \in E$, and the counters are modified according to $e_1, e_2 \in \{-1, 0, +1\}$. If $x \in T$, then the machine was scanning $x$ on the input tape, and the head moves one cell to the right; if $x = \varepsilon$, then the machine performs the transition irrespective of the scanned input symbol, and the reading head does not move.

The symbol $Z$ appears initially on the cells scanned by the storage tape heads and may never appear on any other cell. An integer $t$ can be stored by moving a tape head $t$ cells to the right of $Z$. A stored number can be incremented or decremented by moving the tape head right or left. The machine is capable of checking whether a stored value is *zero* or not by looking at the symbol scanned by the storage tape heads. If the scanned symbol is $Z$, then the value stored in the corresponding counter is *zero*. Note that although we do not allow to explicitly check the non-emptiness of the counters which is allowed in [3], this feature can be simulated: After successfully decrementing and incrementing a counter, the stored value is not altered, but the machine can be sure that the scanned symbol is $B$. A word $w \in T^*$ is accepted by the two counter machine if the input head has read the last non-blank symbol on the input tape, and the machine is in the accepting state $q_F$. Two-counter machines are computationally complete; they are just as powerful as Turing-machines, see [3].

## 3   The Universality of RPC Systems

First we show that RPC systems with $i$-communication are computationally universal; they characterize the class of recursively enumerable languages, even with a limited number of components.

**Theorem 1.** $iRPC_{10}CF = RE$.

*Proof.* We only give the proof of the inclusion $RE \subseteq iRPC_{10}CF$. The reverse inclusion follows from the Church thesis. To this aim, let us consider a recursively enumerable language $L \subseteq T^*$ and a two-counter machine $TCM = (T \cup \{Z, B\}, E, R)$, as presented in the previous section, characterizing the language $L$. We construct an RPC system accepting $L$ using $i$-communication. Let

$$\Pi = (V, \mu, C_{sel}, C_{gen}, C_{ch_1}, C_{S_4}, C_{c_1}, C_{ind_1}, C_{ch_{2,1}}, C_{c_2}, C_{ind_2}, C_{ch_{2,2}}, sel),$$

where if $Ind = \{sel, gen, c_1, c_2, ind_1, ind_2, ch_1, S_4, ch_{2,1}, ch_{2,2}\}$ then the components are $C_\alpha = (M_\alpha, R_\alpha)$ for $\alpha \in Ind$, and the membrane structure is defined as $\mu = [\,[\,[\,[\,]_{S_4}\,[\,[\,]_{ind_1}\,[\,]_{ch_{2,1}}]_{c_1}\,[\,[\,]_{ind_2}\,[\,]_{ch_{2,2}}]_{c_2}\,]_{ch_1}\,]_{gen}\,]_{sel}$.

Let $D = \{[q, x, c_1, c_2, q', e_1, e_2] \mid (q, x, c_1, c_2) \rightarrow (q', e_1, e_2) \in R\}$, and let us define for any $\alpha = [q, x, c_1, c_2, q', e_1, e_2] \in D$, the following notations: $State(\alpha) = q$, $Read(\alpha) = x$, $NextState(\alpha) = q'$, and $Store(\alpha, i) = c_i$, $Action(\alpha, i) = e_i$ for $i = 1, 2$.

The general idea of the simulation is to represent the states and the transitions of $TCM$ with nonterminals of $D$ and the values of the counters by strings of nonterminals containing as many $A$ symbols as the value stored in the given counter. Let $V = N \cup K \cup T$, where $K = \{Q_\alpha \mid \alpha \in Ind\}$, let the set of nonterminals be $N = \{S'_i, F_i, B_i \mid 1 \leq i \leq 6\} \cup \{S_i, C_i \mid 1 \leq i \leq 7\} \cup \{\alpha_i, H_i \mid \alpha \in D, 1 \leq i \leq 8\} \cup \{\alpha'_1, \alpha''_1, \bar{\alpha}'_1, \bar{\alpha}''_1, \underline{\bar{\alpha}}'_1 \mid \alpha \in D\} \cup \{F_{1,i} \mid 1 \leq i \leq 5\} \cup \{J_i \mid 1 \leq i \leq 4\} \cup \{A, \bar{F}'_1, \bar{\underline{F}}'_1, \underline{\bar{F}}'_1, \bar{F}''_1, E, E_1, I, J, S''_6\}$, and let the rules be defined as follows.

$$M_{sel} = \{\{I\}\},$$
$$\begin{aligned}
R_{sel} = &\{I \rightarrow \alpha_1 \mid \alpha \in D, State(\alpha) = q_0\} \cup \\
&\{\alpha_8 \rightarrow \beta_1 \mid \alpha, \beta \in D, NextState(\alpha) = State(\beta)\} \cup \\
&\{\alpha_8 \rightarrow F_1 \mid \alpha \in D, NextState(\alpha) = q_F\} \cup \\
&\{\alpha_i \rightarrow \alpha_{i+1} \mid \alpha \in D, 1 \leq i \leq 7\} \cup \{F_i \rightarrow F_{i+1} \mid 1 \leq i \leq 5\} \cup \\
&\{F_6 \rightarrow Q_{gen}, S'_6 \rightarrow S''_6, S''_6 \rightarrow Q_{gen}, E \rightarrow \varepsilon, \bar{\alpha}''_1 \rightarrow \varepsilon\}.
\end{aligned}$$

This region keeps track of the current state of the simulated two-counter machine and also selects the transition to be simulated. The symbol $I$ is used to initialize the system by introducing one of the initial transition symbols of the form $[q_0, x, c_1, c_2, q', e_1, e_2]_1$ where $q_0$ is the initial state. It also produces the result of the computation when after simulating the entering of the counter machine into the final state (that is, after the appearance of the nonterminal $F_1$), it receives the strings produced in the lower regions and erases the occurrences of the nonterminals $E$ which, if the simulation was successful, produces a terminal word accepted by the two-counter machine.

$$\begin{aligned}
M_{gen} = &\{\{S_1, S'_1\}\}, \\
R_{gen} = &\{S_1 \rightarrow S_2, S_2 \rightarrow Q_{sel}, S'_1 \rightarrow Q_{sel}\} \cup \\
&\{\alpha_1 \rightarrow \alpha'_1, \alpha_2 \rightarrow Q_{sel}, \alpha_i \rightarrow \alpha_{i+1} \mid \alpha \in D, 3 \leq i \leq 6\} \cup
\end{aligned}$$

$$\{\alpha_1' \to \alpha_1'', \alpha_1'' \to S_2', \alpha_7 \to xS_1 \mid \alpha \in D, Read(\alpha) = x\} \cup$$
$$\{F_1 \to F_{1,1}, F_{1,i} \to F_{1,i+1}, F_{1,5} \to Q_{ch_1} \mid 1 \le i \le 4\} \cup$$
$$\{S_i' \to S_{i+1}' \mid 2 \le i \le 4\} \cup \{S_5' \to S_1', A \to \varepsilon, J \to \varepsilon\} \cup$$
$$\{F_2 \to Q_{sel}, F_i \to F_{i+1} \mid 3 \le i \le 4\} \cup \{F_5 \to Q_{ch_1}\} \cup$$
$$\{S_6' \to S_6'', H_7 \to \varepsilon\}.$$

This region generates the string accepted by the counter machine by adding the symbol $Read(\alpha)$ for each $\alpha \in D$ chosen in the selector region. After the appearance of the nonterminal $F_1$ in the system, this region will append the words from the checking region $ch_1$ to its own string and send this string which also contains the generated word to the region $sel$. Then it will receive the word from region $ch_2$ and erase all $A$ and $J$ symbols before forwarding it also to region $sel$.

$$M_{ch_1} = \{\{S_1, S_1'\}\},$$
$$R_{ch_1} = \{S_1 \to S_2, S_2 \to S_3, S_3 \to Q_{gen}, S_1' \to S_2', S_2' \to Q_{gen}\} \cup$$
$$\quad \{\alpha_1'' \to \delta_1 \delta_2 Q_{S_4} \mid \alpha \in D, \delta_j = Q_{c_j} \text{ if } Store(\alpha, j) = Z,$$
$$\quad \text{ or } \delta_j = \varepsilon \text{ otherwise}\} \cup \{S_i \to S_{i+1} \mid 4 \le i \le 6\} \cup$$
$$\quad \{\alpha_1' \to \bar{\alpha}_1', \bar{\alpha}_1' \to \underline{\bar{\alpha}}_1', \underline{\bar{\alpha}}_1' \to S_3', S_3' \to S_4', S_4' \to S_5'\} \cup$$
$$\quad \{F_{1,1} \to \bar{F}_1', \bar{F}_1' \to \underline{\bar{F}}_1', \underline{\bar{F}}_1' \to \underline{\underline{F}}_1', \underline{\underline{F}}_1' \to Q_{c_1} Q_{c_2}\} \cup$$
$$\quad \{S_7 \to S_1, S_5' \to S_1', F_{1,2} \to Q_{S_4}\}, \text{ and}$$
$$M_{S_4} = \{\{S_4\}\},$$
$$R_{S_4} = \emptyset.$$

The region $ch_1$ checks whether the counter contents are zero when they should be zero by collecting the counter strings from regions $c_1, c_2$ when necessary. At the end of the simulation, the collected string is forwarded to the region $gen$ and then to region $sel$, where a terminal string can only be produced if the word originating in region $ch_1$ contains no $A$ symbols.

For $j = 1, 2$, let

$$M_{c_j} = \{\{J, C_1\}\},$$
$$R_{c_j} = \{J \to J_1, J_1 \to J_2, J_2 \to Q_{ch_1}, A \to Q_{ind_j}, J_3 \to Q_{ind_j}, J_4 \to J\} \cup$$
$$\quad \{\bar{\alpha}_1' \to \bar{\alpha}_1'', \bar{\alpha}_1'' \to \delta_\alpha J_3, \underline{\bar{\alpha}}_1' \to C_5 \mid \alpha \in D, \delta_\alpha = A \text{ if } Action(\alpha, j) = 0,$$
$$\quad \delta_\alpha = AA \text{ if } Action(\alpha, j) = +1, \delta_\alpha = \varepsilon \text{ if } Action(\alpha, j) = -1\} \cup$$
$$\quad \{C_i \to C_{i+1}, C_4 \to Q_{ch_1}, C_7 \to C_1 \mid i \in \{1, 2, 3, 5, 6\} \} \cup$$
$$\quad \{\bar{F}_1' \to \bar{F}_1'', \underline{\bar{F}}_1' \to Q_{ch_2, j}, E \to E_1, H_6 \to H_7\}.$$

These regions maintain strings representing the contents of the two counters. After the selection of a transition symbol in the region corresponding to $C_{sel}$, they execute the action required by the chosen transition symbol by adding $AA$, $A$, or $\varepsilon$ to the counter string and then deleting one $A$ and $J_3$ by rewriting it to

**Table 1.** Components of $\Pi$ in the proof of Theorem 1, simulating the behavior of the first counter when using an instruction with $\alpha = [q, x, Z, B, q', +1, -1]$. The region $C_{S_4}$ is omitted since it always contains the string $S_4$.

| | $C_{sel}$ | $C_{gen}$ | $C_{ch_1}$ | $C_{c_1}$ | $C_{ind_1}$ | $C_{ch_{2,1}}$ |
|---|---|---|---|---|---|---|
| 0 | $\beta_8$ | $wS_1, S_1'$ | $E...S_1, S_1'$ | $E...J, C_1$ | $B_1$ | $AEJ...H_1$ |
| 1 | $\alpha_1$ | $wS_2, Q_{sel}$ | $E...S_2, S_2'$ | $E...J_1, C_2$ | $B_2$ | $AEJ...H_2$ |
| 1C | $\alpha_1$ | $wS_2, \alpha_1$ | $E...S_2, S_2'$ | $E...J_1, C_2$ | $B_2$ | $AEJ...H_2$ |
| 2 | $\alpha_2$ | $wQ_{sel}, \alpha_1'$ | $E...S_3, Q_{gen}$ | $E...J_2, C_3$ | $B_3$ | $AEJ...H_3$ |
| 2C | $\alpha_2$ | $w\alpha_2, \alpha_1'$ | $E...S_3, \alpha_1'$ | $E...J_2, C_3$ | $B_3$ | $AEJ...H_3$ |
| 3 | $\alpha_3$ | $wQ_{sel}, \alpha_1''$ | $E...Q_{gen}, \bar{\alpha}_1'$ | $E...Q_{ch_1}, C_4$ | $B_4$ | $AEJ...H_4$ |
| 3C | $\alpha_3$ | $w\alpha_3, \alpha_1''$ | $E...\alpha_1'', \bar{\alpha}_1'$ | $E...\bar{\alpha}_1', C_4$ | $B_4$ | $AEJ...H_4$ |
| 4 | $\alpha_4$ | $w\alpha_4, S_2'$ | $E...Q_{c_1}Q_{S_4}, \bar{\alpha}_1'$ | $E...\bar{\alpha}_1'', Q_{ch_1}$ | $B_5$ | $AEJ...H_5$ |
| 4C | $\alpha_4$ | $w\alpha_4, S_2'$ | $E...\bar{\alpha}_1''S_4, \bar{\alpha}_1'$ | $E...\bar{\alpha}_1'', \bar{\alpha}_1'$ | $B_5$ | $AEJ...H_5$ |
| 5 | $\alpha_5$ | $w\alpha_5, S_3'$ | $E...\bar{\alpha}_1''S_5, S_3'$ | $E...\delta_\alpha J_3, C_5$ | $B_6$ | $AEJ...H_6$ |
| 6 | $\alpha_6$ | $w\alpha_6, S_4'$ | $E...\bar{\alpha}_1''S_6, S_4'$ | $E...Q_{ind_1}J_3, C_6$ | $E$ | $AEJ...H_7$ |
| 6C | $\alpha_6$ | $w\alpha_6, S_4'$ | $E...\bar{\alpha}_1''S_6, S_4'$ | $E...EJ_3, C_6$ | $E$ | $AEJ...H_7$ |
| 7 | $\alpha_7$ | $w\alpha_7, S_5'$ | $E...\bar{\alpha}_1''S_7, S_5'$ | $E...EQ_{ind_1}, C_7$ | $J_4$ | $AEJ...H_8$ |
| 7C | $\alpha_7$ | $w\alpha_7, S_5'$ | $E...\bar{\alpha}_1''S_7, S_5'$ | $E...EJ_4, C_7$ | $J_4$ | $AEJ...H_8$ |
| 8 | $\alpha_8$ | $wxS_1, S_1'$ | $E...\bar{\alpha}_1''S_1, S_1'$ | $E...EJ, C_1$ | $B_1$ | $AEJQ_{c_1}H_1$ |
| 8C | $\alpha_8$ | $wxS_1, S_1'$ | $E...\bar{\alpha}_1''S_1, S_1'$ | $E...EJ, C_1$ | $B_1$ | $AEJ...H_1$ |

$Q_{ind_j}$. The simulation can only be successful if exactly one $A$ and the symbol $J_3$ is rewritten. This is ensured by region $C_{ind_j}$. If there is a string obtained after the two queries which contain only a number of $A$ or $E$ symbols and one $J_4$ symbol, then the simulation of the actions required by the chosen transition was successful. If a counter is empty, this construction also forbids the successful execution of the decrement instruction since this would introduce $E_1$ in the counter strings.

The rules of the region supporting the work of the counters $C_{c_j}$, $j = 1, 2$, are defined as follows.

$$M_{ind_j} = \{\{B_1\}\},$$
$$R_{ind_j} = \{B_i \rightarrow B_{i+1} \mid 1 \leq i \leq 5\} \cup \{B_6 \rightarrow E, E \rightarrow J_4, J_4 \rightarrow B_1\},$$

and

$$M_{ch_{2,j}} = \{\{H_1\}\},$$
$$R_{ch_{2,j}} = \{H_i \rightarrow H_{i+1} \mid 1 \leq i \leq 7\} \cup \{H_8 \rightarrow Q_{c_j}H_1\}.$$

Instead of giving a detailed proof of the correctness of our construction, we demonstrate the work of the system in Table 1 and Table 2 by indicating a possible transition sequence of $\Pi$ while simulating an instruction of the two-counter machine $TCM$, and by presenting the terminating part of the simulation. Note that the cells of the tables contain only some of the strings produced by the regions, those which are interesting from the point of view of the simulation.

Let us first look at Table 1. The simulated instruction is represented by a non-terminal $\alpha_1 = [q, x, Z, *, q', +1, -1]_1$ chosen in region $C_{sel}$ in the first step. This indicates that the first counter should be empty which requirement is satisfied since region $C_{c_1}$ contains a string containing zero $A$ symbols. In the following few steps, the indexed versions of $\alpha$ reach the regions $C_{gen}, C_{ch_1}, C_{c_j}, j \in \{1, 2\}$, and each of these regions executes its part of the simulation. $C_{gen}$ generates the letter read by the two-counter machine, $C_{ch_1}$ queries the regions simulating the counters in the case when their contents should be zero, and this way collects a "checker" string. If this string contains the nonterminal $A$, then the simulation is not correct. $C_{c_j}$ maintain the contents of the counters by adding or deleting $A$-s. Its work is aided by $C_{ind_1}$ and $C_{ch_{2,j}}$. The region $C_{ch_{2,j}}$ collects the counter strings at the end of each simulating cycle. The simulation was successful if and only if this collected string only contains $A, E$ or $J$ symbols.

The terminating phase of the simulation is presented on Table 2. When $G_{sel}$ selects the symbol $F$, the system prepares to finish its work. The variously indexed versions of $F$ travel through the system and result in the transfer of the word generated in $G_{gen}$ and the checker string of region $C_{ch_1}$ to region $C_{sel}$. There the symbol $A$ cannot be erased, so a terminal word can only be produced if the checker string does not contain this symbol. Meanwhile, the other checker strings are transferred from $C_{ch_{2,j}}$ to region $C_{gen}$ where the $A$ and $J$ symbols can be erased, but nothing else, so when later also this string is transferred to $C_{sel}$, a terminal string can only be produced if the behavior of the counter simulating regions were correct in each step of the simulation. The last row of the table represents the situation when the erasing process begins. When all $A$ and $J$ have disappeared from the string in region $C_{gen}$, then $S_6'$ can be changed to a query symbol transferring the result to $C_{sel}$, where all remaining symbols can be erased, in the case when the simulation was correct.  □

Next we prove that any RPC system using $i$-communication can be simulated with an RPC system using $p$-communication.

**Theorem 2.** $iRPC_nCF \subseteq pRPC_{3n}CF$, for any $n \geq 1$.

*Proof.* Let $\Pi = (V, \mu, (M_1, R_1), \ldots, (M_n, R_n), 1)$ be a system of degree $n$ with $V = N \cup K \cup T$. We construct $\Pi' = (V', \mu', (M_1', R_1'), \ldots, (M_{3n}', R_{3n}'), 1)$ of degree $3n$, such that $L_p(\Pi') = L_i(\Pi)$.

Let $\mu'$ be defined by adding two new regions $[\ [\ ]_{2n+i}\ ]_{n+i}$ inside every region $i$. This way, $n + i \in neighbor_{\mu'}(i)$, and $2n + i \in neighbor_{\mu'}(n + i)$ for all $1 \leq i \leq n$.

Let $V' = N' \cup K' \cup T$ where $N' = N \cup \{S_1, S_2, S_3, S_4\}$, and let the rules of $\Pi'$ be defined as

$$M_i' = M_i \cup \{\{S_1, S_2\}\},$$
$$R_i' = R_i \cup \{S_1 \to Q_i, S_2 \to Q_{n+i}\},$$

and

$$M_{n+i}' = \{\{S_1, S_2, S_3, S_4\}\}, \quad R_{n+i}' = \{S_3 \to Q_{n+i}, S_4 \to Q_{2n+i}\},$$
$$M_{2n+i}' = \{\{S_1, S_2, S_3, S_4\}\}, \quad R_{2n+i}' = \{S_1 \to Q_{n+i}, S_2 \to Q_{2n+i}\}$$

for all $1 \leq i \leq n$.

**Table 2.** Components of $\Pi$ in the proof of Theorem 1, simulating the terminating phase of the system. We only present the components which are also included in Table 1.

| | $C_{sel}$ | $C_{gen}$ | $C_{ch_1}$ | $C_{c_1}$ | $C_{ind_1}$ | $C_{ch_{2,1}}$ |
|---|---|---|---|---|---|---|
| 0 | $\beta_8$ | $wS_1, S_1'$ | $E...S_1, S_1'$ | $E...J, C_1$ | $B_1$ | $AEJ...H_1$ |
| 1 | $F_1$ | $wS_2, Q_{sel}$ | $E...S_2, S_2'$ | $E...J_1, C_2$ | $B_2$ | $AEJ...H_2$ |
| 1C | $F_1$ | $wS_2, F_1$ | $E...S_2, S_2'$ | $E...J_1, C_2$ | $B_2$ | $AEJ...H_2$ |
| 2 | $F_2$ | $wQ_{sel}, F_{1,1}$ | $E...S_3, Q_{gen}$ | $E...J_2, C_3$ | $B_3$ | $AEJ...H_3$ |
| 2C | $F_2$ | $wF_2, F_{1,1}$ | $E...S_3, F_{1,1}$ | $E...J_2, C_3$ | $B_3$ | $AEJ...H_3$ |
| 3 | $F_3$ | $wQ_{sel}, F_{1,2}$ | $E...Q_{gen}, F_1'$ | $E...Q_{ch_1}, C_4$ | $B_4$ | $AEJ...H_4$ |
| 3C | $F_3$ | $wF_3, F_{1,2}$ | $E...F_{1,2}, F_1'$ | $E...F_1', C_4$ | $B_4$ | $AEJ...H_4$ |
| 4 | $F_4$ | $wF_4, F_{1,3}$ | $E...Q_{S_4}, \bar{F}_1'$ | $E...\bar{F}_1'', Q_{ch_1}$ | $B_5$ | $AEJ...H_5$ |
| 4C | $F_4$ | $wF_4, F_{1,3}$ | $E...S_4, \bar{F}_1'$ | $E...\bar{F}_1'', \bar{F}_1'$ | $B_5$ | $AEJ...H_5$ |
| 5 | $F_5$ | $wF_5, F_{1,4}$ | $E...S_5, \underline{\bar{F}}_1'$ | $E...\bar{F}_1''$<br>$Q_{ch_{2,1}}$ | $B_6$ | $AEJ...H_6$ |
| 5C | $F_5$ | $wF_5, F_{1,4}$ | $E...S_5, \underline{\bar{F}}_1'$ | $E...\bar{F}_1''$<br>$AEJ...H_6$ | $B_6$ | $AEJ...H_6$ |
| 6 | $F_6$ | $wQ_{ch_1}$<br>$F_{1,5}$ | $E...S_6$<br>$Q_{c_1}Q_{c_2}$ | $E...\bar{F}_1''$<br>$AEJ...H_7$ | $E$ | $AEJ...H_7$ |
| 6C | $F_6$ | $wE...S_6$<br>$F_{1,5}$ | $E...S_6$<br>$AEJ......H_7$ | $E...\bar{F}_1''$<br>$AEJ...H_7$ | $E$<br>$E$ | $AEJ...H_7$<br>$AEJ...H_7$ |
| 7 | $Q_{gen}$ | $wE...S_6'$<br>$Q_{ch_1}$ | $E...S_7$<br>$AEJ......H_7$ | $E...\bar{F}_1''$<br>$AEJ...H_7$ | $J_4$ | $AEJ...H_8$ |
| 7C | $wE...S_6'$ | $wE...S_6'$<br>$AEJ...H_7$ | $E...S_7$<br>$AEJ......H_7$ | $E...\bar{F}_1''$<br>$AEJ...H_7$ | $J_4$ | $AEJ...H_8$ |

The additional membranes of $\Pi'$ work as "suppliers" of symbols. In each step, each region $i$ rewrites $S_1$ and $S_2$ to query itself, and the region $n+i$. From itself it "receives" the strings it contains besides $S_1, S_2$, from region $n+i$ it receives $S_1, S_2$, so the same behavior can be repeated in the next step. This self query mechanism is used in each region to keep a copy of its contents even in the case when it is requested by some other region. This way, $\Pi'$ simulates the communication behavior of $\Pi$.  □

By Theorems 1 and 2, we obtain the immediate corollary.

**Corollary 3.** $iRPC_{10} = pRPC_{30}CF = RE$.

## 4    Closing Remarks

We proved that in the case of string rewriting P systems, communication according to dynamically emerging requests leads to computational completeness not only in tissue-like P systems (MPC systems in short, their computational completeness was shown in [2]), but also in standard P systems (RPC systems), even in the case of systems with bounded size parameters.

There have remained several open problems for further study. For example, it is not known whether the obtained size bounds are sharp or not, and whether or not the sharp bounds are different for MPC systems and RPC systems.

Apart from the problem of the sharpness of the bounds, another natural research direction would be to examine the impact of the number of different query symbols on the power of these systems. In RPC systems, for example, $maxc_\mu + 2$ where $maxc_\mu$ is the maximal number of children of any region in the membrane structure $\mu$ (plus *one* for the parent region and *one* for the region itself) is a trivial upper bound for the number of different query symbols necessary. What happens if we decrease the number of symbols, for example, by forbidding "self"-queries, by allowing only one symbol for any of the child membranes of each region of the system, or by using the "extreme" restriction of having only one query symbol type available for use by all the regions? Can these decreases in the "query-complexity" be compensated by more complicated membrane structures?

# References

1. Csuhaj-Varjú, E., Dassow, J., Kelemen, J., Păun, G.: Grammar Systems. A Grammatical Approach to Distribution and Cooperation. Gordon and Breach, London (1994)
2. Csuhaj-Varjú, E., Păun, G., Vaszil, G.: Tissue-like P systems communicating by request. In: Krithivasan, K., Rama, R. (eds.) Formal Language Aspects of Natural Computing. Ramanujan Mathematical Society Lecture Notes Series, vol. 3, pp. 143–153, Ramanujan Mathematical Society (2006)
3. Fischer, P.C.: Turing machines with restricted memory access. Information and Control 9, 364–379 (1966)
4. Martín-Vide, C., Păun, G., Pazos, J., Rodríguez-Patón, A.: Tissue P systems. Theoretical Computer Science 296(2), 295–326 (2003)
5. Păun, G.: Computing with membranes. Journal of Computer and System Sciences 61(1), 108–143 (2000) (and Turku Center for Computer Science-TUCS Report 208 (November 1998), www.tucs.fi)
6. Păun, G.: Membrane Computing: An Introduction. Springer, Berlin (2002)
7. Rozenberg, G., Salomaa, A. (eds.): Handbook of Formal Languages. Springer, Berlin (1997)

# On the Dynamics of PB Systems with Volatile Membranes

Giorgio Delzanno[1] and Laurent Van Begin[2]

[1] Università di Genova, Italy
giorgio@disi.unige.it
[2] Université Libre de Bruxelles, Belgium
lvbegin@ulb.ac.be

**Abstract.** We investigate decision problems like reachability and boundedness for extensions of PB systems with volatile membranes. Specifically, we prove that reachability and boundedness are decidable for PB systems extended with rules for membrane dissolution. For PB systems extended with membrane creation, reachability is still decidable whereas boundedness becomes undecidable. Furthermore, we show that both problems are undecidable for PB systems extended with both dissolution and creation rules. Finally, we prove that reachability and boundedness become decidable for PB systems with dissolution rules and in which only one instance of each type of membrane can be created during a computation. Our work extends the results in [4] obtained by Dal Zilio and Formenti for PB systems with static membrane structure.

## 1 Introduction

The PB systems of Bernardini and Manca [2] are a variant of P-systems [15] in which boundary rules can be used to move multisets of objects across a membrane. As shown, e.g., in [6], PB systems can be applied to model complex interactions among biological membranes. To fully exploit the power of PB systems, it seems important to develop methods for qualitative and quantitative analysis of models specified in this formalism. In this paper we focus our attention on decision problems related to the qualitative analysis of PB systems, and, more precisely, on problems like reachability and boundedness. Some preliminary results on decision problems for PB systems have been obtained in [4]. Specifically, in [4] Dal Zilio and Formenti proved that the reachability problem is decidable for PB systems with symbol objects. The *reachability problem* consists in checking if a given system can evolve into a fixed a priori configuration. The decidability proof in [4] is based on an encoding of PB systems into Petri nets [16], an infinite-state model of concurrent systems for which the reachability problem is decidable [13,10]. A Petri net is a collection of places that contain tokens, and of transitions that define how tokens move from one place to another. The current configuration of a net is called marking. A marking specifies the current number of tokens in each place. A PB system can be encoded as a Petri net in which membranes are modeled as places, symbol objects as tokens, configurations as markings, and internal/boundary rules as transitions. The execution of a rule is simulated then by the firing of the corresponding Petri net transition. The Petri net encoding shows that the reachability problem is decidable in PB systems. The same reduction can be used to decide other properties like *boundedness* [5]. In [4] the authors observe that the aforementioned encoding can be extended to more sophisticated Petri net models so as to deal with dynamically changing membrane structures. As an example, Petri net transitions extended with transfer arcs naturally model the dissolution of a membrane. Indeed, a transfer arc can be used to atomically transfer all tokens from one place to another. This operation can be applied to move the content of a dissolved membrane to its father. Unfortunately, as pointed out in [4], this connection cannot be exploited in order to extend the decidability results obtained for PB systems. Indeed, problems like reachability and boundedness become undecidable in presence of transfer or reset arcs [5]. Thus, the decidability of reachability and boundedness for PB systems with volatile or moving membrane is still an open problem. In this paper we focus our attention on decision problems for extensions of PB systems with dissolution and creation rules. Our technical results are as follows.

We first show that reachability is decidable in PB systems with dissolution rules (PBD systems). Dissolution rules are a peculiar feature of P-systems. Thus, PBD systems represent a natural extension of PB system. Our decidability proof is still based on a reduction to a Petri net reachability problem. Our construction extends the Petri net encoding of [4] in order to weakly simulate the original PBD system. More precisely, from a PBD reachability problem we compute a Petri net that may contain executions that do not correspond to real computations of the corresponding PBD system. Spurious computations can however be eliminated by enforcing special conditions (e.g. requiring a special set of places to be empty) on the initial and target markings used to encode a PBD reachability problem. It is important to notice that our reduction does not require the additional power provided by Petri nets with transfer arcs.

As a second result, we show that reachability is decidable in PB systems extended with creation rules (PBC systems). We consider here creation rules inspired to those proposed for P-systems in [12]. Our proof exploits structural properties of PBC systems that allow us to reduce the reachability of a target configuration $c$, to a reachability problem in a Petri net extracted from both the original PBC system and the configuration $c$.

We consider then a model with both dissolution and creation rules (PBDC systems). For this model, we first give a general negative result for the decidability of reachability, and then study a non-trivial subclass in which reachability becomes decidable. Specifically, we first show that it is sufficient to consider three membranes with the same name to encode a reachability problem for a two counter machine [14] as reachability of a PBDC system. We define then a restricted semantics for PBDC systems in which at most only one copy of each type of membrane can be created during a computation. This semantics is inspired to a view of membrane names/types as bounded resources. Under this semantics, we prove the decidability of PBDC reachability via a reduction

to Petri net reachability. In the encoding we use special places to identify the
membrane structure of the current configuration. The encoding is exponentially
more complex than the encoding used for PBD and PBC systems, since it requires the construction of a Petri net where the number of places is equal to
number of tree structures that can be built upon a finite and fixed a priori set
of membrane names.

As a last analysis, we study the boundedness problem for the aforementioned
extensions of PB systems. Specifically, we first show that boundedness is decidable for PBDC systems with restricted semantics. The proof exploits the theory
of well-quasi ordering [1]. As a consequence, we obtain the decidability of boundedness for PBD systems. Finally, we prove that boundedness is undecidable in
PBC systems. This result is obtained by encoding counter machines as PBC systems. The encoding exploits the possibility of creating several instances of the
same membrane to simulate a counter (i.e. the same encoding cannot be applied
in the restricted semantics of PBDC systems).

To our current knowledge, these are the first decidability/undecidability results obtained for reachability and boundedness in extensions of PB systems with
dissolution and creation rules. Our decidability results extend those obtained for
PB systems by Dal Zilio and Formenti in [4].

**Plan of the Paper.** In Section 2 we recall the main definitions of PB systems
(Petri (P/T) nets and counter machines are defined in appendix). In Section 3,
4, and 5 we study the reachability problem for extensions of PB systems resp.
with dissolution, creation, and both dissolution and creation rules. In Section
6 we study the boundedness problem for the aforementioned extensions of PB
systems. Finally, in Section 7 we address some conclusion and future work.

## 2   Preliminaries

In this section we recall the main definitions for PB systems with symbol objects
taken from [2,4]. We first need some preliminary notions. Let $\mathbb{N}$ be the set of
positive integers. Consider a finite alphabet $\Gamma$ of symbols. A multiset over $\Gamma$ is
a mapping $u : \Gamma \rightsquigarrow \mathbb{N}$. For any $a \in \Gamma$, the value $u(a)$ denotes the multiplicity
of $a$ in $u$ (the number of occurrences of symbol $a$ in $u$). We often use a multiset
as a string $a_1 \cdot \ldots \cdot a_n$ of symbols, i.e., $a_i \in \Gamma$. Furthermore, we use $\epsilon$ to denote
the empty multiset, i.e., such that $\epsilon(a) = 0$ for any $a \in \Gamma$. As an example, for
$\Gamma = \{a, b, c, d\}$, $a \cdot b \cdot c \cdot c$ represents the multiset $u$ such that $u(a) = u(b) =
1, u(c) = 2, u(d) = 0$. We use $\Gamma^{\otimes}$ to denote the set of all possible multisets over
the alphabet $\Gamma$. Given two multisets $u, v$ over $\Gamma$, we write $u \doteq v$ if $u(a) = v(a)$
for all $a \in \Gamma$, and $u \preceq v$ if $u(a) \leq v(a)$ for all $a \in \Gamma$. Furthermore, we use $\oplus$
and $\ominus$ to denote multiset union and difference, respectively. Specifically, for any
$a \in \Gamma$ we have that $(u \oplus v)(a) = u(a) + v(a)$, and $(u \ominus v)(a) = u(a) - v(a)$.

*PB-systems.* A PB system [2] with symbol object is a tuple $\Pi = (\Gamma, M, R, \mu_0)$,
where $\Gamma$ is a finite alphabet of symbols; $M$ is a finite tree representing the
*membrane structure* with membrane names taken from a set $N$, $R$ is a finite

set of rules, and $\mu_0$ is the *initial configuration*, i.e., a mapping from membranes (nodes in $M$) to $\Gamma^\otimes$. Rules can be of the following two forms:

(1)  *Internal Rule* :  $[_i \; u \to [_i \; v$         (2)  *Boundary Rule* :  $u \; [_i \; v \to u' \; [_i \; v'$

where $i \in N$, and $u, u', v, v' \in \Gamma^\otimes$ and we assume that at least one between $u$ and $u'$ is not empty. A configuration $\mu$ of a PB system $\Pi$ is a distribution of objects in $\Gamma$ in the membranes in $M$, i.e., a mapping from $M$ to $\Gamma^\otimes$. A rule of the form (1) is enabled at $\mu$, if $i$ is a membrane in $M$ and $u \preceq \mu(i)$. Its application leads to a new configurations $\nu'$ such that $\nu'(i) = (\nu(i) \ominus u) \oplus v$ and $\nu'(j) = \nu(j)$ for any $j \in N$ s.t. $j \neq i$. Suppose now that membrane $j$ contains as immediate successor in $M$ membrane $i$. A rule of the form (2) is enabled at $\mu$, if $u \preceq \mu(j)$ and $v \preceq \mu(i)$. Its application leads to a new configurations $\nu'$ such that $\nu'(j) = (\nu(j) \ominus u) \oplus u'$ and $\nu'(i) = (\nu(i) \ominus v) \oplus v'$ and $\nu'(k) = \nu(k)$ for any $k \in N$ s.t. $k \neq i, j$. We say that there is a transition $\mu \Rightarrow \mu'$ if $\mu'$ can be obtained from $\mu$ by applying a rule in $R$. A computation with initial configuration $\mu_0$ is a sequence of transitions $\mu_0 \Rightarrow \mu_1 \Rightarrow \ldots$. A configuration $\mu$ is reachable from $\mu_0$ if there exists a sequence of transitions $\mu_0 \Rightarrow \ldots \Rightarrow \mu$.

Given a PB system $\Pi$ with initial configuration $\mu_0$ and a configuration $\mu$, the *reachability problem* consists in checking if $\mu$ is reachable from $\mu_0$. Given a PB system $\Pi$ with initial configuration $\mu_0$, the *boundedness problem* consists in deciding if the set of configurations reachable from $\mu_0$ is finite. Reachability and boundedness are decidable for PB systems with symbol objects. The proof is based on an encoding PB systems into Petri nets defined in [4].

## 3   PB Systems with Dissolution Rules

A PB system with dissolution rules (PBD) provides, in addition to internal and boundary rules, a third kind of rules of the following form:

(3)  *Dissolution Rule* :    $[_i \; u \to [_i \; v \cdot \delta$

where $\delta$ is a symbol not in $\Gamma$. The intuitive meaning of this rule is that after applying the rule $[_i \; u \to [_i \; v$ the membrane $i$ is dissolved and its content (including its sub-membranes) is moved to the membrane $j$ that contains $i$ as immediate successor in the current membrane structure. To make the semantics formal, we make the membrane structure part of the current configuration, $M_0$ being the initial tree. Thus, a configuration is now a pair $c = (M, \mu)$, where $M$ is a tree, and $\mu$ is a mapping from nodes of $M$ to $\Gamma^\otimes$. Rules of type (1) and (2) operate on a configuration $c = (M, \mu)$ without changing the tree structure $M$ and changing $\mu$ as specified in the semantics of PB systems. A dissolution rule like (3) operates on a configuration $c = (M, \mu)$ as follows. For simplicity, we assume that membrane $i$ is not the root of $M$. Suppose now that $i$ is an immediate successor of $j$ in $M$. The rule is enabled if $u \preceq \mu(i)$. Its application leads to a new configuration $c' = (M', \nu')$ such that $M'$ is the tree obtained by removing node $i$ and by letting all successor nodes of $i$ become successors of $j$; $\nu'$ is the mapping defined as $\nu'(j) = \nu(j) \oplus (\nu(i) \ominus u) \oplus v$ and $\nu'(k) = \nu(k)$ for any $k \in M$ s.t. $k \neq i, j$. Notice that rules of type $(1 - 3)$ are enabled at $c = (M, \mu)$ only if the membrane $i$ is in current tree $M$. The definition of sequences of transitions and of reachability problems can naturally be extended to the new type of rules.

## 3.1   Decidability of Reachability in PBD Systems

In this section we prove that the reachability problem is decidable in PB systems with dissolution rules. We assume here that names of membranes are all different. However, the construction we present can be extended to the general case. The starting point of our construction is the reduction of reachability for PB systems to reachability in Petri nets given in [4]. Let $\Pi = (\Gamma, M, R, \mu_0)$ be a PB system. For each membrane $i$ in $M$ and each symbol $a \in \Gamma$, the Petri net $\mathcal{N}$ associated to $\Pi$ makes use of place $a^i$ to keep track of the number of occurrences (multiplicity) of objects of type $a$ in $i$. Transitions associated to internal rules redistribute tokens in the set of places associated to the corresponding membrane. As an example, a rule like $[_i \; a \cdot b \;\rightarrow\; [_i \; c$ is encoded by a Petri net transition that removes one token from place $a^i$ and one token from place $b^i$ and adds one token to place $c^i$. Boundary rules are modeled by Petri net transitions that work on places associated to pairs of membranes. As an example, if membrane $j$ contains $i$, a rule like $a \; [_i \; b \;\rightarrow\; b \; [_i \; a$ is encoded by a Petri net transition that removes one token from place $a^j$ and one token from place $b^i$ and adds one token to place $b^j$ and one token to place $a^i$. For a membrane structure $M$, a configuration $\mu : M \rightsquigarrow \Gamma^\otimes$ is represented by a marking $m_\mu$ such that for every node $i$ in $M$, $a^i$ has $k$ tokens in $m$ iff $a$ has $k$ occurrences in $\mu(i)$. Reachability of a configuration $\mu$ is reduced the to reachability of the marking $m_\mu$ starting from $m_{\mu_0}$ in $\mathcal{N}$.

The Petri net encoding of [4] exploits the property that the membrane structure of a PB system is never changed by the application of a rule. This property does not hold anymore for dissolution rules, since they removes nodes from the current membrane structure. Thus, the size of the membrane structure may decrease in a sequence of transitions. Our decidability proof is still based on a reduction to a Petri net reachability problem. Our reduction exploits the property that the number of applications of dissolution rules is bounded a priori by the size of the initial membrane structure $M_0$. Thus it is enough to associate a flag *present/dissolved* to each membrane occurring in the initial configuration to keep track of the current membrane structure. Special care must be taken in the transfer of objects from a dissolved membrane to its parent. For this task we need to operate in two modes. In *normal* mode we simulate internal and boundary rules. In *dissolving* mode we stop all other operations, move objects one-by-one to the current parent membrane, and non-deterministically go back to the normal mode. Good simulations can be distinguished by bad ones by enforcing places associated to dissolved membranes to be empty in the target configuration.

The formal definition of the Petri net encoding is as follows. Assume a PBD system $\Pi = (\Gamma, M_0, R, \mu_0)$, where $\Gamma = \{a_1, \ldots, a_m\}$, and $M_0$ has the membranes with names in $N = \{n_0, n_1, \ldots, n_k\}$, $n_0 \in N$ being the root node. Given a membrane $i$, let $path(i)$ be the sequence of nodes in the (unique) path from $n_0$ to $i$ in $M_0$. We define the Petri net $\mathcal{N}$ encoding $\Pi$ in several steps. First of all we assume that $\mathcal{N}$ has at least the places $normal, dissolving_1, \ldots, dissolving_k$ that we use to determine the simulation mode as described in the previous paragraphs. We assume here that $normal$ contains one token iff $dissolving_i$ is empty for all $i : 1, \ldots, k$, and $dissolving_i$ contains one token iff $normal$ as well $dissolving_j$ are empty for any $j \neq i$. Furthermore, for each membrane $i$, the Petri net $\mathcal{N}$ has a place $present_i$ and a place $dissolved_i$ and, for any $a \in \Gamma$, a place $a^i$.

*Notation.* In the rest of the paper given a multiset of objects $u$ and a membrane $i$ we use $\pi_i(u)$ to denote the multiset of places in which, for each $a \in \Gamma$, $a^i$ has the same number of occurrences as those of $a$ in $u$.

An internal rule $r = [_i u \to [_i v$ is encoded by a transition $t_r$ that satisfies the following conditions. The pre-set of $t_r$ contains place $normal$ (normal mode), $present_i$ (membrane $i$ is still present), and the multiset of places $\pi_i(u)$. The post-set of $t_r$ contains $normal, present_i$ and the multiset of places $\pi_i(v)$. Thus, the only difference with the encoding of PB system is the condition on the $normal$ and $present_i$ flags (in normal mode internal rules are enabled only when the membrane is not dissolved).

Now let $path(i) = (n_0, n_1, \ldots, n_q, i)$ with $q \geq 0$. A boundary rule $r = u[_i v \to u'[_i v'$ is encoded by a set $B_r = \{b_r^{n_0}, \ldots, b_r^{n_q}\}$ of transitions. The pre-set of transition $b_r^{n_j}$ contains places $normal$ and $present_i$ together with the set of places $D^{n_j} = \{present_{n_j}, dissolved_{n_{j+1}}, \ldots, dissolved_{n_q}\}$, and the multisets $\pi_{n_j}(u)$ and $\pi_i(v)$. The post-set contains places $normal$ and $present_i$ together with the set of places $D^{n_j}$ defined for the pre-set and the multisets $\pi_{n_j}(u')$ and $\pi_i(v')$. The pre-condition $D^{n_j}$ allows us to select the membrane that is the immediate ancestor of $i$ in the current configuration, i.e., a membrane $n_j \in path(i)$ that is not dissolved and such that all the intermediate membranes between $n_j$ and $i$ in $path(i)$ are dissolved. Notice that, by the assumptions we made on the $normal/dissolving$ and $present/dissolved$ flags, in normal mode at most one rule in $B_r$ can be enabled at a given configuration.

Consider now a dissolution rule $r = [_i u \to [_i v \cdot \delta$. We first model the internal rule by the transition $s_r$. The pre-set of $s_r$ contains the places $normal$ and $present_i$ and the multiset $\pi_i(u)$. The post-set contains the place $dissolving_i$ and the multiset $\pi_i(v)$. We model the transfer of the contents of membrane $i$ to its current immediate ancestor via a set of transitions $S_r^a = \{s_a^{n_0}, \ldots, s_a^{n_q}\}$ for each $a \in \Gamma$. The pre-set of transition $s_a^{n_j}$ contains places $dissolving_i$ ($i$ is dissolving) and $a^i$ (the source of a token to be transferred) together with the set of places $D^{n_j}$ defined in the case of boundary rules. The post-set contains places $dissolving_i$ and $a^j$ (the destination of a transferred token), and the set $D^{n_j}$. Finally, we add a transition $d_r^i$ to stop the transfer of tokens and to switch the operating mode back to $normal$. The pre-set of $d_r^i$ contains the place $dissolving_i$ and its post-set contains the places $normal$ and $dissolved_i$. Notice that the simulation phase of a dissolution rule for membrane $i$ can be activated only if $present_i$ is not empty. This implies that once the $dissolving_i$ flag is reset (i.e. the mode goes back to *normal*) it cannot be set in successive executions (a membrane can dissolve at most once).

The Petri net $\mathcal{N}$ is built by taking the union of the places and transitions used in the encoding described before. Let $M$ be a membrane structure with a subset of the nodes in $M_0$ (initial structure of $\Pi$). A configuration $c = (M, \mu)$ is encoded by a marking $m_c$ in which there is one token in *normal*, one token in *present$_i$* for each membrane $i$ in $M$, and one token in *dissolved$_j$* for each $j$ not in $M$. Furthermore, for each membrane $i$ in $c$ and $a \in \Gamma$, place $a^i$ has as many tokens as the number occurrences of $a$ in $\mu(i)$. All the remaining places in $\mathcal{N}$ (*dissolving$_i$* for any $i$ and *present$_j$* for any $j$ not in $M$) are empty.

By construction of $\mathcal{N}$, it is immediate to see that if there is a sequence of transitions from $c_0 = (M_0, \mu_0)$ to $c = (M, \mu)$ passing through the configurations $c_1, \ldots, c_v$ then there is a firing sequence from $m_{c_0}$ to $m_c$ passing through the markings $m_{c_1}, \ldots, m_{c_v}$. Such a firing sequence is obtained by completing all transfers of objects required by the simulation of each dissolution rule (i.e. after the simulation of the dissolution of membrane $i$, the normal mode is reactivate only when the places associated to the objects contained in $i$ are all empty). Vice versa, suppose there exists a firing sequence from $m_{c_0}$ to $m_c$. We first notice that only the markings in which the place *normal* is not empty correspond to configurations of the original PBD system. Furthermore, suppose that during the simulation of the dissolution of membrane $i$, the transfer of objects is stopped when some of the places associated to objects in $i$ are not empty. Let $m$ be the resulting marking. Now we notice that the first step of the simulation of dissolution is to set the *present$_i$* flag to false. This implies that in the marking $m$ place *present$_i$* is empty, while there exists $a \in \Gamma$ such that $a^i$ is not empty (some token has not been transferred). It is easy to check that if $m$ has these two properties, for any marking $m'$ derived from $m$ by applying transitions of $\mathcal{N}$, the content of the place $a^i$ in $m'$ is the same as in $m$. Indeed, transitions that simulate internal, boundary and dissolution rules operating directly on $i$ are no more enabled (the condition *present$_i$* fails). Furthermore, a dissolution rule on a membrane $j$ nested into $i$ in $M_0$ cannot transfer tokens to $i$ since *dissolved$_i$* is checked when searching for the father of $j$ in the current tree structure. In other words, if the simulation of a dissolution rule is not correctly executed, then there exists at least one non-empty place $a^i$ for a dissolved membrane $i$. By definition, however, $m_c$ is the marking in which all places associate to dissolved membranes are empty. Thus, if $m_c$ is reachable from $m_{c_0}$ then the corresponding firing sequence corresponds to a real computation in $\mathcal{N}$. Thus, we have that the reachability of a configuration $c = (M, \mu)$ in $\Pi$ can be encoded as the reachability of the marking $m_c$ in $\mathcal{N}$ from $m_{c_0}$. From the decidability of reachability in Petri nets, we obtain the following theorem.

**Theorem 1.** *The reachability problem is decidable in PBD systems.*

The previous theorem extends the decidability result for PB systems in [4].

# 4    PB Systems with Creation

In this section we consider an extension of PB systems inspired to the membrane creation operation studied in [12]. Let N be a possibly infinite list of membrane names. A PB system with creation rules (PBC) provides, in addition to internal and boundary rules, a third kind of rules of the following form:

$$(4)\ \text{Creation}:\quad a\ \rightarrow [_i\ v\ ]_i$$

where $a \in \Gamma$, $v \in \Gamma^\otimes$, and $i \in \mathsf{N}$. The intuitive meaning of this rule is that after applying the rule $a \rightarrow [_i v]_i$ inside a membrane $j$, object $a$ is replaced by the new membrane $i$ containing the multiset of objects $v$.

To make the semantics formal, we assume that membrane structures are trees whose nodes are labeled with names in N. Furthermore, we make both the set of used names and the membrane structure part of the current configuration, $N_0 \subseteq \mathsf{N}$ being the initial set of used names, and $M_0$ being the initial tree defined over $N_0$. Thus, a configuration is now a triple $c = (N, M, \mu)$ where $N$ is a set of names, $M$ is a tree with nodes labeled in $N$, and $\mu$ is a mapping $M \rightsquigarrow \Gamma^\otimes$. Rules of type (1) and (2) operate on a configuration $c = (N, M, \mu)$ without changing $N$ and $M$. A creation rule like (4) operates on a configuration $c = (N, M, \mu)$ as follows. Suppose that $n$ is a node in $M$. The rule is enabled if $a \in \mu(n)$. Its application leads to a new configurations $c' = (N', M', \nu')$ such that $N' = N \cup \{i\}$; $M'$ is the tree obtained by adding a new node $m$ labeled by $i$ as a successor of node $n$; $\nu'$ is the mapping defined as $\nu'(n) = \nu(n) \ominus a$, $\nu'(m) = v$, and $\nu'(p) = \nu(p)$ for the nodes $p \neq m, n$. Notice that rules of type (4) can be applied in any membrane. Indeed, the only precondition for the application of rule 4 is the existence of object $a$ in a membrane. Furthermore, such an application may create different nodes with the same membrane name. The reachability problem can naturally be reformulated for the extended semantics of rules. Specifically, it consists in checking whether a given target configuration $c$ is reachable from the initial configuration $c_0$.

Notice that in our model we distinguish nodes from membrane names. Thus, different nodes may have the same name. In presence of creation rules the membrane structure can grow in an arbitrary manner both in width and depth. Despite of these powerful features of PBC systems, the reachability problems can still be decided by resorting to an encoding into Petri net reachability as explained in the next section.

## 4.1    Decidability of Reachability in PBC Systems

Differently from the encoding defined for PBD systems in the previous section, the encoding needed here is function of both the initial and the target configuration. Indeed, since PBC rules can only add new nodes, to decide if a configuration $c = (N, M, \mu)$ is reachable from $c_0$ we can restrict our attention to membrane structures of size comprised between the size of $M_0$ and the size of $M$ and with (possibly repeated) labels in $N$. Actually, we can make some simplification that allows us to build a Petri net by considering only the target configuration $M$. Indeed, as shown in [12], with creation rules we can safely consider initial configurations with only the root membrane (we can always add a finite number of creation rules to generate any initial configuration in a preliminary phase of the computation).

Let $\Pi$ be a PBC system with initial configuration $c_0 = (N_0, M_0, \mu_0)$ where $M_0$ is a single node labeled $n_0$. Consider now a target configuration $(N, M, \mu)$ where $N = \{n_0, \ldots, n_k\}$ and $M$ has $m$ nodes with $k \leq m$ and the root node of $M$ is labeled $n_0$. Starting from $\Pi$ and $c$ we build the Petri net $\mathcal{N}$ described next. For each node $n$ in $M$, the Petri net $\mathcal{N}$ has places $used_n$ and $notused_n$ (used as one flip-flop), and $a^n$ for each $a \in \Gamma$ (to model the content of membrane $n$). We assume that $used_n$ is not empty iff the membrane has been created and it is in use, and $notused_n$ is not empty iff the membrane has still to be created. PBC rules are modeled as follows.

For each node $n$ in $M$ with label $i$, an internal rule $r = [_i u \rightarrow [_i v$ is encoded by a transition $t_r^n$ that satisfies the following conditions: The pre-set contains place $used_n$ together with multiset $\pi_n(u)$; The post-set contains $used_n$ together with multiset $\pi_n(u)$. The differences with the encoding of PB/PBD systems is the condition on the $used_n$ flag and the fact that we work on nodes of membrane structures and not directly on membrane names (as said before two different nodes may have the same name). The pre-condition on $used_n$ is needed in order to enable rules operating on node $n$ only after the corresponding creation rule has been fired.

For each node $m$ in $M$ that has an immediate successor $n$ with label $i$, a boundary rule $r = u[_i v \rightarrow u'[_i v'$ is encoded by a transition $b_r^{m,n}$ that satisfies the following conditions. The pre-set contains places $used_n$ and $used_m$ together with the multisets $\pi_m(u)$ and $\pi_n(v)$. The post-set contains places $used_n$ and $used_m$ together with the multisets $\pi_m(u')$ and $\pi_n(v')$. Notice that, differently from the encoding used in PBD, in PBC we do not have to consider paths in the membrane structure $M$.

For each node $m$ in $M$ that has an immediate successor $n$ with label $i$, a creation rule $r = a \rightarrow [_i v]_i$ is encoded by a transition $c_r^{m,n}$ that satisfies the following conditions. The pre-set contains places $used_m$, $a^m$ and $notused_n$. The post-set contains places $used_m$ and $used_n$, together with the multiset $\pi_n(v)$.

The Petri net $\mathcal{N}$ is built by taking the union of the places associated to each membrane and the union of the set of transitions used to encode internal, boundary and creation rules described before. A configuration $c' = (N', M', \mu')$ is encoded by a marking $m_{c'}$ in which for each node $n$ in $M'$ there is one token in $used_n$, and for each $a \in \Gamma$, $a^n$ has as many tokens as the number occurrences of $a$ in $\mu(n)$. Furthermore, for each node $n'$ in $M$ that do not occur in $M'$, we put a token in $notused_{n'}$. All the remaining places are empty.

By construction of $\mathcal{N}$, it is immediate to see that there is a sequence of transitions from $c_0 = (N_0, M_0, \mu_0)$ to $c = (N, M, \mu)$ passing through the configurations $c_1, \ldots, c_v$ if and only if there is a firing sequence from $m_{c_0}$ to $m_c$ passing through the markings $m_{c_1}, \ldots, m_{c_v}$. The creation of a new node $n$ corresponds to the activation of the part of the Petri net $\mathcal{N}$ that models node $n$. Since nodes are created in "cascade", a node $m$ is created only after all ancestors have been created. This property is ensured by the condition on the *used* flag inserted in the transitions modeling creation rules.

Following from the decidability of reachability in Petri nets, we obtain the following theorem.

**Theorem 2.** *The reachability problem is decidable for PBC systems.*

The previous theorem extends the decidability result for PB systems in [4].

## 5   PB Systems with Dissolution and Creation

In this section we consider an extension of PB systems with both dissolution and creation rules (PBDC systems). The semantics is obtained in a natural way by adapting the semantics of dissolution rules to membrane structures with labeled nodes. More precisely, a dissolution rule applied to a configuration $(N, M, \mu)$ modifies $M$ and $\mu$ as specified in Section 3 while it does not modify $N$, i.e., the set $N$ of used names can only grow monotonically. Notice that the reachability problem for PBDC systems allows to determine if, for a PBDC system $\Pi$, it is possible from the initial configuration to build a membrane structure $M$ with a mapping $\mu$ that associates multisets of objects to nodes, whatever the set of names $N$ is. Indeed, the set of possible names for membranes that appear in executions of $\Pi$ is determined by the creation rules of the PBDC system (and its initial configuration). Hence, the number of possibilities for $N$ is finite and the problem can be reduced to a finite number of reachability problems.

In presence of both creation and dissolution the membrane structure can change in an arbitrary manner. This feature gives additional power to PB systems. Indeed, we can reduce the reachability problem for two counter machines (known to be undecidable) to reachability in a PBDC system. For this reduction it is enough to consider dissolution and creation rules working on three membranes with the same name. Specifically, consider a system with initial configuration $c_0 = [_0\ s_0\ [_0\ c_1\ ]_0\ [_0\ c_2\ ]_0\ ]_0$, where $s_0$ represents the initial control state of a two counter machine. Membrane $[_0 c_i]_0$ is used to represent counter $c_i$ with value zero for $i : 1, 2$. Counter $c_i$ with value $k$ is represented by the membrane $[_0 c_i \cdot u]_0$ where $u$ is the multiset with $k$ occurrences of a special object $a$. The increment of counter $c_i$ in control state $s$ and update to control state $s'$ is encoded by the boundary rule $s\ [_0\ c_i \rightarrow s'\ [_0\ c_i \cdot a$ for $i : 1, 2$. The decrement of counter $c_i$ in control state $s$ and update to control state $s'$ is encoded by th e boundary rule $s\ [_0\ c_i \cdot a \rightarrow s'\ [_0\ c_i$ for $i : 1, 2$. The zero test on counter $c_i$ in control state $s$ and update to control state $s'$ is simulated by five rules. We use the two rules $s\ [_0\ c_i \rightarrow aux_s\ [_0\ c_i \cdot dissolve$ and $[_0\ dissolve \rightarrow [_0 \epsilon \cdot \delta$, where $\epsilon$ is the empty multiset, to move to an auxiliary state $aux_s$ and dissolve the membrane containing $c_i$. We then use the rules $[_0\ aux_s \cdot c_i \rightarrow [_0\ aux'_{s',c_i}$ and $aux'_{s',c_i} \rightarrow [_0\ c_i \cdot out_{s'}\ ]_0$ to create a new empty instance of the same membrane containing the objects $c_i$ and $out_{s'}$. Finally, we use the boundary rule $[_0\ c_i \cdot out_{s'} \rightarrow s'\ [_0\ c_i$ to move to the next state $s'$. The effect of the execution of these five rules is that of moving the contents of the counter on which the zero-test is executed to the top level membrane 0. Indeed, the membrane containing $c_i$ is first dissolved and then re-created. If the zero-test is executed when a counter is not zero, some object $a$ will remain inside the topmost membrane in all successive configurations. This feature can be used to distinguish good simulations from bad ones. Specifically, let us consider the reachability problem for a two counter machine in which the initial and the target configurations both coincide with the configuration with a given control state $s$ and both counters set to zero. This problem can be expressed as the reachability of the configuration $c_0$ from $c_0$. Thus, the following property holds.

**Theorem 3.** *The reachability problem is undecidable for PBDC systems in which configurations have at most three different membranes with the same name.*

### 5.1   PBDC Systems with Restricted Semantics

As a final analysis, we consider a restricted semantics for PBDC systems in which newly created membranes must be assigned fresh and unused names. In other words we assume that creation rules can be applied at most once for each *type* of membrane. Another possible view is that membrane names are themselves resources that can be used at most once.

Formally, assume a configuration $c = (N, M, \mu)$. Suppose that $n$ is a node in $M$. In the restricted semantics, the creation rule (4) is enabled if $a \in \mu(n)$ and $i \notin N$, i.e., the name $i$ is *fresh*. Its application leads to a new configurations $c' = (N', M', \nu')$ such that $N' = N \cup \{i\}$, $M'$ is the tree obtained by adding a new node $m$ labeled $i$ as a successor of node $n$, and $\nu'$ is the mapping defined as $\nu'(n) = \nu(n) \ominus a$, $\nu'(m) = v$, and $\nu'(p) = \nu(p)$ for $p \neq m, n$, $p \in N$).

Since with creation rules in the style of [12] the set of rules operating on membranes is fixed and known a priori, we can assume that the number of distinct names is finite (it corresponds to the set of names occurring in internal, boundary, dissolution and creation rules and in the initial configuration). This restriction yields the following key observation.

**Observation 1.** If the set of possible membrane names N is finite and every name in N can be used only once, then starting from a configuration with a single membrane, the number of distinct membrane structures that we can generate is finite. Every such membrane structure has at most |N| nodes.

This property does not imply that the number of configurations is finite. Indeed, there are no restrictions on creation and deletion of *objects* inside membranes. As an example, the PBDC system with the internal rule $[_0 a \rightarrow [_0 a \cdot a$ and the initial membrane $[_0 a]_0$ generates an infinite set of configurations (membrane 0 with any number of repetitions of object $a$).

The aforementioned property can be used to show that reachability is decidable in PBDC Systems with restricted semantics. Let $\Pi$ be a PBDC system defined over a finite set of names $\Lambda$. Suppose that $\Lambda$ has cardinality $K$. Furthermore, assume that the initial configuration $c_0 = (N_0, M_0, \mu_0)$ is such that $M_0$ is a single node. We first build the set $\Theta$ of all possible membrane structures with *at most K* nodes labeled with distinct labels taken from $\Lambda$. As an example, if $\Lambda = \{0, 1, 2\}$, then we will consider all trees with at most three nodes and such that each node has a distinct label taken from $\Lambda$, i.e., $[_0\ ]_0$, $[_1\ ]_1$, $[_1\ [_0\ ]_0\ ]_1$, $[_2\ [_0\ ]_0\ [_1\ ]_1]_2$, and so on. Notice that for a fixed membrane structure $T$ we can always determine the immediate ancestor $j$ of a node $i$ (if it exists) at static time.

Starting from $\Pi$ and $\Theta$, we now define a Petri net $\mathcal{N}$ that satisfies the following conditions. First of all, we associate a place $T$ to each membrane structure $T \in \Theta$. We assume that only one of such places can be non empty during the simulation of the restricted semantics. A non-empty place $T \in \Theta$ corresponds to the current membrane structure. Furthermore, for each $i \in \Lambda$ we add to $\mathcal{N}$ places $used_i$ and $notused_i$ (to model freshness of name $i$), and, for each $a \in \Gamma$, place $a^i$ (to model the content of membrane $i$ in the current membrane structure). Notice that since names are used at most once, we can safely confuse nodes of membrane structure with their labels (each node has a different label in $\Lambda$). PBDC rules are modeled as the finite set of transitions in $\mathcal{N}$ defined as follows.

For each membrane structure $T \in \Theta$ with a membrane $i$, an internal rule $r = [_i u \rightarrow [_i v$ is encoded by a transition $t_r^T$ that satisfies the following conditions. The pre-set contains the places $T$ (the current membrane structure) and $used_i$ ($i$ is in use) together with multiset $\pi_i(u)$. The post-set contains places $T$ and $used_i$ together with multiset $\pi_i(v)$. Thus, only the internal rules defined on the current membrane structure are enabled.

For each membrane structure $T \in \Theta$ with a membrane $j$ with immediate successor $i$, a boundary rule $r = u[_i v \rightarrow u'[_i v'$ is encoded by a transition $b_r^{T,i,j}$ that satisfies the following conditions. The pre-set contains places $T$, $used_i$, $used_j$, and the multisets $\pi_i(u)$ and $\pi_i(v)$. The post-set contains places $T$, $used_i$, $used_j$, and the multisets $\pi_i(u')$ and $\pi_i(v')$. Thus, only boundary rules defined on the current membrane structure are enabled.

For each $T \in \Theta$ such that $i$ does not occur in the set of names in $T$ (the side condition that ensures the freshness of generated membrane names), and for each name $j$ occurring in $T$, a creation rule $r = a \rightarrow [_i v]_i$ is encoded by a transition $c_r^{T,i,j}$ that satisfies the following conditions. The pre-set contains places $T$, $used_j$, $notused_i$, and $a^j$. The post-set contains places $used_i$ and $used_j$, the multiset $\pi_i(v)$, and the place $T_{j+i} \in \Theta$ associated to the membrane structure obtained from $T$ by adding a new node labeled $i$ as immediate successor of $j$.

For each membrane structure $T \in \Theta$ with a membrane $j$ with immediate successor $i$, a dissolution rule $r = [_i u \rightarrow [_i v \cdot \delta$ is encoded by the following set of transitions. We first define a transition $c_r^{T,i,j}$ that starts the dissolution phase of node $i$. The pre-set of $c_r^{T,i,j}$ contains places $T$, $used_i$, $used_j$, and the multiset $\pi_i(u)$. The post-set contains the place $dissolve^{T,i,j}$, and, the multiset $\pi_i(v)$. Notice that, by removing a token from the place $T$, we automatically disable all transitions not involved in the dissolution phase (i.e. $T$ plays the role of flag *normal* used for simulating dissolution rules in PBD systems). Now, for each $a \in \Gamma$, we model the transfer of the content of node $i$ to node $j$ via a transition $m_r^{T,i,j,a}$ that satisfies the following conditions: The pre-set contains the places $dissolve^{T,i,j}$ and $a^i$ (the source of a token to be transferred). The post-set contains the places $dissolve^{T,i,j}$ and $a^j$ (the destination of a transferred token). Finally, let $T_{j-i}$ be the membrane structure obtained by $T$ by removing membrane $i$ and moving all of its sub-membranes into membrane $j$. Then, we add transition $d_r^{T,i,j}$ to non-deterministically stop the transfer of tokens and to update the membrane structure to $T_{j-i}$, i.e., the pre-set of this transition contains the place $dissolve^{T,i,j}$ and its post-set contains the places $T_{j-i}$, $used_i$ and $used_j$. Notice that name $i$ remains marked as $used$ after dissolving the corresponding membrane (i.e. it cannot be used in successive creation rules).

The Petri net $\mathcal{N}$ is built by taking the union of the places and transitions used to encode internal, boundary, creation and dissolution rules described before.

A generic configuration $c = (N, M, \mu)$ is encoded by a marking $m_c$ in which: there is one token in the place associated to the membrane structure $M$, one token in $used_i$ for each $i \in N$, one token in $notused_i$ for each $i \in \Lambda \setminus N$, and, for each $i$ that occurs in $M$ and for each $a \in \Gamma$, as many tokens in $a^i$ as the number of occurrences of object $a$ in $\mu(i)$. All other places are empty.

Notice that after a membrane with name $i$ is introduced by a creation rule (i.e., the place $unused_i$ is emptied while one token is put in $used_i$), no other membranes with the same name can be created (there is no rule that puts a token back to $unused_i$). The membrane $i$ however can dissolve in a successive transition, i.e., in a target configuration $used_i$ can be non-empty (i.e., $i \in N$), even if $i$ does not occur in the current membrane structure. Also notice that in $m_c$ we enforce all places associated to membranes not occurring in $M$ to be empty. The combination of these two properties allows us to distinguish good simulations (i.e., in which after the application of dissolution rules all tokens are transferred to the father membrane) from bad ones (some tokens are left in a place $a^i$, $used_i$ is non empty, but $i$ is no more in the current membrane structure). Following from this observation and from the construction of $\mathcal{N}$, we have that $c = (N, M, \mu)$ is reachable from $c_0$ if and only if the marking $m_c$ is reachable from $m_{c_0}$.

Following from the decidability of reachability in Petri nets, we obtain the following theorem.

**Theorem 4.** *The reachability problem is decidable in PBDC systems with restricted semantics.*

## 6   Boundedness Problem for Extended PB Systems

In [4] Dal Zilio and Formenti exploit the Petri net encoding used for deciding reachability to prove that boundedness is decidable too for PB systems with symbol objects. In this section we investigate the boundedness problem for the different extensions of PB systems proposed in the present paper.

As a first result, we show that the boundedness problem is decidable for PBDC systems with restricted semantics where a membrane name can be used at most once. To prove this property, let us first define the following partial order $\sqsubseteq$ over configurations. Assume two configurations $c_1 = (N_1, M_1, \mu_1)$ and

$c_2 = (N_2, M_2, \mu_2)$. We define $c_1 \sqsubseteq c_2$ if and only if $N_1 = N_2$ and $M_1 = M_2$ (i.e. $c_1$ and $c_2$ have the same tree structure), and $\mu_1(n) \leq \mu_2(n)$ (the multiset associated to $n$ in $c_1$ is contained in that associated to $n$ in $c_2$) for all node $n$ in $M_1$. If we fix an upper bound on the number of possible nodes occurring in a membrane structure along a computation, then $\sqsubseteq$ has the following property.

**Proposition 1.** *Fixed a $k \in \mathbb{N}$, for any infinite sequence of configurations $c_1 c_2 \ldots$ with membrane structure of size at most $k$, there exist positions $i < j$ such that $c_i \sqsubseteq c_j$ (i.e. $\sqsubseteq$ is a* well-quasi ordering*).*

The proof is a straightforward application of composition properties of well-quasi ordering, see e.g. [1]. Now assume an infinite computation $c_0 = (N_0, M_0, \mu_0) c_1 = (N_1, M_1, \mu_1) \ldots$ of a PBDC system with restricted semantics. From Observation 1 it follows that for all $i \geq 0$ the number of nodes in $M_i$ is bounded by the number of possible names. Hence, by Prop. 1 we know that there exist positions $i < j$ such that $c_i \sqsubseteq c_j$. Furthermore, if $c_i \sqsubseteq c_j$ and $c_i \neq c_j$, then $N_i = N_j$, $M_i = M_j$, and $\mu_i \prec \mu_j$. Thus, the transition sequence $\sigma$ from $c_i$ to $c_j$ does not modify the membrane structure but strictly increases the number of objects contained at each of its node. This implies that the application of $\sigma$ can be iterated starting from $c_j$, leading to a infinite strictly increasing, w.r.t $\sqsubseteq$, sequence of configurations.

As a consequence of these properties, the boundedness problem for PBDC systems can be decided by building a computation tree $\mathcal{T}$ such that: the root node $n_0$ of $\mathcal{T}$ is labeled by the initial configuration $c_0$, if $n_0, \ldots, n_k$ is a path in $\mathcal{T}$ such that $n_i$ is labeled with $c_i$ $i : 0, \ldots, k$, and for all $i : 0, \ldots k-1$ $c_i \neq c_k$ and $c_i \not\sqsubseteq c_k$ then we add a node $n'$ labeled $c'$ as successor of $n_k$ if and only if $c_k \Rightarrow c'$. Furthermore, the PBDC system is not bounded if and only if there exists a leaf $n$ labeled with $c$ and a predecessor $n'$ of $n$ labeled with $c'$ in $\mathcal{T}$ such that $c' \sqsubseteq c$. Since $\sqsubseteq$ is a decidable relation and the tree $\mathcal{T}$ is finite (by Observation 1 and Prop. 1), the following property then holds.

**Theorem 5.** *The boundedness problem is decidable for PBDC systems with restricted semantics.*

From Theorem 5, we know that boundedness is decidable for PB systems (consistently with the result in [4]) and for PBD systems (they form a subclass of PBDC systems where the restricted and standard semantics coincides).

The boundedness problem turns out to be undecidable for PBC systems with standard semantics in which there is no limit on the number of instances of a given type of membranes that can be created during a computation. The proof is based on a reduction of two counter machines to PBC systems. The idea is to use nested membranes to model the current value of a counter. E.g., the membrane $[_1 used\ [_1\ used\ [_1\ unused\ [_1\ end\ ]_1]_1]_1]_1$ can be used to encode counter $c_1$ with value 2 (the number of occurrences of symbol $used$). Hence a configuration of a two counter machine with both counters set to zero is encoded as a configuration of the form

$$[_0 \ell \ [_1\ unused\ [_1\ \ldots [_1\ end\ ]_1\ \ldots]_1]_1[_2\ unused\ [_2 \ldots [_2\ end\ ]_2\ \ldots]_2\ ]_2\ ]_0$$

$$A : \begin{bmatrix} \ell \ [_i \ unused \ \rightarrow \ \ell_1 \ [_i \ used \\ \ell \ [_i \ used \ \rightarrow \ \ell'_i \ [_i \ down \\ down \ [_i \ used \ \rightarrow \ down \ [_i \ down \\ down \ [_i \ unused \ \rightarrow \ up \ [_i \ used \\ down \ [_i \ up \ \rightarrow \ up \ [_i \ used \\ \ell'_i \ [_i \ up \ \rightarrow \ \ell_1 \ [_i \ used \end{bmatrix}$$

$$B : \begin{bmatrix} down \ [_i \ end \ \rightarrow \ down \ [_i \ create \\ create \ \rightarrow \ [_i \ exit \cdot end \ ]_i \\ \epsilon \ [_i \ exit \ \rightarrow \ up \ [_i \ \epsilon \end{bmatrix}$$

$$C : \begin{bmatrix} \ell \ [_i \ used \ \rightarrow \ \ell'_i \ [_i \ used_1 \\ used_1 \ [_i \ used \ \rightarrow \ used_1 \ [_i \ used_1 \\ used_1 \ [_i \ unused \ \rightarrow \ used_2 \ [_i \ unused \\ used_1 \ [_i \ end \ \rightarrow \ used_2 \ [_i \ end \\ used_1 \ [_i \ used_2 \ \rightarrow \ used_3 \ [_i \ unused \\ used_1 \ [_i \ used_3 \ \rightarrow \ used_3 \ [_i \ used \\ \ell'_i \ [_i \ used_2 \ \rightarrow \ \ell_1 \ [_i \ unused \\ \ell'_i \ [_i \ used_3 \ \rightarrow \ \ell_1 \ [_i \ used \end{bmatrix}$$

**Fig. 1.** Encoding of increment ($A$ and $B$) and decrement ($C$)

where $\ell$ is a symbol corresponding to the current location of the two counter machine, and membrane with name $i$ encodes counter $i$ for $i : 1, 2$. Increment of counter $i$ in location $\ell$ with $\ell_1$ as successor location is simulated by replacing the first *unused* symbol encountered when descending the tree from the membrane 0 with the symbol *used*. This is implemented by the set of rules $A$ in Fig. 1 used to descend the structure of a membrane of type $i$ in search for the first occurrence of symbol *unused*. Here $\ell'_i, down, up$ are auxiliary symbols ($\ell'_i$ is blocking for the other counter(s), *down* propagate the search down in the tree, *up* propagate the success notification up in the tree). Furthermore, if the current tree has no membrane containing the *unused* symbol (i.e. all membranes contain *used*) then a new membrane is created by applying the set of rules $B$ in Fig. 1. Decrement is simulated by changing the last occurrence of *used* encountered by descending the membrane tree from the root into symbol *unused*. We can safely assume here that the counter is non-zero, i.e., that there is at least one membrane with *used* object. The set of rules $C$ in Fig. 1 implement this idea. Here $\ell'_i, used_1, used_2, used_3$ are auxiliary symbols, $used_1$ is used to mark nodes during the downward search (for *unused*), $used_2$ is used to mark the *used* node to be replaced by *unused*, and $used_3$ is used to replace nodes marked with $used_1$ with *used* during the return from the search. The zero-test on counter $c_i$ in location $\ell$ with successor $\ell_1$ can be implemented by testing if all objects in membranes $i$ are *unused* (or *end*). Note that increment, resp. decrement, of $c_i$ is encoded by a top-down traversal of membranes $i$ until reaching a membrane containing an object *unused*, resp. *used*, which is then replaced by *used*, resp. *unused*. Furthermore, each membrane contains one object. Hence, no membrane $i$ contain a *used* symbols if and only if the top level membrane $i$ has object *unused* or *end*. This can be checked with the following rules:

$$\ell \ [_i \ unused \ \rightarrow \ \ell_1 \ [_i \ unused \qquad \ell \ [_i \ end \ \rightarrow \ \ell_1 \ [_i \ end$$

By construction, we directly have that a two counter machine is bounded if and only if its encoding into PBC systems is bounded. Since boundedness is undecidable for two-counter machines, we obtain the following negative result.

**Theorem 6.** *The boundedness problem is undecidable for PBC systems.*

The undecidability of boundedness proved in Theorem 6 is not in contradiction with the decidability of reachability proved in Theorem 2. Similar results obtained for fragments of process calculi, see e.g. [3], seem to indicate that, in general, the decidability of reachability cannot be used to give an estimation of the expressive power of a computational model.

# 7 Conclusions and Related Work

In this paper we have investigated the decidability of reachability and boundedness in extensions of PB systems with rules that dynamically modify the tree structure of membranes. Concerning related work, as mentioned in the introduction our results extend those obtained for PB-systems in [4]. Decision problems for qualitative analysis of subclasses of P-systems have been studied in [11,8]. A methodology based on P/T nets to characterize the completeness of biological models has been proposed in [7]. The application of this methodology to the extended PB systems studied in this paper is an interesting research line for future work.

## Acknowledgements

The research of the second author was supported by the Belgian National Science Foundation (FNRS).

## References

1. Aziz-Abdulla, P., Nylén, A.: On Efficient Verification of Infinite-State Systems. In: LICS 2000, pp. 132–140 (2000)
2. Bernardini, F., Manca, V., Systems, P.: P Systems with Boundary Rules. In: Păun, G., Rozenberg, G., Salomaa, A., Zandron, C. (eds.) WMC 2002. LNCS, vol. 2597, pp. 107–118. Springer, Heidelberg (2003)
3. Busi, N., Zavattaro, G.: Deciding Reachability in Mobile Ambients. In: Sagiv, M. (ed.) ESOP 2005. LNCS, vol. 3444, pp. 248–262. Springer, Heidelberg (2005)
4. Dal Zilio, S., Formenti, E.: On the Dynamics of PB Systems: A Petri Net View. In: Martín-Vide, C., Mauri, G., Păun, G., Rozenberg, G., Salomaa, A. (eds.) Membrane Computing. LNCS, vol. 2933, pp. 153–167. Springer, Heidelberg (2004)
5. Dufourd, C., Finkel, A., Schnoebelen, P.: Reset Nets Between Decidability and Undecidability. In: Larsen, K.G., Skyum, S., Winskel, G. (eds.) ICALP 1998. LNCS, vol. 1443, pp. 103–115. Springer, Heidelberg (1998)
6. Franco, G., Manca, V.: A Membrane System for the Leukocyte Selective Recruitment. In: Martín-Vide, C., Mauri, G., Păun, G., Rozenberg, G., Salomaa, A. (eds.) Membrane Computing. LNCS, vol. 2933, pp. 181–190. Springer, Heidelberg (2004)
7. Frisco, P.: P-systems, Petri Nets and Program Machines. In: Freund, R., Păun, G., Rozenberg, G., Salomaa, A. (eds.) WMC 2005. LNCS, vol. 3850, pp. 209–223. Springer, Heidelberg (2006)

8. Ibarra, O.H., Dang, Z., Egecioglu, Ö.: Catalytic P Systems, Semilinear Sets, and Vector Addition Systems. TCS 312(2-3), 379–399 (2004)
9. Karp, R.M., Miller, R.E.: Parallel Program Schemata. JCSS 3, 147–195 (1969)
10. Kosaraju, S.R.: Decidability of Reachability in Vector Addition Systems. STOC, 267–281 (1982)
11. Li, C., Dang, Z., Ibarra, O.H., Yen, H.-C.: Signaling P Systems and Verification Problems. In: Caires, L., Italiano, G.F., Monteiro, L., Palamidessi, C., Yung, M. (eds.) ICALP 2005. LNCS, vol. 3580, pp. 1462–1473. Springer, Heidelberg (2005)
12. Martin-Vide, C., Păun, G., Rodriguez-Paton, A.: On P Systems with Membrane Creation Comp. Sc. J. of Moldova 9(2), 134–145 (2001)
13. Mayr, E.W.: An Algorithm for the General Petri Net Reachability Problem. SIAM J. Comput. 13(3), 441–460 (1984)
14. Minsky, M.: Computation: Finite and Infinite Machines. Prentice-Hall, Englewood Cliffs (1967)
15. Păun, G.: Computing with Membranes. JCSS 61(1), 108–143 (2000)
16. Petri, C.A.: Kommunikation mit Automaten. Ph.D. Thesis. U. of Bonn (1962)

# A    Petri (P/T) Nets and Counter Machines

A *Petri net* [16] is a pair $(P, T, m_0)$ where $P$ is a finite set of *places*, $T$ is a finite set of *transitions* $(T \subseteq P^\otimes \times P^\otimes)$, and $m_0$ is the initial marking. A *transition* $t$ is defined by the pre-set ${}^\bullet t$ and by the post-set $t^\bullet$, two multisets of places in $P$. A marking is a multiset over $P$. Given a marking $m$ and a place $p$, we say that the place $p$ contains $m(p)$ *tokens*. A transition $t$ is enabled at the marking $m$ if ${}^\bullet t \preceq m$. When enabled, the firing of $t$ at $m$ produces a marking $m'$ (written $m \xrightarrow{t} m'$) defined as $(m \ominus {}^\bullet t) \oplus t^\bullet$. A firing sequence is a sequence of markings $m_0 m_1 \ldots$ such that $m_i$ is obtained from $m_{i-1}$ by firing a transition in $T$ at $m_i$. Finally, we say that $m'$ is reachable from $m_0$ if there exists a firing sequence from $m_0$ passing through $m'$. The reachability problem is decidable for Petri nets [10].

A *two counter machine* [14] is a finite automaton extended with two counters taking their value into the non-negative integers. When the automaton moves from one location to another it can access one counter by using an operation/test of the following three forms: increment, decrement and test for zero. These operations/tests have their usual semantics. In particular, decrement is blocking when the counter is equal to 0.

# A Logarithmic Bound for Solving Subset Sum with P Systems

Daniel Díaz-Pernil, Miguel A. Gutiérrez-Naranjo,
Mario J. Pérez-Jiménez, and Agustín Riscos-Núñez

Research Group on Natural Computing
University of Sevilla, Spain
{sbdani,magutier,marper,ariscosn}@us.es

**Abstract.** The aim of our paper is twofold. On one hand we prove the ability of polarizationless P systems with dissolution and with division rules for non-elementary membranes to solve **NP**-complete problems in a polynomial number of steps, and we do this by presenting a solution to the Subset Sum problem. On the other hand, we improve some similar results obtained for different models of P systems by reducing the number of steps and the necessary resources to be of a logarithmic order with respect to $k$ (recall that $n$ and $k$ are the two parameters used to indicate the size of an instance of the Subset Sum problem).

As the model we work with does not allow cooperative rules and does not consider the membranes to have an associated polarization, the strategy that we will follow consists on using objects to represent the weights of the subsets through their multiplicities, and comparing the number of objects against a fixed number of membranes. More precisely, we will generate $k$ membranes in $\log k$ steps.

## 1 Introduction

This paper is the continuation of a series of results on Complexity Classes in Membrane Computing that are trying to establish the relevance, in terms of computing power, of each one of the possible features of a P system (see [3]).

The Subset Sum problem is a well-known **NP**-complete problem which can be formulated as follows: *Given a finite set $A$, a weight function, $w : A \to \mathbb{N}$, and a constant $k \in \mathbb{N}$, determine whether or not there exists a subset $B \subseteq A$ such that $w(B) = k$.* It has been a matter of study in Membrane Computing several times, being mainly used to prove the ability of different P system models in order to solve problems from the **NP** class in a polynomial time.

This speed-up is achieved by trading space for time, in the sense that the considered models allow that an exponential amount of membranes can be produced by a P system in a polynomial number of steps. For example, solutions to the Subset Sum problem working in a number of steps which is linear with respect to the parameters $n$ and $k$ have been designed using P systems with active membranes [9], using tissue P systems with cell division [2], and using P systems with membrane creation [4].

G. Eleftherakis et al. (Eds.): WMC8 2007, LNCS 4860, pp. 257–270, 2007.
© Springer-Verlag Berlin Heidelberg 2007

In this paper we work with P systems using division of non-elementary membranes and dissolution rules. Our aim goes beyond adding this P system model to the above mentioned list; we improve previous complexity results by solving the Subset Sum problem in a linear number of steps with respect to $n$ and $\log k$. We also improve the pre-computation process, as the initial resources are also bounded by $\log k$.

The paper is structured as follows: in the next section we present the formal framework, i.e., we recall the definition of recognizing P systems, the P system model used along the paper is settled and the class $\mathbf{PMC}_{\mathcal{AM}^0(+d,+ne)}$ is presented. In Section 3, our design of the solution of the Subset Sum problem is presented and some conclusions are given in the last section.

## 2   Formal Framework

In this paper we are using cellular systems for attacking the resolution of decision problems. This means that for each instance of a problem that we try to solve, we are only interested in obtaining a Boolean answer (*Yes* or *No*). Therefore, the P system can behave as a *black box* to which the user supplies an input and from which an affirmative or negative answer is received. This is indeed the motivation for defining the concept of *recognizing P systems* (introduced in [13]).

### 2.1   Recognizing P Systems

Let us recall that a decision problem, $X$, is a pair $(I_X, \theta_X)$ where $I_X$ is a language over an alphabet whose elements are called *instances* and $\theta_X$ is a total Boolean function over $I_X$. If $u$ is an instance of the problem $X$ such that $\theta_X(u) = 1$ (respectively, $\theta_X(u) = 0$), then we say that the answer to the problem for the instance considered is *Yes* (respectively, *No*).

Keeping this in mind, recognizing P systems are defined as a special class of membrane systems that will be used to solve decision problems, in the framework of the complexity classes theory. Note that this definition is stated informally, and it can be adapted for any kind of membrane system paradigm.

A recognizing P system is a P system with input and with external output having two distinguished objects **yes** and **no** in its working alphabet such that:

- All computations halt.
- If $\mathcal{C}$ is a computation of $\Pi$, then either the object **yes** or the object **no** (but not both) must have been released into the environment, and only in the last step of the computation.

### 2.2   The P System Model

The power of membrane division as a tool for efficiently solving **NP** problems in Membrane Computing has been widely proved. Many examples of designs of P systems solving **NP**-complete problems have been proposed in the framework of P systems with active membranes with two polarizations and three polarizations and in the framework of P systems with non-elementary membrane division. The key of such solutions is the creation of an exponential amount of workspace (membranes) in a polynomial time.

In the literature, one can find two quite different rules for performing membrane division. On the one hand, in [7], P systems with active membranes were presented. In this model new membranes were obtained through the process of *mitosis* (membrane division). In these devices membranes have polarizations, one of the "electrical charges" $0, -, +$, and several times the problem was formulated whether or not these polarizations are necessary in order to obtain polynomial solutions to **NP**–complete problems. The last result is that from [1], where one proves that two polarizations suffice.

P systems with active membranes have been successfully used to design (uniform) solutions to well-known **NP**–complete problems, such as SAT [13], *Subset Sum* [9], *Knapsack* [10], *Bin Packing* [11], *Partition* [5], and the *Common Algorithmic Problem* [12].

The syntactic representation of membrane division rule is

$$[\,a\,]_h^{e_1} \rightarrow [\,b\,]_h^{e_2} \,[\,c\,]_h^{e_3} \tag{1}$$

where $h$ is a label, $e_1, e_2$ and $e_3$ are electrical charges and $a, b$ and $c$ are objects. The interpretation is well-known: An elementary membrane can be divided into two membranes with the same label, possibly transforming some objects and changing the electrical charge. All objects present in the membrane except the object triggering the rule are copied into both new membranes.

In [6], a variant of this rule was used in which the polarization was dropped:

$$[\,a\,]_h \rightarrow [\,b\,]_h \,[\,c\,]_h. \tag{2}$$

In both cases (with and without polarizations) the key point is that the membranes are always elementary membranes. In the literature, there also exist rules for the division of non-elementary polarizationless membranes, as

$$[\,[\,]_{h_1} \,[\,]_{h_2}\,]_{h_0} \rightarrow [\,[\,]_{h_1}\,]_{h_0} \,[\,[\,]_{h_2}\,]_{h_0} \tag{3}$$

where $h_0, h_1$ and $h_2$ are labels. There exists an important difference with respect to elementary membrane division: in the case of (3), the rule is not triggered by the occurrence of an object inside a membrane, but by the membrane structure instead. This point has a crucial importance in the design of solutions, since a membrane can be divided by the corresponding rule even if there are no objects inside it.

According to the representation (3), the membrane $h_0$ divides into two new membranes also with label $h_0$ and all the information (objects and membranes) different from membranes $h_1$ and $h_2$ inside is duplicated.

In this paper we use a type of membrane division which is syntactically equivalent to (2)

$$[\,a\,]_h \rightarrow [\,b\,]_h \,[\,c\,]_h, \tag{4}$$

but we will consider a semantic difference; the dividing membrane can be elementary or non-elementary and after the division, all the objects and membranes inside the dividing membrane are duplicated, except the object $a$ that triggers the rule, which appears in the new membranes possibly modified (represented as objects $b$ and $c$).

In this paper we work with a variant of P systems with active membranes which we call with weak division, and that does not use polarizations.

**Definition 1.** *A P system with active membranes with weak division is a P system with $\Gamma$ as working alphabet, with $H$ as the finite set of labels for membranes, and where the rules are of the following forms:*

(a) $[a \rightarrow u]_h$ *for $h \in H$, $a \in \Gamma$, $u \in \Gamma^*$. This is an object evolution rule, associated with a membrane labelled with $h$: an object $a \in \Gamma$ belonging to that membrane evolves to a multiset $u \in \Gamma^*$.*

(b) $a[\ ]_h \rightarrow [b]_h$ *for $h \in H$, $a, b \in \Gamma$. An object from the region immediately outside a membrane labeled with $h$ is introduced in this membrane, possibly transformed into another object.*

(c) $[a]_h \rightarrow b[\ ]_h$ *for $h \in H$, $a, b \in \Gamma$. An object is sent out from membrane labeled with $h$ to the region immediately outside, possibly transformed into another object.*

(d) $[a]_h \rightarrow b$ *for $h \in H$, $a, b \in \Gamma$: A membrane labeled with $h$ is dissolved in reaction with an object. The skin is never dissolved.*

(e) $[a]_h \rightarrow [b]_h [c]_h$ *for $h \in H$, $a, b, c \in \Gamma$. A membrane can be divided into two membranes with the same label, possibly transforming some objects. The content of the membrane is duplicated. The membrane can be elementary or not.*

These rules are applied according to the following principles:

- All the rules are applied in parallel and in a maximal manner. In one step, one object of a membrane can be used by only one rule (chosen in a non–deterministic way), but any object which can evolve by one rule of any form, must evolve.

- If at the same time a membrane labeled with $h$ is divided by a rule of type (e) and there are objects in this membrane which evolve by means of rules of type (a), then we suppose that first the evolution rules of type (a) are used, and then the division is produced. Of course, this process takes only one step.

- The rules associated with membranes labeled with $h$ are used for all copies of this membrane. At one step, a membrane can be the subject of *only one* rule of types (b)-(e).

Let us note that in this framework we work without cooperation, without priorities, with weak division, and without changing the labels of membranes.

In this paper we work within the model of *polarizationless P systems using weak division of non-elementary membranes and dissolution*. Let $\mathcal{AM}^0(+d, +ne)$ be the class of such systems.

## 2.3  The Class PMC$_{\mathcal{AM}^0(+d,+ne)}$

**Definition 2.** *We say that a decision problem $X = (I_X, \theta_X)$ is solvable in polynomial time by a family $\mathbf{\Pi} = \{\Pi(n) : n \in \mathbb{N}\}$ of recognizing P systems from $\mathcal{AM}^0(+d, +ne)$ if the following holds:*

- *The family $\mathbf{\Pi}$ is* polynomially uniform *by Turing machines, that is, there exists a deterministic Turing machine working in polynomial time which constructs the system $\Pi(n)$ from $n \in \mathbb{N}$.*
- *There exists a pair $(cod, s)$ of polynomial-time computable functions over $I_X$ such that:*
  - *for each instance $u \in I_X$, $s(u)$ is a natural number and $cod(u)$ is an input multiset of the system $\Pi(s(u))$;*
  - *the family $\mathbf{\Pi}$ is* polynomially bounded *with regard to $(X, cod, s)$, that is, there exists a polynomial function $p$, such that for each $u \in I_X$ every computation of $\Pi(s(u))$ with input $cod(u)$ is halting and, moreover, it performs at most $p(|u|)$ steps;*
  - *the family $\mathbf{\Pi}$ is* sound *with regard to $(X, cod, s)$, that is, for each $u \in I_X$, if there exists an accepting computation of $\Pi(s(u))$ with input $cod(u)$, then $\theta_X(u) = 1$;*
  - *the family $\mathbf{\Pi}$ is* complete *with regard to $(X, cod, s)$, that is, for each $u \in I_X$, if $\theta_X(u) = 1$, then every computation of $\Pi(s(u))$ with input $cod(u)$ is an accepting one.*

In the above definition we have imposed to every P system $\Pi(n)$ a *confluent* condition, in the following sense: every computation of a system with the *same* input multiset must always give the *same* answer. The pair of functions $(cod, s)$ is called a *polynomial encoding* of the problem in the family of P systems.

We denote by **PMC**$_{\mathcal{AM}^0(+d,+ne)}$ the set of all decision problems which can be solved by means of recognizing polarizationless P systems using division of non-elementary membranes and dissolution in polynomial time.

## 3  Designing the Solution to Subset Sum

In this section we address the resolution of the problem following a brute force algorithm, implemented in the framework of recognizing P systems from the $\mathcal{AM}^0(+d, +ne)$ class. The idea of the design is better understood if we divide the solution to the problem into several stages:

- *Generation stage*: for every subset of $A$, a membrane labeled by $e$ is generated via membrane division.
- *Calculation stage*: in each membrane the weight of the associated subset is calculated (using the auxiliary membranes $e_0, \ldots, e_n$).
- *Checking stage*: in each membrane it is checked whether the weight of its associated subset is exactly $k$ (using the auxiliary membranes $ch$).
- *Output stage*: the system sends out the answer to the environment, according to the result of the checking stage.

Let us now present a family of recognizing P systems from the $\mathcal{AM}^0(+d, +ne)$ class that solves Subset Sum, according to Definition 2.

We shall use a tuple $(n, (w_1, \ldots, w_n), k)$ to represent an instance of the Subset Sum problem, where $n$ stands for the size of $A = \{a_1, \ldots, a_n\}$, $w_i = w(a_i)$, and $k$ is the constant given as input for the problem. Let $g : \mathbb{N} \times \mathbb{N} \to \mathbb{N}$ be a function defined by

$$g(n, k) = \frac{(n + k)(n + k + 1)}{2} + n$$

This function is primitive recursive and bijective between $\mathbb{N} \times \mathbb{N}$ and $\mathbb{N}$ and computable in polynomial time. We define the polynomially computable function $s(u) = g(n, k)$.

We shall provide a family of P systems where each P system solves all the instances of the Subset Sum problem with the same size. Let us consider the binary decomposition of $k$, $\Sigma_{i \in I} 2^i = k$, where the indices $i \in I$ indicate the positions of the binary expression of $k$ where a 1 occurs. Let $I' = \{1, \ldots, \lfloor \log k \rfloor\} - I$ be the complementary set, that is, the positions where a 0 occurs. This binary encoding of $k$, together with the weight function $w$ of the concrete instance, will be provided via an input multiset determined by the function $cod$ as follows:

$$cod(u) = cod_1(u) \cup cod_2(u),$$

$$\text{where } cod_1(u) = \{\{b_i^{w_i} : 1 \leq i \leq n\}\} \text{ and}$$
$$cod_2(u) = \{\{c_j : j \in I\}\} \cup \{\{c_j' : j \in I'\}\}$$

Next, we shall provide a family $\mathbf{\Pi} = \{\Pi(g(n, k)) : n, k \in \mathbb{N}\}$ of recognizing P systems which solve the Subset Sum problem in a number of steps being of $O(n + \log k)$ order. We shall indicate for each system of the family its initial configuration and its set of rules. We shall present the list of rules divided by groups, and we shall provide for each of them some comments about the way their rules work.

Let us consider an arbitrary pair $(n, k) \in \mathbb{N} \times \mathbb{N}$. The system $\Pi(g(n, k))$ is determined by the tuple $(\Gamma, \Sigma, \mu, M, \mathcal{R}, i_{in}, i_0)$, that is described next:

- Alphabet:

$$\Gamma = \Sigma \cup \{b_i^+, b_i^-, b_i^=, d_i, d_i^+, d_i^-, p_i, q_i : i = 1, \ldots, n\}$$
$$\cup \{g_0, \ldots, g_{2\lfloor \log k \rfloor + 2}, h_0, \ldots, h_{2\lfloor \log k \rfloor + 2n + 8}, l_0, \ldots, l_{2\lfloor \log k \rfloor + 2n + 10}\}$$
$$\cup \{v_0, \ldots, v_{2\lfloor \log k \rfloor + 2n + 12}\}$$
$$\cup \{w_0, \ldots, w_{2\lfloor \log k \rfloor + 2n + 18}\}$$
$$\cup \{x_0, \ldots, x_{2\lfloor \log k \rfloor + 2n + 15}, z_0, \ldots, z_{2\lfloor \log k \rfloor + 2n + 7}\}$$
$$\cup \{s, \mathsf{yes}, \mathsf{no}, Trash\}$$

- Input alphabet: $\Sigma(n, k) = \{b_1, \ldots, b_n, c_0, \ldots, c_{\lfloor \log k \rfloor}, c_0', \ldots, c_{\lfloor \log k \rfloor}'\}$.

The initial configuration consists of $n + \lfloor \log k \rfloor + 9$ membranes, arranged as shown in Figure 1. Formally, the membrane structure $\mu$ is

$$[[[[[[\cdot^n_\cdot \cdot [[[[[]_{ch} \cdots []_{ch}]_{a_1}]_{a_2}]_{e_0}]_{e_1} \cdot^n_\cdot \cdot]_{e_n}]_{a_3} []_c]_{a_4}]_e]_f]_{skin}$$

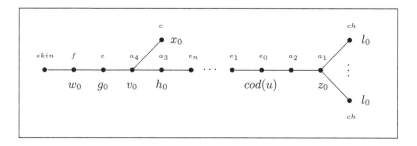

**Fig. 1.** Initial Configuration

where there are exactly $\lfloor \log k \rfloor + 1$ copies of membrane $[\,]_{ch}$.

Roughly speaking (more precise explanations will be given for the rules), we can classify the membranes according to their role as follows:

- $n+2$ membranes that take care of the generation stage, namely those labeled by $e_0, e_1, \ldots e_n$ and $e$.
- $\lfloor \log k \rfloor + 3$ membranes that take care of preparing and implementing the checking stage, namely those labeled by $ch$, $a_1$ and $a_2$.
- 4 membranes that take care of the answer stage, handling and synchronizing the results of the checking, namely those labeled by $a_3, a_4, c$ and $f$.

• The initial multisets are:

$$M(f) = \{\{w_0\}\};\ M(e) = \{\{g_0\}\};\ M(a_4) = \{\{v_0\}\};\ M(a_3) = \{\{h_0\}\};$$

$$M(c) = \{\{x_0\}\};\ M(a_1) = \{\{z_0\}\};\ M(ch) = \{\{l_0\}\}$$

$$M(skin) = M(a_2) = M(e_0) = \cdots = M(e_n) = \emptyset$$

• The input membrane is $i_{in} = e_0$, and the output region is the environment ($i_0 = env$).

**First task: generate $k$ membranes $ch$.** At the beginning of the computation, $k$ membranes $ch$ will be generated inside the innermost region of the structure. The strategy works as follows:

1. Initially, there are $\lfloor \log k \rfloor$ membranes $ch$ in the region $a_1$, and the input multiset is located in region $e_0$ (recall that $cod_2(u)$ consists of $\lfloor \log k \rfloor$ objects $c_i$ or $c'_i$ representing the binary encoding of $k$).
2. In the first $\lfloor \log k \rfloor$ steps, the objects from $cod_2(u)$ get into membrane $a_2$ (the objects enter one by one membrane $a_2$). Simultaneously, the counter $z_i$ is evolving inside membrane $a_1$ and dissolves it at the $\lfloor \log k \rfloor$ step.
3. Thus, in the next step each element from $cod_2(u)$ will go inside a membrane $ch$ (all objects go in parallel into different membranes in a one-to-one manner).

4. Objects $c_i'$ will dissolve the membranes where they enter, while each object $c_i$ will generate by division $2^i$ membranes $ch$.
5. After at most $\lfloor \log k \rfloor$ further steps all divisions have been completed, and the number of membranes $ch$ is exactly $k$.

Membrane $a_2$ will not be divided until the generation and weight calculation stages have been completed, acting as a separator between objects from $cod_1(u)$ and membranes $ch$.

**Set (A1).**
$$\left.\begin{array}{l} c_i[\,]_{a_2} \to [c_i]_{a_2} \\ c_i'[\,]_{a_2} \to [c_i']_{a_2} \\ \\ c_i[\,]_{ch} \to [c_i]_{ch} \\ c_i'[\,]_{ch} \to [c_i']_{ch} \\ [c_i']_{ch} \to Trash \end{array}\right\} \text{ for } i \in \{0,\ldots,\lfloor \log k \rfloor\}.$$

**Set (A2).**
$$\begin{array}{ll} [c_0 \to Trash]_{ch} & \\ [c_i]_{ch} \to [c_{i-1}]_{ch}\,[c_{i-1}]_{ch} & \text{for } i = 1,\ldots,\lfloor \log k \rfloor \\ [z_i \to z_{i+1}]_{a_1} & \text{for } i = 0,\ldots,\lfloor \log k \rfloor - 1 \\ [z_{\lfloor \log k \rfloor}]_{a_1} \to z_{\lfloor \log k \rfloor+1} & \\ [g_i \to g_{i+1}]_e & \text{for } i = 0,\ldots,2\lfloor \log k \rfloor + 1 \\ [g_{2\lfloor \log k \rfloor+2} \to d_1 s]_e & \end{array}$$

In the last step of this stage, the counter $g_i$ produces the objects $d_1$ and $s$ which will trigger the beginning of the next stage.

**Set(B).**
$$\left.\begin{array}{l} [w_i \to w_{i+1}]_f \\ [v_i \to v_{i+1}]_{a_4} \\ [h_i \to h_{i+1}]_{a_3} \\ [x_i \to x_{i+1}]_c \\ [l_i \to l_{i+1}]_{ch} \end{array}\right\} \text{ for } i \in \{0,\ldots,2\lfloor \log k \rfloor + 2\}.$$

$[z_i \to z_{i+1}]_{a_2}$  for $i \in \{\lfloor \log k \rfloor + 1,\ldots,2\lfloor \log k \rfloor + 2\}$.

The rest of the counters simply increase their indices in this stage. (See Fig. 2.)

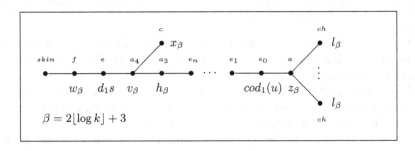

**Fig. 2.** TIME  $2\lfloor \log k \rfloor + 3$

**Second task: generate $2^n$ membranes $e$.** Objects $d_i$ residing inside membrane(s) $e$ will produce $n$ consecutive divisions, thus yielding $2^n$ copies of membrane $e$. To each one of them, a subset of $A$ is associated in the following way: after each division, the membranes where object $p_i$ occurs correspond to subsets of $A$ containing $a_i$, and conversely, membranes where $q_i$ occurs will be associated with subsets not containing $a_i$.

**Set (C).**  $\begin{aligned} &[d_i]_e \to [d_i^+]_e [d_i^-]_e & &\text{for } i = 1, \ldots n \\ &[d_i^+ \to p_i d_{i+1}]_e & &\text{for } i = 1, \ldots n - 1 \\ &[d_i^- \to q_i d_{i+1}]_e & &\text{for } i = 1, \ldots n - 1 \\ &[d_n^+ \to p_n]_e \\ &[d_n^- \to q_n]_e \end{aligned}$

Membrane divisions take place every two steps, so in the $(2\lfloor \log k \rfloor + 2n + 2)$-th step there will be $2^n$ membranes $e$.

**Set (D).**  $\begin{aligned} &s\,[\,]_{a_i} \to [s]_{a_i} & &\text{for } i = 3, 4 \\ &s\,[\,]_{e_i} \to [s]_{e_i} & &\text{for } i = 0, \ldots, n \\ &[s]_{e_0} \to Trash \\ &p_j\,[\,]_{a_i} \to [p_j]_{a_i} & &\text{for } i = 3, 4 \quad j = 1, \ldots, n \\ &p_j\,[\,]_{e_i} \to [p_j]_{e_i} & &\text{for } j = 1, \ldots, n \quad i = j, \ldots, n \\ &[p_i \to q_i]_{e_i} & &\text{for } i = 1, \ldots, n \\ &q_j\,[\,]_{a_i} \to [q_j]_{a_i} & &\text{for } i = 3, 4 \quad j = 1, \ldots, n \\ &q_j\,[\,]_{e_i} \to [q_j]_{e_i} & &\text{for } j = 1, \ldots, n \quad i = j, \ldots, n \\ &[q_i]_{e_i} \to Trash & &\text{for } i = 1, \ldots, n \end{aligned}$

While the divisions are being carried out, objects $s$, $p_j$ and $q_j$, for $j = 1, \ldots, n$, travel into inner membranes (recall that whenever membrane $e$ gets divided, the internal nested structure of membranes $e_i$ is duplicated). In the $(2\lfloor \log k \rfloor + n + 2)$-th step, an object $s$ arrives to every membrane $e_0$. This object dissolves the membrane in the next step, and therefore in the $(2\lfloor \log k \rfloor + n + 3)$-th step we find inside every membrane $e_1$ the multiset $cod_1(u)$, and in this moment the weight calculation stage begins (see rules in **Set (E)**).

As we said before, objects $p_j$ and $q_j$ are traveling into inner membranes, until they reach $e_j$. This is done in such a way that in the $(2\lfloor \log k \rfloor + n + 3)$-th step there is in each membrane $e_1$ either an object $p_1$ or an object $q_1$, in addition to the multiset $cod_1(u)$.

Before going on, let us state two points. First, recall that in the input multiset, introduced in $e_0$ at the beginning of the computation, there are $w(a_i)$ copies of $b_i$, for $i = 1, \ldots, n$. Second, let us note that objects $q_i$ dissolve membrane $e_i$ immediately after arriving to it, while objects $p_i$ take two steps to dissolve membrane $e_i$ (first they are transformed into $q_i$ and in the next step the dissolution takes place).

**Set (E).**    $[b_1 \rightarrow b_1^+]_{e_1}$

$[b_{i+1} \rightarrow b_{i+1}^-]_{e_i}$     for $i = 1, \ldots, n-1$

$[b_{i+2} \rightarrow b_{i+2}^=]_{e_i}$     for $i = 1, \ldots, n-2$

$[b_{i+3} \rightarrow b_{i+3}^=]_{e_i}$     for $i = 1, \ldots, n-3$

$[b_i^+ \rightarrow b_0]_{e_i}$     for $i = 1, \ldots, n$

$[b_i^+ \rightarrow Trash]_{e_j}$     for $i = 1, \ldots, n-1, \quad j = i+1$

$[b_i^- \rightarrow b_i^+]_{e_i}$     for $i = 2, \ldots, n$

$[b_{i+1}^- \rightarrow b_{i+1}^+]_{e_i}$     for $i = 1, \ldots, n-1$

$[b_i^= \rightarrow b_i^+]_{e_i}$     for $i = 3, \ldots, n$

$[b_{i+1}^= \rightarrow b_{i+1}^-]_{e_i}$     for $i = 2, \ldots, n-1$

$[b_{i+2}^= \rightarrow b_{i+2}^-]_{e_i}$     for $i = 1, \ldots, n-2$

$[b_n^+ \rightarrow Trash]_{a_3}$

The basic strategy consists on allowing objects $b_i$ to get transformed into objects $b_0$ only if the element $a_i \in A$ belongs to the associated multiset.

Let us summarize informally the evolution of objects $b_i$ for all possible cases. Recall that in the $(2\lfloor \log k \rfloor + 2)$-th step, the counter $g_i$ produces an object $s$ in membrane $e$:

- At step $t = 2\lfloor \log k \rfloor + 3$ object $s$ enters in $e_n$ and either $d_1^+$ or $d_1^-$ appear in each one of the two existing copies of membrane $e$.
- At step $t = 2\lfloor \log k \rfloor + 4$ object $s$ enters in $e_{n-1}$ and either $p_1$ or $q_1$ appear in membranes $e$.
- At step $t = 2\lfloor \log k \rfloor + 5$, after the second division has been carried out, there are 4 membranes labeled by $e$. Object $s$ enters in $e_{n-2}$ (this happens in all 4 copies) and $p_1$ or $q_1$ get into $e_n$ (there are two of each).
- ...
- At step $t = 2\lfloor \log k \rfloor + n + 3$ object $s$ arrives into $e_0$, and $p_1$ or $q_1$ enter in $e_2$.
- At step $t = 2\lfloor \log k \rfloor + n + 4$ object $s$ dissolves $e_0$ (and hence objects $b_i$ are moved to $e_1$), and $p_1$ or $q_1$ arrive into $e_1$.
- At step $t = 2\lfloor \log k \rfloor + n + 5$ objects $b_1$, $b_2$ and $b_3$ have been transformed in $b_1^+$, $b_2^-$ and $b_3^=$, respectively, and they will be located either in $e_1$ (if the membrane contained an object $p_1$) or in $e_2$ (if there was an object $q_1$ in $e_1$). Besides, in the same step $p_2$ or $q_2$ get into $e_2$.
- At step $t = 2\lfloor \log k \rfloor + n + 6$
  - Objects $b_1^+$ evolve to $b_0$ (if they were in $e_1$) or to $Trash$ (if they were in $e_2$).
  - Objects $b_2^-$ evolve to $b_2^+$.
  - Objects $b_3^=$ have been transformed into $b_3^-$ (both those that were in $e_2$ and those in $e_1$).
  - All the objects $b_i^\alpha$ ($i = 1, \ldots, n$ and $\alpha \in \{+, -, =\}$) will be located either in membrane $e_2$ (if the latter contained an object $p_2$) or in $e_3$ (if there was an object $q_2$ in $e_2$).
  - Besides, in this moment $p_3$ or $q_3$ get into $e_3$.

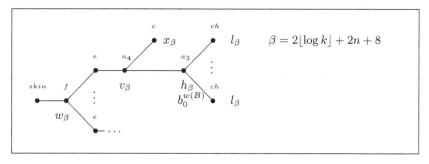

**Fig. 3.** TIME $2\lfloor \log k \rfloor + 2n + 8$

The design has been adjusted in such a way that in the moment when objects $p_i$ and $q_i$ arrive into membranes $e_i$ it happens that the objects $b_j^\alpha$ ($j = i, \ldots, n$ and $\alpha \in \{+, -, =\}$) are located in $e_i$ in half of the membranes or in $e_{i+1}$ in the rest of membranes. In the next step there will be objects $b_i^+$ in $e_i$ only for those cases where there was an object $p_i$, and hence the weight of element $a_i \in A$ should be added to the weight of the associated multiset (that is, $w(a_i)$ copies of $b_0$ will be produced in those membranes).

**Set (F).**
$$\left. \begin{array}{l} [w_i \to w_{i+1}]_f \\ [v_i \to v_{i+1}]_{a_4} \\ [h_i \to h_{i+1}]_{a_3} \\ [x_i \to x_{i+1}]_c \\ [z_i \to z_{i+1}]_{a_2} \\ [l_i \to l_{i+1}]_{ch} \end{array} \right\} \text{ for } i \in \{2\lfloor \log k \rfloor + 3, \ldots, 2\lfloor \log k \rfloor + 2n + 6\}.$$

$$[z_{2\lfloor \log k \rfloor + 2n + 7}]_{a_2} \to Trash$$

The rest of the counters simply increase their indices during this stage. At the end of the stage, in the $(2\lfloor \log k \rfloor + 2n + 7)$-th step, $z_i$ will dissolve all membranes $a_2$. Therefore, in the next step we have $2^n$ membranes labeled by $e$, and inside them (more precisely, inside membranes $a_3$) we have multisets of objects $b_0$ encoding the weights of all possible subsets $B \subseteq A$ (each membrane encodes a different subset) and also exactly $k$ copies of membrane $ch$, see Fig. 3.

**Third task: compare $k$ to the weight of each subset.** We shall focus next on the checking stage. That is, the system has to check in all membranes $a_3$ if the number of objects $b_0$ (encoding the weight of the associated subset) matches or not the parameter $k$ (represented as the number of membranes $ch$). This task is performed by the following set of rules (for the sake of simplicity, we denote $\beta = 2\lfloor \log k \rfloor + 2n + 8$):

**Set (G).**
$$b_0 [\,]_{ch} \to [c^*]_{ch}$$
$$[b_0 \to u_1]_{a_4}$$
$$[c^*]_{ch} \to Trash$$
$$[h_\beta]_{a_3} \to Trash$$

At the step $t = \beta$, objects $b_0$ get into membranes $ch$, and simultaneously membrane $a_3$ is dissolved. There are three possible situations:

1. There are exactly $k$ objects $b_0$. In this case at step $t = \beta + 1$ there will not be any object $b_0$ remaining, and all membranes $ch$ have been dissolved.
2. The number of objects $b_0$ is lower than $k$. In this case at step $t = \beta + 1$ there will not be any object $b_0$ remaining, but there will be some membranes $ch$ that have not been dissolved (because no object $b_0$ entered them).
3. The number of objects $b_0$ is greater than $k$. In this case there are some objects $b_0$ that could not get inside a membrane $ch$ (recall that the rules are applied in a maximal parallel way, but for each membrane only one object can cross it at a time).

In the second case, inside each membrane $ch$ that has not been dissolved the rules $[l_{\beta+1} \to l_{\beta+2}]_{ch}$ and $[l_{\beta+2}]_{ch} \to u_2$ are applied in the steps $t = \beta + 1$ and $t = \beta + 2$, respectively. Hence at step $t = \beta + 3$ there will be an object $u_2$ in $a_4$.

In the third case, the exceeding objects $b_0$ may, nondeterministically, either get into a membrane $ch$ (avoiding that the dissolution rule is applied to that membrane) or evolve into object $u_1$. Irrespectively of the nondeterministic choice, we know that there will be no more objects $b_0$ in $a_4$ at step $t = \beta + 2$.

Of course, during this stage the rest of the counters continue evolving:

**Set (H).**  $[l_{\beta+i-1} \to l_{\beta+i}]_{ch}$ for $i = 0, 1, 2$

$[v_{\beta+i-1} \to v_{\beta+i}]_{a_4}$ for $i = 0, \dots, 4$

$[x_{\beta+i-1} \to x_{\beta+i}]_c$ for $i = 0, \dots, 7$

$[w_{\beta+i-1} \to w_{\beta+i}]_f$ for $i = 0, \dots, 10$

The next set of rules guarantees that in every membrane where the weight of the associated subset was different from $k$ (and only in such membranes) there will be some objects $u_3$.

**Set (I1).**  $[u_i \to u_{i+1}]_{a_4}$ for $i = 1, 2$

$[l_{\beta+2}]_{ch} \to u_2$

$[l_{\beta+2} \to u_3]_{a_4}$

$[c^* \to u_3]_{a_4}$

These objects $u_3$, being in membrane $a_4$, will go into membranes $c$ and dissolve them. We have here a similar situation as before, as there may be several objects $u_3$ willing to go into a membrane $c$. The counter $v_i$ takes care of dissolving membrane $a_4$ so that any exceeding object $u_3$ will be moved to membrane $e$ and subsequently transformed into $Trash$.

**Set (I2).**  $u_3 [\,]_c \to [u_4]_c$

$[v_{\beta+4}]_{a_4} \to Trash$

$[u_3 \to Trash]_e$

$[u_4 \to u_5]_c$

$[u_5]_c \to Trash$

**Final task: answer stage.** Therefore, only in the branches where the number of objects $b_0$ were equal to $k$ we have a membrane $c$ inside membrane $e$ at step $\beta + 7$. Besides, we also have a counter $w_i$ evolving in membrane $f$:

- If the instance of the Subset Sum problem has an affirmative answer, i.e., if there exists a subset of $A$ whose weight is $k$, then in the step $\beta + 7$ there will be a membrane $e$ with a membrane $c$ inside and an object $x_{\beta+7}$ in it. This object will produce an object **yes** which will dissolve his way out to the environment.

  On the contrary, if the instance has a negative answer, then there will not exist any membrane $c$ in the system in the step $\beta + 7$ and the object **yes** will not be produced. Hence, the membrane $f$ will not be dissolved by **yes** and when the counter $w_i$ reaches $w_{\beta+10}$, an object **no** will appear and will be sent to the environment.

The set of rules is the following one:

**Set (J).** $\quad [x_{\beta+7}]_c \rightarrow yes$
$\qquad\qquad [yes]_e \rightarrow yes$
$\qquad\qquad [yes]_f \rightarrow yes$
$\qquad\qquad [yes]_{skin} \rightarrow yes\,[\,]_{skin}$
$\qquad\qquad [w_{\beta+10}]_f \rightarrow no$
$\qquad\qquad [no]_{skin} \rightarrow no\,[\,]_{skin}$

Consequently, if the answer is affirmative the P system halts after $\beta+11$ steps and otherwise after $\beta + 12$ steps.

## 4   Conclusions

In this paper we have combined different techniques for designing P systems in order to get a uniform family of P systems that solves the Subset Sum problem in the framework of P systems with weak division, with dissolution and without polarization. The main contribution of this paper is related to the Complexity Theory of P systems. The best solution of the **NP**-complete problem Subset Sum in any P system model up to now was linear in both input parameters $n$ and $k$. In this paper we show that the dependency on $k$ can be significantly reduced, since we show a solution where the resources and the number of steps are of a logarithmic order with respect to $k$.

## Acknowledgement

The authors acknowledge the support of the project TIN2006-13425 of the Ministerio de Educación y Ciencia of Spain, cofinanced by FEDER funds, and the support of the project of excellence TIC-581 of the Junta de Andalucía.

# References

1. Alhazov, A., Freund, R., Păun, G.: P Systems with Active Membranes and Two Polarizations. In: Păun, G., Riscos-Núñez, A., Romero-Jiménez, A., Sancho-Caparrini, F. (eds.) Proc. Second Brainstorming Week on Membrane Computing, Report RGNC 01/04, pp. 20–35 (2004)
2. Díaz-Pernil, D., Gutiérrez-Naranjo, M.A., Pérez-Jiménez, M.J., Riscos-Núñez, A.: Solving Subset Sum in Linear Time by Using Tissue P Systems with Cell Division. LNCS, vol. 4527, pp. 170–179 (2007)
3. Gutiérrez-Naranjo, M.A., Pérez-Jiménez, M.J., Riscos-Núñez, A., Romero-Campero, F.J.: Computational Efficiency of Dissolution Rules in Membrane Systems. International Journal of Computer Mathematics 83(7), 593–611 (2006)
4. Gutiérrez Naranjo, M.A., Pérez Jiménez, M.J., Romero-Campero, F.J.: A Linear Solution of Subset Sum Problem by Using Membrane Creation. In: Mira, J.M., Álvarez, J.R. (eds.) IWINAC 2005. LNCS, vol. 3561, pp. 258–267. Springer, Heidelberg (2005)
5. Gutiérrez-Naranjo, M.A., Pérez-Jiménez, M.J., Riscos-Núñez, A.: A Fast P System for Finding a Balanced 2-Partition. Soft Computing 9(9), 673–678 (2005)
6. Gutiérrez-Naranjo, M.A., Pérez-Jiménez, M.J., Riscos-Núñez, A., Romero-Campero, F.: On the Power of Dissolution in P Systems with Active Membranes. In: Freund, R., Păun, G., Rozenberg, G., Salomaa, A. (eds.) WMC 2005. LNCS, vol. 3850, pp. 373–394. Springer, Heidelberg (2006)
7. Păun, G.: Computing with Membranes: Attacking **NP**–complete Problems. In: Antoniou, I., Calude, C., Dinneen, M.J. (eds.) UMC 2000. Unconventional Models of Computation, pp. 94–115. Springer, Berlin (2000)
8. Păun, G.: Membrane Computing. An Introduction. Springer, Berlin (2002)
9. Pérez-Jiménez, M.J., Riscos-Núñez, A.: Solving the Subset-Sum problem by P Systems with Active Membranes. New Generation Computing 23(4), 367–384 (2005)
10. Pérez-Jiménez, M.J., Riscos-Núñez, A.: A Linear–time Solution to the Knapsack Problem Using P Systems with Active Membranes. In: Martín-Vide, C., Mauri, G., Păun, G., Rozenberg, G., Salomaa, A. (eds.) Membrane Computing. LNCS, vol. 2933, pp. 250–268. Springer, Heidelberg (2004)
11. Pérez-Jiménez, M.J., Romero-Campero, F.J.: Solving the Bin Packing Problem by Recognizer P Systems with Active Membranes. In: Păun, G., Riscos-Núñez, A., Romero-Jiménez, A., Sancho-Caparrini, F. (eds.) Proc. Second Brainstorming Week on Membrane Computing, Report RGNC 01/04, University of Seville, pp. 414–430 (2004)
12. Pérez-Jiménez, M.J., Romero–Campero, F.J.: Attacking the Common Algorithmic Problem by Recognizer P Systems. In: Margenstern, M. (ed.) MCU 2004. LNCS, vol. 3354, pp. 304–315. Springer, Heidelberg (2005)
13. Pérez-Jiménez, M.J., Romero-Jiménez, A., Sancho-Caparrini, F.: A Polynomial Complexity Class in P Systems Using Membrane Division. In: DCFS 2003. Proc. 5th Workshop on Descriptional Complexity of Formal Systems, pp. 284–294 (2003)

# A Formal Framework
# for Static (Tissue) P Systems

Rudolf Freund[1] and Sergey Verlan[2]

[1] Faculty of Informatics, Vienna University of Technology
Favoritenstr. 9, 1040 Vienna, Austria
rudi@emcc.at
[2] LACL, Département Informatique
UFR Sciences et Technologie, Université Paris XII
61, av. Général de Gaulle, 94010 Créteil, France
verlan@univ-paris12.fr

**Abstract.** The formalism of P systems is known for many years, yet just recently new derivation modes and halting conditions have been proposed. For developing comparable results, a formal description of their functioning, in particular, of the derivation step is necessary. We introduce a formal general framework for static membrane systems that aims to capture most of the essential features of (tissue) P systems and to define their functioning in a formal way.

## 1 Introduction

P systems were introduced by Gh. Păun (see [8], [14]) as distributed parallel computing devices, based on inspiration from biochemistry, especially with respect to the structure and the functioning of a living cell. The cell is considered as a set of compartments enclosed by membranes; the membranes are nested one in another and contain objects and evolution rules. The basic model neither specifies the nature of these objects nor the nature of the rules. Specifying these two parameters, a lot of different models of computing have been introduced, see [20] for a comprehensive bibliography. Tissue P systems, first considered by Gh. Păun and T. Yokomori in [18] and [19], also see [11], use the graph topology in contrast to the tree topology used in the basic model of P systems.

In this paper, we design a general class of multiset rewriting systems containing, in particular, P systems and tissue P systems. We recall that any P system may be seen at the most abstract level as a multiset rewriting system with only one compartment, encoding the membrane as part of the object representation. However, this approach completely ignores the inner structure of the system because all structural information is hidden (by an encoding) which makes it difficult do deduce any compartment-related information or to model (processes in) biological systems. At a lower level of abstraction, a P system may be seen as a network of cells (compartments) evolving with multi-cell multiset rewriting rules. At the lowest level, the graph/tree structure appears as well as a specialization of rules which are of a very particular form. This last level is usually used

G. Eleftherakis et al. (Eds.): WMC8 2007, LNCS 4860, pp. 271–284, 2007.

in the area of P systems because it permits to easily specify the system and to incorporate different new types of rules.

It is worth noting that in the definition of membrane systems the application of rules often is defined in a quite informal way. This is related to the fact that for a long time only the maximally parallel derivation mode was considered and a P system was supposed to work only in this mode. Recent developments in P systems area have revealed that other derivation modes as the minimally parallel derivation mode might be considered [5] and allow for many interesting new results, yet depending on specific interpretations of this notion. Moreover, different halting conditions have been investigated (see [10], [1]), too. All these articles have shown that there is a need for a formal definition of part of the semantics of membrane systems as the derivation step and the halting procedure like it was done for splicing test tube systems [6] or networks of language processors [7]. In particular, this is important for a classification of P systems as well as for their implementation. For approaches to find operational and logic based semantics for P systems we refer to [4] and [2]; a Petri net semantics for membrane systems is discussed in [12].

This article is an attempt to fulfill the goal of formally defining a procedural semantics for a quite large number of well-known variants of (tissue) P systems considered so far in the literature, but, of course, we do not at all claim to have captured all the variants having already appeared in the literature. In order to be quite general, we place our reasoning at the abstract level of *networks of cells*, already considered in a slightly different way in [3]. We adapt an implementational point of view and also give a formal definition of the derivation step, the halting condition and the procedure for obtaining the result of a computation. Moreover, we give examples of applying our concepts to some well-known variants of P systems.

## 2    Preliminaries

We recall some of the notions and the notations we use (for further details see [8] and [17]). Let $V$ be a (finite) alphabet; then $V^*$ is the set of all strings (a language) over $V$, and $V^+ = V^* - \{\lambda\}$ where $\lambda$ denotes the empty string. $FIN$ ($FIN(V)$)denotes the set of finite languages (over the alphabet $V$), and $RE$, $REG$, and $MAT^\lambda$ denote the families of recursively enumerable and regular languages as well as matrix languages, respectively. For any family of string languages $F$, $PsF$ denotes the family of Parikh sets of languages from $F$ and $NF$ the family of Parikh sets of languages from $F$ over a one-letter alphabet. By $\mathbb{N}$ we denote the set of all non-negative integers, by $\mathbb{N}^k$ the set of all vectors of non-negative integers.

Let $V$ be a (finite) set, $V = \{a_1, ..., a_k\}$. A *finite multiset* $M$ over $V$ is a mapping $M : V \longrightarrow \mathbb{N}$, i.e., for each $a \in V$, $M(a)$ specifies the number of occurrences of $a$ in $M$. The size of the multiset $M$ is $|M| = \sum_{a \in V} M(a)$. A multiset $M$ over $V$ can also be represented by any string $x$ that contains exactly $M(a_i)$ symbols $a_i$ for all $1 \leq i \leq k$, e.g., by $a_1^{M(a_1)}...a_k^{M(a_k)}$, or else by the set $\left\{ a_i^{M(a_i)} \mid 1 \leq i \leq k \right\}$.

The support of $M$ is the set $supp(M) = \{a \in V \mid f(a) \geq 1\}$. For example, the multiset over $\{a, b, c\}$ defined by the mapping $a \rightarrow 3, b \rightarrow 1, c \rightarrow 0$ can be specified by $a^3 b$ or $\{a^3, b\}$, its support is $\{a, b\}$.

The set of all finite multisets over the set $V$ is denoted by $\langle V, \mathbb{N} \rangle$. We may also consider mappings $M$ of form the $M : V \longrightarrow \mathbb{N}_\infty$ where $\mathbb{N}_\infty = \mathbb{N} \cup \{\infty\}$, i.e., elements of $M$ may have an infinite multiplicity; we shall call such multisets where $M(a_i) = \infty$ for at least one $i$, $1 \leq i \leq k$, *infinite multisets*. The set of all such multisets $M$ over $V$ with $M : V \longrightarrow \mathbb{N}_\infty$ is denoted by $\langle V, \mathbb{N}_\infty \rangle$. For $W \subseteq V$, $W^\infty$ denotes the infinite multiset with $W(a) = \infty$ for all $a \in W$.

Let $x$ and $y$ be two multisets over $V$, i.e., from $\langle V, \mathbb{N} \rangle$ or $\langle V, \mathbb{N}_\infty \rangle$. Then $x$ is called a submultiset of $y$, written $x \leq y$ or $x \subseteq y$, if and only if $x(a) \leq y(a)$ for all $a \in V$; if, moreover, $x(a) < y(a)$ for some $a \in V$, then $x$ is called a strict multiset of $y$. Observe that for all $n \in \mathbb{N}$, $n + \infty = \infty$, and $\infty - n = \infty$. The sum of $x$ and $y$, denoted by $x + y$ or $x \cup y$, is a multiset $z$ such that $z(a) = x(a) + y(a)$ for all $a \in V$. The difference of two multisets $x$ and $y$, denoted by $x - y$, provided that $y \subseteq x$, is the multiset $z$ with $z(a) = x(a) - y(a)$ for all $a \in V$. Observe that in the following, when taking the sum or the difference of two multisets $x$ and $y$ from $\langle V, \mathbb{N}_\infty \rangle$, we shall always assume $\{x(a), y(a)\} \cap \mathbb{N} \neq \emptyset$.

If $X = (x_1, \ldots, x_m)$ and $Y = (y_1, \ldots, y_m)$ are vectors of multisets over $V$, then $X \leq Y$ if and only if $x_j \subseteq y_j$ for all $j$, $1 \leq j \leq m$; in the same way, sum and difference of vectors of multisets are defined by taking the sum and the difference, respectively, in each component.

## 3 Networks of Cells

In this section we consider a general framework for describing membrane systems with a static membrane structure. We consider membrane systems as a collection of interacting cells containing multisets of objects [3].

**Definition 3.1.** *A network of cells of degree* $n \geq 1$ *is a construct*

$$\Pi = (n, V, w, Inf, R), \text{ where:}$$

1. *n is the number of cells;*
2. *V a finite alphabet;*
3. *$w = (w_1, \ldots, w_n)$ where $w_i \in \langle V, \mathbb{N} \rangle$, for all $1 \leq i \leq n$, is the finite multiset initially associated to cell i;*
4. *$Inf = (Inf_1, \ldots, Inf_n)$ where $Inf_i \subseteq V$, for all $1 \leq i \leq n$, is the set of symbols occurring infinitely often in cell i (in most of the cases, only one cell, called* the environment, *will contain symbols occurring with infinite multiplicity);*
5. *R is a finite set of interaction rules of the form*

$$(X \rightarrow Y; P, Q)$$

*where $X = (x_1, \ldots, x_n)$, $Y = (y_1, \ldots, y_n)$, and $x_i, y_i \in \langle V, \mathbb{N} \rangle$, $1 \leq i \leq n$, are vectors of multisets over V as well as $P = (p_1, \ldots, p_n)$, $Q = (q_1, \ldots, q_n)$,* *and $p_i, q_i$, $1 \leq i \leq n$, are finite sets of multisets over $V$. We will also use the notation*

$$((x_1, 1) \ldots (x_n, n) \to (y_1, 1) \ldots (y_n, n) ; (p_1, 1) \ldots (p_n, n), (q_1, 1) \ldots (q_n, n))$$

*for a rule $(X \to Y; P, Q)$; moreover, if some $p_i$ or $q_i$ is an empty set or some $x_i$ or $y_i$ is equal to the empty multiset, $1 \leq i \leq n$, then we may omit it from the specification of the rule.*

A network of cells consists of $n$ cells, numbered from 1 to $n$, that contain (possibly infinite) multisets of objects over $V$; initially cell $i$ contains $w_i \cup Inf_i^\infty$. Cells can interact with each other by means of the rules in $R$. An interaction rule

$$((x_1, 1) \ldots (x_n, n) \to (y_1, 1) \ldots (y_n, n) ; (p_1, 1) \ldots (p_n, n), (q_1, 1) \ldots (q_n, n))$$

rewrites objects $x_i$ from cells $i$ into objects $y_j$ in cells $j$, $1 \leq i, j \leq n$, if every cell $k$, $1 \leq k \leq n$, contains all multisets from $p_k$ and does not contain any multiset from $q_k$. In other words, the first part of the rule specifies the rewriting of symbols, the second part of the rule specifies permitting conditions and the third part of the rule specifies the forbidding conditions. In the next section we give an even more detailed precise definition for the application of an interaction rule.

For an interaction rule $r$ of the form above, the set

$$\{i \mid x_i \neq \lambda \text{ or } y_i \neq \emptyset \text{ or } p_i \neq \emptyset \text{ or } q_i \neq \lambda\}$$

induces a relation between the interacting cells. However, this relation need not give rise to a *structure* relation like a tree as in P systems or a graph as in tissue P systems (e.g., see [15] for definitions of P systems and tissue P systems), though most models of membrane systems with a static membrane structure can be seen as special variants of networks of cells, and moreover, a lot of important features of membrane systems, in particular the derivation step and the halting condition, may be described at the level of networks of cells.

## 4   Systems with a Static Structure

In this section we consider networks of cells having a static structure, i.e., the number of cells does not change during the evolution of the system. We first define configurations, transition steps, and then halting conditions.

**Definition 4.1.** *Consider a network of cells $\Pi = (n, V, w, Inf, R)$. A configuration $C$ of $\Pi$ is an $n$-tuple of multisets over $V$ $(u_1', \ldots, u_n')$ with $u_i' \in \langle V, \mathbb{N}_\infty \rangle$, $1 \leq i \leq n$; in the following, $C$ will also be described by its finite part $C^f$ only, i.e., by $(u_1, \ldots, u_n)$ satisfying $u_i' = u_i \cup Inf_i^\infty$ and $u_i \cap Inf_i = \emptyset$, $1 \leq i \leq n$.*

In the sense of the preceding definition, the *initial configuration* of $\Pi$, $C_0$, is described by $w$, i.e., $C_0^f = w = (w_1, \ldots, w_n)$, whereas $w_i' = w_i \cup Inf_i^\infty$, $1 \leq i \leq n$, is the initial contents of cell $i$, i.e., $C_0 = w \cup Inf^\infty$.

**Definition 4.2.** *We say that an interaction rule $r = (X \to Y; P, Q)$ is eligible for the configuration $C$ with $C = (u_1, \dots, u_n)$ if and only if for all $i$, $1 \le i \le n$, the following conditions hold true:*

- *for all $p \in P_i$, $p \subseteq u_i$ (every $p \in P_i$ is a submultiset of $u_i$),*
- *for all $q \in q_i$, $q \not\subseteq u_i$ (no $q \in q_i$ is a submultiset of $u_i$), and*
- *$x_i \subseteq u_i$ ($x_i$ is a submultiset of $u_i$).*

*Moreover, we require that $x_j \cap (V - inf_j) \ne \emptyset$ for at least one $j$, $1 \le j \le n$. This last condition ensures that at least one symbol appearing only in a finite number of copies is involved in the rule. The set of all rules eligible for $C$ is denoted by $Eligible\,(\Pi, C)$.*

*The marking algorithm.* Let $C = (v_1, \dots, v_n)$ be a configuration of a network of cells $\Pi$ and $C^f$ its finite description; moreover, let $R'$ be a finite multiset of rules from $R$ consisting of the (copies of) rules $r_1, \dots, r_k$, where for each $i$, $1 \le i \le k$, we have $r_i = (X_i \to Y_i; P_i, Q_i) \in Eligible\,(\Pi, C)$, $X_i = (x_{i,1}, \dots, x_{i,n})$, $Y_i = (y_{i,1}, \dots, y_{i,n})$. Moreover, let $X_i'$ and $Y_i'$, $1 \le i \le k$, be the the vectors of finite multisets from $\langle V, \mathbb{N} \rangle$ with $X_{i,j} = X_{i,j}' \cup Inf_j^\infty$ and $X_{i,j}' \cap Inf_j = \emptyset$, $1 \le j \le n$.
Then:

1. consider the vector of multisets $Mark_0\,(\Pi, C, R') = (\lambda, \dots, \lambda)$ of size $n$ and let $i = 1$;
2. if $X_i' \le C^f - Mark_{i-1}\,(\Pi, C, R')$, then set

$$Mark_i\,(\Pi, C, R') = \left(C^f - Mark_{i-1}\,(\Pi, C, R')\right) - X_i',$$

   otherwise, end the algorithm and return **false**;
3. if $i = k$ then end the algorithm and return **true**, otherwise set $i$ to $i+1$ and return to step 2.

If the marking algorithm returns **true** for the pair $(C, R')$ then we say that the configuration $C$ may be *marked* by $R'$, and we define $Mark\,(\Pi, C, R') = Mark_k\,(\Pi, C, R')$.

**Definition 4.3.** *Consider a configuration $C$ and let $R'$ be a multiset of rules from $Eligible\,(\Pi, C)$ (i.e., a multiset of eligible rules). We say that the multiset of rules $R'$ is applicable to $C$ if the marking algorithm as described above returns **true** and $Mark\,(\Pi, C, R')$. The set of all multisets of rules applicable to $C$ is denoted by $Appl\,(\Pi, C)$.*

**Definition 4.4.** *Consider a configuration $C$ and a multiset of rules $R' \in Appl\,(\Pi, C)$. According to the marking algorithm described above, we define the configuration being the result of applying $R'$ to $C$ as*

$$Apply\,(\Pi, C, R') = (C - Mark\,(\Pi, C, R')) + \Sigma_{1 \le i \le k} Y_i'.$$

We remark that $Apply(R', C)$ is again a configuration.

For the specific *derivation modes* to be defined in the following, the selection of multisets of rules applicable to a configuration $C$ has to be a specific subset of $Appl\,(\Pi, C)$.

**Definition 4.5.** *For the derivation mode $\vartheta$, the selection of multisets of rules applicable to a configuration $C$ is denoted by $Appl\,(\Pi, C, \vartheta)$.*

**Definition 4.6.** *For the* asynchronous *derivation mode (asyn),*

$$Appl\,(\Pi, C, asyn) = Appl\,(\Pi, C)\,,$$

*i.e., there are no particular restrictions on the multisets of rules applicable to $C$.*

**Definition 4.7.** *For the* sequential *derivation mode (sequ),*

$$Appl\,(\Pi, C, sequ) = \{R' \mid R' \in Appl\,(\Pi, C) \ \ and \ \ |R'| = 1\}\,,$$

*i.e., any multiset of rules $R' \in Appl\,(\Pi, C, sequ)$ has size 1.*

The most important derivation mode considered in the area of P systems from the beginning is the *maximally parallel* derivation mode where we only select multisets of rules $R'$ that are not extensible, i.e., there is no other multiset of rules $R'' \supsetneq R'$ applicable to $C$.

**Definition 4.8.** *For the* maximally parallel *derivation mode (max),*

$$Appl\,(\Pi, C, max) = \{R' \mid R' \in Appl\,(\Pi, C) \ \ and \ there \ is$$
$$no \ R'' \in Appl\,(\Pi, C) \ \ with \ R'' \supsetneq R'\}\,.$$

For the *minimally parallel* derivation mode, we need an additional feature for the set of rules $R$, i.e., we consider a partition of $R$ into disjoint subsets $R_1$ to $R_h$. Usually, this partition of $R$ may coincide with a specific assignment of the rules to the cells. For any set of rules $R' \subseteq R$, let $\|R'\|$ denote the number of sets of rules $R_j$, $1 \le j \le h$, with $R_j \cap R' \ne \emptyset$.

There are several possible interpretations of this minimally parallel derivation mode which in an informal way can be described as applying multisets such that from every set $R_j$, $1 \le j \le h$, at least one rule – if possible – has to be used (e.g., see [5]). We start with the basic variant where in each derivation step we only choose a multiset of rules $R'$ from $Appl\,(\Pi, C, asyn)$ that cannot be extended to $R'' \in Appl\,(\Pi, C, asyn)$ with $R'' \supsetneq R'$ as well as $(R'' - R') \cap R_j \ne \emptyset$ and $R' \cap R_j = \emptyset$ for some $j$, $1 \le j \le h$, i.e., extended by a rule from a set of rules $R_j$ from which no rule has been taken into $R'$.

**Definition 4.9.** *For the* minimally parallel *derivation mode (min),*

$$Appl\,(\Pi, C, min) = \{R' \mid R' \in Appl\,(\Pi, C, asyn) \ \ and$$
$$there \ is \ no \ R'' \in Appl\,(\Pi, C, asyn)$$
$$with \ R'' \supsetneq R', \ (R'' - R') \cap R_j \ne \emptyset$$
$$and \ R' \cap R_j = \emptyset \ for \ some \ j, \ 1 \le j \le h\}.$$

In the following we also consider further restricting conditions on the four basic modes defined above, especially interesting for the minimally parallel derivation mode, thus obtaining some new combined derivation modes.

A derivation mode closely related to the maximally parallel one, yet not considered so far in the literature is the following one, where we not only demand that the chosen multiset $R'$ is not extensible, but also contains the maximal number of rules among all applicable multisets:

**Definition 4.10.** *For any basic derivation mode $\delta \in \{asyn, sequ, max, min\}$, we define the* maximal in rules $\delta$ derivation mode *($max_{rule}\delta$) by setting*

$$Appl\,(\Pi, C, max_{rule}\delta) = \{R' \mid R' \in Appl\,(\Pi, C, \delta) \ \ and$$
$$there \ is \ no \ R'' \in Appl\,(\Pi, C, \delta)$$
$$with \ |R''| > |R'|\}.$$

In the case of the minimally parallel derivation mode, we have two more very interesting variants of possible interpretations, the first one maximizing the sets of rules involved in a multiset to be applied ($max_{set}min$), and the second one demanding that all sets of rules that could contribute should contribute ($all_{aset}min$). The corresponding restricting conditions are based on a partition of the rules which usually will be the same as that one given in the definition for the minimally parallel derivation mode. In general, we define these variants for any of the basic derivation modes as follows:

**Definition 4.11.** *For any basic derivation mode $\delta \in \{asyn, sequ, max, min\}$, we define the* maximal in sets $\delta$ derivation mode *($max_{set}\delta$) by setting*

$$Appl\,(\Pi, C, max_{set}\delta) = \{R' \mid R' \in Appl\,(\Pi, C, \delta) \ \ and$$
$$there \ is \ no \ R'' \in Appl\,(\Pi, C, \delta)$$
$$with \ \|R''\| > \|R'\|\}.$$

**Definition 4.12.** *For any basic derivation mode $\delta \in \{asyn, sequ, max, min\}$, we define the* using all applicable sets $\delta$ derivation mode *($all_{aset}\delta$) by setting*

$$Appl\,(\Pi, C, all_{aset}\delta) = \{R' \mid R' \in Appl\,(\Pi, C, \delta) \ \ and$$
$$for \ all \ j, \ 1 \leq j \leq h,$$
$$R_j \cap \bigcup_{X \in Appl(\Pi,C)} X \neq \emptyset$$
$$implies \ R_j \cap R' \neq \emptyset\}.$$

We should like to mention that, for example, the derivation modes $max_{set}\delta$ and $all_{aset}\delta$ with $\delta \in \{asyn, sequ, max, min\}$ could be extended by the constraint that a maximal number of rules has to be used, too, thus yielding derivation modes $max_{rule}max_{set}\delta$ and $max_{rule}all_{aset}\delta$. Yet we do not consider such combinations of restricting conditions in this paper. Moreover, there are several other derivation modes considered in the literature, for instance, we may apply (at most) $k$ rules in parallel in every derivation step, but we leave the task to define such derivation modes in the general framework elaborated in this paper to the reader.

For all the derivation modes defined above, we now can define how to obtain a next configuration from a given one by applying an applicable multiset of rules according to the constraints of the underlying derivation mode:

**Definition 4.13.** *Given a configuration $C$ of $\Pi$ and a derivation mode $\vartheta$, we may choose a multiset of rules $R' \in Appl\,(\Pi, C, \vartheta)$ in a non-deterministic way and apply it to $C$. The result of this transition step from the configuration $C$ with applying $R'$ is the configuration $Apply\,(\Pi, C, R')$, and we also write $C \Longrightarrow_{(\Pi,\vartheta)} C'$. The reflexive and transitive closure of the transition relation $\Longrightarrow_{(\Pi,\vartheta)}$ is denoted by $\Longrightarrow^*_{(\Pi,\vartheta)}$.*

Looking carefully into the definitions for all the (basic and combined) derivation modes defined above, we immediately infer the following equalities which do not depend on the kind of rules at all (observe that the restricting conditions for the combined modes using the condition $max_{rule}$ or $max_{set}$ are defined with respect to the underlying basic mode, which, for example, immediately implies the equalities for the sequential mode):

**Lemma 4.1.** *The following equalities for derivation modes hold true in general for all kinds of networks of cells:*

$$max_{rule}sequ = max_{set}sequ = sequ,$$
$$max_{rule}asyn = max_{rule}min = max_{rule}max.$$

*Provided that the partitions with respect to min as well as $all_{aset}$ and $max_{set}$ are the same, also the following equalities hold true:*

$$max_{set}min = max_{set}asyn,$$
$$all_{aset}min = all_{aset}asyn.$$

On the other hand, it is well known from the literature that the four basic derivation modes may yield different application results. Moreover, the following simple example shows most of the incomparability between combined and/or basic derivation modes:

**Example 4.1.** *Consider the network of cells*

$$\Pi = \left(4, \{a, b\}, \left(b^3, a^3, b, b\right), (\emptyset, \emptyset, \emptyset, \emptyset), R\right)$$

*with the following rules in R:*

1. $(b, 1)(a, 2) \rightarrow (a, 1)(b, 2)$
2. $(a, 2)(b, 3) \rightarrow (b, 2)(a, 3)$
3. $(aa, 2)(b, 4) \rightarrow (b, 2)(aa, 4)$

*In fact, $\Pi$ can be interpreted as a P system with antiport rules, 3 membranes with membrane $i$ represented by cell $i + 1$, as well as cell 1 representing the environment (see subsection 4.4 as well as Figure 1). Due to the availability*

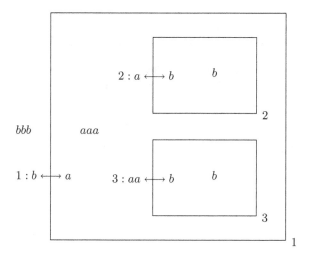

**Fig. 1.** Network of cells depicted as P system with antiport rules

*of objects in the four cells, only the following multisets of rules (represented as strings) are applicable to the initial configuration $C_0$, $C_0 = \left(b^3, a^3, b, b\right)$:*

$$Appl\left(\Pi, C_0, asyn\right) = Appl\left(\Pi, C_0\right) = \left\{1, 1^2, 1^3, 1^2 2, 12, 13, 2, 23, 3\right\}$$

*Assuming the partition for the minimally parallel derivation mode to be the partition into the three single rules (which corresponds to assigning the rule $i$ to cell $i + 1$ – corresponding to membrane $i$ in a membrane system – and no rule to cell 1 which represents the environment), we obtain the following sets of multisets of rules applicable to $C_0$ according to the different derivation modes:*

$Appl\left(\Pi, C_0, \delta_1\right) = \left\{1, 2, 3\right\}$ *for* $\delta_1 \in \left\{sequ, max_{rule}sequ, max_{set}sequ\right\}$,
$Appl\left(\Pi, C_0, min\right) = \left\{1^3, 1^2 2, 12, 13, 23\right\}$,
$Appl\left(\Pi, C_0, max\right) = \left\{1^3, 1^2 2, 13, 23\right\}$,
$Appl\left(\Pi, C_0, max_{rule}\delta_2\right) = \left\{1^3, 1^2 2\right\}$, $\delta_2 \in \left\{asyn, min, max\right\}$,
$Appl\left(\Pi, C_0, max_{set}\delta_3\right) = \left\{1^2 2, 12, 13, 23\right\}$, $\delta_3 \in \left\{asyn, min\right\}$,
$Appl\left(\Pi, C_0, max_{set}max\right) = \left\{1^2 2, 13, 23\right\}$,
$Appl\left(\Pi, C_0, all_{aset}\delta_4\right) = \emptyset$, $\delta_4 \in \left\{sequ, asyn, min, max\right\}$.

*All these sets of multisets of rules listed above are different which shows the incomparability of the corresponding derivation modes (observe that the inherent equalities for the modes $\delta_1$, $\delta_2$, and $\delta_3$ follow from Lemma 4.1). Due to the competition of the rules for the objects in cell 2, we get $Appl\left(\Pi, C_0, all_{aset}\delta_4\right) = \emptyset$ for all basic derivation modes $\delta_4 \in \left\{sequ, asyn, min, max\right\}$.*

*Omitting cell 4 in the network of cells above as well as the corresponding rule 3, we obtain a network of cells $\Pi'$,*

$$\Pi' = \left(3, \{a, b\}, \left(b^3, a^3, b\right), \left(\emptyset, \emptyset, \emptyset\right), R'\right),$$

*with only the rules 1 and 2 in $R'$ as well as*

$$\emptyset \subsetneqq Appl\left(\Pi', C_0', all_{aset}min\right) = \left\{12, 1^2 2\right\} \subsetneqq \left\{1^2 2\right\} = Appl\left(\Pi', C_0', all_{aset}max\right).$$

Especially for computing and accepting devices, the notion of determinism is of major importance. For networks of cells, determinism can be defined as follows:

**Definition 4.14.** *A configuration $C$ is said to be* accessible *in $\Pi$ with respect to the derivation mode $\vartheta$ if and only if $C_0 \Longrightarrow_{(\Pi, \vartheta)}^* C$ ($C_0$ is the initial configuration of $\Pi$). The set of all accessible configurations in $\Pi$ is denoted by $Acc\left(\Pi\right)$.*

**Definition 4.15.** *A network of cells $\Pi$ is said to be* deterministic *with respect to the derivation mode $\vartheta$ if and only if $\left|Appl\left(\Pi, C, \vartheta\right)\right| \leq 1$ for any accessible configuration $C$.*

## 4.1 Halting Conditions

A halting condition is a predicate applied to an accessible configuration. The system halts according to the halting condition if this predicate is true for the current configuration. In such a general way, the notion halting with final state or signal halting can be defined as follows:

**Definition 4.16.** *An accessible configuration $C$ is said to fulfill the* signal halting *condition or* final state halting *condition (S) if and only if $C \in S\left(\Pi, \vartheta\right)$ where*

$$S\left(\Pi, \vartheta\right) = \left\{C' \mid C' \in Acc\left(\Pi\right) \text{ and } State\left(\Pi, C', \vartheta\right) = \textbf{true}\right\}.$$

Here $State\left(\Pi, C', \vartheta\right)$ means a decidable feature of the underlying configuration $C'$, e.g., the occurrence of a specific symbol (signal) in a specific cell.

The most important halting condition used from the beginning in the P systems area is the *total halting*, usually simply considered as *halting*:

**Definition 4.17.** *An accessible configuration $C$ is said to fulfill the* total halting *condition (H) if and only if no multiset of rules can be applied to $C$ with respect to the derivation mode anymore, i.e., if and only if $C \in H\left(\Pi, \vartheta\right)$ where*

$$H\left(\Pi, \vartheta\right) = \left\{C' \mid C' \in Acc\left(\Pi\right) \text{ and } Appl\left(\Pi, C', \vartheta\right) = \emptyset\right\}.$$

The adult halting condition guarantees that we still can apply a multiset of rules to the underlying configuration, yet without changing it anymore:

**Definition 4.18.** *An accessible configuration $C$ is said to fulfill the* adult halting *condition (A) if and only if $C \in A\left(\Pi, \vartheta\right)$ where*

$$A\left(\Pi, \vartheta\right) = \{C' \mid C' \in Acc\left(\Pi\right), \; Appl\left(\Pi, C', \vartheta\right) \neq \emptyset \text{ and}$$
$$Apply\left(\Pi, C', R'\right) = C' \text{ for every } R' \in Appl\left(\Pi, C', \vartheta\right)\}.$$

We should like to mention that we could also consider $A\left(\Pi, \vartheta\right) \cup H\left(\Pi, \vartheta\right)$ instead of $A\left(\Pi, \vartheta\right)$.

For introducing the notion of partial halting, we have to consider a partition of $R$ into disjoint subsets $R_1$ to $R_h$ as for the minimally parallel derivation mode (eventually, this partition for partial halting might also be different from the partition used for the minimally parallel derivation mode). We then say that we are not halting only if there still is a multiset of rules $R'$ from $Appl\left(\Pi, C\right)$ with $R' \cap R_j \neq \emptyset$ for all $j$, $1 \leq j \leq h$:

**Definition 4.19.** *An accessible configuration $C$ is said to fulfill the* partial halting *condition (h) if and only if $C \in h\left(\Pi, \vartheta\right)$ where*

$$h\left(\Pi, \vartheta\right) = \{C' \mid C' \in Acc\left(\Pi\right) \text{ and there is}$$
$$\text{no } R' \in Appl\left(\Pi, C'\right) \text{ with } R' \cap R_j \neq \emptyset \text{ for all } j, \ 1 \leq j \leq h\}.$$

## 4.2  Computation, Goal and Result of a Computation

A computation in a network of cells $\Pi$, $\Pi = (n, V, w, Inf, R)$, starts with the initial configuration $C_0$, $C_0 = w \cup Inf^\infty$, and continues with transition steps according to the chosen derivation mode until the halting condition is met.

The computations with a network of cells may have different goals, e.g., to generate *(gen)* a (vector of) non-negative integers in a specific output cell (membrane) or to accept *(acc)* a (vector of) non-negative integers placed in a specific input cell at the beginning of a computation. Moreover, the goal can also be to compute *(com)* an output from a given input or to output yes or no to decide *(dec)* a specific property of a given input.

The results not only can be taken as the number $(N)$ of objects in a specified output cell, but, for example, also be taken modulo a terminal alphabet $(T)$ or by subtracting a constant from the result $(-k)$.

Such different tasks of a network of cells may require additional parameters when specifying its functioning, e.g., we may have to specify the output/input cell(s) and/or the terminal alphabet.

We shall not go into the details of such definitions here, we just mention that the goal of the computations $\gamma \in \{gen, acc, com, dec\}$ and the way to extract the results $\rho$ (usually taken from halting computations) are two other parameters to be specified and clearly defined when defining the functioning of a network of cells or a membrane system.

## 4.3  Taxonomy of Networks of Cells and (Tissue) P Systems

For a particular variant of networks of cells or especially P systems/tissue P systems we have to specify the derivation mode, the halting condition as well as the procedure how to get the result of a computation, but also the specific kind of rules that are used, especially some complexity parameters.

For networks of cells, we shall use the notation

$$O_m C_n\left(\vartheta, \phi, \gamma, \rho\right) [\text{parameters for rules}]$$

to denote the family of sets of vectors obtained by networks of cells $\Pi = (n, V, w, Inf, R)$ with $m = |V|$, as well as $\vartheta, \phi, \gamma, \rho$ indicating the derivation mode, the halting condition, the goal of the computations, and the way how to get results, respectively; the *parameters for rules* describe the specific features of the rules in $R$. If any of the parameters $m$ and $n$ is unbounded, we replace it by $*$.

For P systems, with the interaction between the cells in the rules of the corresponding network of cells allowing for a tree structure as underlying interaction graph, we shall use the notation

$$O_m P_n \left( \vartheta, \phi, \gamma, \rho \right) [\text{parameters for rules}].$$

Observe that usually the environment is not counted when specifying the number of membranes in P systems, but this often hides the important role that the environment takes in the functioning of the system.

For tissue P systems, with the interaction between the cells in the rules of the corresponding network of cells allowing for a graph structure as underlying interaction graph, we shall use the notation

$$O_m t P_n \left( \vartheta, \phi, \gamma, \rho \right) [\text{parameters for rules}].$$

As a special example, let us now consider *symport/antiport P systems*.

## 4.4   A Specific Example: P Systems with Symport/Antiport Rules

For definitions and results concerning P systems with symport/antiport rules, we refer to the original paper [13] as well as to the overview given in [16]. An *antiport rule* is a rule of the form $((x, i) (u, j) \to (x, j) (u, i))$ usually written as $(x, out; u, in)$, $xu \neq \lambda$, where $j$ is the region outside the membrane $i$ in the underlying graph structure. A *symport rule* is of the form $((x, i) \to (x, j))$ or $((u, j) \to (u, i))$.

The weight of the antiport rule $(x, out; u, in)$ is defined as $\max \{|x|, |u|\}$. Using only antiport rules with weight $\leq k$ induces the type of rules $\alpha$ usually written as $anti_k$. The weight of a symport rule $(x, out)$ or $(u, in)$ is defined as $|x|$ or $|u|$, respectively. Using only symport rules with weight $\leq k$ induces the type of rules $\alpha$ usually written as $sym_k$. If only antiport rules $(x, out; u, in)$ of weight $\leq 2$ and with $|x| + |u| \leq 3$ as well as symport rules of weight 1 are used, we shall write $anti_{2'}$.

As is well known, $O_* P_2 \left( max, H, gen, N \right) [anti_{2'}] = NRE$.

Observe that we only need one membrane separating the environment and the skin region, but this means that two regions corresponding to two cells are involved.

## 4.5   A General Result

For any network of cells using multiset rewriting rules of type $\alpha$, with a derivation mode $\vartheta \in \{all_{aset} min, max_{set} min, asyn, sequ\}$ and partial halting, we only get Parikh sets of matrix languages (regular sets of non-negative integers):

**Theorem 4.2** *For every* $\vartheta \in \{all_{aset}min, max_{set}min, asyn, sequ\}$,

$$O_*C_* (\vartheta, h, gen, T) [\alpha] \subseteq PsMAT^\lambda \text{ and } O_*C_* (\vartheta, h, gen, N) [\alpha] \subseteq NREG$$

*provided that the partitions for the derivation modes* $all_{aset}min$, $max_{set}min$ *and the partial halting* $h$ *are the same.*

The proof follows the ideas of a similar result proved for a general variant of P systems with permitting contexts in [1] for $\vartheta \in \{all_{aset}min, asyn, sequ\}$ and therefore is omitted; we just have to mention that the results are still valid if we take the derivation mode $max_{set}min$ instead of $all_{aset}min$, because when using partial halting we always have to take at least one rule from every set of rules (provided that the partitions for the derivation modes $all_{aset}min$, $max_{set}min$ and the partial halting $h$ all are the same). On the other hand, we do not know whether these results also hold true for the derivation modes $min$ and/or $max_{rule}min$.

## 5  Conclusions

The main purpose of this paper is to elaborate a general framework for static P systems and tissue P systems, but there are many variants of membrane systems not yet covered by this general framework, especially dynamic changes of the number of cells cannot be handled with the current version. Yet we have already started to extend our approach to such dynamic variants like P systems with active membranes. Moreover, also spiking neural P systems require some efforts for being captured within this framework. Our approach aims at formalizing the main features of membrane systems in such a way that derivation modes and halting conditions can be defined in a clear and unambiguous way to avoid that different interpretations of notions and concepts in the P systems area yield incomparable results (as a special example consider the variants described for the minimally parallel derivation mode). Moreover, specifying the marking algorithm in a procedural way should allow for easier and unambiguous implementations. Considering variants of (tissue) P systems at such a high level of abstraction allows for establishing quite general results.

## Acknowledgements

The authors gratefully especially acknowledge the useful suggestions and remarks from Artiom Alhazov and Markus Beyreder (elaborating Example 4.1) as well as all the interesting discussions with the participants of WMC 8 in Thessaloniki.

## References

1. Alhazov, A., Freund, R., Oswald, M., Verlan, S.: Partial versus total halting in P systems. In: Gutiérrez-Naranjo, M.A., Păun, G., Romero-Jiménez, A., Riscos-Núñez, A. (eds.) Proc. Fifth Brainstorming Week on Membrane Computing, Sevilla, pp. 1–20 (2007)

2. Andrei, O., Ciobanu, G., Lucanu, D.: A rewriting logic framework for operational semantics of membrane systems. Theoretical Computer Science 373(3), 163–181 (2007)
3. Bernardini, F., Gheorghe, M., Margenstern, M., Verlan, S.: Networks of Cells and Petri Nets. In: Gutiérrez-Naranjo, M.A., Păun, G., Romero-Jiménez, A., Riscos-Núñez, A. (eds.) Proc. Fifth Brainstorming Week on Membrane Computing, Sevilla, pp. 33–62 (2007)
4. Ciobanu, G., Andrei, O., Lucanu, D.: Structural operational semantics of P systems. In: [9], 1–23
5. Ciobanu, G., Pan, L., Păun, G., Pérez-Jiménez, M.J.: P systems with minimal parallelism. Theoretical Computer Science 378(1), 117–130 (2007)
6. Csuhaj-Varjú, E., Kari, L., Păun, G.: Test tube distributed systems based on splicing. Computers and AI 15(2–3), 211–232 (1996)
7. Csuhaj-Varjú, E.: Networks of Language Processors. Current Trends in Theoretical Computer Science, 771–790 (2001)
8. Dassow, J., Păun, G.: On the power of membrane computing. Journal of Universal Computer Science 5(2), 33–49 (1999)
9. Freund, R., Lojka, G., Oswald, M.: Gh. Păun (Eds.): WMC 2006. Pre-Proceedings of Sixth International Workshop on Membrane Computing, Vienna (June 18-21, 2005)
10. Freund, R., Oswald, M.: P systems with partial halting (accepted, 2007)
11. Freund, R., Păun, G., Pérez-Jiménez, M.J.: Tissue-like P systems with channel states. Theoretical Computer Science 330, 101–116 (2005)
12. Kleijn, J., Koutny, M., Rozenberg, G.: Towards a Petri net semantics for membrane systems. In: [9] 439–460
13. Păun, A., Păun, G.: The power of communication: P systems with symport/ antiport. New Generation Computing 20(3), 295–306 (2002)
14. Păun, G.: Computing with membranes. J. of Computer and System Sciences 61, 108–143 (2000), TUCS Research Report 208 (1998), http://www.tucs.fi
15. Păun, G.: Membrane Computing. An Introduction. Springer, Berlin (2002)
16. Rogozhin, Y., Alhazov, A., Freund, R.: Computational power of symport/antiport: history, advances, and open problems. In: Freund, R., Păun, G., Rozenberg, G., Salomaa, A. (eds.) WMC 2005. LNCS, vol. 3850, pp. 1–30. Springer, Heidelberg (2006)
17. Rozenberg, G., Salomaa, A. (eds.): Handbook of Formal Languages, vol. 3, Springer, Berlin (1997)
18. Păun, G., Sakakibara, Y., Yokomori, T.: P systems on graphs of restricted forms. Publicationes Matimaticae 60 (2002)
19. Păun, G., Yokomori, T.: Membrane computing based on splicing. In: Winfree, E., Gifford, D.K. (eds.) DNA Based Computers V. DIMACS Series in Discrete Mathematics and Theoretical Computer Science, vol. 54, pp. 217–232. American Mathematical Society, Providence, RI (1999)
20. The P Systems Web Page: http://psystems.disco.unimib.it

# Conformon-P Systems with Negative Values

Pierluigi Frisco

School of Mathematical and Computer Sciences
Heriot-Watt University, Edinburgh, EH14 4AS, UK
pier@macs.hw.ac.uk

**Abstract.** Some initial results on the study of conformon-P systems with negative values are reported.

One model of these conformon-P systems is proved to be computationally universal while another is proved to be at least as powerful as partially blind program machines.

## 1  Introduction

The subdivision of a cell into compartments delimited by membranes inspired G. Păun to define a new class of (distributed and parallel) models of computation called *membrane systems* [8]. The hierarchical structure, the locality of interactions, the inherent parallelism, and also the capacity (in less basic models) for membrane division, represent the distinguishing hallmarks of membrane systems. Research on membrane systems, also called 'P systems' (where 'P' stays for 'Păun'), has really flourished [9].

One of the lines of research within membrane systems deals with the study of the generative power of models of these systems.

Recent results [3,4] obtained with the use of Petri nets and P/T systems [10] show that the study of the generative variants of computing systems based on symbol objects (membrane systems, program machines, brane calculi, etc.) can be facilitated if someone considers the number of unbounded elements present in these systems. In the present paper we do not introduce the notation of Petri net and P/T systems but only one result obtained with their use. These information can be found in the just mentioned publications.

In particular [Corollary 2] from [4] indicates:

*A P/T system with two unbounded elements has computational power equivalent to the one of program machines;*

*A P/T system with only unbounded number of tokens has computational power equivalent to the one of partially blind program machines;*

*A P/T system with only unbounded number of places has computational power equivalent to the one of restricted program machines (in this case restrictions in the composition of building blocks are present).*

There *unbounded elements* refers to some components of the P/T systems (as, for instance, number of places and tokens) that are present in unbounded quantity. In [4] it is also proved that maximal parallelism is equivalent to the

G. Eleftherakis et al. (Eds.): WMC8 2007, LNCS 4860, pp. 285–297, 2007.

presence of an unbounded number of places. The results proved in [4] indicate that in the study of a computing system the number and kind of unbounded elements can give an indication (upper bounds and precise characterization) of the computing power of the system.

The research reported in the present paper does not have the level of generality (i.e., the use of Petri nets) used in [4]. It refers to our initial results on the study of conformon-P systems having one 'extended' unbounded element: the value of the conformons ranges from $-\infty$ to $+\infty$, differently from previous studies in which it was ranging from 0 to $+\infty$.

## 2    Basic Definitions

We assume the reader to have familiarity with basic concepts of formal language theory [6] and program machines [7]. We indicate with $\mathbb{N}$ the set of positive integers, $\mathbb{N}_0 = \{0\} \cup \mathbb{N}$ and $\mathbb{Z} = \mathbb{N}_0 \cup \{-i \mid i \in \mathbb{N}\}$ indicates the set of all integers (positive, negative and zero).

### 2.1    Program Machines

A *program machine* (also known as *(multi)counter machines, multipushdown machines, register machines* and *counter automata*) with $n$ counters ($n \in \mathbb{N}$) is defined as $M = (S, R, s_0, s_d)$, where $S$ is a finite set of *states*, $s_0, s_d \in S$ are respectively called the *initial* and *final* states, $R$ is the finite set of *instructions* of the form $(s_i, l_-, s_g, s_u)$ or $(s_i, l_+, s_q)$, with $s_i, s_g, s_u, s_q \in S$, $s_i \neq s_d, 1 \leq l \leq n$.

A *configuration* of a program machine $M$ with $n$ counters is given by an element in the $n+1$-tuples $(s_j, \mathbb{N}_0^n)$, $s_j \in S$. Given two configurations $(s_i, l_1, \ldots, l_n)$ and $(s'_j, l'_1, \ldots, l'_n)$ we define a *computational step* as $(s_i, l_1, \ldots, l_n) \vdash (s_j, l'_1, \ldots, l'_n)$:

- if $(s_i, l_-, s_g, s_u)$, $l = l_p$ and $l_p \neq 0$, then $s_j = s_g$, $l'_p = l_p - 1$, $l'_k = l_k$, $k \neq p$, $1 \leq k \leq n$;
  if $l = l_p$ and $l_p = 0$, then $s_j = s_u$, $l'_k = l_k$, $1 \leq k \leq n$;
  (informally: in state $s_i$ if the content of counter $l$ is greater than 0, then subtract 1 from that counter and change state into $s_g$, otherwise change state into $s_u$);
- if $(s_i, l_+, s_q)$, $l = l_p$, then $s_j = s_q, l'_p = l_p + 1$, $l'_k = l_k$, $k \neq p$, $1 \leq k \leq n$;
  (informally: in state $s_i$ add 1 to counter $l$ and change state into $s_q$).

The reflexive and transitive closure of $\vdash$ is indicated by $\vdash^*$.

A *computation* is a finite sequence of transitions between configurations of a program machine $M$ starting from the initial configuration $(s_0, l_1, \ldots, l_n)$ with $l_1 \neq 0$, $l_k = 0$, $2 \leq k \leq n$. If the last of such configurations has $s_d$ as state, then we say that $M$ *accepted* the number $l_1$. The set of numbers accepted by $M$ is defined as $L(M) = \{l_1 \mid (s_0, l_1, \cdots, l_n) \vdash^* (s_d, l''_1, \cdots, l''_n)\}$. For every program machine it is possible to create another one accepting the same set of numbers and having all counters empty in the final state.

*Partially blind program machines* (also known as *partially blind multicounter machines*) were introduced in [5] and defined as program machines without test on zero. The only allowed operations are increase and decrease of one unit per time of the counters indicated as $(s_i, l_+, s_q)$ and $(s_i, l_-, s_q)$ respectively. In case the machine tries to subtract from a counter having value zero it stops in a non final state. In [5] it is also proved that such machines are strictly less powerful than non blind ones.

## 2.2  Conformon-P System with Negative Values

A *conformon-P system with negative values* has conformons, a name-value pair, as objects. If $V$ is an alphabet (a finite set of letters), then we can define a conformon as $[\alpha, a]$, where $\alpha \in V$ and $a \in \mathbb{Z}$ (in our previous works on conformons, see for instance [1,2], we considered $a \in \mathbb{N}_0$). We say that $\alpha$ is the *name* and $a$ is the *value* of the conformon $[\alpha, a]$. If, for instance, $V = A, B, C, \ldots$, then $[A, 5], [C, 0], [Z, -14]$ are conformons, while $[AB, 21]$ and $[D, 0.5]$ are not.

Two conformons can interact according to an *interaction rule*. An interaction rule is of the form $r : \alpha \xrightarrow{n} \beta$, where $r$ is the label of the rule (a kind of name, it makes easier to refer to the rule) $\alpha, \beta \in V$ and $n \in \mathbb{N}_0$, and it says that a conformon with name $\alpha$ can give $n$ from its value to the value of a conformon having name $\beta$. If, for instance, there are conformons $[G, 5]$ and $[R, 9]$ and the rule $r : G \xrightarrow{3} R$, one application of $r$ leads to $[G, 2]$ and $[R, 12]$, another application of $r$ (to $[G, 2]$ and $[R, 12]$) leads to $[G, -1]$ and $[R, 15]$.

The compartments (membranes) present in a conformon-P system have a label (again, a kind of name which makes it easier to refer to a compartment), every label being different. Compartments can be unidirectionally connected to each other and for each connection there is a *predicate*. A predicate is an element of the set $\{\geq n, \leq n \mid n \in \mathbb{Z}\}$. Examples of predicates are: $\geq 5, \leq -2$, etc.

If, for instance, there are two compartments (with labels) $m_1$ and $m_2$ and there is a connection from $m_1$ to $m_2$ having predicate $\geq 4$, then conformons having value greater or equal to 4 can pass from $m_1$ to $m_2$. In a time unit any number of conformons can move between two connected membranes as long as the predicate on the connection is satisfied. Notice that we have *unidirectional connections* that is: $m_1$ connected to $m_2$ does not imply that $m_2$ is connected to $m_1$. Moreover, each connection has its own predicate. If, for instance, $m_1$ is connected to $m_2$ and $m_2$ is connected to $m_1$, the two connections can have different predicates. It is possible to have multiple connections (with different predicates) between compartments.

The interaction with another conformon and the passage to another membrane are the only *operations* that can be performed by a conformon.

Formally, a *conformon-P system with negative values* of degree $m, m \geq 1$, is a construct $\Pi = (V, \mu, \alpha_a, ack, L_1, \ldots, L_m, R_1, \ldots, R_m)$, where $V$ is an alphabet; $\mu = (N, E)$ is a *directed labeled graph* underlying $\Pi$. The set $N$ contains *vertices* (the membrane compartments), while the set $E$ defines directed labeled *edges* (the connections) between vertices.

In $\alpha_a$ the value of $\alpha$ can either be *input* or *output*, in the former case $\Pi$ is an accepting device, in the latter case $\Pi$ is a generating device, while $a \in \{1, \ldots, m\}$

indicates the input or output membrane, respectively. $ack \in N$ indicates the *acknowledgment membrane*.

The multisets $L_i$ contain conformons associated to region $i$; $R_i$ are finite sets of rules for conformons interaction associated to region $i$.

A *configuration* of $\Pi$ is an $m$-tuple indicating the multisets of conformons present in each membrane of the system. A *transition* is the passage from one configuration to another as the consequence of the application of operations.

A *computation* is a finite sequence of transitions between configurations of a system $\Pi$ starting from $(L_1, \ldots, L_m)$, the *initial configuration* characterized by the fact that no conformon is present in the acknowledgment membrane. If used as a generating device, then the result of a computation is given by the multisets of conformons associated to membrane $a$ when any conformon is associated to membrane $ack$. When this happens the computation is halted, that is no other operation is performed even if it could. When a conformon is associated to the acknowledge membrane the number of conformons (counted with their multiplicity) associated to membrane $a$ defines the *number generated* by $\Pi$.

If used as an accepting device, then the input is given by the multiset of conformons associated to $a$ in the initial configuration. If $\Pi$ reaches a configuration with any conformon in $ack$, then no other operation is performed even if it could and $\Pi$ accepts the input.

Some of the conformon-P systems considered in this paper work under *maximal parallelism*: in every configuration the maximum number of operations that can be performed is performed. If in one configuration some operations are conflicting (so that they cannot be executed together as they involve the same conformons), then any maximum number of non conflicting operations is performed. The passage of two conformons through the same connection is considered as two different operations (similarly for the interactions of two different pairs of conformons due to one rule).

## 2.3    Some Modules for Conformon-P Systems

In the following we use the concept of *module*: a group of membranes with conformons and interaction rules in a conformon-P system able to perform a specific task.

An example of module is a *splitter* [1]: a module that, when a conformon $[X, x]$ with $x \in \{x_1, \ldots, x_h\}, x_i < x_{i+1}, 1 \le i \le h - 1$ is associated with a specific membrane of it, it may pass such a conformon to other specific membranes according to its value $x$. A detailed splitter is depicted in Figure 1.a. Vertices outgoing a module representation of a splitter, Figure 1.b, have as predicates elements in the set $\{= n \mid \mathbb{N}_0\}$, this is a shorthand indicating the function performed by this module.

It should be clear that if a splitter is part of a conformon-P system with maximal parallelism, then the number of steps required to a conformon to pass from the $u_{h+1}$ membrane to any other of the $u$ membrane depends on the value of the conformon. If we consider the splitter depicted in Figure 1.a, a conformon

(a)

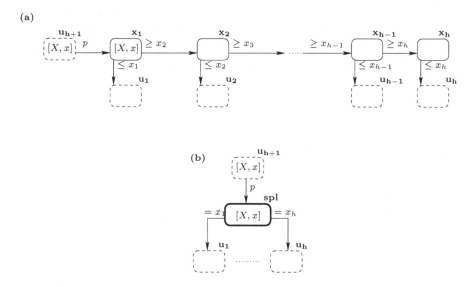

(b)

**Fig. 1.** A detailed splitter (a) and its module representation (b)

present in membrane $u_{h+1}$ requires only two steps to pass to membrane $u_1$ but it requires $h + 1$ steps to pass to membrane $u_h$.

In order to have this time constant (equal to $h+1$ in the example), then delays (i.e., sequences of membranes) have to be introduced. We make this assumption for all the splitters considered in the proof of Theorem 1.

## 2.4  Figures in This Paper

The representation of the conformon-P systems considered in this paper follows some rules aimed to a more concise representation an to an easier understanding of them.

The label of each membrane (a number) is indicated in **bold** on the top right corner of each compartment. Splitters are depicted by a thicker line, their label (also in **bold**) starts with **spl**, and their edges have '=' as predicate. The module representation of a splitter is depicted in Figure 1.b.

Oval compartments with a label inside are shortcuts for membranes or modules.

Conformons present in the initial configuration of the system are depicted in **bold**, the remaining conformons are the ones that could be present in the membrane during the computation.

The predicate associated to an edge is indicated close to the edge.

Some predicates and the value of some conformons contain a slash (/). This is a shorthand for multiple predicates or values. For instance, the conformon $[A, 3/4]$ indicates that in a membrane the conformons $[A, 3]$ and $[A, 4]$ can be present. If there is a connection from membrane $m_1$ to membrane $m_2$ and the connection has predicate $\leq 0/ \geq 5$, then this is equivalent to two connections from $m_1$ to $m_2$, one with predicate $\leq 0$ and the other with predicate $\geq 5$.

If the conformon $[A, a]$ is present in $m$ copies in a certain membrane, then this is indicated with $([A, a], m)$, where an unbounded number of copies is indicated with $+\infty$.

# 3    Results

**Theorem 1.** *The class of numbers accepted by conformon-P systems with negative values and with maximal parallelism coincides with the one accepted by program machines.*

*Proof.* This proof follows from the one of [Theorem 2] from [1] (where priorities between interaction rules are present) and from [Theorem 1] from [4].

Figure 1 represents such a conformon-P system used as accepting device simulating a program machine. During this proof we refer to this figure.

For each state $s_i$ of the simulated program machine there is a conformon with name $s_i$. For each instructions of the kind $(s_i, l_+, s_q) \in R$ there is a conformon with name $s'_{q,l}$; for each instruction of the kind $(s_i, l_-, s_g, s_u) \in R$ there are conformons with name $s''_{g,l}$ and $\bar{s}_{g,l}$. For the final state $s_d \in S$ there is one conformon with name $s'''_d$.

The initial configuration of the conformon-P system with priorities has all conformons with name $s'_{q,l}$, $s''_{g,l}$ and $s'''_d$ and 0 as value in membrane 1; all the ones with name $\bar{s}_{g,l}$ and 1 as value in membrane 17; all the ones with name $s_i$ and 0 as value in membrane 11 except the one with name of the initial state $s_0$ that is in membrane 1 with value 9 (in Figure 2 the generic conformon $[s_i, 9]$ is present in membrane 1); conformons $[a, 8]$ and $[c, 0]$ are initially present in membrane 6 and 13 respectively. Moreover for each counter $l$ of the simulated machine there are an unbounded number of occurrences of the conformons $[l, 0]$ in membrane 8, while the input membrane (membrane 14 in the figure) contains as many copies of such conformons as the values $k_l$ of the counters at the initial configuration of the simulated machine.

The addition of one unit to a counter $l$ is simulated by moving one occurrence of the conformon $[l, 0]$ from membrane 8 to membrane 14; the subtraction of one unit is simulated with the passage of one occurrence of the same conformon from membrane 14 to membrane 8.

For each instruction of the type $(s_i, l_+, s_q) \in R$ there is in membrane 1 the rule $s_i \xrightarrow{6} s'_{q,l}$; for each instruction of the kind $(s_i, l_-, s_g, s_u) \in R$, there is in membrane 1 the rule $s_i \xrightarrow{7} s''_{g,l}$; for $s_d \in S$ there is in membrane 1 the rule $s_i \xrightarrow{8} s'''_d$.

Only one conformons of the kind $[s_i, 9]$ may be associated to membrane 1. When such a conformon is present in membrane 1, then one of the interaction rules indicated above can occur.

Let us consider now the case that the rule $s_i \xrightarrow{8} s'''_d$ is applied. As there is only one instance of $[s'''_d, 0]$ then the newly created $[s'''_d, 8]$ passes to $spl_1$ and from

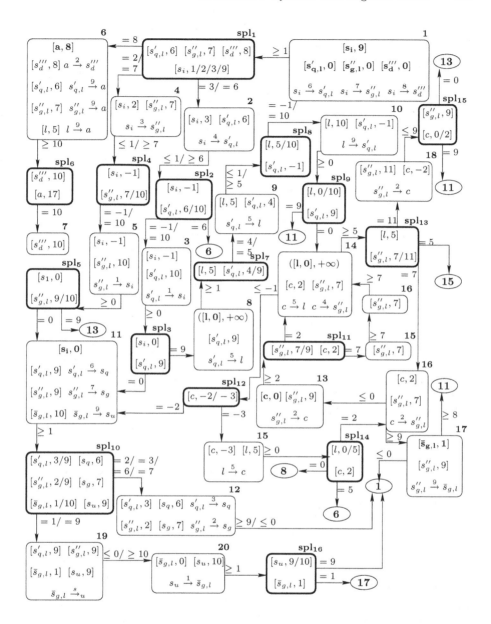

**Fig. 2.** The conformons-P system related to Theorem 1

here to membrane 6. After the interaction two sets of rules are applicable: one with a second interaction of $s_i$ and $s_d'''$ and another with the passage of $[s_d''', 8]$ and $[s_i, 1]$ to $spl_1$. Maximal parallelism forces this last set to be applied.

In membrane 6 $[s_d''', 8]$ interacts with $[a, 8]$ such that $[s_d''', 10]$ is created. When this happens this conformon passes first to $spl_6$ and then to membrane 7, the acknowledgment membrane, halting in this way the computation.

It is important to notice that the presence of $[a, 8]$ in membrane 6 is necessary for the halting of the computation. If in a configuration the conformon $[s_d''', 8]$ passes in membrane 6 but $[a, 8]$ is not there, then the simulation does not halt.

If instead the interaction in membrane 1 involves the conformon with name $s_i$ and either $s_{q,l}'$ or $s_{g,l}''$ (due to either $(s_i, l_+, s_q) \in R$ or $(s_i, l_i, s_g, s_u) \in R$), then the following sets of applicable operations are.

1. the conformon with name $s_i$ can interact again with the same instance of either $s_{q,l}'$ or $s_{g,l}''$;
2. if there is either $(s_i, l_+, s_{q'}) \in R$ or $(s_i, l_1, s_{g'}, s_{u'}) \in R$, then the conformon with name $s_i$ can interact with an instance of either $s_{q',l}'$ or $s_{g',l}''$ and the conformon created in the previous interaction (either $[s_{q,l}', 6]$ or $[s_{g,l}'', 7]$) can pass to $spl_1$;
3. both $s_i$ and either $[s_{q,l}', 6]$ or $[s_{g,l}'', 7]$ can pass to $spl_1$.

Because of maximal parallelism only the second and third set of operations in the previous list can take place (as they contain two elements while the first set contains only one element).

If the second set occurs, then the system never reaches an halting configuration. This can be seen if, for instance, we consider the conformon $[s_{q,l}', 6]$. Once in $spl_1$ this conformons passes to membrane 2, then to $spl_2$ and from here to membrane 6 where it interacts with $[a, 8]$. As a consequence of this interaction the conformon with name $a$ passes to $spl_6$ so that the system does never halt.

The role of $[a, 8]$ in membrane 6 is just this: if in any stage during the computation the system performed an operation that does not follow the simulation of the program machine, then a conformon passes to membrane 6 and interacts with $[a, 8]$ making it unavailable for $[s_d''', 8]$.

The creation of $[s_i, 3]$ and $[s_{q,l}', 6]$ in membrane 1 starts the simulation of the instruction $(s_i, l_+, s_q) \in R$. As we said, the simulation of the addition of 1 to the value of the counter is performed with the passage of one instance of $[l, 0]$ from membrane 8 to membrane 14. When in membrane 2 $[s_i, 3]$ and $[s_{q,l}', 6]$ interact, $[s_i, -1]$ and $[s_{q,l}', 10]$ are created and they pass to $spl_2$ (in case $[s_{q,l}', 6]$ passes to $spl_2$, then the system never halts). From $spl_2$ $[s_i, -1]$ and $[s_{q,l}', 10]$ pass together to membrane 3 where they interact, $[s_i, 0]$ and $[s_{q,l}', 9]$ are created and pass to $spl_3$. From here $[s_i, 0]$ passes to membrane 11 while $[s_{q,l}', 9]$ passes to membrane 8.

The membrane-splitter-membrane-splitter sequence that we just described (membrane 2 - $spl_2$ - membrane 3 - $spl_3$) is present in other parts of the system. This sequence allows to control the interaction of two conformons in a very precise way and to discard the outcome (i.e., conformons) of undesired interactions.

Once in membrane 8 the conformon with name $s_{q,l}'$ goes under another membrane-splitter sequence. In membrane 8 $[s_{q,l}', 9]$ interacts with an instance of $[l, 0]$. After this interaction three sets of applicable operations are possible, this situation is similar to the one described before for membrane 1. In this case the undesired computation sees a conformons $[l, 5]$ passing from $spl_8$ to membrane 6, while a proper simulation sees a conformon $[l, 0]$ passing to membrane 14 and $[s_{q,l}', 9]$ passing to membrane 11.

In membrane 11 $[s'_{q,l}, 9]$ interacts with $[s_q, 0]$ such that $[s'_{q,l}, 3]$ and $[s_q, 6]$ are created. Again a membrane-splitter sequence allows to create $[s'_{q,l}, 0]$ and $[s_q, 9]$ and let them pass to membrane 1. The instruction $(s_i, l_+, s_q) \in R$ has been performed and the system simulates the program machine being in state $q$.

The simulation of the instruction $(s_i, l_-, s_g, s_u) \in R$ starts with the interaction of $[s_i, 9]$ and $[s''_{g,l}, 0]$. If when this happens no $[l, 0]$ conformon is present in membrane 14, then the conformon $[s_u, 9]$ passes to membrane 1, otherwise an occurrence of $[l, 0]$ passes from membrane 14 to membrane 8 and the conformon $[s_g, 9]$ passes to membrane 1.

One interaction of $[s_i, 9]$ and $[s''_{g,l}, 0]$ in membrane 1 creates $[s_i, 2]$ and $[s''_{g,l}, 7]$ and, similarly to what described before, they can follow a membrane-splitter sequence at the end of which $[s_i, 2]$ is in membrane 11 and $[s''_{g,l}, 9]$ is in membrane 13.

In this last membrane $[s''_{g,l}, 9]$ interacts with $[c, 0]$ so that $[s''_{g,l}, 7]$ and $[c, 2]$ are created, then these two conformons pass to $spl_{11}$. From here $[c, 2]$ passes to membrane 14. The conformon $[s''_{g,l}, 7]$ also passes to this membrane but only after two steps (in the meantime it goes in membrane 15 and 16).

If in membrane 14 there is at least an occurrence of $[l, 0]$, then $[c, 2]$ interacts with any of these so that $[c, -3]$ and $[l, 5]$ are created (at the same time $[s''_{g,l}, 7]$ pass to membrane 17). In this configurations a few things can happen. Similarly to the second and third set of operations indicated in the list above $[c, -3]$ can either remain in membrane 14 and interact with another instance of $[l, 0]$ (if present) or it can pass to $spl_{12}$ and from here to membrane 15. In any case $[l, 5]$ passes to $spl_{13}$ and from here to membrane 15. If $[c, -3]$ is not present in this membrane when $[l, 5]$ is present, then this last conformon passes to $spl_{14}$ and from here to membrane 6 (and here it interacts with $[a, 8]$ such that the system never halts).

It should be clear now that if $[c, -3]$ and $[l, 5]$ do not move together out of membrane 14 the system never halts. If they do so, then they pass to membrane 15 at the same time. Here $[l, 5]$ can either pass to $spl_{14}$ (and then to membrane 6) or it can interact with $[c, -3]$ so that $[l, 0]$ and $[c, 2]$ are created and they pass together to $spl_{14}$. From here $[l, 0]$ passes to membrane 8 and $[c, 2]$ to membrane 16 (where it waits until $[s''_{g,l}, 7]$ arrives).

When $[s''_{g,l}, 7]$ passes to membrane 14 it can be that the conformon with name $c$ is there or not. This last conformon can be in membrane 14 for two reasons: either no occurrence of $[l, 0]$ was in that membrane, or one occurrence of $[l, 0]$ was there and $[c, -2]$ did not pass to $spl_{12}$. We know from the above that in this last case the system does not halt (because an $[l, 2]$ is heading membrane 6), so we are not going to discuss the consequences of the interaction between $[s''_{g,l}, 7]$ and the conformon with name $c$ when this last has a negative value.

If $[c, 2]$ is present in membrane 14 when $[s''_{g,l}, 7]$ arrives there, too, then two things can happen: the two conformons interact or not. In this last case $[s''_{g,l}, 7]$ passes to $spl_{13}$ and from here to membrane 16 and no operation can happen in the system. If instead the two conforms interact, then $[s''_{g,l}, 11]$ and $[c, -1]$ are created and they pass to membrane 18 (through $spl_{13}$ and $spl_{12}$, respectively).

Here either $[c, -2]$ passes to $spl_{15}$ and no further operation is applied, or the two conformons interact so to create $[s''_{g,l}, 9]$ and $[c, 0]$ and then these two conformons pass to membrane 11 and 13, respectively. So when $[s''_{g,l}, 9]$ is present in $spl_{15}$, then in the simulation the counter $l$ was empty.

What happens to the conformon with name $s''_{g,l}$ in membrane 11 is similar to what discussed for the conformon with name $s'_{q,l}$ earlier on. The result of these operations is that $[s_g, 9]$ and $[s''_{g,l}, 0]$ are created and pass to membrane 1.

We still have to discuss the case in which no conformon with name $c$ is present in membrane 14 when $[s''_{g,l}, 7]$ arrives. Here maximal parallelism forces this conformon to pass to $spl_{13}$ and from here to membrane 16 where $[c, 2]$ is also present. This means that if $[c, 2]$ and $[s''_{g,l}, 7]$ are present in membrane 16, then the simulation of the subtraction of 1 from counter $l$ has been performed.

When in membrane 16 $[c, 2]$ and $[s''_{g,l}, 7]$ interact so that $[c, 0]$ and $[s''_{g,l}, 9]$ are created and pass to membrane 13 and 17, respectively. In this last membrane $[s''_{g,l}, 9]$ interacts with $[\bar{s}_{g,l}, 1]$ so that $[s''_{g,l}, 0]$ and $[\bar{s}_{g,l}, 10]$ are created. Because of maximal parallelism these last two conformons pass to membrane 1 and 11, respectively.

What happens to the conformon with name $\bar{s}_{g,l}$ in membrane 11 is similar to what discussed for the conformon with name $s'_{q,l}$ earlier on. The result of these operations is that $[s_u, 9]$ and $[\bar{s}_{g,l}, 1]$ are created and pass to membrane 1 and 17 respectively.

If on a given input the program machine reaches an halting state, then the simulating conformon-P system can reach a final configuration.

The assumption that a program machine can simulate any such conformon-P system derives from the Turing-Church thesis.                                    □

In the figure related to the following proof some conformons have a parametric value of the kind $a + bn$, $a, b \in \mathbb{N}, n \geq 0$. This indicates all the possible values that a conformons can have as a consequence of interactions. If, for instance, the conformons $[A, a], [C, c]$ and the interaction rule $C \xrightarrow{b} A$ are present in the same membrane, then the value of the $A$ conformon can change into $a + bn$, where $n$ indicates the number of interactions between the $A$ and the $C$ conformon.

**Theorem 2.** *Conformon-P systems with negative values and without maximal parallelism can simulate partially blind program machines.*

*Proof.* This proof follows from the one of [Theorem 1] from [1].

Figure 2 represents such a conformon-P system used as an accepting device simulating a program machine. During this proof we refer to this figure.

For each state $s_i$ of the simulated program machine there is a conformon with name $s_i$. For each instruction of the kind $(s_i, l_+, s_q) \in R$ there is a conformon with name $s'_{q,l}$; for each instruction of the kind $(s_i, l_-, s_g) \in R$ there is a conformon with name $s''_{g,l}$. For the final state $s_d \in S$ there is one conformon with name $s'''_d$.

The initial configuration of the conformon-P system has all conformons with name $s'_{q,l}$, $s''_{g,l}$ and $s'''_d$ and 0 as value in membrane 1; all the ones with name $s_i$ and 0 as value in membrane 8 except the one with name of the initial state $s_0$

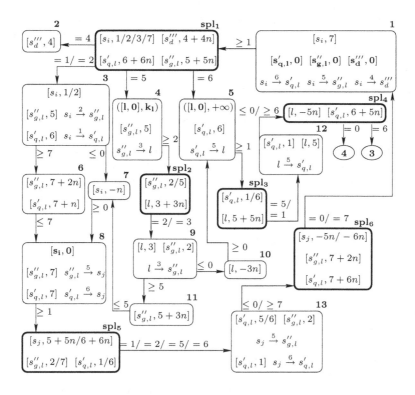

**Fig. 3.** The conformons-P system related to Theorem 2

that is in membrane 1 with value 7 (in Figure 3 the generic conformon $[s_i, 7]$ is present in membrane 1).

Moreover, for each counter $l$ of the simulated machine there are an unbounded number of occurrences of the conformons $[l, 0]$ in membrane 5, while the input membrane (membrane 4 in the figure) as many copies of such conformons as the values $k_l$ of the counters at the initial configuration of the simulated machine. The addition of one unit to one counter $l$ is simulated moving one occurrence of the conformon $[l, 0]$ from membrane 5 to membrane 4; the subtraction of one unit is simulated with the passage of one occurrence of the same conformon from membrane 4 to membrane 5.

For each instruction of the type $(s_i, l_+, s_q) \in R$ there is in membrane 1 the rule $s_i \xrightarrow{6} s'_{q,l}$; for each instruction of the kind $(s_i, l_-, v) \in R$, there is in membrane 1 the rule $s_i \xrightarrow{5} s''_{g,l}$; for $s_d \in S$ there is in membrane 1 the rule $s_i \xrightarrow{4} s'''_d$.

For any configuration of the conformon-P system only one conformon of the kind $[s_i, 7]$ may be associated to membrane 1. As we already said, initially this conformon is the one related to the initial state of the program machine.

When a conformon of the kind $[s_i, 7]$ is present in membrane 1, then one of the interaction rules indicate above can occur.

Let us consider now the case that the rule $s_i \xrightarrow{8} s_d'''$ can be applied. Of course this rule can be applied more than once, if this happens the value of the $s_i$ conformon goes below 0 and the one of the $s_d'''$ conformon is $4n$. If $n$ is at least 1, then the $s_d'''$ conformon can pass to $spl_1$, but only if $n = 1$, then $[s_d''', 4]$ passes to membrane 2, the acknowledgment membrane, halting in this way the computation.

This process of 'filtering out' (with splitters) conformons with an undesired value is present in many places in this conformon-P system. If two conformons over interacted, then the system never reaches an halting configuration.

If instead a rule of the kind $s_i \xrightarrow{6} s_{q,r}'$ is applied, then $[s_{q,r}', 6 + 6n]$ can pass to $spl_1$, but only $[s_{q,r}', 6]$ can pass to membrane 5. If in membrane 1 $[s_i, 1]$ is produced, then it passes to membrane 3 (through $spl_1$).

In membrane 5 two things can happen:

1. $s_{q,l}'$ interacts several times with the same $l$ conformon;
2. $s_{q,l}'$ interacts with different $l$ conformons.

The only case such that $[s_{q,l}', 1]$ is produced and passes to membrane 12 (through $spl_3$) is when $[s_{q,l}', 6]$ interacts only once with one $[l, 0]$ conformons. In all the other cases the value of the $s_{q,l}'$ conformon becomes negative and such a conformon does not pass to membrane 12 (in this way the system never halts).

If $[s_{q,l}', 1]$ is produced, then also one $[l, 5]$ is produced and, once in membrane 12, these two conformons interact so that $[s_{q,l}', 6]$ and $[l, 0]$ are recreated (because of $spl_4$, an over interaction in membrane 12 leads the resulting conformons to remain in that splitter) and they can pass to membrane 3 and 4, respectively.

The passage of an instance of $[l, 0]$ from membrane 5 to membrane 4 simulates the subtraction of one unit from the $l$ counter. If no $[l, 0]$ conformon is present in membrane 5 when a $s_{q,l}'$ conformon gets there, then the system never reaches an halting configuration.

In membrane 3 $s_i$ and $s_{q,l}'$ can interact such that $[s_i, -n]$ and $[s_{q,l}', 7+n], n \geq 0$ are produced and they pass to membrane 7 and 6 respectively. Only if $n = 0$ (which means that only one interaction took place), then $[s_i, 0]$ and $[s_{q,l}', 7]$ pass to membrane 8. Here $s_{q,l}'$ interacts with $s_j$ and, in a way similar to what described until now, this can lead to the production of $[s_j, 7]$ and $[s_{q,l}', 0]$ and these two conformons can pass to membrane 1.

The simulation of an instruction of the kind $(s_i, l_-, v)$ is performed in a similar way.

If on a given input the partially blind program machine reaches an halting state, then the simulating conformon-P system can reach a final configuration.
□

## 4   Final Remarks

The results reported in this paper leave us somewhat confused. How shall we interpret these results in terms of unbounded elements [4]? Must the range of values (from $-\infty$ to $+\infty$) be regarded as one or two unbounded elements or as no unbounded elements at all? (because the two infinities 'cancel' each other)

In this last case [Corollary 2] from [4] is confirmed as Theorem 1 has two unbounded elements (number of conformons and maximal parallelism) and Lemma 2 only one (number of conformons). This also implies that it is possible to prove that a partially blind program machine can simulate any conformon-P system with negative values without maximal parallelism.

In case the range of values is regarded as one unbounded element, then it should be possible to prove that the computational power of conformon-P system with negative values is equivalent to the one of program machines. This implies that Theorem 1 is redundant (because it uses three unbounded elements).

In case the range of values is regarded as two unbounded elements, then some of the results reported in [4] should be extended and made more general.

We believe that the range of values should be regarded an one unbounded element.

Another problem on this line of research regards the characterization of blind program machines in terms of unbounded elements. *Blind program machines* (also known as *blind multicounter machines*) [5] are program machines that cannot sense (so neither halt or check) if the value of a counter is zero. It is know that these machines are strictly less computationally powerful then partially blind program machines.

# References

1. Frisco, P.: The conformon-P system: A molecular and cell biology-inspired computability model. Theoretical Computer Science 312(2-3), 295–319 (2004)
2. Frisco, P.: Infinite hierarchies of conformon-P systems. In: Hoogeboom, H.J., Păun, G., Rozenberg, G., Salomaa, A. (eds.) WMC 2006. LNCS, vol. 4361, pp. 395–408. Springer, Heidelberg (2006)
3. Frisco, P.: P systems, Petri nets, and Program machines. In: Freund, R., Păun, G., Rozenberg, G., Salomaa, A. (eds.) WMC 2005. LNCS, vol. 3850, pp. 209–223. Springer, Heidelberg (2006)
4. Frisco, P.: An hierarchy of recognising computational processes, as Tech. Rep. 0047 (submitted, 2007), also available at
   http://www.macs.hw.ac.uk:8080/techreps/build_table.jsp
5. Greibach, S.A.: Remarks on blind and partially blind one-way multicounter machines. Theoretical Computer Science 7, 311–324 (1978)
6. Hopcroft, J.E., Ullman, D.: Introduction to Automata Theory, Languages, and Computation. Addison-Wesley, Reading (1979)
7. Minsky, M.L.: Computation: Finite and Infinite Machines. Automatic computation. Prentice-Hall, Englewood Cliffs (1967)
8. Păun, G.: Computing with membranes. Journal of Computer and System Sciences 1(61), 108–143 (2000)
9. Păun, G.: Membrane Computing. An Introduction. Springer, Berlin (2002)
10. Reisig, W., Rozenberg, G. (eds.): Lectures on Petri Nets I: Basic Models. LNCS, vol. 1491. Springer, Heidelberg (1998)

# Optimizing Evolution Rules Application and Communication Times in Membrane Systems Implementation

Jorge A. Tejedor, Abraham Gutiérrez, Luis Fernández, Fernando Arroyo,
Ginés Bravo, and Sandra Gómez

Natural Computing Group
Escuela Universitaria de Informática, Universidad Politécnica de Madrid
Crta. de Valencia Km. 7, 28031 Madrid, Spain
{jtejedor,abraham,setillo,farroyo,gines,sgomez}@eui.upm.es
http://www.eui.upm.es

**Abstract.** Several published time analyses in P systems implementation
have proved that there is a very strong relationship between communication and evolution rules application time in membranes of the system.
This work shows how to optimize the evolution rule application and communication times using two complementary techniques: the improvement
of evolution rules algorithms and the usage of compression schema.

On the one hand, this work uses the concepts of competitiveness relationship among active rules and competitiveness graph. For this, it takes
into account the fact that some active rules in a membrane can consume
disjoint object sets. Based on these concepts, we present a new evolution rules application algorithm that improves throughput of active rules
elimination algorithms (sequential and parallel).

On the other hand, this work presents an algorithm for compressing
information related to multisets and evolution rules, based on the assumption that algorithmic complexity of the operations performed over
multisets, in evolution rules application algorithms, is determined by the
representation of multiset information of these rules. This representation
also affects the communication phase among membranes phase.

## 1   Introduction

Computation with membranes was introduced by Gheorghe Păun in 1998 [14]
through a definition of transition P systems. This new computational paradigm
is based on the observation of biochemical processes. The region defined by a
membrane contains chemical elements (multisets) which are subject to chemical
reactions (evolution rules) to produce other elements. Transition P systems are
hierarchical, as the region defined by a membrane may contain other membranes.
Multisets generated by evolution rules can be moved to adjacent membranes
(parent and children). This multiset transfer feeds back into the system so that
new products are consumed by further chemical reactions in the membranes.

G. Eleftherakis et al. (Eds.): WMC8 2007, LNCS 4860, pp. 298–319, 2007.

These systems perform computations through transition between two consecutive configurations. Each transition or evolution step goes through two steps: rules application and objects communication. First, the evolution rules are applied simultaneously to the multiset in each membrane. This process is performed by all membranes at the same time. Then, also simultaneously, all membranes communicate with their destinations.

Nowadays, membrane systems have been sufficiently characterized from a theoretical point of view. Their computational power has been settled – many variants are computationally complete. Among their most relevant characteristics appears the fact that they can solve non polynomial time problems in polynomial time, but this is achieved by the consumption of an exponential number of resources, in particular, the number of membranes that evolve in parallel.

There are available several membrane systems simulators, [?]. An overview of membrane computing software can be found in [2]. However, the way in which these models can be implemented is a persistent problem today, because "the next generation of simulators may be oriented to solve (at least partially) the problems of information storage and massive parallelism by using parallel language programming or by using multiprocessor computers" [2]. In this sense, information storage in membrane computation implementation is an example of Parkinson's First Law [13]: "storage and transmission requirements grow double than storage and transmission improvements".

The objectives of this paper are: first to present an improvement of the algorithm of active rules elimination [20] used in the rules application step, and second to present a compression algorithm that allows us to compress information without penalizing evolution time in P systems implementation.

To achieve this, the paper is structured as follows: first, related works are presented; then, the basic ideas of the active rules elimination algorithm are summarized, which is followed by a definition of the concept of competition between rules and the optimization of the algorithm is specified. Next sections present requirements for information compression in membrane systems and the proposed compression schema; then we analyze the obtained results for a set of tests for a well known P system. Finally, some conclusions are presented.

## 2  Related Works

The first works over massively parallel implementation for P systems started with Syropoulos [18] and Ciobanu [3] who in their distributed implementations of P systems use Java Remote Method Invocation (RMI) and the Message Passing Interface (MPI) respectively, on a cluster of PC connected by Ethernet. These authors do not carry out a detailed analysis about the importance of the time used during communication phase in the total time of P system evolution; although Ciobanu stated that "the response time of the program has been acceptable. There are however executions that could take a rather long time due to unexpected network congestion".

Recently, in [19] and [1] one presents analyses for distributed architectures that are technology independent, based on: the allocation of several membranes in the same processor; the use of proxies for communication among processors; and, token passing in the communication. These solutions avoid communication collisions, and reduce the number and length for communication among membranes. All these allow to obtain a better step evolution time than in others suggested architectures congested quickly by the network collisions when the number of membranes grows. Table 1 summarizes minimum times $(T_{min})$, optimal amount of processors and membranes located in each processor ($P_{opt}$ and $K_{opt}$) to reach those minimum times, and the throughput obtained with corresponding processors and communications ($Th_{proc}$ and $Th_{com}$) for the architecture. This analysis considers the P system number of membranes $(M)$ that would evolve, the maximum time used by the slowest membrane in applying its rules $(T_{apl})$, and the maximum time used by the slowest membrane for communication $(T_{com})$.

**Table 1.** Distributed architecture parameters depending on application rules time $(T_{apl})$, communication time $(T_{com})$ and number of membranes $(M)$

| Distributed Architecture [19] | Distributed Architecture [1] |
|---|---|
| $T_{min} = 2\sqrt{2\,M\,T_{apl}\,T_{com}} - 2\,T_{com}$ <br> $P_{opt} = \sqrt{\dfrac{M\,T_{apl}}{2\,T_{com}}}$ <br> $K_{opt} = \sqrt{\dfrac{2\,M\,T_{com}}{T_{apl}}}$ <br> $Th_{proc} \sim 50\%$ <br> $Th_{com} \sim 50\%$ | $T_{min} = 2\sqrt{M\,T_{apl}\,T_{com}} + T_{com}$ <br> $P_{opt} = \sqrt{\dfrac{M\,T_{apl}}{T_{com}}}$ <br> $K_{opt} = \sqrt{\dfrac{M\,T_{com}}{T_{apl}}}$ <br> $Th_{proc} \sim 50\%$ <br> $Th_{com} \sim 100\%$ |

From all these, we may conclude that to reach minimum times over distributed architectures, there should be a balance between the time dedicated to evolution rules application and the time used for communication among membranes. So, depending on the existing relation between both times, and on the number of membranes in the P system, it is possible to determine the number of processors and the number of membranes that will be located at each of them to obtain the evolution minimum time.

The difference between these architectures lies on the different topology for the processors net and the policy for token passing. Thus, [1] reaches a throughput near to a 100% of the communication line, an increment in the parallelism level by the increment of a 40% in the processors amount involved in the architecture and a reduction to reach the 70% of the evolution time. Both works conclude that, for a specific number of membranes $M$, if it is possible that:

1. For $T_{apl}$ to be $N$ times faster, the number of membranes that would be hosted in a processor would be multiplied by $\sqrt{N}$, the number of required processors would be divided by the same factor and the time required to perform an evolution step would improve approximately with the same factor $\sqrt{N}$.

2. For $T_{com}$ to be $N$ times faster, the number of required processors would be multiplied by $\sqrt{N}$, the number of membranes that would be hosted in a processor would be divided by the same factor and the time required to perform an evolution step would improve approximately by the same factor $\sqrt{N}$.

Table 2 summarizes the importance of reducing $T_{apl}$ and $T_{com}$ over the distributed architectures parameters (minimum evolution time, optimum number of processors and optimum number of membranes per processor).

**Table 2.** Repercussion on distributed architecture parameters depending on $T_{apl}$ and $T_{com}$

| Conditions | $T_{\min}$ | $P_{opt}$ | $K_{opt}$ |
|---|---|---|---|
| $T_{apl}$ be N faster and $T_{com}$ be equal | $\frac{T_{\min}}{\sqrt{N}}$ | $\frac{P_{opt}}{\sqrt{N}}$ | $K_{opt} \cdot \sqrt{N}$ |
| $T_{apl}$ be equal and $T_{com}$ be N' faster | $\frac{T_{\min}}{\sqrt{N'}}$ | $P_{opt} \cdot \sqrt{N'}$ | $\frac{K_{opt}}{\sqrt{N'}}$ |
| $T_{apl}$ be N faster and $T_{com}$ be N' faster | $\frac{T_{\min}}{\sqrt{N'} \cdot \sqrt{N}}$ | $\frac{P_{opt} \cdot \sqrt{N'}}{\sqrt{N}}$ | $\frac{P_{opt} \cdot \sqrt{N}}{\sqrt{N'}}$ |

These architectures need to know the time required to perform rules application to be able to optimally distribute membranes among processors. Analysis of the rules application algorithms published to date shows that only the execution time of active rules elimination algorithm [20](and its parallel version [8]) can be known beforehand. These two algorithms enable prior determination of the maximum execution time, since this value depends on the number of rules rather than on the cardinality of the multiset to which they are applied, as it is reported in other algorithms [3], [6], [7]. In addition, these algorithms are the fastest in their category (sequential and parallel).

This paper describes how to optimize evolution rule application and communication times by means of two strategies. On one hand, the active rules elimination algorithm modification taking into account the fact that some active rules in a membrane can consume disjoint object sets will improve the evolution rule application time. On the other hand, the use of a compression schema for multisets and evolution rules presented in membranes will improve both times.

# 3    Optimization of Active Rules Elimination Algorithm

This section first presents the main ideas about active rules elimination algorithm, second it introduces the concepts of competitive rules and competitiveness graph and finally, three optimizations of the algorithm are carried out taking into account the competitiveness among rules and the features of the algorithm itself.

### 3.1   Active Rules Elimination Algorithm

The general idea of this algorithm is to eliminate, one by one, the rules from the set of active rules. Each step of rule elimination performs two consecutive actions:

1. Iteratively, any rule other than that which is to be eliminated is applied for a randomly selected number of times in an interval from 0 to the maximum applicability threshold. This action ensures the non-determinism inherent to P systems.
2. The rule to be eliminated is applied a number of times which is equal to its maximum applicability threshold, thus making it no longer applicable and resulting in its disappearance from the set of active rules.

We assume that:

1. The object multiset to which active rules are applied is $\omega$.
2. The active rules set is transformed to an indexed sequence $R$ in which the order of rules is not relevant.
3. The object multiset resulting from application of active rules is $\omega'$.
4. The multiset of applied rules that constitute the algorithm output is $\omega_R$.
5. Operation $|R|$ determines the number of rules in the indexed sequence $R$.
6. Operation $\Delta_{R[Ind]} \lceil \omega' \rceil$ calculates the maximum applicability threshold of the rule $R[Ind]$ over $\omega'$.
7. The operation $input\,(R[Ind]) \cdot K$ performs the scalar product of the antecedent of rules by a natural number.

The algorithm is as follows:

$$
\begin{aligned}
&(1) \quad \omega' \leftarrow \omega \\
&(2) \quad \omega_R \leftarrow \emptyset_{M_{R(O,T)}} \\
&(3) \quad \textbf{FOR } Last = |R| \textbf{ DOWNTO } 1 \\
&(4) \quad\quad \textbf{BEGIN} \\
&(5) \quad\quad\quad \textbf{FOR } Ind = 1 \textbf{ TO } Last - 1 \textbf{ DO} \\
&(6) \quad\quad\quad\quad \textbf{BEGIN} \\
&(7) \quad\quad\quad\quad\quad Max \leftarrow \Delta_{R[Ind]} \lceil \omega' \rceil \\
&(8) \quad\quad\quad\quad\quad K \leftarrow random(0, Max) \\
&(9) \quad\quad\quad\quad\quad \omega_R \leftarrow \omega_R + \left\{ R[Ind]^K \right\} \\
&(10) \quad\quad\quad\quad\;\; \omega' \leftarrow \omega' - input\,(R[Ind]) \cdot K \\
&(11) \quad\quad\quad\quad \textbf{END} \\
&(12) \quad\quad\quad Max \leftarrow \Delta_{R[Last]} \lceil \omega' \rceil \\
&(13) \quad\quad\quad\;\; \omega_R \leftarrow \omega_R + \left\{ R[Last]^{Max} \right\} \\
&(14) \quad\quad\quad\;\; \omega' \leftarrow \omega' - input\,(R[Last]) \cdot Max \\
&(15) \quad\quad \textbf{END}
\end{aligned}
$$

Remember that if rule $R[i]$ is no longer applicable in the elimination step for $R[j]$, it is no longer necessary to perform the elimination step for $R[i]$, and thus the algorithm is greatly improved, as shown in [20].

In each iteration of the algorithm of actives rules elimination, the maximum applicability threshold of a rule is calculated and then the rule is applied. The number of iterations executed at worst is:

$$\#iterations = \sum_{i=1}^{q} i = \frac{q \cdot (q+1)}{2}$$

Let $q$ be the cardinality of the indexed sequence of active rules. Thus, this algorithm allows one to know how long it takes to be executed in the worst case, with knowledge of the rules set of a membrane.

It is important to note that, in general, it is essential to perform the first action in each elimination step of a rule. This action is necessary to ensure that any possible result of the rules application to the multiset is produced by the algorithm. In case the action is not performed, the eliminated rule (applied as many times as the value of its maximum applicability threshold) may consume the objects necessary so that any other rule can be applied. However, the latter does not always occur and the first action in each elimination step can be simplified. For the sake of illustration, let us assume that the antecedents of a set of active rules are shown in Figure 1.

```
input(r1) = a        input(r2) = ab
input(r3) = cd       input(r4) = d
```

**Fig. 1.** Antecedent of the Active Rules Set

In this case, in the elimination step of the rule $r_1$ only the first action with the rule $r_2$ has to be taken, as $r_1$ and $r_2$ are the only rules with the object $a$ in its antecedents. The same is the case with rules $r_3$ and $r_4$, as these two compete for the object $d$. Thus, taking into account the competition between rule antecedents, one can adjust the rule elimination algorithm to perform only 6 iterations in the worst case, rather than 10 (2 to eliminate $r_1$, 1 to eliminate $r_2$, 2 to eliminate $r_3$, 1 to eliminate $r_4$) as shown in Figure 2.

## 3.2 Definition of Competitiveness Between Rules

Let $R$ be a set of active rules, $R = \{r_1, r_2, ..., r_q\}$ with $q > 0$, and let $C$ be a binary relation defined over the set $R$ such that

$$\forall x, y \in R, \; x \neq y \quad x \, C \, y \; \Leftrightarrow \; input(x) \cap input(y) \neq \emptyset$$

This binary relation can be represented by a non-directed graph $CG = (R, C)$ called a competitiveness graph, where the rules are related to each other if and

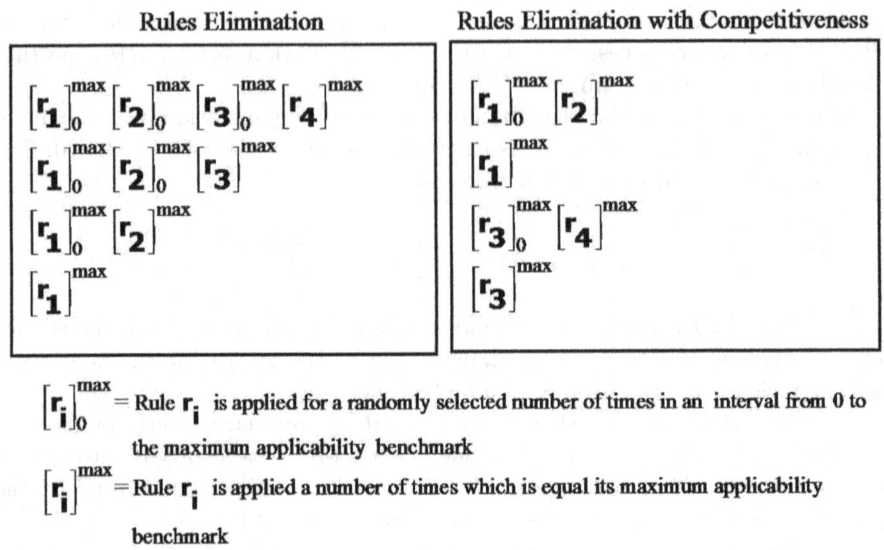

| Rules Elimination | Rules Elimination with Competitiveness |
|---|---|
| $\left[\mathbf{r_1}\right]_0^{max} \left[\mathbf{r_2}\right]_0^{max} \left[\mathbf{r_3}\right]_0^{max} \left[\mathbf{r_4}\right]^{max}$ | $\left[\mathbf{r_1}\right]_0^{max} \left[\mathbf{r_2}\right]^{max}$ |
| $\left[\mathbf{r_1}\right]_0^{max} \left[\mathbf{r_2}\right]_0^{max} \left[\mathbf{r_3}\right]^{max}$ | $\left[\mathbf{r_1}\right]^{max}$ |
| $\left[\mathbf{r_1}\right]_0^{max} \left[\mathbf{r_2}\right]^{max}$ | $\left[\mathbf{r_3}\right]_0^{max} \left[\mathbf{r_4}\right]^{max}$ |
| $\left[\mathbf{r_1}\right]^{max}$ | $\left[\mathbf{r_3}\right]^{max}$ |

$\left[\mathbf{r_i}\right]_0^{max}$ = Rule $\mathbf{r_i}$ is applied for a randomly selected number of times in an interval from 0 to the maximum applicability benchmark

$\left[\mathbf{r_i}\right]^{max}$ = Rule $\mathbf{r_i}$ is applied a number of times which is equal its maximum applicability benchmark

**Fig. 2.** Execution trace of Rules Elimination and Rules Elimination with competitiveness algorithms

only if their antecedents have an object in common. For example, given the rules inputs shown in Figure 3, the competitiveness graph generated by these rules taking into account the relation $C$ will be as shown in Figure 4.

$input(r_1) = a$        $input(r_2) = a$        $input(r_3) = b$
$input(r_4) = bc$       $input(r_5) = c$        $input(r_6) = cd$
$input(r_7) = de$       $input(r_8) = ef$       $input(r_9) = dg$
                        $input(r_{10}) = fg$

**Fig. 3.** Antecedents of an active rules set

Consider a competitiveness graph $CG = (R, C)$, a rule $x \in R$, and a set $R' \subseteq R$. The subgraph resulting from elimination of rule $x$ is defined as

$$CSG = (R - \{x\}, C \cap R - \{x\} \times R - \{x\})$$

and the competitiveness subgraph induced by the subset $R'$ is the graph

$$CSG = (R', C \cap R' \times R').$$

For a competitiveness graph $CG = (R, C)$, a competitiveness chain is defined as an ordered sequence of rules pertaining to $R$

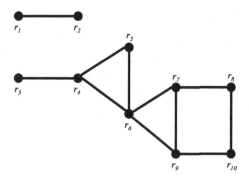

**Fig. 4.** Competitiveness graph

$$s_1, s_2, \ldots, s_n \quad s_i \in R,$$

satisfying:

$$s_i \, C \, s_{i+1} \quad \forall i \in \{1, \ldots, n-1\}$$

By definition, there is always a competitiveness chain composed of a single rule.

For a competitiveness graph $CG = (R, C)$, the accessible rule relation $(A)$ is defined as:

$$x, y \in R \quad x \, A \, y \iff \exists \text{ a competitiveness chain } s_1, \ldots, s_n \,|\, s_1 = x \wedge s_n = y$$

This is an equivalence relation which divides the rule set $R$ into equivalence classes.

Let $E$ be an equivalence class produced by $A$. The connected component of $CG$ is defined as the graph induced by the nodes pertaining to the equivalence class $E$. Then $CG$ is called connected if and only if it has a connected component.

For a competitiveness graph $CG = (R, C)$ and a rule $x \in R$, it is said that $x$ is an articulation of $CG$ if and only if the subgraph resulting from the elimination of rule $x$ has more connected components than $CG$.

### 3.3   The Algorithm Based on Rules Competitiveness

Based on the rules competitiveness relation, one can improve the algorithm of elimination of active rules. To do this, an analysis must be made of the evolution rules of each membrane prior to P system evolution. The analysis will determine the order of active rules elimination and what rules set are used in the first action of each elimination step of a given rule. The following optimizations can be made of the algorithm of rule elimination:

**First optimization.** The idea of this optimization is based on the fact that in the elimination step of a rule, the first action of the algorithm must be applied to the rules in the same connected component of the competitiveness graph. This can be done because the antecedents of rules in different connected components do not compete for common objects of the multiset.

The analysis prior to the execution of each P system calculates the competitiveness graph of each membrane. Then the connected components of the graph are calculated. The algorithm of active rule elimination will be applied independently to the rules of each of the connected components, with no need for any change in its codification.

In the worse case of the example in Figure 4, the sequential version of this algorithm will need to perform 3 iterations in the connected component consisting of the rules $\{r_1, r_2\}$ and 36 iterations in the connected component consisting of the rules $\{r_3, r_4, r_5, r_6, r_7, r_8, r_9, r_{10}\}$. Therefore, this example has gone from 55 iterations in the worst case of the algorithm of active rules elimination to 39 iterations (Figure 5), that is, it has been reduced by 71% compared to the active rules elimination algorithm.

Making a parallel version of the algorithm is quite simple. One needs only to apply the algorithm of active rules elimination in parallel to the rules of each connected component on the competitiveness graph. The parallel version would require only 36 iterations ($maximum(36, 3)$) in the worst case, as shown in Figure 6), therefore it has been reduced by 65% compared to the active rules elimination algorithm.

**Second optimization.** This optimization is applied in each connected component of the competitiveness graph. If the competitiveness graph of a membrane has articulations, the algorithm can be used to eliminate these rules first and

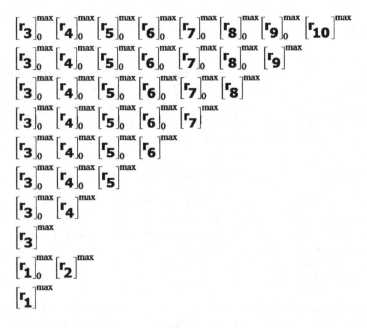

$$[r_3]_0^{max} [r_4]_0^{max} [r_5]_0^{max} [r_6]_0^{max} [r_7]_0^{max} [r_8]_0^{max} [r_9]_0^{max} [r_{10}]^{max}$$
$$[r_3]_0^{max} [r_4]_0^{max} [r_5]_0^{max} [r_6]_0^{max} [r_7]_0^{max} [r_8]_0^{max} [r_9]^{max}$$
$$[r_3]_0^{max} [r_4]_0^{max} [r_5]_0^{max} [r_6]_0^{max} [r_7]_0^{max} [r_8]^{max}$$
$$[r_3]_0^{max} [r_4]_0^{max} [r_5]_0^{max} [r_6]_0^{max} [r_7]^{max}$$
$$[r_3]_0^{max} [r_4]_0^{max} [r_5]_0^{max} [r_6]^{max}$$
$$[r_3]_0^{max} [r_4]_0^{max} [r_5]^{max}$$
$$[r_3]_0^{max} [r_4]^{max}$$
$$[r_3]^{max}$$
$$[r_1]_0^{max} [r_2]^{max}$$
$$[r_1]^{max}$$

**Fig. 5.** Execution trace of $1^{st}$ sequential optimization

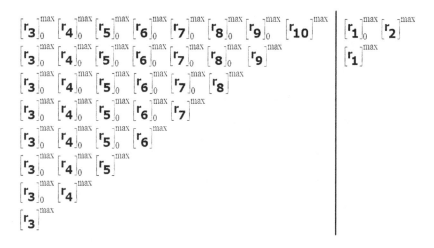

**Fig. 6.** Execution trace of $1^{st}$ parallel optimization

cause the appearance of new connected components. Thus, if rule $r_6$ is eliminated in our example (Figure 4) the connected component splits in two: the one composed of $\{r_3, r_4, r_5\}$ and the one composed of $\{r_7, r_8, r_9, r_{10}\}$.

When a connected component has no articulations, elimination of more than one rule can break it into more than one connected component. Continuing with the example we have proposed, if we first remove from connected component $\{r_7, r_8, r_9, r_{10}\}$ rules $r_7$ and $r_{10}$ in two elimination steps, it then splits into two connected components consisting of the rules $r_8$ and $r_9$, respectively.

To perform this optimization, a slight change must be made in the sequential algorithm of active rules elimination. Now, each step of elimination of a rule must eliminate a specific rule. Moreover, there is a certain partial order in the elimination steps of a rule. Whereas order is irrelevant in previous versions of the active rules elimination algorithm, it is decisive in this version. The set of rules used and the rule being eliminated in each elimination step is calculated for each membrane in the analysis prior to the evolution of the P system; as a result, the calculation does not penalize the execution time of the algorithm.

Figure 7 shows the order in which evolution rules are eliminated and the set of rules used in each elimination step for the example in Figure 4. The number of iterations of this algorithm in the worst case is 25, so it has been reduced by 45% compared to the active rules elimination algorithm.

The parallel version of the algorithm involves applying the sequential version to each of the connected components that are either in the original competitiveness graph or that are generated as a result of the elimination of a rule.

The execution trace of the parallel algorithm used with the set of rules of the example in Figure 4 is shown in Figure 8. It may be noted that the number of iterations in the worst case is 16 ($maximum(8, 2) + maximum(3, 4, 1) + maximum(1, 1, 3) + maximum(1, 1)$) using 5 processes. Hence, the number of iterations is reduced by 29% compared to the active rules elimination algorithm.

$$[r_4]_0^{max} \ [r_3]_0^{max} \ [r_5]_0^{max} \ [r_7]_0^{max} \ [r_{10}]_0^{max} \ [r_8]_0^{max} \ [r_9]_0^{max} \ [r_6]^{max}$$

$$[r_5]_0^{max} \ [r_3]_0^{max} \ [r_4]^{max}$$

$$[r_5]^{max}$$

$$[r_3]^{max}$$

$$[r_8]_0^{max} \ [r_9]_0^{max} \ [r_{10}]_0^{max} \ [r_7]^{max}$$

$$[r_8]_0^{max} \ [r_9]_0^{max} \ [r_{10}]^{max}$$

$$[r_8]^{max}$$

$$[r_9]^{max}$$

$$[r_1]_0^{max} \ [r_2]^{max}$$

$$[r_1]^{max}$$

**Fig. 7.** Execution trace of $2^{nd}$ sequential optimization

**Third optimization.** This last optimization is based on an analysis of the execution trace of the $2^{nd}$ optimization. It can occasionally be observed that the elimination step of one rule $r_j$ also eliminates one or more additional rules $r_i$. This can occur either because $r_i$ is applied a number of times that coincides with the maximum applicability threshold, or the rules applied prior to $r_i$ consume the objects needed to continue being active. This can be used in three ways to improve the execution time of the algorithm:

1. There is no need to execute the elimination step of the rule $r_i$ eliminated in a previous step. Bearing in mind the execution trace in Figure 7, if the elimination step of rule $r_6$ also eliminates rule $r_4$, then it would no longer be necessary to execute the elimination step of $r_4$, thus allowing execution of the algorithm to save 3 iterations.
2. The rule $r_i$ is not to be applied in the elimination steps of subsequent rules. Bearing in mind the execution trace in Figure 7, if the elimination step of the rule $r_6$ also eliminates rule $r_8$, it is therefore unnecessary in elimination steps of the rules $r_7$ and $r_{10}$ to try to apply $r_8$, thus allowing execution of the algorithm to save 2 iterations.
3. Elimination of the rule $r_i$ causes a change in the composition and order of the subsequent elimination steps. Keeping in mind the execution trace in Figure 7, if the elimination step of rule $r_6$ also eliminates rule $r_8$, then it is beneficial for the execution of the algorithm that $r_9$ be the next rule to be eliminated. This is the case because once $r_6$, $r_8$ and $r_9$ have been eliminated, $r_7$ and $r_{10}$ can be eliminated in a single iteration in their elimination step since they do not share objects. Here, 3 iterations would be saved.

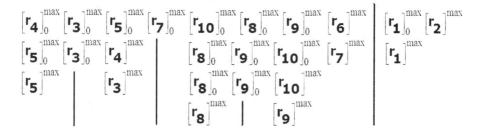

**Fig. 8.** Execution trace of $2^{nd}$ parallel optimization

To implement this optimization, a determination is necessary of what rules continue to be active whenever an elimination step is performed, and this information is used to calculate the next optimal elimination step to be taken. Logically, calculation of the next optimal elimination step would severely penalize the execution time of the algorithm. Hence, a different solution must be sought. This solution involves making an analysis prior to the execution of each P system, in which we can calculate all the possible active rule sets and assign them the next optimal step of rule elimination. All this information would be reflected in a director graph of the algorithm, the definition of which is as follows.

Let $R$ be a set of active rules. The director graph of the algorithm of rule application is composed of a triple $DG = (Q, A, T)$ where:

1. $Q$ is the node set of the graph, composed of a subset of parts of $R$, that is:
   $\forall q \in Q,, q \in \mathcal{P}(R)$
2. $A$ is a correspondence whose initial set is $Q$ and whose final set is a set of sequences of rules composed of rules from the origin element of $Q$. Thus, each set of active rules has one or more sequences of rules. Each sequence of rules indicates the order in which elimination step rules are applied. So, a state can have several elimination steps associated in the analysis prior the evolution of each P system.
   $A : Q \rightarrow E$ *where $E$ is the set of possible sequences with elements in $Q$*
3. $T$ is a set of transitions. Each transition is composed of a triad $\langle q_i, A(q_i), q_f \rangle$ where $q_i$, $q_f \in Q$ are the initial and final state, respectively, of the transition and $A(q_i)$ are the elimination step (s) of rules associated to state $q_i$, which, after being executed, means that active rules are those of state $q_f$.

The execution of the sequential algorithm of application of competitive rules will involve making a loop that ends when it reaches a state with no active rules. In each iteration, there are three steps:

1. The elimination steps associated to the state are executed.
2. Active rules are calculated.
3. The state represented by active rules is transited.

Execution of the parallel algorithm of application of competitive rules will be similar to the sequential one. The difference is that execution of several elimination steps associated to a state is performed in a parallel way. At worst, the third optimization performs the same iterations as the second optimization.

# 4  Multisets and Rules Compression

Algorithmic complexity of the operations over multisets used in the evolution rules application algorithms is determined by the representation of multiset information of these rules. This representation also affects the communication between membranes. So the use of a suitable compression schema can have a positive influence over the reduction of evolution rule application and communication times.

## 4.1  Compression Requirements

First, unlike other environments, where it is admissible a non lossless information system (i.e., multimedia contents transmission), in our environment it is essential that our compression system has no information loss.

Almost all the compression methods require two phases: the first one for analysis followed by a second one for conversion. First, an initial analysis of the information is done to identify repeated strings. From this analysis, an equivalences table is created to assign short codes to those strings. In a second phase, information is transformed using equivalent codes for repeated strings. Besides, this table is required with the information for its future compression/decompression. On the other hand, we must realize that a higher compression without any information loss will take more processing time. *Bitrate* is always variable and it is used mainly in text compression [17]. Because all of this, in spite of the fact that there are compression systems that are able to reach entropy limit - highest limit for data compression (e.g., Huffman codes) - they are not the ideal candidates for our system because of the following reasons:

1. Table storage will increase the needs for memory resources and would decrease compression goal.
2. Time for the phase of evolution rules application is penalized with compression/decompression processes when accessing compressed information on the P system. This reduces parallelism level from distributed systems and increases evolution time.
3. And also, despite of the fact that communication phase time will be reduced because a lowest amount of information is transmitted, this will be counteracted by the time needed for decompression in the destination.

In this way compression schema for information from P system should accomplish the following requirements:

1. there should be no information loss;
2. it should use the lowest amount of space for storage and transmission;
3. it should not penalize time for rules application phase and communication among membranes while processing compressed information. Thus, this means that the system should:

(a) encode information for a direct manipulation in both phases without having to use coding/decoding processes,

(b) do the compression in a previous stage to the P system evolution,

(c) therefore, abandon entropy limit to be able to maintain parallelism level and evolution time reached in previous research works.

## 4.2  Compression Schema

The second goal of this work pretends to compress the information from multisets that are present in regions and rules antecedents and consequents from each rule of a P system. But it does not address the compression of another kind of information, such as priorities, membrane targets in rule consequents nor dissolving rule capability.

Representation for multisets information in related literature is Parikh's vector [4]. Data compression is directly associated with its representation. A compression schema is presented here in three consecutive steps beginning with Parikh's vector codification over the P system alphabet.

**Parikh's vector over P system alphabet.** Each region of a membrane can potentially host an unlimited number of objects, represented by the symbols from a given alphabet $V$. We use $V^*$ to denote the set of all strings over the alphabet $V$ (we consider only finite alphabets). For $a \in V$ and $x \in V^*$ we denote by $|x|_a$ the number of occurrences of $a$ in $x$. Then, for $V = \{a_1, ..., a_n\}$, the Parikh vector associated with $V$ is the mapping on $V^*$ denoted by $\psi_V(x) = (|x|_{a_i}, \cdots, |x|_{a_n})$ for each $x \in V^*$. The byte's order reflects the order of the objects within the alphabet and consequently, the position directly indicates which symbol's multiplicity is being stored.

**Parikh's vector for each membrane's alphabet.** First step in compression considers only the alphabet subset that may exist in each of the regions for the membrane system, whatever are the possible configurations for the P system evolution. This subset may be calculated by a static analysis, previous to P system evolution time. Rules to consider when determining each membrane's alphabet in a given P system are:

1. Every object present at the region for the P system initial configuration belongs to its membrane's alphabet.

2. Every object present at the consequent for a membrane's evolution rule with target "here" belongs to its membrane's alphabet.

3. Every object present at the consequent for a membrane's evolution rule with target "in" to another membrane, belongs to the target membrane's alphabet.

4. Every object present at the consequent for a membrane's evolution rule with target "out", belongs to its father alphabet.

5. Every object present at any membrane alphabet with an evolution rule with dissolution capability belongs to its father alphabet.

**Parikh's vector without null values.** Next compression step is an alteration over the *Run Length Encoding (RLE)* algorithm [11], used mainly to compress FAX transmissions. In this lossless codification, data sequences with same value (usually zeros) are stored as a unique value plus its count. RLE compression factor is, approximately:

$$\frac{E\left(X\right)}{E\left\{\log_2 x\right\}}$$

where X is a discrete random variable that represents the number of successive zeros between two ones and E(X) is its expected value (average). Compression value stands between 20% and 30%.

In our case, what we pretend is to eliminate all the null values in Parikh's vector, that is, to eliminate all the references to the alphabet elements in a membrane that do not appear in its multiset. This information may be considered as redundant because it may be obtained from the new coded information. In a formal way, let $V = \{a_1, a_2, \ldots, a_n\}$ be an ordered finite alphabet, for $x \in V^*$ the encoded Parikh vector associated with $V$ is defined by $\Psi_V^E(x) = \{(|x|_{a_i}, i) \mid |x|_{a_i} \neq 0\}$.

At this point we should remark an important factor that is the variable or constant character for the multiset multiplicities. For the cases with multisets present at a membrane region, independently from the initial configuration, its multiplicities values are variable depending on the evolution that takes the membrane system in a non deterministic way. On the other hand, for the cases with multisets present at the evolution rules antecedents and consequents, its multiplicities values are constant and known previously to the P system evolution.

According to this situation, the compression second step encodes without null values just the information that belongs to constant multisets present at evolution rules. Thus, we get a more compressed (and lossless) representation. The reason that does this representation possible is the fact that the absence of these null values multiplicities does not affect none of the multisets operations (addition, subtraction, applicability, scalar product, ... ).

**Storage unit compression.** Last compression step concerns storage unit size for each of the P system information values. Depending on the storage unit size (measured in bits), we will be able to codify a greater or smaller range of values. In membrane computing, that does not allow negative values, given $t$ bits for the storage unit, the range for possible values will vary from 0 to $2^t - 1$.

In this section, we will have to take into account multisets present in the regions separately from the ones present in evolution rules. For the first case, storage unit size depends on the value range we want to reach during evolution without having an overflow. Instead of this, for the second case, we have to take into account, as it was shown in previous sections, that each membrane's ordered alphabet and their multiplicities are constant. Thus, an analysis previous to the P system evolution allows calculating the value ranges that are present in constant multisets for evolution rules and, so, the size that is needed to get their codification:

1. value range for multiplicities present at the antecedents and consequents for each membrane,
2. value range for Parikh's vector positions over the ordered alphabet for each membrane.

## 5    Analysis of Results

In this section we present the analysis of results obtained from the improvement of evolution rules algorithms and the usage of compression schema. First we analyze the impact that algorithms and compression have over the time required for evolution rules application. Second, we analyze the impact that compression has over the time needed for communication among membranes. Afterward, we analyze the global impact over distributed architectures parameters: evolution minimum time, optimum number of processors and membranes in each processor. Finally, we analyze the schema compression itself and its benefits over viable architectures for P systems implementations.

For the analysis of the following sections, we examine some P systems considered in [14] and [15]. Table 3 describes these P Systems.

**Table 3.** P System used for testing

| P System | Task | Reference |
|----------|------|-----------|
| A. | First example | [14] |
| B. | Decidability: n mod k = 0 | [14] |
| C. | Generating: $n^2, n \geq 1(1^{st}$ version$)$ | [14] |
| D. | Generating: $n^2, n \geq 1$ ($2^{nd}$ version) | [15] |

### 5.1    Impact Analysis for Evolution Rules Application Time

The algorithms for evolution rules application that have been referred to in this paper, are based upon a limited set of primitive operations over multisets. These are computation of: applicability, maximum applicability, antecedent/consequent addition and subtraction over its region multiset and the scalar product of an antecedent/consequent.

Table 4 shows the number of operations over multisets performed at worst by the algorithms:

- Actives rules elimination (ARE) [20]
- Sequential version of competitive rules with $2^{nd}$ optimization (SCR)
- Delimited massively parallel (DMP) [8]
- Parallel version of competitive rules with $2^{nd}$ optimization (PCR)

applied to P systems mentioned in Table 3.

**Table 4.** Number of operations over multisets performed at worst

| P System | Sequential | | | Parallel | | |
|---|---|---|---|---|---|---|
| | ARE | SCR | SCR/ARE | DMP | PCR | PCR/DMP |
| A. | 18 | 12 | 66,6% | 18 | 9 | 50% |
| B. | 9 | 9 | 100% | 10 | 9 | 100% |
| C. | 18 | 12 | 66,6% | 18 | 9 | 50% |
| D. | 18 | 12 | 66,6% | 18 | 9 | 50% |
| Average | 15.75 | 11.25 | 75% | 16 | 9 | 60% |

According to these empirical values, SCR algorithm decreases its execution time 75% against ARE. Consequently, evolution rules application time will be approximately 1.33 times faster. With the parallel algorithms we have that PCR algorithm decreases its execution time 60% against DMP. Consequently, evolution rules application time will be approximately 1.67 times faster.

The algorithmic complexity of the operations over multisets used in the evolution rules application algorithms is determined by the representation of multiset information of these rules. At worst, using representation through Parikh's vector over the P system alphabet, complexity will be equal to the alphabet cardinality. On the other hand, using representation through the proposed compression schema, complexity at worst will be equal to the multiset support that is present at the evolution rule antecedent/consequent. Table 5 presents, for each of the P systems in table 3, its alphabet support, the average support for multisets present in its evolution rules and a percentage based relation among both cardinalities. Last row presents these cardinalities average values and their relation.

**Table 5.** Alphabet cardinality and support average from P systems of Table 3

| P System | $\mid V \mid$ | $\mid\mid$ support(w) $\mid\mid$ | % |
|---|---|---|---|
| A. | 4 | 1.05 | 26.3% |
| B. | 4 | 1.50 | 37.5% |
| C. | 5 | 1.13 | 22.6% |
| D. | 5 | 1.13 | 22.6% |
| Average | 4.5 | 1.20 | 27.25% |

According to these empirical values, each of the primitive operations previously mentioned will decrease its execution time approximately until a 27.25%. Consequently, evolution rules application time will be approximately 3.67 times faster.

Taking into account both factors (decrease number of operations and decrease time per operation) we can affirm that the evolution rules application time with SRC algorithm will be approximately 4.88 times faster than ARE algorithm and PRC algorithm will be approximately 6.09 times faster than DMP algorithm.

## 5.2 Impact Analysis for Communication Time Among Membranes

Communication among membranes addresses submission of multisets present at the applied application rules consequents and, in case of dissolution, the region multiset itself. Depending on information representation, the data packet size to transmit will be smaller or bigger. Table 6 shows, for each of the P systems shown in Table 3, information compression rate for its communication for different storage units sizes. Last row presents compression rates average.

**Table 6.** Compression degree for communication units from P systems of Table 3

|          | Storage unit size | | | |
|----------|--------|---------|---------|---------|
| P System | 8 bits | 16 bits | 32 bits | 64 bits |
| A.       | 55.0%  | 45.0%   | 40.0%   | 37.5%   |
| B.       | 60.0%  | 50.0%   | 45.0%   | 42.5%   |
| C.       | 54.0%  | 44.0%   | 39.0%   | 36.5%   |
| D.       | 53.3%  | 44.4%   | 40.0%   | 37.8%   |
| Average  | 55.6%  | 45.9%   | 41.0%   | 38.6%   |

According to these empirical values, a reduction until a 55.6% of the information to transmit among membranes may be reached in the worst case. Considering that communication is a linear process that depends upon the amount of information to transmit, communication time among membranes will be approximately 1.8 times faster.

## 5.3 Global Impact Analysis

At this point, we present an impact analysis of the optimization of the evolution rule application and communication times over distributed architecture parameters. In particular, we examine, following the criteria shown in Table 2, the implication in optimum number of processors and membranes per processor and minimum evolution time.

On one hand, time reduction for evolution rules application increases the number of membranes per processor. It also decreases the number of processors and evolution time.

Using the compression schema with SCR algorithm the following results are obtained: the evolution rules application time will be approximately 4,88 times faster so we get an increment of a 120.9% for membranes per processor and a reduction until a 45.26% for number of processors and for evolution time.

Using the compression schema with PCR algorithm the following results are obtained: the evolution rules application time will be approximately 6.09 times faster so we get an increment of a 146.78% for membranes per processor and a reduction until a 40.52% for number of processors and for evolution time.

On the other hand, time reduction for communication among membranes increases the number of processors. It also decreases the number of membranes per processor and evolution time. According to the previous empirical data, from a communication time 1.80 times faster, we get, for the worst case, a 34.2% increment for number of processors and a reduction until a 74.5% for the number of membranes per processor and for evolution time.

Taking into account both factors, reduction for application and communication time, counteract their effects over the number of processors and the number of membranes per processor.

Using the compression schema with SCR algorithm the following results are obtained: we get a reduction of a **60.7%** for the number of processors, an increment of a **64.65%** for the number of membranes per processor and a reduction until a **33,74%** for evolution time.

Using the compression schema with PCR algorithm the following results are obtained: we get a reduction of a **54.36%** for the number of processors, an increment of a **83.93%** for the number of membranes per processor and a reduction until a **30.2%** for evolution time.

### 5.4   Compression Schema Analysis

Table 7 shows compression rates reached for each P system from Table 4, considering different storage unit sizes. Last row presents average compression rates for each storage size.

**Table 7.** Compression degree for P System from Table 3

|  | Storage unit size | | | |
|---|---|---|---|---|
| P System | 8 bits | 16 bits | 32 bits | 64 bits |
| A. | 59.8% | 37.8% | 26.8% | 21.3% |
| B. | 75.0% | 47.7% | 34.1% | 27.3% |
| C. | 51.1% | 32.1% | 22.8% | 18.1% |
| D. | 52.2% | 33.3% | 23.9% | 19.2% |
| Average compression degree | 59.5% | 37.7% | 26.9% | 21.5% |

Considering the worst case for this compression schema (8 bits for all the storage units), at least, we reach a compression rate of 75,0%, which implies an increase of a 33,3% for memory availability to store information. For average compression rate (59,5%), it is reached an increase of 68,0% of memory availability. So we attenuate the storage problem for information in distributed architectures implemented with low storage capacity microcontrollers based technologies. Using this compression schema, it will be possible to allocate more membranes in each microcontroller and so, it will be possible to reach minimum times at the same time that we are maximizing resources.

On the other hand, it has to be underlined that the compression process is done by an analysis previous to the P system evolution. Thus, evolution is not penalized with compression/decompression processes while phases for evolution rules application or communication among membranes.

## 6 Conclusions

Several published time analyses have proved that there is a very strong relationship between communication and evolution rules application times during membranes evolution in P systems implementation. This relation determines the number of membranes that can be allocated per processor in order to obtain the minimum evolution time for the P system. This work shows how to optimize the evolution rule application and communication times using two complementary techniques: the improvement of evolution rules algorithms and the usage of compression schema.

On the one hand, this paper introduces the concept of a competitiveness relationship among active rules. Based on this concept, a new way of parallelism has been opened toward the massively parallel character needed in rules application in P systems. Moreover, the sequential version of this algorithm performs a lower number of operations in execution than in other sequential algorithms published to date. Both the sequential and the parallel versions of the algorithm carry out a limited number of operations, thus allowing for prior knowledge of the execution time. This characteristic makes both versions of the proposed algorithm appropriate for being used in viable distributed architectures for P systems implementations. This is very important because architectures require determining the distribution of the number of membranes to be located in each processor of the architecture in order to obtain minimal evolution step times with minimal resources.

On the other hand, this work has presented a schema for compressing multisets and evolution rules for P system. The schema gets the possible highest compression level for the information without penalizing compression and decompression time with cost-consuming operations.The whole compression process is performed by mean of a previous static analysis to the P system execution. These facts, thanks to the chosen representation of information, improve the system performance reducing evolution rule application and communication times, what is very important because it implies a direct reduction of the evolution time in system execution.

An additional advantage obtained by the new algorithm and compression scheme is applied to hardware solutions and architectures based on microprocessors nets. In these cases the amount of information that has to be stored and transmitted is very important. In the first case, the main problem is due to the low storage capacity of microcontrollers. So, reducing this amount of information needed to represent a membrane, means to be able to extend the variety of problems that can be solved with this technology. In the second case, reducing the amount of information to transmit means to minimize the bottleneck in processor communication and so, increase the parallelism level.

# References

1. Bravo, G., Fernández, L., Arroyo, F., Tejedor, J.: Master-Slave Parallel Architecture for Implementing P Systems. In: MCBE 2007. The 8th WSEAS International Conference on Mathematics and Computers in Business and Economics, Vancouver (Canada) (June 2007)
2. Ciobanu, G., Păun, G., Pérez-Jiménez, M. (eds.): Applications of Membrane Computing. Natural Computing Series. Springer, Heidelberg (2006)
3. Ciobanu, G., Wenyuan, G.: A P System running on a Cluster of Computers. In: Martín-Vide, C., Mauri, G., Păun, G., Rozenberg, G., Salomaa, A. (eds.) Membrane Computing. LNCS, vol. 2933, pp. 123–150. Springer, Heidelberg (2004)
4. Dassow, J.: Parikh Mapping and Iteration. In: Calude, C.S., Pun, G., Rozenberg, G., Salomaa, A. (eds.) Multiset Processing. LNCS, vol. 2235, pp. 85–102. Springer, Heidelberg (2001)
5. Fernández, L., Martínez, V.J., Arroyo, F., Mingo, L.F.: A Hardware Circuit for Selecting Active Rules in Transition P Systems. In: Workshop on Theory and Applications of P Systems, Timisoara (Romania) (September 2005)
6. Fernández, L., Arroyo, F., Castellanos, J., Tejedor, J.A., García, I.: New Algorithms for Application of Evolution Rules based on Applicability Benckmarks. In: BIOCOMP 2006. International Conference on Bioinformatics and Computational Biology, Las Vegas (EEUU) (July 2006)
7. Fernández, L., Arroyo, F., Tejedor, J.A., Castellanos, J.: Massively Parallel Algorithm for Evolution Rules Application in Transition P Systems. In: WMC 2006, pp. 337–343 (July 2006)
8. Gil, F.J., Fernández, L., Arroyo, F., Tejedor, J.A.: Delimited Massively Parallel Algorithm based on Rules Elimination for Application of Active Rules in Transition P Systems. In: i.TECH-2007. Fifth International Conference Information Research and Applications, Varna (Bulgary) (June 2007)
9. Gutiérrez, A., Fernández, L., Arroyo, F., Martínez, V.: Design of a Hardware Architecture based on Microcontrollers for the Implementation of Membrane Systems. In: SYNASC 2006. 8th International Symposium on Symbolic and Numeric Algorithms for Scientific Computing, (September 26-29, 2006), Timisoara, Romania (2006)
10. Gutiérrez, A., Fernández, L., Arroyo, F., Alonso, S.: Hardware and Software Architecture for Implementing Membrane Systems: A case of study to Transition P Systems. In: DNA13 2007. 13th International Meeting on DNA Computing Memphis, EEUU (June 4-8, 2007)
11. Lelewer, D.A., Hirschberg, D.S.: Data Compression. ACM Computing, 8902-0069 (1987)
12. Martínez, V., Fernández, L., Arroyo, F., Gutiérrez, A.: A Hardware Circuit for the Application of Active Rules in a Transition P Systems Region. In: Fourth Inter. Conference Information Research and Applications, (June 20-25, 2006), Bulgaria, Varna (2006)
13. Parkinson, C.N.: Parkinson's Law, or the Pursuit of Progress. John Murray (1957)
14. Păun, G.: Computing with Membranes. Journal of Computer and System Sciences 61 (2000), Turku Center of Computer Science-TUCS Report 208 (1998)
15. Păun, G., Rozenberg, G.: A Guide to Membrane Computing. Theoretical Computer Science 287, 73–100 (2000)
16. Petreska, B., Teuscher, C.: A Reconfigurable Hardware Membrane System. In: Alhazov, A., Martín-Vide, C., Paun, G. (eds.) Preproceedings of the Workshop on Membrane Computing, Tarragona, July 17-22 2003, pp. 343–355 (2003)

17. Salomon, D.: Data Compression: The Complete Reference. Springer, Heidelberg (2004)
18. Syropoulos, A., Mamatas, E.G., Allilomes, P.C., Sotiriades, K.T.: A Distributed Simulation of P Systems. In: Preproceedings of the Workshop on Membrane Computing, Tarragona, pp. 455–460 (2003)
19. Tejedor, J.A., Fernández, L., Arroyo, F., Bravo, G.: An Architecture for Attacking the Bottleneck Communication in P System. In: AROB 2007. XII International Symposium on Artificial Life and Robotics, Oita, JAPAN (January 25-27, 2007)
20. Tejedor, J.A., Fernández, L., Arroyo, F., Gutiérrez, A.: Algorithm of Active Rule Elimination for Application of Evolution Rules. In: MCBE 2007. The 8th WSEAS International Conference on Mathematics and Computers in Business and Economics, Vancouver (Canada) (June 2007)

# Hill Kinetics Meets P Systems:
# A Case Study on Gene Regulatory Networks as Computing Agents *in silico* and *in vivo*

Thomas Hinze[1], Sikander Hayat[2], Thorsten Lenser[1],
Naoki Matsumaru[1], and Peter Dittrich[1]

[1] Friedrich-Schiller-Universität Jena, Bio Systems Analysis Group
Ernst-Abbe-Platz 1-4, D-07743 Jena, Germany
{hinze,thlenser,naoki,dittrich}@minet.uni-jena.de
[2] Universität des Saarlandes, Computational Biology Group
Center for Bioinformatics, P.O. Box 15 11 50, D-66041 Saarbrücken, Germany
s.hayat@bioinformatik.uni-saarland.de

**Abstract.** Modeling and simulation of biological reaction networks is an essential task in systems biology aiming at formalization, understanding, and prediction of processes in living organisms. Currently, a variety of modeling approaches for specific purposes coexists. P systems form such an approach which owing to its algebraic nature opens growing fields of application. Here, emulating the dynamical system behavior based on reaction kinetics is of particular interest to explore network functions. We demonstrate a transformation of Hill kinetics for gene regulatory networks (GRNs) into the P systems framework. Examples address the switching dynamics of GRNs acting as NAND gate and RS flip-flop. An adapted study *in vivo* experimentally verifies both practicability for computational units and validity of the system model.

## 1 Introduction

Along with the development of systems biology, a variety of modeling techniques for biological reaction networks have been established during the last years [1]. Inspired by different methodologies, three fundamental concepts emerged mostly independent of each other: *analytic*, *stochastic*, and *algebraic* approaches. Each paradigm specifically emphasizes certain modeling aspects. Analytic approaches, primarily adopted from chemical reaction kinetics, enable a macroscopic view on species concentrations in many-body systems. Based on differential equations considering generation and consumption rates of species, deterministic monitoring and prediction of temporal or spatial system behavior is efficiently expressed by continuous average concentration gradients. In contrast, stochastic approaches reflect aspects of uncertainty in biological reaction networks by incorporating randomness and probabilities. So, ranges of possible scenarios and their statistical distribution can be studied facilitating a direct comparison with wetlab experimental data. Statistical tools help in discovering correlations between network components. Furthermore, algebraic approaches appear as flexible instruments regarding the level of abstraction for system description. Due to

G. Eleftherakis et al. (Eds.): WMC8 2007, LNCS 4860, pp. 320–335, 2007.
© Springer-Verlag Berlin Heidelberg 2007

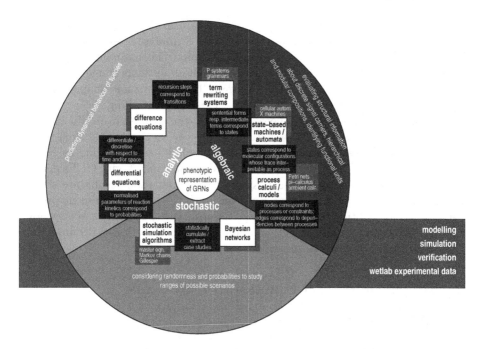

**Fig. 1.** Modeling approaches for biological reaction networks and bridges between them. Algorithmic strategies behind these bridges allow model transformations. Stochastic, analytic, and algebraic approaches form fundamental paradigms, classified into subclasses (white highlighted) and transformational concepts (black highlighted).

their discrete principle of operation, they work by embedding as well as evaluating structural information, modularization, molecular tracing, and hierarchical graduation of provided system information.

Combining advantages of several paradigms comes more and more into the focus of research. On the one hand, heterogeneous models subsume elements from different approaches into an extended framework. On the other hand, transformation strategies aim to model shifting between approaches, see Figure 1. Thus, specific analysis tools as well as advanced techniques for classification, simplification, comparison, and unification can become applicable more easily. This is additionally motivated by the fact that all three paradigms are independently known to be capable of constructing Turing complete models for computation.

In general, P systems represent term rewriting mechanisms, hence algebraic constructs [15,16]. Substantiated by the progress in proteomics, investigating the dynamical behavior of biological reaction networks is essential to understand their function. Although P systems containing appropriate kinetics are useful, reaction kinetics is mostly defined for analytic models. In this paper, we contribute to bridging this gap for GRNs.

Related work addresses corresponding P systems for phenotypic representations of some biological network classes. While metabolic P systems [11] and P systems for cell signalling [9,14] have already been equipped with mass-action kinetics derived from underlying reaction mechanism [5], P systems for GRNs [2,4] and for quorum sensing [3] are restricted to formulate inhibiting or activating effects qualitatively. In order to introduce a homogeneous quantitative model, we decided to incorporate Hill kinetics [12] to the P systems framework by describing the cooperativity in GRNs dynamically using sigmoid-shaped transfer functions that are more precise than two-stage on/off switching.

The paper is organized as follows: Based on the definition of Hill kinetics, we present a method for discretization that leads to P systems $\Pi_{\text{Hill}}$ whose properties are discussed briefly. A case study includes GRNs acting as NAND gate and RS flip-flop. For each logic gate, its GRN in concert with ODEs derived from Hill kinetics, corresponding P system, and simulation results are shown. Finally, we verify that a reporter gene encoding the green fluorescent protein (*gfp*) with transcription factors N-acyl homoserine lactone (AHL) and isopropyl-$\beta$D-thiogalactopyranoside (IPTG) can mimic the aforementioned RS flip-flop *in vivo*. Here, *gfp* expression is quantified using flow cytometry.

## 2    Transforming Hill Kinetics to P Systems

### 2.1    Hill Kinetics

Hill kinetics [12] represents a homogeneous analytic approach to model cooperative and competitive aspects of interacting biochemical reaction networks dynamically. It formulates the relative intensity of gene regulations by sigmoid-shaped threshold functions h of degree $m \in \mathbb{N}_+$ and threshold $\Theta > 0$ such that $x \geq 0$ specifies the concentration level of a transcription factor that activates resp. inhibits gene expression. Function value h then returns the normalized change in concentration level of the corresponding gene product:

activation (upregulation) $\rightarrow$:    $h^+(x, \Theta, m) = \frac{x^m}{x^m + \Theta^m}$

inhibition (downregulation) $\perp$: $h^-(x, \Theta, m) = 1 - h^+(x, \Theta, m)$

Functions $h^+$ and $h^-$ together with a proportional factor $c_1$ quantify the production rate of a certain gene product *GeneProduct*. Here we assume a linear spontaneous decay with rate $c_2[GeneProduct]$ such that the differential equation takes the form $\frac{d[GeneProduct]}{dt} = ProductionRate - c_2[GeneProduct]$. Different activation and inhibition rates are simply multiplied as in the following example illustrated in Figure 2.1 ($c_1, c_2 \in \mathbb{R}_+$):

$$\frac{d[GeneProduct]}{dt} = c_1 \cdot h^+(A_1, \Theta_{A_1}, m) \cdot \ldots \cdot h^+(A_n, \Theta_{A_n}, m) \cdot$$
$$\left(1 - h^+(I_1, \Theta_{I_1}, m) \cdot \ldots \cdot h^+(I_p, \Theta_{I_p}, m)\right) - c_2[GeneProduct]$$

For simplicity, each differential coefficient $\frac{dy}{dt}$ is subsequently denoted as $\dot{y}$.

By coupling gene regulatory units we obtain GRNs. Here, gene products can act as transcription factors for other genes within the network. Additional complex formation among gene products allows conjunctive composition of transcription factors and the introduction of further nonlinearities. Thus, an

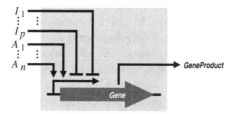

**Fig. 2.** Gene regulatory unit. Repetitive expression of a *Gene* leads to generation of a specific *GeneProduct*, a protein whose amino acid sequence is encoded by the DNA sequence of the *Gene*. Transcription factors (specific single proteins or complexes) quantitatively control the expression rate by their present concentration. Two types of transcription factors can be distinguished: Inhibitors, here symbolized by $I_1, \ldots, I_p$, repress *Gene* expression by downregulation while activators $A_1, \ldots, A_n$ cause the opposite amplifying effect by upregulation.

effective signal transduction and combination between different network elements becomes feasible.

## 2.2 Discretization

The analytic nature of Hill kinetics based on continuous concentrations requires a discretization with respect to value and time in order to derive a homologous term rewriting mechanism. Following the intention to approximate continuous concentrations by absolute particle numbers, we assume a large but finite pool of molecules. The application of a reaction rule in terms of a rewriting process removes a number of reactant particles from this pool and simultaneously adds all products. Therefore, selection and priorization of reaction rules to apply are controlled by an underlying iteration scheme with temporally stepwise operation.

Since Hill kinetics is characterized by variable reaction rates due to the sigmoid-shaped functions h, this variability should also be reflected in the term rewriting mechanism. For this reason, we introduce dynamic stoichiometric factors resulting in time dependent reaction rules. Let $\Delta\tau > 0$ be the constant time discretization interval (step length), the gene regulatory unit depicted in Figure 2.1 consists of two reaction rules with variable stoichiometric factors $s$ and $u$:

$$s \; Gene \longrightarrow s \; GeneProduct + s \; Gene \Big|_{A_1,\ldots,A_n,\neg I_1,\ldots,\neg I_p} \quad \text{where}$$

$$s = \lfloor \Delta\tau \cdot c_1 \cdot [Gene] \cdot$$
$$h^+(A_1, \Theta_{A_1}, m) \cdot \ldots \cdot h^+(A_n, \Theta_{A_n}, m) \cdot$$
$$\left(1 - h^+(I_1, \Theta_{I_1}, m) \cdot \ldots \cdot h^+(I_p, \Theta_{I_p}, m)\right) \rfloor$$

$$u \; GeneProduct \longrightarrow \emptyset \quad \text{where} \quad u = \lfloor \Delta\tau \cdot c_2 \cdot [GeneProduct] \rfloor$$

Here, the upper reaction formulates the generation of *GeneProduct* particles with regard to the limiting resource of available *Gene* objects. Reaction conditions coming from the presence of activators $A_1, \ldots, A_n$ and absence ($\neg$) of inhibitors $I_1, \ldots, I_p$ affect the stoichiometric factor $s$. The notation of indexes after the vertical bar declares the elements which occur in the h-components $(\mathrm{h}^+, \mathrm{h}^-)$ of the function regulating the rule. In order to map normalized concentrations from Hill kinetics into absolute particle numbers, we introduce the factor term $[Gene]$ which represents the total number of $Gene$ objects present in the reaction system. Accordingly, the decay (consumption) of $GeneProduct$ is expressed by the lower transition rule.

The rounding regulation ($\lfloor\ \rfloor$) provides for integer numbers as stoichiometric factors. This is necessary for handling the discrete manner of term rewriting. Nevertheless, a discretization error can occur and propagate over the time course. The higher the total number of particles in the reaction system is initially set, the more this inaccuracy can be reduced.

Now, we incorporate the reaction system obtained by discretization into the P systems framework. Therefore, we firstly define some syntactical conventions with respect to formal languages and multisets.

### 2.3   Formal Language and Multiset Prerequisites

We denote the empty word by $\varepsilon$. Let $A$ be an arbitrary set and $\mathbb{N}$ the set of natural numbers including zero. A multiset over $A$ is a mapping $F : A \longrightarrow \mathbb{N} \cup \{\infty\}$. $F(a)$, also denoted as $[a]_F$, specifies the multiplicity of $a \in A$ in $F$. Multisets can be written as an elementwise enumeration of the form $\{(a_1, F(a_1)), (a_2, F(a_2)), \ldots\}$ since $\forall (a, b_1), (a, b_2) \in F : b_1 = b_2$. The support $\mathrm{supp}(F) \subseteq A$ of $F$ is defined by $\mathrm{supp}(F) = \{a \in A \mid F(a) > 0\}$. A multiset $F$ over $A$ is said to be empty iff $\forall a \in A : F(a) = 0$. The cardinality $|F|$ of $F$ over $A$ is $|F| = \sum_{a \in A} F(a)$. Let $F_1$ and $F_2$ be multisets over $A$. $F_1$ is a subset of $F_2$, denoted as $F_1 \subseteq F_2$, iff $\forall a \in A : (F_1(a) \leq F_2(a))$. Multisets $F_1$ and $F_2$ are equal iff $F_1 \subseteq F_2 \wedge F_2 \subseteq F_1$. The intersection $F_1 \cap F_2 = \{(a, F(a)) \mid a \in A \wedge F(a) = \min(F_1(a), F_2(a))\}$, the multiset sum $F_1 \uplus F_2 = \{(a, F(a)) \mid a \in A \wedge F(a) = F_1(a) + F_2(a)\}$, and the multiset difference $F_1 \ominus F_2 = \{(a, F(a)) \mid a \in A \wedge F(a) = \max(F_1(a) - F_2(a), 0)\}$ form multiset operations. The term $\langle A \rangle = \{F : A \longrightarrow \mathbb{N} \cup \{\infty\}\}$ describes the set of all multisets over $A$ while $\mathcal{P}(A)$ denotes the power set of $A$.

### 2.4   Transformation: Definition of the Corresponding P System

The general form of a P system $\Pi_{\mathrm{Hill}}$ emulating the dynamical behavior of GRNs using Hill kinetics is a construct

$$\Pi_{\mathrm{Hill}} = (V_{\mathrm{Genes}}, V_{\mathrm{GeneProducts}}, \Sigma, [_1]_1, L_0, r_1, \ldots, r_k, \mathrm{f}_1, \ldots, \mathrm{f}_k, \Delta\tau, m)$$

where $V_{\mathrm{Genes}}$ denotes the alphabet of genes, $V_{\mathrm{GeneProducts}}$ the alphabet of gene products (without loss of generality $V_{\mathrm{Genes}} \cap V_{\mathrm{GeneProducts}} = \emptyset$), and $\Sigma \subseteq V_{\mathrm{GeneProducts}}$ represents the output alphabet. $\Pi_{\mathrm{Hill}}$ does not incorporate inner membranes, so the only membrane is the skin membrane $[_1]_1$. The single membrane property results from the spatial globality of GRNs within an organism: Gene expression is located in the cell nuclei flanked by a receptor-controlled intercellular transduction and combination of transcription factors. Resulting GRNs form independent network structures of high stability within living organisms.

Let $V = V_{\text{Genes}} \cup V_{\text{GeneProducts}}$. The multiset $L_0 \in \langle V \rangle$ over $V$ holds the initial configuration of the system.

Initial reaction rules[1] $r_i \in \langle E_{i,0} \rangle \times \langle P_{i,0} \rangle \times \mathcal{P}(TF_i)$ with multiset of reactants $E_{i,0} \subseteq V \times \mathbb{N}$, multiset of products $P_{i,0} \subseteq V \times \mathbb{N}$ and set of involved transcription factors $TF_i \in V_{\text{GeneProducts}}$, $i = 1, \ldots, k$, define the potential system activity at time point 0. A function $f_i : \mathbb{R}_+ \times \langle V \rangle \times \mathbb{N}_+ \to \mathbb{N}$ is associated with each initial reaction rule $r_i$. This function adapts the stoichiometric factors according to the discretized Hill kinetics as described above.

Furthermore, we introduce two global parameters. The time discretization interval $\Delta\tau \in \mathbb{R}_+$ corresponds to the length of a time step between discrete time points $t$ and $t+1$. The degree $m \in \mathbb{N}_+$ is used for all embedded sigmoid-shaped functions.

Finally, the dynamical behavior of the P system is specified by an iteration scheme updating both the system configuration $L_t$ and the stoichiometric factors of reaction rules $r_i$ starting from $L_0$ where $i = 1, \ldots, k$:

$$L_{t+1} = L_t \ominus Reactants_t \uplus Products_t \quad \text{with}$$

$$Reactants_t = \biguplus_{i=1}^{k} (E_{i,t+1} \cap L_t)$$

$$Products_t = \begin{cases} \displaystyle\biguplus_{i=1}^{k} P_{i,t+1} & \text{iff } Reactants_t = \displaystyle\biguplus_{i=1}^{k} E_{i,t+1} \\ \emptyset & \text{else} \end{cases}$$

$$E_{i,t+1} = \{(e, a') \mid (e, a) \in E_{i,t} \wedge a' = f_i(\Delta\tau, L_t, m)\} \tag{1}$$

$$P_{i,t+1} = \{(q, b') \mid (q, b) \in P_{i,t} \wedge b' = f_i(\Delta\tau, L_t, m)\} \tag{2}$$

Informally, the specification of $E_{i,t+1}$ and $P_{i,t+1}$ means that all reactants $e$ and products $q$ remain unchanged over the time course. Just their stoichiometric factors are updated from value $a$ to $a'$ (reactants) and from $b$ to $b'$ (products) according to functions $f_i$. These functions may utilize the numbers of copies for all $|V|$ types of particles recently present in the system. The cardinality $|L_t \cap \{(w_j, \infty)\}|$ then identifies this amount for any $w_j \in V$.

In terms of computational devices, P systems $\Pi_{\text{Hill}}$ carry an output providing the outcome of a calculation. For this purpose, the multiplicity of those gene products listed in the output alphabet is suitable. We define an output function output $: \mathbb{N} \to \mathbb{N}$ by

$$\text{output}(t) = |L_t \cap \{(w, \infty) \mid w \in \Sigma\}|.$$

For better readability, we subsequently write a reaction rule $r_i = \Big(\{(e_1, a_1), \ldots, (e_h, a_h)\}, \{(q_1, b_1), \ldots, (q_v, b_v)\}, \{tf_1, \ldots, tf_c\}\Big)$ with $\text{supp}(E_{i,t}) = \{e_1, \ldots, e_h\}$ and $\text{supp}(P_{i,t}) = \{q_1, \ldots, q_v\}$ as well as $TF_i = \{tf_1, \ldots, tf_c\}$ by using the chemical denotation $r_i : a_1 e_1 + \ldots + a_h e_h \longrightarrow b_1 q_1 + \ldots + b_v q_v \big|_{tf_1, \ldots, tf_c}$.

---

[1] Note that in our case the stoichiometry of reaction rules changes over time which is used to implement time-varying reaction rates.

As a first example, $\Pi_{\text{Hill}}$ of the gene regulatory unit shown in Figure 2.1 reads:

$$\Pi_{\text{Hill,GRNunit}} = (V_{\text{Genes}}, V_{\text{GeneProducts}}, \Sigma, [_1]_1, L_0, r_1, r_2, f_1, f_2, \Delta\tau, m)$$

$$V_{\text{Genes}} = \{Gene\}$$

$$V_{\text{GeneProducts}} = \{A_1, \ldots, A_n, \neg I_1, \ldots, \neg I_p, GeneProduct\}$$

$$\Sigma = \{GeneProduct\}$$

$$L_0 = \{(Gene, g), (A_1, a_1), \ldots, (A_n, a_n), (\neg I_1, i_1), \ldots, (\neg I_p, i_p)\}$$

$$r_1 \ : \ s_1 \ Gene \longrightarrow s_1 \ GeneProduct + s_1 \ Gene\big|_{A_1, \ldots, A_n, \neg I_1, \ldots, \neg I_p}$$

$$r_2 \ : \ s_2 \ GeneProduct \longrightarrow \emptyset$$

$$f_1(\Delta\tau, L_t, m) = \lfloor \Delta\tau \cdot |L_t \cap \{(Gene, \infty)\}| \cdot$$

$$\frac{|L_t \cap \{(A_1, \infty)\}|^m}{|L_t \cap \{(A_1, \infty)\}|^m + \Theta_{A_1}^m} \cdots \frac{|L_t \cap \{(A_n, \infty)\}|^m}{|L_t \cap \{(A_n, \infty)\}|^m + \Theta_{A_n}^m} \cdot$$

$$\left(1 - \frac{|L_t \cap \{(\neg I_1, \infty)\}|^m}{|L_t \cap \{(\neg I_1, \infty)\}|^m + \Theta_{\neg I_1}^m} \cdots \frac{|L_t \cap \{(\neg I_p, \infty)\}|^m}{|L_t \cap \{(\neg I_p, \infty)\}|^m + \Theta_{\neg I_p}^m}\right)\rfloor$$

$$f_2(\Delta\tau, L_t, m) = \lfloor \Delta\tau \cdot |L_t \cap \{(GeneProduct, \infty)\}| \rfloor$$

$$\Delta\tau \in \mathbb{R}_+$$

$$m \in \mathbb{N}_+$$

Note that $s_1$ at time point $t+1$ is equal to $f_1(\Delta\tau, L_t, m)$ at time point $t$ or holds its initialization value at time point 0. Respectively, $s_2$ at time point $t+1$ is equal to $f_2(\Delta\tau, L_t, m)$ at time point $t$ or holds its initialization value at time point 0, see equations (1) and (2).

At low molecular concentrations, deterministic application of Hill functions can conflict between different functions which want to update the system configuration. This is the case if the amount of reactants is too small to satisfy the needs of all functions. Since the number of multiset elements always remains nonnegative (see definition of $\ominus$), the system can violate mass conservation by satisfying these needs. A system extension based on stochastic rewriting mechanisms and/or priorization of reaction rules can overcome this insufficiency.

## 2.5   System Classification, Properties and Universality

$\Pi_{\text{Hill}}$ belongs to P systems with symbol objects and time varying transition rules whose evolution is based on conditional rewriting by quantitative usage of promoters and inhibitors. Thus, the dynamical behavior formulated in Hill kinetics is time- and value-discretely approximated by a stepwise adaptation. This leads to a deterministic principle of operation.

From the view on computational completeness, there are several indicators for Turing universality. On the one hand, we will demonstrate within the next section how NAND gates and compositions of NAND gates can be emulated by

P systems of the form $\Pi_{\text{Hill}}$. Arbitrarily extendable circuits consisting of coupled NAND gates can be seen as computational complete. On the other hand, the multiplicity of each symbol object within the system may range through the whole recursively enumerable set of natural numbers. So, copies of a gene product expressed by a dedicated gene are able to represent the register value of a random access machine. Autoactivation loops keep a register at a certain value while external activation increases the amount of gene product (increment operation) and external inhibition decreases respectively (decrement operation). Incrementing and decrementing transcription factors always form complexes with program counter objects. The interplay of those specific transcription factors manages the program control.

# 3    Case Study: Computational Units and Circuits

Synthetic GRNs have been instrumental in elucidating basic principles that govern the dynamics and consequences of stochasticity in the gene expression of naturally occurring GRNs. The realization as computational circuits infers inherent evolutionary fault tolerance and robustness to these modular units.

In a case study, we introduce two synthetic GRNs for logic gates (NAND gate, RS flip-flop) and describe their dynamical behavior quantitatively by an ordinary differential equation model using Hill kinetics and by corresponding P systems $\Pi_{\text{Hill}}$.

A variety of distinguishable transcription factors given by their concentration over the time course enables communication between as well as coupling of computational units. Thus, circuit engineering becomes feasible.

## 3.1    NAND Gate

input:    concentration levels of transcription factors $x$ (input1), $y$ (input2)
output:    concentration level of gene product $z$.

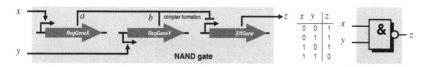

| $x$ | $y$ | $z$ |
|---|---|---|
| 0 | 0 | 1 |
| 0 | 1 | 1 |
| 1 | 0 | 1 |
| 1 | 1 | 0 |

**Ordinary Differential Equations**

$\dot{a} = \text{h}^+(x, \Theta_x, m) - a$
$\dot{b} = \text{h}^+(y, \Theta_y, m) - b$
$\dot{z} = 1 - \text{h}^+(a, \Theta_a, m) \cdot \text{h}^+(b, \Theta_b, m) - z$

**Simulation Result** (Copasi [10], ODE solver)

dynamical behavior depicted for $m = 2$, $\Theta_j = 0.1$, $j \in \{x, y, a, b\}$

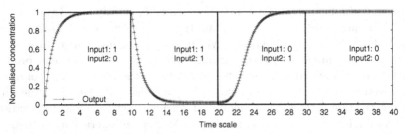

## Corresponding P System

$$\Pi_{\text{Hill,GRNnand}} = (V_{\text{Genes}}, V_{\text{GeneProducts}}, \Sigma, [_1]_1, L_0, r_1, \ldots, r_6, f_1, \ldots, f_6, \Delta\tau, m)$$

$$V_{\text{Genes}} = \{RegGeneX, RegGeneY, EffGene\}$$

$$V_{\text{GeneProducts}} = \{x, y, z, \neg a, \neg b\}$$

$$\Sigma = \{z\}$$

$$L_0 = \{(RegGeneX, rgx), (RegGeneY, rgy), (EffGene, eg),$$
$$(x, x_0), (y, y_0), (z, z_0), (\neg a, a_0), (\neg b, b_0)\}$$

$$r_1 : \quad s_1 \; RegGeneX \longrightarrow s_1 \; \neg a + s_1 \; RegGeneX \Big|_x$$

$$r_2 : \quad s_2 \; \neg a \longrightarrow \emptyset$$

$$r_3 : \quad s_3 \; RegGeneY \longrightarrow s_3 \; \neg b + s_3 \; RegGeneY \Big|_y$$

$$r_4 : \quad s_4 \; \neg b \longrightarrow \emptyset$$

$$r_5 : \quad s_5 \; EffGene \longrightarrow s_5 \; z + s_5 \; EffGene \Big|_{\neg a, \neg b}$$

$$r_6 : \quad s_6 \; z \longrightarrow \emptyset$$

$$f_1(\Delta\tau, L_t, m) = \left\lfloor \Delta\tau \cdot |L_t \cap \{(RegGeneX, \infty)\}| \cdot \frac{|L_t \cap \{(x, \infty)\}|^m}{|L_t \cap \{(x, \infty)\}|^m + \Theta_x^m} \right\rfloor$$

$$f_2(\Delta\tau, L_t, m) = \left\lfloor \Delta\tau \cdot |L_t \cap \{(\neg a, \infty)\}| \right\rfloor$$

$$f_3(\Delta\tau, L_t, m) = \left\lfloor \Delta\tau \cdot |L_t \cap \{(RegGeneY, \infty)\}| \cdot \frac{|L_t \cap \{(y, \infty)\}|^m}{|L_t \cap \{(y, \infty)\}|^m + \Theta_y^m} \right\rfloor$$

$$f_4(\Delta\tau, L_t, m) = \left\lfloor \Delta\tau \cdot |L_t \cap \{(\neg b, \infty)\}| \right\rfloor$$

$$f_5(\Delta\tau, L_t, m) = \Big\lfloor \Delta\tau \cdot |L_t \cap \{(EffGene, \infty)\}| \cdot$$
$$\left(1 - \frac{|L_t \cap \{(\neg a, \infty)\}|^m}{|L_t \cap \{(\neg a, \infty)\}|^m + \Theta_{\neg a}^m} \cdot \frac{|L_t \cap \{(\neg b, \infty)\}|^m}{|L_t \cap \{(\neg b, \infty)\}|^m + \Theta_{\neg b}^m}\right)\Big\rfloor$$

$$f_6(\Delta\tau, L_t, m) = \left\lfloor \Delta\tau \cdot |L_t \cap \{(z, \infty)\}| \right\rfloor$$

$$\Delta\tau \in \mathbb{R}_+$$

$$m \in \mathbb{N}_+$$

## Simulation Result (MATLAB, P system iteration scheme)

dynamical behavior depicted for $m = 2$, $\Delta\tau = 0.1$, $\Theta_j = 500$, $j \in \{x, y, \neg a, \neg b\}$

$rgx = 10,000$, $rgy = 10,000$, $eg = 10,000$, $x_0 = 0$, $y_0 = 0$, $z_0 = 0$, $a_0 = 0$, $b_0 = 0$

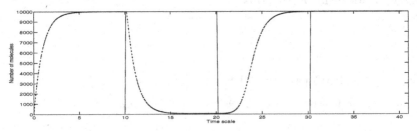

## 3.2 RS Flip-Flop

input:    concentration levels of transcription factors $\overline{S}, \overline{R}$
output:   concentration level of gene product $Q$

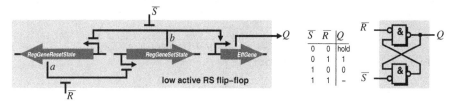

### Ordinary Differential Equations

$$\dot{a} = 1 - \mathrm{h}^+(b, \Theta_b, m) \cdot \mathrm{h}^-(\overline{S}, \Theta_{\overline{S}}, m) - a$$
$$\dot{b} = 1 - \mathrm{h}^+(a, \Theta_a, m) \cdot \mathrm{h}^-(\overline{R}, \Theta_{\overline{R}}, m) - b$$
$$\dot{Q} = \mathrm{h}^+(b, \Theta_b, m) \cdot \mathrm{h}^-(\overline{S}, \Theta_{\overline{S}}, m) - Q$$

### Simulation Result (Copasi, ODE solver)

dynamical behavior depicted for $m = 2$, $\Theta_j = 0.1$, $j \in \{a, b, \overline{R}, \overline{S}\}$

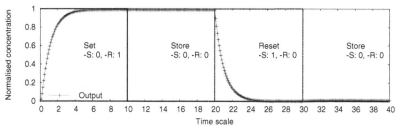

### Corresponding P System

$$\Pi_{\mathrm{Hill,GRNrsff}} = (V_{\mathrm{Genes}}, V_{\mathrm{GeneProducts}}, \Sigma, [_1]_1, L_0, r_1, \ldots, r_6, \mathrm{f}_1, \ldots, \mathrm{f}_6, \Delta\tau, m)$$
$$V_{\mathrm{Genes}} = \{RegGeneResetState, RegGeneSetState, EffGene\}$$
$$V_{\mathrm{GeneProducts}} = \{Q, \neg\overline{S}, \neg\overline{R}, \neg a, \neg b\}$$
$$\Sigma = \{Q\}$$
$$L_0 = \{(RegGeneResetState, rgr), (RegGeneSetState, rgs),$$
$$(EffGene, eg), (Q, q_0), (\neg\overline{S}, ss_0), (\neg\overline{R}, rs_0), (\neg a, a_0), (\neg b, b_0)\}$$

$r_1$ :  $s_1 \ RegGeneResetState \longrightarrow s_1 \ \neg a + s_1 \ RegGeneResetState|_{\neg\overline{S}, \neg b}$

$r_2$ :  $s_2 \ \neg a \longrightarrow \emptyset$

$r_3$ :  $s_3 \ RegGeneSetState \longrightarrow s_3 \ \neg b + s_3 \ RegGeneSetState \ |_{\neg\overline{R}, \neg a}$

$r_4$ :  $s_4 \ \neg b \longrightarrow \emptyset$

$r_5$ :  $s_5 \ EffGene \longrightarrow s_5 \ Q + s_5 \ EffGene \ |_{\neg\overline{S}, \neg b}$

$r_6$ :  $s_6 \ Q \longrightarrow \emptyset$

$$\mathrm{f}_1(\Delta\tau, L_t, m) = \left\lfloor \Delta\tau \cdot |L_t \cap \{(RegGeneResetState, \infty)\}| \cdot \right.$$
$$\left. \left( 1 - \frac{|L_t \cap \{(\neg b, \infty)\}|^m}{|L_t \cap \{(\neg b, \infty)\}|^m + \Theta_{\neg b}^m} \cdot \left( 1 - \frac{|L_t \cap \{(\neg\overline{S}, \infty)\}|^m}{|L_t \cap \{(\neg\overline{S}, \infty)\}|^m + \Theta_{\neg\overline{S}}^m} \right) \right) \right\rfloor$$
$$\mathrm{f}_2(\Delta\tau, L_t, m) = \lfloor \Delta\tau \cdot |L_t \cap \{(\neg a, \infty)\}| \rfloor$$

$$f_3(\varDelta\tau, L_t, m) = \left\lfloor \varDelta\tau \cdot |L_t \cap \{(RegGeneSetState, \infty)\}| \cdot \right.$$
$$\left. \left(1 - \frac{|L_t\cap\{(\neg a,\infty)\}|^m}{|L_t\cap\{(\neg a,\infty)\}|^m + \Theta_{\neg a}^m} \cdot \left(1 - \frac{|L_t\cap\{(\neg\overline{R},\infty)\}|^m}{|L_t\cap\{(\neg\overline{R},\infty)\}|^m + \Theta_{\neg\overline{R}}^m}\right)\right)\right\rfloor$$
$$f_4(\varDelta\tau, L_t, m) = \lfloor \varDelta\tau \cdot |L_t \cap \{(\neg b, \infty)\}| \rfloor$$
$$f_5(\varDelta\tau, L_t, m) = \left\lfloor \varDelta\tau \cdot |L_t \cap \{(EffGene, \infty)\}| \cdot \right.$$
$$\left. \frac{|L_t\cap\{(\neg b,\infty)\}|^m}{|L_t\cap\{(\neg b,\infty)\}|^m + \Theta_{\neg b}^m} \cdot \left(1 - \frac{|L_t\cap\{(\neg\overline{S},\infty)\}|^m}{|L_t\cap\{(\neg\overline{S},\infty)\}|^m + \Theta_{\neg\overline{S}}^m}\right)\right\rfloor$$
$$f_6(\varDelta\tau, L_t, m) = \lfloor \varDelta\tau \cdot |L_t \cap \{(Q, \infty)\}| \rfloor$$
$$\varDelta\tau \in \mathbb{R}_+$$
$$m \in \mathbb{N}_+$$

**Simulation Result** (MATLAB, P system iteration scheme)

dynamical behavior depicted for $m = 2$, $\varDelta\tau = 0.1$, $\Theta_j = 500$, $j \in \{\neg a, \neg b, \neg\overline{R}, \neg\overline{S}\}$
$rgr = 10,000$, $rgs = 10,000$, $eg = 10,000$, $q_0 = 0$, $ss_0 = 0$, $rs_0 = 0$, $a_0 = 0$, $b_0 = 0$

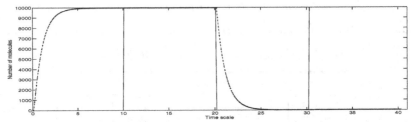

A homologous analytic model of a bistable toggle switch was introduced in [6]. In case of the forbidden input signalling $\overline{S} = 1$, $\overline{R} = 1$, the normalized concentrations of both inhibitors $\neg a$ and $\neg b$ converge to 0.5. By setting or resetting input signalling, the flip-flop restores.

## 4     RS Flip-Flop Validation *in vivo*

In addition to prediction and simulation of GRNs acting as logic gates, we demonstrate the practicability of the RS flip-flop by an experimental study *in vivo*. Resulting output protein data measured over the time course can validate the system model. Following the pioneering implementation of a bistable toggle switch [6], we could confirm its function in a previous study [8]. Two extensions were investigated: Firstly, the effects of IPTG and AHL as appropriate intercellular inducers for flip-flop setting were shown. Secondly, flow cytometry was used to quantitatively measure protein concentrations within the flip-flop implementation. We give a brief overview of experimental setup and results.

### 4.1     Biological Principles and Prerequisites

#### Quorum Sensing and Autoinduction via AHL

In quorum sensing, bacterial species regulate gene expression based on cell-population density [13]. An alteration in gene expression occurs when an intercellular signalling molecule termed autoinducer, produced and released by the bacterial cells reaches a critical concentration. Termed as quorum sensing or autoinduction, this fluctuation in autoinducer concentration is a function of bacterial cell-population density. *Vibrio fischeri*, a well studied bacterium, colonizes the light organs of a variety of marine fishes and squids, where it occurs at very high densities ($10^{10} \frac{cells}{ml}$) and produces light. The two genes essential for cell density regulation of luminescence are: luxI, which codes for an autoinducer synthase; and luxR, which codes for an autoinducer-dependent activator of the luminescence genes. The luxR and luxI genes are adjacent and divergently transcribed, and luxI is the first of seven genes in the luminescence or lux operon. LuxI-type proteins direct AHL synthesis while LuxR-type proteins function as transcriptional regulators that are capable of binding AHL signal molecules. Once formed, LuxR-AHL complexes bind to target promoters of quorum-regulated genes. Quorum sensing is now known to be widespread among both Gram-positive and Gram-negative bacteria.

**Bioluminescence in** *Vibrio fischeri*
Bioluminescence in general is defined as an enzyme catalyzed chemical reaction in which the energy released is used to produce an intermediate or product in an electronically excited state, which then emits a photon. It differs from fluorescence or phosphorescence as it is not depended on light absorbed. The mechanism for gene expression and the structure of the polycistronic message of the lux structural genes in *Vibrio fischeri* have been thoroughly characterized [7]. Briefly, there are two substrates, luciferin, which is a reduced flavin mononucleotide ($FMNH_2$), and a long chain ($7-16$ carbons) fatty aldehyde (RCHO). An external reductant acts via flavin mono-oxygenase oxidoreductase to catalyze the reduction of FMN to $FMNH_2$, which binds to the enzyme and reacts with $O_2$ to form a 4a-peroxy-flavin intermediate. This complex oxidizes the aldehyde to form the corresponding acid (RCOOH) and a highly stable luciferase-hydroxyflavin intermediate in its excited state, which decays slowly to its ground state emitting blue-green light $h\nu$ with a maximum intensity at about 490nm:

$$FMNH_2 + RCHO + O_2 \xrightarrow{\text{lucif.}} FMN + H_2O + RCOOH + h\nu$$

**Transcription Control by LacR and $\lambda$CI Repressor Proteins**
*Escherichia coli* cells repress the expression of the lac operon when glucose is abundant in the growth medium. Only when the glucose level is low and the lactose level is high, the operon is fully expressed. The Lac repressor LacR is a 360 residue long protein that associates into a homotetramer. It contains a helix-turn-helix (HTH) motif through which it interacts with DNA. This interaction represses transcription by hindering association with RNA polymerase and represents an example of combinatorial control widely seen in prokaryotes and eukaryotes. The CI repressor of bacteriophage lambda is the key regulator in lambda's genetic switch, a bistable switch that underlies the phage's ability to efficiently use its two modes of development.

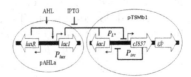

**Fig. 3.** A schematic diagram of an AHL biosensor module interfaced with the genetic toggle switch adapted from [8]. The transgenic artificial GRN consists of a bistable genetic toggle switch [6] which is interfaced with genes from the lux operon of the quorum sensing signalling pathway of *Vibrio fischeri*.

### Flow Cytometry

Flow cytometry refers to the technique where microscopic particles are counted and examined as they pass in a hydro-dynamically focused fluid stream through a measuring point surrounded by an array of detectors. Previously, flow cytometry analyzes were performed by us using a BD LSRII flow cytometer equipped with 405nm, 488nm and 633nm lasers. 488nm laser was used for *gfp* and yellow fluorescent protein (*yfp*) quantification.

### 4.2    Experimental Setup and Implementation

We have shown that an *in vivo* system [8] can potentially be used to mimic a RS flip-flop and have quantified its performance using flow cytometry. The presence or absence of the inducers IPTG or AHL in combination with temperature shift acts as an input signal, see Figure 3. The toggle switch comprising of structural genes for reporter/output proteins fused to promoter regions that are regulated by input signals is visualized as a RS flip-flop. This design endows cells with two distinct phenotypic states: where the $\lambda$CI activity is high and the expression of lacI is low (referred to as high or 1 state), or where the activity of LacR is high and the expression of $\lambda$CI is low (referred to as low or 0 state). *gfp* is expressed only in the high $\lambda$CI/low LacR state.

### 4.3    Results and Discussion

For co-relational purposes, all experiments were conducted with both BL21 and Top10 strains of *Escherichia coli*. The concentration of IPTG used in all the experiments was 2mM and that of AHL was $1\mu$M. Experiments conducted without the use of inducers, lead to an unreliable shifting of the states, signifying the use if inducers in a tightly, mutually regulated circuit. Further experiments conducted to understand the switching dynamics of the circuit revealed that in the current scenario, it was easier to switch from a high to a low state than vice versa. This discrepancy in switching behavior is attributed to the differing modes of elimination of LacR and $\lambda$CI repressor proteins. While switching from low to high state, the repression due to IPTG-bound Lac repressor needs to be overcome by cell growth. Switching from high to low state is effected by immediate thermal degradation of the temperature-sensitive $\lambda$CI. Experiments were

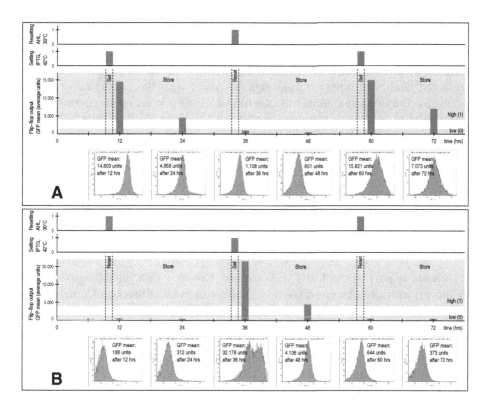

**Fig. 4.** Inducer-dependent switching. Repeated activation and deactivation of the toggle switch based on inducers and temperature. Temperature was switched every 24 hours. Cells were incubated with inducers for 12 hours, followed by growth for 12 hours without inducers, initially kept at 30°C (**A**) and 42°C (**B**). The cells successfully switched states thrice.

also conducted to test the sustainability of states. The plug and play property of the circuit was examined by employing *yfp* as the reporter gene instead of *gfp*. As shown in Figure 4, the circuit could mimic a RS flip-flop. A massive parallelism permissible by the use of large quantities of cells can compensate for the slow speed of switching. Further tests are to be performed to confirm this hypothesis.

# 5   Conclusions

The dynamical behavior of GRNs is able to emulate information processing in terms of performing computations. In order to formalize this capability, we have introduced P systems of the form $\Pi_{\text{Hill}}$ incorporating cooperativity and competitivity between transcription factors based on Hill kinetics. Its transformation to a dedicated iteration scheme for a discrete term rewriting mechanism with variable stoichiometric factors in $\Pi_{\text{Hill}}$ provides a homogeneous approach that allows to compose GRNs towards functional units like computing agents. Examples address computational units (NAND gate, RS flip-flop), each defined by GRN, its ODE model, and the corresponding P system. Simulations of the dynamical behavior quantitatively show the switching characteristics as well as the expected quality of binary output signals. Along with the prediction of GRNs acting as computational units, an experimental study *in vivo* demonstrates their practicability. Although the measurement of the dynamic switching behaviour was condensed to 12 points in time, they approximate the expected course. At the crossroad of modelling, simulation, and verification of biological reaction networks, the potential of amalgamating analytic, stochastic, and algebraic approaches into the P systems framework seems promising for applications in systems biology to explore network functions.

## Acknowledgements

This work is part of the ESIGNET project (Evolving Cell Signalling Networks *in silico*), which has received research funding from the European Community's Sixth Framework Programme (project no. 12789). Further funding from the Federal Ministry of Education and Research (BMBF, grant 0312704A) and from German Research Foundation (DFG, grant DI852/4-1) is acknowledged. We are very grateful to J.J. Collins for providing us with the plasmids and their sequences; to W. Pompe, G. Rödel, K. Ostermann, and L. Brusch from Dresden University of Technology for their scientific support and V. Helms from Saarland University for administrative support.

## References

1. Alon, U.: An Introduction to Systems Biology. Chapman & Hall, Sydney, Australia (2006)
2. Barbacari, N., et al.: Gene Regulatory Network Modelling by Membrane Systems. In: Freund, R., Păun, G., Rozenberg, G., Salomaa, A. (eds.) WMC 2005. LNCS, vol. 3850, pp. 162–178. Springer, Heidelberg (2006)
3. Bernardini, F., et al.: Quorum Sensing P Systems. Theor. Comp. Sci. 371, 20–33 (2007)
4. Busi, N., et al.: Computing with Genetic Gates, Proteins, and Membranes. In: Hoogeboom, H.J., Păun, G., Rozenberg, G., Salomaa, A. (eds.) WMC 2006. LNCS, vol. 4361, pp. 233–249. Springer, Heidelberg (2006)
5. Fontana, F., et al.: Discrete Solutions to Differential Equations by Metabolic P Systems. Theor. Comput. Sci. 372(1), 165–182 (2007)
6. Gardner, T.S., et al.: Construction of a Genetic Toggle Switch in *Escherichia coli*. Nature 403, 339–342 (2000)
7. Hastings, J., et al.: Bacterial Bioluminescence. Annu. Rev. Microbiol. 31, 549–595 (1977)
8. Hayat, S., et al.: Towards *in vivo* Computing: Quantitative Analysis of an Artificial Gene Regulatory Network Behaving as a RS Flip-Flop. In: Proc. Bionetics (2006)

9. Hinze, T., et al.: A Protein Substructure Based P System for Description and Analysis of Cell Signalling Networks. In: Hoogeboom, H.J., Păun, G., Rozenberg, G., Salomaa, A. (eds.) WMC 2006. LNCS, vol. 4361, pp. 409–423. Springer, Heidelberg (2006)
10. Hoops, S., et al.: Copasi: a COmplex PAthway SImulator. Bioinf. 22, 3067–3074 (2006)
11. Manca, V.: Metabolic P Systems for Biomolecular Dynamics. Progress in Natural Sciences 17(4), 384–391 (2006)
12. Mestl, T., et al.: A Mathematical Framework for Describing and Analysing Gene Regulatory Networks. J. Theor. Biol. 176, 291–300 (1995)
13. Miller, M., et al.: Quorum Sensing in Bacteria. Annu. Rev. Microbiol. 55, 165–199 (2001)
14. Păun, A., et al.: Modeling Signal Transduction Using P Systems. In: Hoogeboom, H.J., Păun, G., Rozenberg, G., Salomaa, A. (eds.) WMC 2006. LNCS, vol. 4361, pp. 100–122. Springer, Heidelberg (2006)
15. Păun, G.: Computing with Membranes. J. Comp. Syst. Sci. 61(1), 108–143 (2000)
16. Păun, G.: Membrane Computing: An Introduction. Springer, Berlin (2002)

# Solving Numerical NP-Complete Problems with Spiking Neural P Systems

Alberto Leporati, Claudio Zandron,
Claudio Ferretti, and Giancarlo Mauri

Dipartimento di Informatica, Sistemistica e Comunicazione
Università degli Studi di Milano – Bicocca
Via Bicocca degli Arcimboldi 8, 20126 Milano, Italy
{leporati,zandron,ferretti,mauri}@disco.unimib.it

**Abstract.** Starting from an extended nondeterministic spiking neural P system that solves the SUBSET SUM problem in a constant number of computation steps, recently proposed in a previous paper, we investigate how different properties of spiking neural P systems affect the capability to solve numerical **NP**–complete problems. In particular, we show that by using maximal parallelism we can convert any given integer number from the usual binary notation to the unary form, and thus we can initialize the above P system with the required (exponential) number of spikes in polynomial time. On the other hand, we prove that this conversion cannot be performed in polynomial time if the use of maximal parallelism is forbidden. Finally, we show that if we can choose whether each neuron works in the nondeterministic vs. deterministic and/or in the maximal parallel vs. sequential way, then there exists a *uniform* family of spiking neural P systems that solves the SUBSET SUM problem.

## 1   Introduction

*Spiking neural P systems* (SN P systems, for short) have been introduced in [6] as a new class of distributed and parallel computing devices. They were inspired by *membrane systems* (also known as *P systems*) [11,12,15], in particular by tissue–like P systems [10], and are based on the neurophysiological behavior of neurons sending electrical impulses (*spikes*) along axons to other neurons.

In SN P systems the processing elements are called *neurons*, and are placed in the nodes of a directed graph, called the *synapse graph*. The contents of each neuron consist of a number of copies of a single object type, namely the *spike*. Neurons may also contain *firing* and/or *forgetting* rules. The *firing rules* allow a neuron to send information to other neurons in the form of electrical impulses (also called spikes) which are accumulated at the target cell. The application of the rules depends on the contents of the neuron; in the general case, applicability is determined by checking the contents of the neuron against a regular set associated with the rule. As inspired from biology, when a neuron sends out spikes it becomes "closed" (inactive) for a specified period of time, that reflects the refractory period of biological neurons. During this period, the neuron does not

G. Eleftherakis et al. (Eds.): WMC8 2007, LNCS 4860, pp. 336–352, 2007.
© Springer-Verlag Berlin Heidelberg 2007

accept new inputs and cannot "fire" (that is, emit spikes). Another important feature of biological neurons is that the length of the axon may cause a time delay before a spike arrives at the target. In SN P systems this delay is modeled by associating a delay parameter to each rule which occurs in the system. If no firing rule can be applied in a neuron, there may be the possibility to apply a *forgetting rule*, that removes from the neuron a predefined number of spikes.

Formally, an SN P system of degree $m \geq 1$, as defined in [7], is a construct of the form

$$\Pi = (O, \sigma_1, \sigma_2, \ldots, \sigma_m, syn, in, out),$$

where:

1. $O = \{a\}$ is the singleton alphabet ($a$ is called *spike*);
2. $\sigma_1, \sigma_2, \ldots, \sigma_m$ are *neurons*, of the form $\sigma_i = (n_i, R_i)$, with $1 \leq i \leq m$, where:
   (a) $n_i \geq 0$ is the *initial number of spikes* contained in $\sigma_i$;
   (b) $R_i$ is a finite set of *rules* of the following two forms:
      (1) $E/a^c \to a; d$, where $E$ is a regular expression over $a$, and $c \geq 1$, $d \geq 0$ are integer numbers; if $E = a^c$, then it is usually written in the following simplified form: $a^c \to a; d$;
      (2) $a^s \to \lambda$, for $s \geq 1$, with the restriction that for each rule $E/a^c \to a; d$ of type (1) from $R_i$, we have $a^s \notin L(E)$ (where $L(E)$ denotes the regular language defined by $E$);
3. $syn \subseteq \{1, 2, \ldots, m\} \times \{1, 2, \ldots, m\}$, with $(i, i) \notin syn$ for $1 \leq i \leq m$, is the directed graph of *synapses* between neurons;
4. $in, out \in \{1, 2, \ldots, m\}$ indicate the *input* and the *output* neurons of $\Pi$.

The rules of type (1) are called *firing* (also *spiking*) *rules*, and they are applied as follows. If the neuron $\sigma_i$ contains $k \geq c$ spikes, and $a^k \in L(E)$, then the rule $E/a^c \to a; d \in R_i$ can be applied. The execution of this rule removes $c$ spikes from $\sigma_i$ (thus leaving $k-c$ spikes), and prepares one spike to be delivered to all the neurons $\sigma_j$ such that $(i, j) \in syn$. If $d = 0$, then the spike is immediately emitted, otherwise it is emitted after $d$ computation steps of the system. (Observe that, as usually happens in membrane computing, a global clock is assumed, marking the time for the whole system, hence the functioning of the system is synchronized.) If the rule is used in step $t$ and $d \geq 1$, then in steps $t, t+1, t+2, \ldots, t+d-1$ the neuron is *closed*, so that it cannot receive new spikes (if a neuron has a synapse to a closed neuron and tries to send a spike along it, then that particular spike is lost), and cannot fire new rules. In the step $t+d$, the neuron spikes and becomes open again, so that it can receive spikes (which can be used starting with the step $t+d+1$) and select rules to be fired.

Rules of type (2) are called *forgetting* rules, and are applied as follows: if the neuron $\sigma_i$ contains *exactly* $s$ spikes, then the rule $a^s \to \lambda$ from $R_i$ can be used, meaning that all $s$ spikes are removed from $\sigma_i$. Note that, by definition, if a firing rule is applicable then no forgetting rule is applicable, and vice versa.

In each time unit, if a neuron $\sigma_i$ can use one of its rules, then a rule from $R_i$ must be used. Since two firing rules, $E_1 : a^{c_1} \to a; d_1$ and $E_2 : a^{c_1} \to a; d_2$, can have $L(E_1) \cap L(E_2) \neq \emptyset$, it is possible that two or more rules can be applied in a neuron. In such a case, only one of them is nondeterministically chosen. Thus, the rules are used in the sequential manner in each neuron, but neurons function in parallel with each other.

The *initial configuration* of the system is described by the numbers $n_1, n_2, \ldots,$ $n_m$ of spikes present in each neuron, with all neurons being open. During the computation, a configuration is described by both the number of spikes present in each neuron and by the number of steps to wait until it becomes open (this number is zero if the neuron is already open). A *computation* in a system as above starts in the initial configuration. A positive integer number is given in input to a specified *input neuron*. This number may be encoded in many different ways, for example as the interval of time steps elapsed between the insertion of two spikes into the neuron (note that this is a unary encoding). Other possible encodings are discussed below. To pass from a configuration to another one, for each neuron a rule is chosen among the set of applicable rules, and is executed. The computation proceeds in a sequential way into each neuron, and in parallel among different neurons. Generally, a computation may not halt. However, in any case the output of the system is usually considered to be the time elapsed between the arrival of two spikes in a designated *output cell*. Defined in this way, SN P systems compute functions of the kind $f : \mathbb{N} \to \mathbb{N}$; they can also indirectly compute functions of the kind $f : \mathbb{N}^k \to \mathbb{N}$ by using a bijection from $\mathbb{N}^k$ to $\mathbb{N}$.

As discussed in [7], there are other possibilities to encode natural numbers read from and/or emitted to the environment by SN P systems; for example, we can consider the number of spikes contained in the input and in the output neuron, respectively, or the number of spikes read/produced in a given interval of time. Also, an alternative way to compute a function $f : \mathbb{N}^k \to \mathbb{N}$ is to introduce $k$ natural numbers $n_1, n_2, \ldots, n_k$ in the system by "reading" from the environment a binary sequence $z = 0^b 10^{n_1} 10^{n_2} 1 \ldots 10^{n_k} 10^g$, for some $b, g \geq 0$; this means that the input neuron of $\Pi$ receives a spike in each step corresponding to a digit 1 from the string $z$. Note that we input exactly $k + 1$ spikes, and that this is again a unary encoding. Sometimes we may need to impose that the system outputs exactly two spikes and halts (sometimes after the second spike) hence producing a spike train of the form $0^{b'} 10^r 10^{g'}$, for some $b', g' \geq 0$ and with $r = f(n_1, n_2, \ldots, n_k)$. In what follows we will also consider systems which have $k$ input neurons. For these systems, the input values $n_1, n_2, \ldots, n_k$ will arrive *simultaneously* to the system, each one entering through the corresponding input neuron. Moreover, the input numbers will be sometimes encoded in *binary* form, using the same number of bits in order to synchronize the different parts of the system. For further details, we refer the reader to section 3.

If we do not specify an input neuron (hence no input is taken from the environment) then we use SN P systems in the *generative* mode; we start from the initial configuration, and the distance between the first two spikes of the output neuron (or the number of spikes, etc.) is the result of the computation. Note that generative SN P systems are inherently nondeterministic, otherwise they would always reproduce the same sequence of computation steps, and hence the same output. Dually, we can neglect the output neuron and use SN P systems in the *accepting* mode; for $k \geq 1$, the natural number $n_1, n_2, \ldots, n_k$ are read in input and, if the computation halts, then the numbers are accepted.

In [6] it was shown that generative SN P systems are universal, that is, can generate any recursively enumerable set of natural numbers. Moreover, a characterization of semilinear sets was obtained by spiking neural P systems with a bounded number of spikes in the neurons. These results can also be obtained with even more restricted forms of spiking P systems; for example, [5] shows that at least one of these features can be avoided while keeping universality: time delay (refractory period) greater than 0, forgetting rules, outdegree of the synapse graph greater than 2, and regular expressions of complex form. These results have been further extended in [3], where it is shown that universality is kept even if we remove some combinations of two of the above features. Finally, in [13] the behavior of spiking neural P systems on infinite strings and the generation of infinite sequences of 0 and 1 was investigated, whereas in [1] spiking neural P systems were studied as language generators (over the binary alphabet $\{0, 1\}$).

We define the *description size* of an SN P system $\Pi$ as the number of bits which are necessary to describe it. Since the alphabet $O$ is fixed, no bits are necessary to define it. In order to represent *syn* we need at most $m^2$ bits, whereas we can represent the values of *in* and *out* by using $\log m$ bits each. For every neuron $\sigma_i$ we have to specify a natural number $n_i$ and a set $R_i$ of rules. For each rule we need to specify its type (firing or forgetting), which can be done with 1 bit, and in the worst case we have to specify a regular expression and two natural numbers. If we denote by $N$ the maximum natural number that appears in the definition of $\Pi$, $R$ the maximum number of rules which occur in its neurons, and $S$ the maximum size required by the regular expressions that occur in $\Pi$ (more on this later), then we need a maximum of $\log N + R(1 + S + 2 \log N)$ bits to describe every neuron of $\Pi$. Hence, to describe $\Pi$ we need a total of $m^2 + 2 \log m + m\big(\log N + R(1 + S + 2 \log N)\big)$ bits. Note that this quantity is polynomial with respect to $m$, $R$, $S$ and $\log N$. Since the regular languages determined by the regular expressions that occur in the system are *unary* languages, the strings of such languages can be bijectively identified with their lengths. Hence, when writing the regular expression $E$, instead of writing unions, concatenations and Kleene closures among strings we can do the same by using the lengths of such strings. (Note that, when concatenating two languages $L_1$ and $L_2$ represented in this way, the lengths in $L_1$ are *summed* with the lengths of $L_2$ by combining them in all possible ways). In this way we obtain a representation of $E$ which is *succinct*, that is, exponentially more compact than the usual representation of regular expressions. As we have seen in [8], this succinct representation yields some difficulties when we try to simulate a deterministic accepting SN P system that contains general regular expressions, by a deterministic Turing machine. However, as shown in [5] and [3], it is possible to restrict our attention to particularly simple regular expressions, without loosing computational completeness. For these expressions, the membership problem (is a given string into the language generated by the regular expression?) is polynomial also when representing the instances in succinct form, and thus they do not yield problems when simulating the system with a deterministic Turing machine.

In what follows it will be convenient to consider also the following slightly extended version of SN P systems. Precisely, we will allow rules of the type $E/a^c \to a^p; d$, where $c \geq 1$, $p \geq 0$ and $d \geq 0$ are integer numbers. The semantics of this kind of rules is similar to that of standard rules, the difference being that now $p$ spikes (instead of one) are emitted by the neuron. As before, a closed neuron does not receive spikes from other neurons, and does not apply any rule. If $p = 0$, then we obtain a forgetting rule as a particular case of our general rules. Also in the extended SN P systems it may happen that, given two rules $E_1/a^{c_1} \to a^{p_1}; d_1$ and $E_2/a^{c_2} \to a^{p_2}; d_2$, if $L(E_1) \cap L(E_2) \neq \emptyset$ then for some contents of the neuron both the rules can be applied. In such a case, one of them is nondeterministically chosen. Note that we do not require that forgetting rules are applied only when no firing rule can be applied. We say that the system is *deterministic* if, for every neuron that occurs in the system, any two rules $E_1/a^{c_1} \to a^{p_1}; d_1$ and $E_2/a^{c_2} \to a^{p_2}; d_2$ in the neuron are such that $L(E_1) \cap L(E_2) = \emptyset$. This means that, for any possible contents of the neuron, at most one of the rules that occur in the neuron may be applied.

Input and output encodings are defined just like in traditional SN P systems. Finally, the *description size* of an extended SN P system is defined exactly as we have done for standard systems, the only difference being that now we require (at most) three natural numbers to describe a rule.

In [8] we have shown that by using our extended version of SN P systems, it is possible to solve any given instance of SUBSET SUM in *constant* time by exploiting the power of nondeterminism. The solution is given in the so called *semi–uniform* setting, that is, for every fixed instance of SUBSET SUM a specific SN P system that solves it is built. In particular, the rules of the system and the number of spikes which occur in the initial configuration depend upon the instance to be solved. A drawback of this solution is that in general the number of spikes needed to initialize the system is exponential with respect to the usually agreed instance size of SUBSET SUM. However, in this paper we show that this preparation can be performed in polynomial time by traditional SN P systems that, endowed with the power of maximal parallelism, read from the environment the $k$-bit integer numbers $v_1, v_2, \ldots, v_n$ encoded in binary and produce $v_1, v_2, \ldots, v_n$ spikes, respectively, in $n$ specified neurons. We also prove that this operation cannot be performed in polynomial time if the use of maximal parallelism is forbidden. Then we design an SN P system that performs the opposite conversion: it takes a given ($k$-bit) natural number $N$ of spikes occurring in a certain neuron, and produces the coefficients of the binary encoding of $N$ in $k$ predefined neurons. Thanks to these two modules, that allow us to move from binary to unary encoding and back, we finally design a *uniform* family $\{\Pi(\langle n, k \rangle\}_{n,k \in \mathbb{N}}$ of SN P systems, where $\Pi(\langle n, k \rangle)$ solves all possible instances $(\{v_1, v_2, \ldots, v_n\}, S)$ of SUBSET SUM such that all $v_i$ and $S$ are $k$-bit natural numbers. As we will see, the construction of $\Pi(\langle n, k \rangle)$ relies upon the assumption that different subsystems can work under different regimes: deterministic vs. nondeterministic, and sequential vs. maximally parallel.

The rest of this paper is organized as follows. In section 2 we recall from [8] how the **NP**–complete problem SUBSET SUM can be solved in constant time by exploiting nondeterminism in our extended SN P systems. In section 3 we convert natural numbers from binary notation to the unary form through maximally parallel SN P systems, and we use such a translation as an initialization stage to solve SUBSET SUM. In section 4 we perform also the opposite conversion, and we design a family of SN P systems that solves SUBSET SUM in a uniform way (according to the above definition). Section 5 concludes the paper.

## 2   Solving Numerical NP–Complete Problems with Extended Spiking Neural P Systems

Let us start by recalling the nondeterministic extended SN P system introduced in [8] to solve the **NP**–complete problem SUBSET SUM in a *constant* number of computation steps. The SUBSET SUM problem can be defined as follows.

*Problem 1.* NAME: SUBSET SUM.

- INSTANCE: a (multi)set $V = \{v_1, v_2, \ldots, v_n\}$ of positive integer numbers, and a positive integer number $S$.
- QUESTION: is there a sub(multi)set $B \subseteq V$ such that $\sum_{b \in B} b = S$?

If we allow to nondeterministically choose among the rules which occur in the neurons, then the extended SN P system depicted in Figure 1 solves any given instance of SUBSET SUM in a constant number of steps. We emphasize the fact that such a solution occurs in the *semi-uniform* setting, that is, for every instance of SUBSET SUM we build an SN P system that specifically solves that instance.

Let $(V = \{v_1, v_2, \ldots, v_n\}, S)$ be the instance of SUBSET SUM to be solved. In the initial configuration of the system, the leftmost neurons contain (from top to bottom) $v_1, v_2, \ldots, v_n$ spikes, respectively, whereas the rightmost neurons contain zero spikes each. In the first step of computation, in each of the leftmost neurons it is nondeterministically chosen whether to include or not the element $v_i$ in the (candidate) solution $B \subseteq V$; this is accomplished by nondeterministically choosing among one rule that forgets $v_i$ spikes (in such a case, $v_i \notin B$) and one rule that propagates $v_i$ spikes to the rightmost neurons. At the beginning of the second step of computation a certain number $N = |B|$ of spikes, that corresponds to the sum of the $v_i$ which have been chosen, occurs in the rightmost neurons. We have three possible cases:

- $N < S$: in this case neither the rule $a^*/a^S \to a; 0$ nor the rule $a^*/a^{S+1} \to a; 1$ (which occur in the neuron at the top and at the bottom of the second layer, respectively) fire, and thus no spike is emitted to the environment;
- $N = S$: only the rule $a^*/a^S \to a; 0$ fires, and emits a single spike to the environment. No further spikes are emitted;

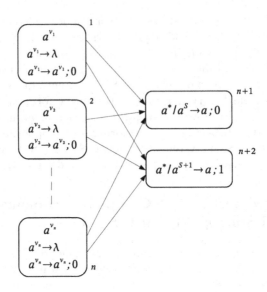

**Fig. 1.** A nondeterministic extended SN P system that solves the SUBSET SUM problem in constant time

- $N > S$: both the rules $a^*/a^S \to a; 0$ and $a^*/a^{S+1} \to a; 1$ fire. The first rule immediately sends one spike to the environment, whereas the second rule sends another spike at the next computation step (due to the delay associated with the rule).

Hence, by counting the number of spikes emitted to the environment at the second and third computation steps we are able to read the solution of the given instance of SUBSET SUM: the instance is positive if and only if a single spike is emitted.

The proposed system is generative; its input (the instance of SUBSET SUM to be solved) is encoded in the initial configuration. We stress once again that the ability to solve SUBSET SUM in constant time derives from the fact that the system is nondeterministic. As it happens with Turing machines, nondeterminism can be interpreted in two ways: (1) the system "magically" chooses the correct values $v_i$ (if they exist) that allow to produce a single spike in output, or (2) at least one of the possible computations produces a single spike in output.

The formal definition of the extended (generative) SN P system depicted in Figure 1 is as follows:

$$\Pi = (\{a\}, \sigma_1, \ldots, \sigma_{n+2}, syn, out),$$

where:

- $\sigma_i = (v_i, \{a^{v_i} \to \lambda, \ a^{v_i} \to a^{v_i}; 0\})$ for all $i \in \{1, 2, \ldots, n\}$;
- $\sigma_{n+1} = (0, \{a^*/a^S \to a; 0\})$;
- $\sigma_{n+2} = (0, \{a^*/a^{S+1} \to a; 1\})$;

- $syn = \bigcup_{i=1}^{n}\{(i, n + 1), (i, n + 2)\}$;
- $out = 0$ indicates that the output is sent to the environment.

However, here we are faced with a problem that we have already met in [9], and that we will meet again in the rest of the paper. In order to clearly expose the problem, let us consider the following algorithm that solves SUBSET SUM using the well known Dynamic Programming technique [2]. In particular, the algorithm returns 1 on positive instances, and 0 on negative instances.

SUBSET SUM($\{v_1, v_2, \ldots, v_n\}, S$)

```
for j ← 0 to S
    do M[1, j] ← 0
M[1, 0] ← M[1, v₁] ← 1
for i ← 2 to n
    do for j ← 0 to S
        do M[i, j] ← M[i − 1, j]
            if j ≥ vᵢ and M[i − 1, j − vᵢ] > M[i, j]
                then M[i, j] ← M[i − 1, j − vᵢ]
return M[n, S]
```

In order to look for a subset $B \subseteq V$ such that $\sum_{b \in B} b = S$, the algorithm uses an $n \times (S + 1)$ matrix $M$ whose entries are from $\{0, 1\}$. It fills the matrix by rows, starting from the first row. Each row is filled from left to right. The entry $M[i, j]$ is filled with 1 if and only if there exists a subset of $\{v_1, v_2, \ldots, v_i\}$ whose elements sum up to $j$. The given instance of SUBSET SUM is thus a positive instance if and only if $M[n, S] = 1$ at the end of the execution.

Since each entry is considered exactly once to determine its value, the time complexity of the algorithm is proportional to $n(S + 1) = \Theta(nS)$. This means that the difficulty of the problem depends on the value of $S$, as well as on the magnitude of the values in $V$. In fact, let $K = max\{v_1, v_2, \ldots, v_n, S\}$. If $K$ is polynomially bounded with respect to $n$, then the above algorithm works in polynomial time. On the other hand, if $K$ is exponential with respect to $n$, say $K = 2^n$, then the above algorithm may work in exponential time and space. This behavior is usually referred to in the literature by telling that SUBSET SUM is a *pseudo–polynomial* **NP**–complete problem.

The fact that in general the running time of the above algorithm is not polynomial can be immediately understood by comparing its time complexity with the instance size. The usual size for the instances of SUBSET SUM is $\Theta(n \log K)$, since for conciseness every "reasonable" encoding is assumed to represent each element of $V$ (as well as $S$) using a string whose length is $O(\log K)$. Here all logarithms are taken with base 2. Stated differently, the size of the instance is usually considered to be the number of bits which must be used to represent in binary $S$ and all the integer numbers which occur in $V$. If we would represent such numbers using the unary notation, then the size of the instance would be $\Theta(nK)$. But in this case we could write a program which first converts the instance in binary form and then uses the above algorithm to solve the problem in polynomial time with respect to the new instance size. We can thus conclude that the difficulty of a numerical **NP**–complete problem depends also on the measure of the instance size we adopt.

The problem we mentioned above concerning the SN P system depicted in Figure 1 is that the rules $a^{v_i} \to \lambda$ and $a^{v_i} \to a^{v_i}; 0$ which occur in the leftmost neurons, as well as those that occur in the rightmost neurons, check for the existence of a number of spikes which may be exponential with respect to the usually agreed instance size of SUBSET SUM. Moreover, to initialize the system the user has to place a number of objects which may also be exponential. This is not fair, because it means that the SN P system that solves the **NP**–complete problem has in general an exponential size with respect to the binary string which is used to describe it; an exponential effort is thus needed to build and initialize the system, that easily solves the problem by working in unary notation (hence in polynomial time with respect to the size of the system, but not with respect to its *description size*). This problem is in some aspects similar to what has been described in [9], concerning traditional P systems that solve numerical **NP**–complete problems.

# 3   Solving SUBSET SUM with Inputs Encoded in Binary

Similarly to what we have done in [9], in this section we show that the ability of the SN P system depicted in Figure 1 to solve SUBSET SUM does not derive from the fact that the system is initialized with an exponential number of spikes, at least if we allow the application of rules in the maximal parallel way.

In this paper, maximal parallelism is intended exactly as in traditional P systems. Since in SN P systems we have only one kind of objects (the spike), this means that at every computation step the (multi)set of rules to be applied in a neuron is determined as follows. Let $k$ denote the number of spikes contained in the neuron. First, one rule is nondeterministically chosen among those which can be applied. If such a rule consumes $c$ spikes, then the selection process is repeated to the remaining $k - c$ spikes, until no rule can be applied. Note that a rule may eventually be chosen many times, and thus at the end of the process we obtain a multiset of rules. A little technical difficulty is given by the fact that the chosen rules may have different delays; hence, we define the delay associated with a multiset of rules as the maximum of the delays that appear in the rules. However, for our purposes it will suffice to consider maximally parallel neurons that contain just one rule. Hence, the process with which the neuron chooses the rules to be applied is unimportant, and no problems arise with the delays: at every computation step the only existing rule is chosen, and is applied as many times as possible (i.e., maximizing the number of spikes which are consumed).

Consider the SN P system depicted in Figure 2, in which all the neurons work in the maximal parallel way. Assume that a sequence of spikes comes from the environment, during $k$ consecutive time steps. Such spikes can be considered as the binary encoding of a $k$-bit natural number $N$, by simply interpreting as 1 (resp., 0) the presence (resp., absence) of a spike in each time step. The system

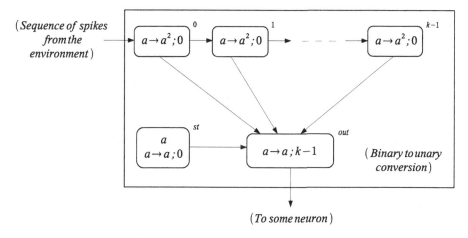

**Fig. 2.** A maximally parallel SN P system that converts a binary encoded positive integer number to unary form

works as follows. In the first step, the most significant bit of $N$ enters into the neuron labeled with 0. Simultaneously, neuron $st$ fires and sends a spike to neuron $out$, that will contain the resulting unary encoding of $N$. This is done in order to close such a neuron, so that it does not receive the intermediate results produced by neurons $0, 1, \ldots, k-1$ during the conversion. During the next $k-1$ steps, all subsequent bits of $N$ enter into the system. Neurons $0, 1, \ldots, k-1$ act as a shift register, and they duplicate every spike before sending both copies to the neighboring neuron. In this way, since rules are applied in the maximal parallel way, at the end of the $k$-th step each neuron $j$, with $j \in \{0, 1, \ldots, k-1\}$, will contain $2^j$ spikes if the $j$-th bit of $N$ is 1, otherwise it will contain 0 spikes. At the $(k+1)$-th step, neuron $out$ becomes open again, and receives exactly $N$ spikes. Two little annoying details are that this neuron emits a "spurious" spike at the $(k+1)$-th computation step, and that it becomes again closed for further $k-1$ time steps. The first spike emitted from the subsystem has obviously to be ignored, whereas during the $(2k)$-th step neuron $out$ emits the $N$ spikes we are interested in. Note that this module can be used only once, since neuron $st$ initially contains just one spike. By making neuron $st$ work in the sequential mode (instead of the maximally parallel mode), and slightly complicating the structure of the system, we can also convert a sequence of $n$ numbers arriving from the environment in $n \cdot k$ consecutive time steps.

By looking at Figure 3, we can see that for any instance $(\{v_1, v_2, \ldots, v_n\}, S)$ of SUBSET SUM it is possible to build a maximally parallel nondeterministic SN P system that solves it as follows. During the first $k$ computation steps, the system reads $n$ sequences of spikes, each one encoding in binary the natural number $v_i$. Each sequence goes to an SN subsystem which performs the conversion from binary to unary, as illustrated in Figure 2. Thus in the $(2k)$-th step, for all $i \in \{1, 2, \ldots, n\}$, $v_i$ spikes reach the neuron labeled with $v_i$. At the next step, each of these neurons *nondeterministically* decides whether to propagate the

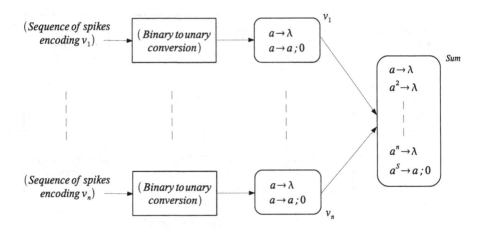

**Fig. 3.** A nondeterministic SN P system that solves the SUBSET SUM problem by working in the maximal parallel way (but for the neuron $Sum$)

spikes it has received, or to delete them. Hence, the rules of neurons $v_i$ are applied not only in the maximal parallel way, but also in a nondeterministic way (in the sense that one of the two rules is nondeterministically chosen, and then is applied in the maximal parallel way). In step $2k + 2$, the neuron labeled with $Sum$ checks whether the number of spikes it has gathered is equal to $S$; if so, it fires one spike to the environment, thus signalling that the given instance of SUBSET SUM is positive. Conversely, the instance is negative if and only if no spike is emitted from the system during the $(2k + 2)$-nd computation step. The forgetting rules which occur in neuron $Sum$ are needed so that at step $k + 2$ all the spurious spikes that (eventually) reach the neuron (coming from the modules that have performed the conversions from binary to unary) are removed from the system, and are not added to the spikes that arrive at step $2k+1$. Of course, here we are assuming that $S > n$; if this is not the case, then the rules must be modified accordingly. Note that neuron $Sum$ is deterministic, and works in the sequential way. We also observe that, if desired, we can use two neurons instead of one in the last layer of the system, as we have done in Figure 1. The first neuron would be just like $Sum$, the only difference being that the rule $a^S \to a; 0$ becomes $a^*/a^S \to a; 0$. The second neuron would contain the same forgetting rules as $Sum$, and the firing rule $a^*/a^{S+1} \to a; 1$ instead of $a^S \to a; 0$. In this way, the instance would be signalled as positive if and only if a single spike is emitted during the steps $2k + 2$ and $2k + 3$.

This solution to the SUBSET SUM problem is still semi–uniform: a single system is able to solve all the instances that have the same value of $S$, and in which all $v_i$ are $k$-bit numbers. A way to make the system uniform would be to read from the environment also the value of $S$, encoded in binary form, and send a corresponding number of spikes to a predefined neuron. The problem would thus reduce to comparing with $S$ the number of spikes obtained by nondeterministically choosing some of the $v_i$. In the next section we will operate in a similar way; however, instead of comparing the contents of two neurons, expressed in unary form, we will operate as follows: we will keep $S$ in binary form, and we will convert the sum of $v_i$ from unary to binary. In this way, the problem to compare $S$ with the sum of $v_i$ is reduced to a bit-by-bit comparison.

Before doing all this, let us show that the conversion from binary to unary of a given natural number cannot be performed in polynomial time without using maximal parallelism. Let $\Pi$ be a deterministic SN P system that works in the sequential way: all the neurons compute in parallel with respect to each other, but in each neuron only one rule is chosen and applied at every computation step. To be precise, even if the contents of the neuron would allow to apply the chosen rule many times (such as it happens, for example, with the rule $a^*/a \rightarrow a^2; 0$ and five spikes occurring in the neuron), only one instance of the rule is applied (in the example, one spike is consumed and two spikes are produced). Without loss of generality, we can assume that the regular expressions that occur in $\Pi$ have the form $a^i$ with $i \leq 3$ or $a(aa)^+$, which suffice to obtain computationally complete SN P systems [5]. Let $m$ be the number of neurons in $\Pi$, and let $t(k)$ be the polynomial number of steps needed by $\Pi$ to convert the $k$-bit natural number $N$ given in input from the binary to the unary form. Moreover, let $Q$ be the maximum number of spikes produced by any rule of $\Pi$. Since in the worst case every neuron is connected with every other neuron, the total number of spikes occurring in the system is incremented by at most $mQ$ units during each computation step. If we denote by $M$ the number of spikes occurring in the initial configuration, then after $t(k)$ computation steps the number of spikes in the system will be at most $M + mQt(k)$. This quantity is polynomial with respect to both the number of steps and the description size of $\Pi$, and thus it cannot cover the exponential gap that exists between the number of objects needed to represent $N$ in binary and in unary form.

## 4   A Uniform Family of SN P Systems for SUBSET SUM

Let us present now a uniform family $\{\Pi(\langle n, k \rangle)\}_{n,k \in \mathbb{N}}$ of SN P systems such that for every $n$ and $k$ in $\mathbb{N}$, the system $\Pi(\langle n, k \rangle)$ solves all possible instances $(\{v_1, v_2, \ldots, v_n\}, S)$ of SUBSET SUM in which $v_1, v_2, \ldots, v_n$ and $S$ are all $k$-bit natural numbers.

As told in the previous section, we first need a subsystem that allows to convert natural numbers from the unary to the binary form. Consider the system depicted in Figure 4. All the neurons work in the maximal parallel way. Initially, neuron $in$ contains $N$ spikes, where $N$ is the $k$-bit natural number we want to convert. In the first computation step, all the spikes contained in neuron $in$ are sent to neuron 0 (thus entering into the subsystem), thanks to the rule $a \rightarrow a; 0$ applied in the maximal parallel way. In the second step, rule $a^2 \rightarrow a; 0$ in neuron 0 halves the number of spikes (indeed, computing an integer division by 2) and sends the result to neuron 1. If the initial number of spikes was even, then in neuron 0 no spikes are left; instead, if the initial number of spikes was odd, then exactly one spike will remain in neuron 0. Hence, the number of spikes remaining

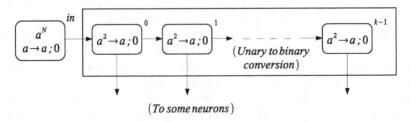

**Fig. 4.** A maximally parallel SN P system that converts a unary encoded positive integer number to binary form

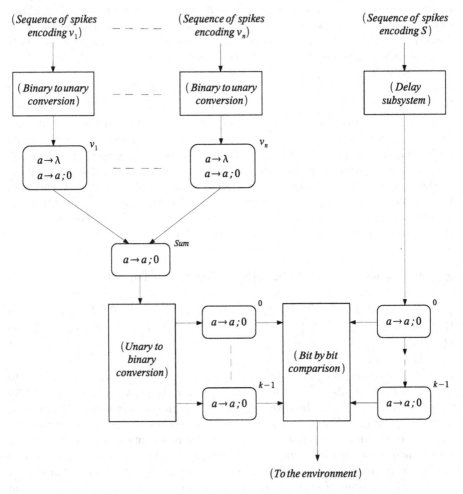

**Fig. 5.** The uniform SN P system $\Pi(\langle n, k \rangle)$ that solves all instances of SUBSET SUM composed by $k$-bit natural numbers

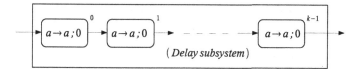

**Fig. 6.** An SN P system that delays of $k$ steps the sequence of spikes given in input

in neuron 0 is equal to the value of the least significant bit of the binary encoding of $N$. The computation proceeds in a similar way during the next $k-1$ steps; in each step, the next bit (from the least significant to the most significant) of the binary encoding of $N$ is computed. Note that the bits that have already been computed are unaffected by subsequent computation steps. After $k$ computation steps, the neurons labeled with $0, 1, \ldots, k-1$ contain all the bits of the binary encoding of $N$. In order to use such bits, we can connect these neurons to other $k$ neurons, which should be kept closed during the conversion by means of a trick similar to that used in Figure 2.

The SN P system $\Pi(\langle n, k \rangle)$ that solves all the instances $(\{v_1, v_2, \ldots, v_n\}, S)$ of SUBSET SUM which are composed by $k$-bit natural numbers is depicted (in a schematic way) in Figure 5. The sequences of spikes that encode $v_1, v_2, \ldots, v_n$ and $S$ in binary form arrive simultaneously from the environment, and enter into the system from the top. The values $v_1, \ldots, v_n$ are first converted to unary and then some of them are summed, as before; the sequence of bits in $S$, instead, is just delayed (using the subsystem depicted in Figure 6) so that it arrives in the "Bit by bit comparison" subsystem simultaneously with the binary representation of the sum of the $v_i$. Such a binary representation is obtained through the subsystem depicted in Figure 4. The bit-by-bit comparison subsystem (depicted in Figure 7) emits a spike if and only if all the bits of the two integer numbers given in input match, that is, if and only if the two numbers are equal. If we denote by $x = \sum_{i=0}^{k-1} x_i 2^i$ and $y = \sum_{i=0}^{k-1} y_i 2^i$ the numbers to be compared, the subsystem computes the following boolean function:

$$\text{COMPARE}(x_0, \ldots, x_{k-1}, y_0, \ldots, y_{k-1}) = \bigwedge_{i=0}^{k-1} \left( \neg (x_i \oplus y_i) \right) = \neg \left( \bigvee_{i=0}^{k-1} (x_i \oplus y_i) \right)$$

where $\oplus$ denotes the logical XOR operation. The subsystem works as follows. Bits $x_i$ and $y_i$ are XORed by the neurons depicted on the top of Figure 7. The neuron labeled with $\vee$ computes the logical OR of its inputs: precisely, it emits one spike if and only if at least one spike enters into the neuron. Neuron *out* receives the output produced by $\vee$ and computes its logical negation (NOT). In order to be able to produce one spike if no spikes come from *out*, we use two auxiliary neurons that send to *out* one spike at every computation step. The number of neurons, as well as the total number of rules, used by $\Pi(\langle n, k \rangle)$ is polynomial with respect to $n$ and $k$.

We conclude by observing that the output of the SN P system depicted in Figure 5 has to be observed exactly after $3k + 6$ computation steps. First of

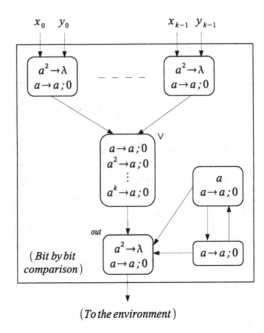

**Fig. 7.** A standard SN P system that compares two $k$-bit natural numbers

all, the bit-by-bit comparison subsystem emits one spike at every computation step if the two $k$-bit integer numbers given in input are both zero. Moreover, the conversion from binary to unary of $v_1, v_2, \ldots, v_n$ produces some spurious spikes before emitting the result. These spurious spikes are added in neuron *sum*, and the result of this addition is first converted to binary and then sent to the comparison subcircuit. Hence, we have to carefully calibrate the delay subsystem so that this value does not interfere with the bits of $S$, that will arrive to the comparison subsystem only later. From a direct inspection of the system in Figure 5, it is easily seen that the correct delay to be applied is equal to $3k + 2$ steps.

## 5    Conclusions

In this paper we have continued the study concerning the computational power of SN P systems, started in [8]. In particular, by slightly extending the original definition of SN P systems given in [6] and [7] we have shown that by exploiting nondeterminism it is possible to solve numerical **NP**–complete problems such as SUBSET SUM and PARTITION (which can be considered as a particular case of SUBSET SUM).

However, a drawback of this solution is that the system may require to specify an exponential number of spikes both when defining the rules and when describing the contents of the neurons in the initial configuration. Hence, we have shown that the numbers $v_1, v_2, \ldots, v_n$ occurring in the instance of SUBSET SUM can be given to the system in binary form, and subsequently converted to the unary form in polynomial time. In this way we have proved that the capability of the above system to solve SUBSET SUM does not derive from the fact that it requires an exponential effort to be initialized.

The new SN P system thus obtained still provides a *semi–uniform* solution, since for each instance of the problem we need to build a specifically designed SN P system to solve it. Thus, we have finally proposed a family $\{\Pi(\langle n, k \rangle)\}_{n,k \in \mathbb{N}}$ of SN P systems such that for all $n, k \in \mathbb{N}$, $\Pi(\langle n, k \rangle)$ solves all the instances $(\{v_1, v_2, \ldots, v_n\}, S)$ of SUBSET SUM such that $v_1, v_2, \ldots, v_n$ and $S$ are all $k$-bit natural numbers. This solution assumes that for each neuron (or, at least, for each subsystem) it is possible to choose whether such a neuron (resp, subsystem) works in a deterministic vs. nondeterministic way, and in the sequential vs. the maximally parallel way.

In [8] we have also studied the computational power of *deterministic* accepting SN P systems working in the sequential way. In particular, we have shown that they can be simulated by deterministic Turing machines with a polynomial slowdown. This means that they are not able to solve **NP**–complete problems in polynomial time unless $\mathbf{P} = \mathbf{NP}$, a very unlikely situation. In future work, we will address the study of the computational power of deterministic accepting SN P systems working in the maximally parallel way.

## Acknowledgments

We gratefully thank Gheorghe Păun for introducing the authors to the stimulating subject of spiking neural P systems, and for asking us a "Milano theorem" (in the spirit of [14]) about their computational power, during the Fifth Brainstorming Week on Membrane Computing, held in Seville from January 29th to February 2nd, 2007.

We are also truly indebted with Mario de Jesús Pérez-Jiménez, as well as with an anonymous referee, for stimulating observations and suggestions made on a previous version of this paper.

This research was partially funded by Università degli Studi di Milano–Bicocca — FIAR 2006.

## References

1. Chen, H., Freund, R., Ionescu, M., Păun, G., Pérez-Jiménez, M.J.: On String Languages Generated by Spiking Neural P Systems. In: Gutiérrez-Naranjo, M.A., Păun, G., Riscos-Núñez, A., Romero-Campero, F.J. (eds.) Fourth Brainstorming Week on Membrane Computing, vol. I, RGCN Report 02/2006, Research Group on Natural Computing, Sevilla University, pp. 169–194. Fénix Editora (2006)
2. Cormen, T.H., Leiserson, C.H., Rivest, R.L.: Introduction to Algorithms. MIT Press, Boston (1990)

3. García-Arnau, M., Peréz, D., Rodríguez-Patón, A., Sosík, P.: Spiking Neural P Systems: Stronger Normal Forms. In: Gutiérrez-Naranjo, M.A., Păun, G., Romero-Jiménez, A., Riscos-Núñez, A. (eds.) Fifth Brainstorming Week on Membrane Computing, RGCN Report 01/2007, Research Group on Natural Computing, Sevilla University, pp. 157–178. Fénix Editora (2007)
4. Garey, M.R., Johnson, D.S.: Computers and Intractability. A Guide to the Theory on NP–Completeness. W.H. Freeman and Company, New York (1979)
5. Ibarra, O.H., Păun, A., Păun, G., Rodríguez-Patón, A., Sosík, P., Woodworth, S.: Normal Forms for Spiking Neural P Systems. Theoretical Computer Science 372(2-3), 196–217 (2007)
6. Ionescu, M., Păun, G., Yokomori, T.: Spiking Neural P Systems. Fundamenta Informaticae 71(2-3), 279–308 (2006)
7. Ionescu, M., Păun, A., Păun, G., Pérez-Jiménez, M.J.: Computing with Spiking Neural P Systems: Traces and Small Universal Systems. In: Mao, C., Yokomori, T. (eds.) DNA Computing. LNCS, vol. 4287, pp. 1–16. Springer, Heidelberg (2006)
8. Leporati, A., Zandron, C., Ferretti, C., Mauri, G.: On the Computational Power of Spiking Neural P Systems. In: Gutiérrez-Naranjo, M.A., et al. (eds.) Fifth Brainstorming Week on Membrane Computing, Research Group on Natural Computing, Sevilla University, pp. 227–245. Fénix Editora (2007)
9. Leporati, A., Zandron, C., Gutiérrez-Naranjo, M.A.: P Systems with Input in Binary Form. International Journal of Foundations of Computer Science 17(1), 127–146 (2006)
10. Martín-Vide, C., Pazos, J., Păun, G., Rodríguez-Patón, A.: A New Class of Symbolic Abstract Neural Nets: Tissue P Systems. In: Ibarra, O.H., Zhang, L. (eds.) COCOON 2002. LNCS, vol. 2387, pp. 290–299. Springer, Heidelberg (2002)
11. Păun, G.: Computing with Membranes. Journal of Computer and System Sciences 61, 108–143 (2000), see also Turku Centre for Computer Science — TUCS Report No. 208 (1998), available at:
    http://www.tucs.fi/Publications/techreports/TR208.php
12. Păun, G.: Membrane Computing. An Introduction. Springer, Berlin (2002)
13. Păun, G., Pérez-Jiménez, M.J., Rozenberg, G.: Infinite Spike Trains in Spiking Neural P Systems (submitted for publication)
14. Zandron, C., Ferretti, C., Mauri, G.: Solving NP–Complete Problems Using P Systems with Active Membranes. In: Antoniou, I., Calude, C.S., Dinneen, M.J. (eds.) Unconventional Models of Computation, pp. 289–301. Springer, London (2000)
15. The P systems Web page: http://psystems.disco.unimib.it/

# Towards a Complete Covering of SBML Functionalities

Tommaso Mazza

'Magna Græcia' University of Catanzaro, Italy
t.mazza@unicz.it

**Abstract.** The complexity of biological systems is at times made worse by the diversity of ways in which they are described: the organic evolution of the science over many years has led to a myriad of conventions. This confusion is reflected by the in-silico representation of biological models, where many different computational paradigms and formalisms are used in a variety of software tools.

The Systems Biology Markup Language (SBML) is an attempt to overcome this issue and aims to simplify the exchange of information by imposing a standardized way of representing models. The success of the idea is attested to by the fact that more than 110 software tools currently support SBML in one form or another.

This work focuses on the translation of the Cyto-Sim simulation language (based on a discrete stochastic implementation of P systems) to SBML. We consider the issues both from the point of view of the employed software architecture and from that of the mapping between the features of the Cyto-Sim language and those of SBML.

## 1 Introduction

Nowadays, very few common exchange formats exist. We face difficulties to exchange models among different analysis and simulation tools. Therefore, taking advantage of the different tools power and capabilities is the main issue among scientists.

To overcome this issue, in March 2001, a first step was taken. During the *First International Symposium on Computational Cell Biology*, (Massachussetts, USA), Michael Hucka presented a new simple, well-supported and with textual substrate (XML) language adding components that reflect the natural conceptual constructs used by modelers in the domain, **SBML: Systems Biology Markup Language**. SBML was intended to be a common exchange format for transferring network models among tools, even if it may not capture everything represented by every tool (lossy transformation).

Inspired to *CellML*, the SBML community immediately begun a joint work with the CellML community with the aim to bring both markup languages together. The fruit beard by this effort has been *CellML2SBML* [32], a suite of XSLT stylesheets that, when applied consecutively, convert models expressed in CellML into SBML without significant loss of information. In the following years, many new and supporting initiatives and tools have been developed

G. Eleftherakis et al. (Eds.): WMC8 2007, LNCS 4860, pp. 353–366, 2007.

in favor of SBML by its growing community. SBML and BioPAX, a common exchange format for databases of pathways, teams worked together to define linkages between both representations. The *libsbml* library had been designed to help modelers to read, write, manipulate, translate, and validate SBML files and data streams. *The Systems Biology Workbench [2] and Systems Biology Markup Language* [8], a project funded by *Japan Science and Technology Corporation ERATO program*, started in the summer 2000. The project goal was to provide software infrastructure which (i) enables sharing of simulation/analysis software and models and (ii) enables collaboration between software developers. Subsequently, a plethora of many applications and databases based on SBML were born: *MathSBML* [33], a *Mathematica* package that manipulates SBML models; *SBMLToolbox* [18] a toolbox that facilitates importing and exporting models represented in SBML in and out of the MATLAB environment. Other two useful tools were *SBMLSupportLayout* and *SBWAutoLayout* [5], supporting reading, creating, manipulating and writing layout information for biochemical models. Yet other smart tools have been developed in the past: *SBML ODE Solver Library* [23] (SOSlib), a programming library for symbolic and numerical analysis of chemical reactions network models encoded in SBML; *SBML-PET* [36], a tool designed to enable parameter estimation for biological models including signalling pathways, gene regulation networks and metabolic pathways. A tool to translate SBML into pi-calculus [6] was presented in 2006 by Eccher and Priami and later also *SBMLR*[1], a tool able to link R to libsbml for SBML parsing and output converting SBML to R graph objects, and more; *SemanticSBML*[2]: a suite of tools to facilitate merging of SBML models for systems biology starting from all elements in the SBML files described by MIRIAM-type annotations and *SBMLeditor* [30] a very simple, low level editor of SBML files.

As many tools have been implemented all around SBML just to highlight the trust of developers on the standardizing initiatives related to the software biological infrastructures towards commons exchange formats. In particular, it is an undeniable fact the increasing and unison consensus among developers in favor of SBML. In fact, several languages have been recently developed to overcome these kind of problems (integrations, standardizing, reuse of biological models) [22], [12], [7], [34], [35], [10], [24], [1], [15], [14]. However, only two XML-based formats are suitable for representing compartmental reaction network models with sufficient mathematical depth that the descriptions can be used as direct input to simulation software. The two are CellML [4], [11] and SBML [13]. The latter is becoming a de-facto standard for a common representation supporting basic biochemical models. In fact, today, SBML is supported by over 110 software systems. As a consequence, many SBML models of gene regulatory networks and metabolic pathways that code a considerably body of biological knowledge have been accumulated in repositories.

Among all databases, I recall (i) the *PANTHER* Classification System, [26], an unique resource that classifies genes by their functions, using published

---

scientific experimental evidence and evolutionary relationships to predict function even in the absence of direct experimental evidence; (ii) *KEGG* [17], a knowledge base for systematic analysis of gene functions, linking genomic information with higher order functional information; (iii) *JWS Online* [28], a Systems Biology tool for simulation of kinetic models from a curated model database; (iv) *Reactome* [16], a curated resource of core pathways and reactions in human biology and (v) *BioModels database* [21], an annotated resource of quantitative models of biomedical interest.

Therefore, with the constant focus on SBML, in this paper I am going to inspect in section 2 how Cyto-Sim can *speak* and *understand* SBML, in the section 3, I am going to show the software facilities employed and the software packages implemented to build a pure Java library to handle SBML documents. In the section 4, I am going to test the software package implemented on real SBML files taken from different data sources and, in the last section, I am going to delineate the future works.

## 2    SBML ⇔ Cyto-Sim

As deeply shown, SBML is a powerful and well defined language for modeling biological interactive systems in standard way. The aim of my work has been to make Cyto-Sim able to *speak* and *understand* SBML.

Cyto-Sim [3] is a stochastic simulator of biochemical processes in hierarchical compartments which may be isolated or may communicate via peripheral and integral membrane proteins. It is available online as a Java applet [27] and as standalone application. For security issue, although the functionalities of the applet has been reduced, it fully and correctly works. By means of it, it is possible to model: (i) interacting species; (ii) compartmental hierarchies; (iii) species localizations inside compartments and membranes and (iv) rules and their and correlated velocity formulas which govern the dynamics of the system to be simulated, as chemical equations.

Some real biological systems have already been successfully simulated in the past by means of Cyto-Sim. Now I am going to try to explain at first how to translate a Cyto-Sim model into SBML (and vice-versa) and later I will test the quality of the translation comparing the simulations available in literature against those obtained by Cyto-Sim about the same models.

### 2.1    Speaking SBML

The conversion process from the Cyto-Sim syntax to the SBML one is quite straightforward. In Cyto-Sim, users must declare the species present into the system writing something like this:

```
/* Object Declaration */
object speciesA, speciesB, speciesC
```

This line of code corresponds to the following SBML chunk of code:

```
<listOfSpecies>
    <species id="compartmentA_0_speciesA" name="speciesA"
        compartment="compartmentA" initialAmount="0.0"/>
    <species id="compartmentB_0_speciesB" name="speciesB"
        compartment="compartmentB" initialAmount="1.0"/>
    <species id="compartmentC_2_speciesC" name="speciesC"
        compartment="compartmentB" initialAmount="2.0"/>
</listOfSpecies>
```

Not all the information in this XML code can be retrieved by the previous objects specification[3]. In fact, the compartment, membrane and initial amount related to one species are reached both from the following code:

```
/* Compartments Declarations */
compartment compartmentA [ruleA]
compartment compartmentB [compartmentA, ruleB, ruleC,
    speciesB, 100 speciesB@7000 : |2 speciesC|]
system compartmentB
```

and from this:

```
/* Rules Declarations */
rule ruleA {
    speciesA k1-> *
    || + speciesA k2-> speciesA + ||
}
rule ruleB speciesB k3-> speciesC
rule ruleC |speciesC| k4-> || + speciesC
```

From the code related to the compartments we take information about (i) the compartment hierarchy, (ii) which rule happens and in what compartment, (iii) the declared initial quantities (species not declared in this context will not still exist as default at the beginning of the simulation) and (iv) eventual re-feeding events at specified evolution times. Considering now that a reaction occurring in a compartment acts only on the species within it, looking to the localization of a species we can infer the localization of any reactions acting on it. Moreover it is possible to notice that the rule *ruleC* acts on the species *speciesC* inside the membrane[4] (membrane number 2) of the compartment *compartmentB*.

In SBML each compartment is quadruplicated to easily handle membranes.

```
<listOfCompartments>
    <compartment id="compartmentA_0" compartmentType="compartmentA"
```

---

[3] The figure between the compartment and the species names within the string assigned to each species id corresponds to the membrane in which a species sits. For more information about the syntax, look at [3].

[4] Recall that in this context a compartment is surrounded by a membrane with a not negligible thickness, therefore a compartment is logically divided into the internal (membrane 0), internal and superficial (membrane 1), intra (membrane 2), external and superficial (membrane 3) and external (membrane 4) membranes.

```
            outside="compartmentA_1"/>
    <compartment id="compartmentA_1" compartmentType="compartmentA"
            outside="compartmentA_2"/>
    <compartment id="compartmentA_2" compartmentType="compartmentA"
            outside="compartmentA_3"/>
    <compartment id="compartmentA_3" compartmentType="compartmentA"
            outside="compartmentB_0"/>
    <compartment id="compartmentB_0" compartmentType="compartmentB"
            outside="compartmentB_1"/>
    <compartment id="compartmentB_1" compartmentType="compartmentB"
            outside="compartmentB_2"/>
    <compartment id="compartmentB_2" compartmentType="compartmentB"
            outside="compartmentB_3"/>
    <compartment id="compartmentB_3" compartmentType="compartmentB"
            outside="system_0"/>
    <compartment id="system_0" compartmentType="system"
            outside="system_1"/>
    <compartment id="system_1" compartmentType="system"
            outside="system_2"/>
    <compartment id="system_2" compartmentType="system"
            outside="system_3"/>
    <compartment id="system_3" compartmentType="system"/>
</listOfCompartments>
```

Then a single compartment generates four independent concentric compartments, as a Matrioska doll toy, related to the same compartment but enclosing different spatial areas and then species. To keep conceptually linked these compartments, a compartment type specification is provided.

```
<listOfCompartmentTypes>
    <compartmentType id="compartmentA"/>
    <compartmentType id="compartmentB"/>
    <compartmentType id="system"/>
</listOfCompartmentTypes>
```

The previously seen reactions are easily translated into SBML differentiating the names of the grouped rules (e.g. the ruleA group contains two reactions. Their names will become: ruleA.0 and ruleA.1). Moreover, the kinetic formulas just touched (k1, k2, etc) before are expressed by MathML expressions inside <kineticLaw> tags.

```
<listOfReactions>
    [...]
    <reaction id="ruleA.1" name="compartmentA_0_ruleA.1">
        <listOfReactants>
            <speciesReference species="compartmentA_0_speciesA"
                stoichiometry="1.0"/>
        </listOfReactants>
        <!--listOfProducts>No Products</listOfProducts-->
        <kineticLaw>
```

```
                <math xmlns="http://www.w3.org/1998/Math/MathML">
                   <apply>
                      <times/>
                      <cn>k1_value</cn>
                      <ci>compartmentA_0_speciesA</ci>
                   </apply>
                </math>
             </kineticLaw>
          </reaction>
</listOfReactions>
```

Cyto-Sim also requires the specification of a range of evolution times and of the species whose quantities have to be plotted on the screen.

```
evolve 0 - 1000
plot compartmentA[speciesA], compartmentB[speciesB:|speciesC|]
```

This information can be encoded in SBML by the use of an annotation which is auto-explicative.

```
<annotation>
    <Cyto-Sim xmlns:cytosim="http://www.sbml.org/2001/ns/cytosim">
        <plot>
            <species>compartmentA_0_speciesA</species>
            <species>compartmentB_0_speciesB</species>
            <species>compartmentB_2_speciesC</species>
        </plot>
        <evolve>
            <from>0</from>
            <to>1000</to>
        </evolve>
    </Cyto-Sim>
</annotation>
```

## 2.2  Understanding SBML

The process to make an existing SBML file comprehensible to Cyto-Sim is more complex than the opposite step. Keeping in mind the correspondences among structures before shown, during this kind of translation we have to check some restrictions and to guarantee some constraints which are now explained.

**Parameters:** SBML optionally carries global parameters, visible everywhere in the file and local ones with more restricted scope. During the parsing time of an SBML file, Cyto-Sim loads all global parameters putting them into a global HashMap. In the case of local parameters inside kineticLaw of reactions, Cyto-Sim considers local and global parameters together taking care to overwrite eventual global parameters with the same name of local ones.

**Species Quantities:** SBML provides optional size for compartments. Cyto-Sim handles quantities and not concentration for species, then each concentration (if any) has to be converted into quantity. To do that, Cyto-Sim requires the size specification for each compartment if there are any specification of the species concentrations inside it.

**Assignments:** Cyto-Sim handles assignment rules at the moment of parsing and use them to replace eventual existing fixed values specified for species quantity, compartment size or parameters value. Up to now, it does not understand initial assignments, rate rules and algebraic rules. These features will be made available soon.

**Functions:** Cyto-Sim does not still handle $\lambda$-functions.

**Units and Constraints:** Cyto-Sim does not still make use of units of measurements and constraints.

## 3   Binding to the SBML Schema

After having conceptually explained how Cyto-Sim converts SBML in its own language and vice-versa, now I am going to show which software architecture gives it the possibility to do that. I used two well known tools for this aim: the Java Architecture for XML Binding (JAXB) package and the XML DOM parser, both build-in the latest release of Java (Java Mustang).

JAXB [9] simplifies access to an XML document from a Java program by presenting the XML document to the program in a Java format. The first step in this process is to bind the schema for the XML document into a set of Java classes that represents the schema. Binding a schema means generating a set of Java classes that represents the schema. All JAXB implementations provide a tool called *binding compiler* in order to bind a schema. In response, the binding compiler generates a set of interfaces and a set of classes that implement the interface. I obtained Java classes for each available XML levels and versions. I mean SBML level 1 version 1, level 1 version 2, level 2 version 1 and level 2 version 2. Later, I compiled and packaged them into just one package. The second step is to unmarshal an SBML document. Unmarshalling means creating a tree of content objects that represents the content and the organization of the document. The content tree is not a DOM-based tree. In fact, content trees produced through JAXB can be more efficient in terms of memory usage than DOM-based trees. The content objects are instances of the classes produced by the binding compiler. In addition to providing a binding compiler, JAXB provides runtime APIs for JAXB-related operations such as marshalling. It is possible to validate source data against an associated schema as part of the unmarshalling operation. If the data is found to be invalid (that is, it doesn't conform to the schema) the JAXB implementation can report it and might take further action. JAXB providers have a lot of flexibility here. The JAXB specification mandates that all provider implementations report validation errors when the errors are encountered, but the implementation does not have to stop processing the data. Some provider implementations might stop processing when the first error is found, others might stop even if many errors are found. In other

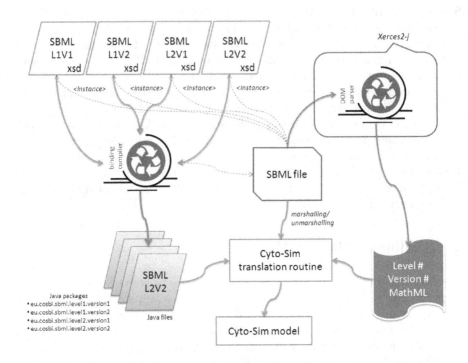

**Fig. 1.** Software Architecture for SBML Binding

words, it is possible for a JAXB implementation to successfully unmarshal an invalid XML document, and build a Java content tree. However, the result will not be valid. The main requirement is that all JAXB implementations must be able to unmarshal valid documents. I unmarshal and validate each SBML file at runtime.

*The W3C Document Object Model (DOM) is a platform and language-neutral interface that allows programs and scripts to dynamically access and update the content, structure, and style of a document.* The XML DOM is the tool to define (i) a standard set of objects for XML, (ii) a standard way to access XML documents; (iii) a standard way to manipulate XML documents. Cyto-Sim uses the DOM parser contained into xerces2-j [29] built into the Java Mustang release. The DOM parser is used to:

**Check Levels and Versions:** Cyto-Sim preliminary opens SBML files and checks levels and versions (delegating validation and comprehension to JAXB). It acquires knowledge about which JAXB context instantiating or, more clearly, which SBML schema considering for binding, unmarshalling and validation;

**Parse MathML expression:** Due to intrinsic limitations of JAXB to handle recursively nested xml tags, Cyto-Sim makes use of DOM to explore MathML expressions and parse their components.

## 4   Experimental Tests

The capability of Cyto-Sim to understand all currently existing SBML levels and versions has been tested on almost all official existing SBML files available on the web. I successfully imported all SBML files generated by Gepasi [25], a software package for modeling biochemical systems and the most part of the models stored into the BioModels database [21]. Gepasi makes available 9 SBML level 1 version 1 files [5] while BioModels has 70 curated and 43 not curated models exported as SBML level 2 version 1 files. I have also tested models from the PAN-THER (130 SBML level 1 version 2 files) Classification System, and from KEGG (77 SBML level 2 version 1 files). All SBML files were converted from KEGG by using a conversion script *kegg2sbml*. Moreover, I retrieved some interesting models among all 238 CellML models and tested them. To do that, I had to manually convert from the CellML format to SBML by means of CellML2SBML [32] and later import and simulate them with Cyto-Sim. All imported files have been successfully parsed by Cyto-Sim. This testifies the quality of the conversion routines and of the architecture employed. Summarizing, I retrieved 567 models from the most known and famous biological model containers available in SBML (or in formats having reference to SBML), and tested them. *I obtained a successful test, when Cyto-Sim had been able to correctly parse the inferred model.* In particular, now I am going to show a couple of examples which Cyto-Sim has been able not only to correctly parse, but also to simulate and get the same results shown in literature.

The first test is related to the model *BIOMD0000000010* picked up from the BioModels database. It concerns the functional organization of signal transduction into protein phosphorylation cascades and in particular the mitogen-activated protein kinase (MAPK) cascades. It greatly enhances the sensitivity of cellular targets to external stimuli [19]. In this paper it is demonstrated that a negative feedback loop combined with intrinsic ultrasensitivity of the MAPK cascade can bring about sustained oscillations in MAPK phosphorylation. The conversion of the SBML file produces the following model with 1 compartment, 8 species and 10 reactions.

```
object MKKK, MKKK_P, MKK, MKK_P, MKK_PP, MAPK, MAPK_P, MAPK_PP

rule J0 MKKK ((1.0*2.5*MKKK)/((1+((MAPK_PP/9.0)^1.0))*(10.0+MKKK)))-> MKKK_P
rule J1 MKKK_P ((1.0*0.25*MKKK_P)/(8.0+MKKK_P))-> MKKK
rule J2 MKK ((1.0*0.025*MKKK_P*MKK)/(15.0+MKK))-> MKK_P
rule J3 MKK_P ((1.0*0.025*MKKK_P*MKK_P)/(15.0+MKK_P))-> MKK_PP
rule J4 MKK_PP ((1.0*0.75*MKK_PP)/(15.0+MKK_PP))-> MKK_P
rule J5 MKK_P ((1.0*0.75*MKK_P)/(15.0+MKK_P))-> MKK
rule J6 MAPK ((1.0*0.025*MKK_PP*MAPK)/(15.0+MAPK))-> MAPK_P
rule J7 MAPK_P ((1.0*0.025*MKK_PP*MAPK_P)/(15.0+MAPK_P))-> MAPK_PP
rule J8 MAPK_PP ((1.0*0.5*MAPK_PP)/(15.0+MAPK_PP))-> MAPK_P
```

---

[5] among all, a very large model representing a set of 100 yeast cells in a liquid culture whose dynamics is represented by means of 2000 reactions.

```
rule J9 MAPK_P ((1.0*0.5*MAPK_P)/(15.0+MAPK_P))-> MAPK

compartment uVol[J0, J1, J2, J3, J4, J5, J6, J7, J8, J9, 280.0 MAPK,
    10.0 MKK_P, 10.0 MKK_PP, 10.0 MKKK_P, 10.0 MAPK_PP, 280.0 MKK,
    10.0 MAPK_P, 90.0 MKKK]
system uVol

evolve 0-33000
plot uVol[MAPK,MAPK_PP]
```

In the figure 2, on the left is shown the simulation result coming from the literature and on the right that one obtained with Cyto-Sim. The graphs are identical.

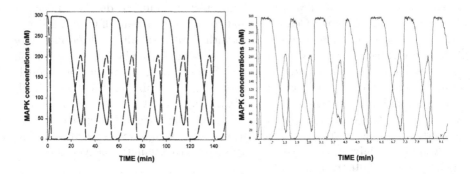

**Fig. 2.** Sustained oscillations in MAPK cascade

The second test is related to the glucose transport by the Bacterial Phosphoenolpyruvate [31] whose model has been found in JWS Online. The resulting model has 1 compartment, 17 species and 10 reactions.

```
object EI, PyrPI, EIP, HPr, EIPHPr, HPrP, EIIA, HPrPIIA, EIIAP, EIICB,
    EIIAPIICB, EIICBP, EIICBPGlc, PEP, Pyr, GlcP, Glc

rule v1 PEP + EI ((1960.0*PEP*EI)-(480000.0*PyrPI))-> PyrPI
rule v2 PyrPI ((108000.0*PyrPI)-(294.0*Pyr*EIP))-> EIP + Pyr
rule v3 HPr + EIP ((14000.0*EIP*HPr)-(14000.0*EIPHPr))-> EIPHPr
rule v4 EIPHPr ((84000.0*EIPHPr)-(3360.0*EI*HPrP))-> HPrP + EI
rule v5 HPrP + EIIA ((21960.0*HPrP*EIIA)-(21960.0*HPrPIIA))-> HPrPIIA
rule v6 HPrPIIA ((4392.0*HPrPIIA)-(3384.0*HPr*EIIAP))-> EIIAP + HPr
rule v7 EIICB + EIIAP ((880.0*EIIAP*EIICB)-(880.0*EIIAPIICB))-> EIIAPIICB
rule v8 EIIAPIICB ((2640.0*EIIAPIICB)-(960.0*EIIA*EIICBP))-> EIICBP +EIIA
rule v9 EIICBP + Glc ((260.0*EIICBP*Glc)-(389.0*EIICBPGlc))-> EIICBPGlc
rule v10 EIICBPGlc ((4800.0*EIICBPGlc)-(0.0054*EIICB*GlcP))-> EIICB +GlcP

compartment compartment_cyto_sim[v1, v2, v3, v4, v5, v6, v7, v8, v9, v10,
    0.0 EIICBPGlc, 5.0 EIICBP, 25.0 HPrP, 2.0 EIP, 20.0 EIIA, 5.0 EIICB,
    25.0 HPr, 2800.0 PEP, 0.0 PyrPI, 0.0 EIPHPr, 50.0 GlcP, 900.0 Pyr,
```

```
      0.0 HPrPIIA, 20.0 EIIAP, 500.0 Glc, 0.0 EIIAPIICB, 3.0 EI]
system compartment_cyto_sim

evolve 0-10000
plot compartment_cyto_sim[HPrP,EIIAPIICB,HPrPIIA]
```

In the figure 3 it is possible to notice that both graphs represent the same behavior. The differences are due to the deterministic (on the left) or stochastic (on the right) nature of the simulations.

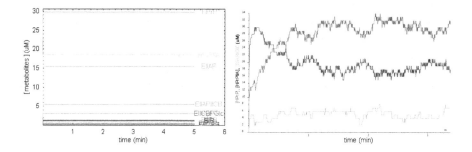

**Fig. 3.** Glucose Transport by the Bacterial Phosphoenolpyruvate

At the end, I tested the whole human reactome derived from Reactome. The actual release of the human reactome I used is an SBML file containing 28 compartments (even including internal membranes of the same compartment), 3054 species (in all their forms) and 1979 interactions represented by means of reactions. Cyto-Sim is able to parse and even to simulate it, although at the moment it cannot have meaning because of the lack of quantitative parameters (reaction rates and initial species quantities).

## 5   Conclusion

kosmopolitês (citizen of the world), has been used to describe a wide variety of important views in moral and socio-political philosophy. The nebulous core shared by all cosmopolitan views is the idea that all human beings, regardless of their political affiliation, do (or at least can) belong to a single community, and that this community should be cultivated. Different versions of cosmopolitanism envision this community in different ways, some focusing on political institutions, others on moral norms or relationships, and still others focusing on shared markets or forms of cultural expression [20].

In the context of the present work, a citizen of the world is anyone who speaks and understands a common language, who can travel to the ends of the earth without worrying about misunderstanding or being misunderstood. Limited comprehension of language is the greatest barrier for people who need to spread information and ideas. This is exactly the case for scientists who wish to share their results and models with the widest possible audience.

In this paper I have presented a quick overview of the SBML story, continually remarking on the increasing interest of scientists both to support and write their biological models in SBML. I have shown how Cyto-Sim converts SBML model into its own syntax and vice-versa and I presented a possible software architectural arrangement to allow simple binding to SBML schemas and correct unmarshalling of SBML files. Finally, I presented tests performed on two models coming from separate databases. I demonstrated the correctness of the translation routines and highlighted the similarities of the obtained simulation results.

Today there are more than 600 models written in SBML, ready to be more accurately studied, confirmed or refuted. Challenging existing knowledge is the means to increase understanding and therefore to *grow* knowledge. The best way to achieve this is to maximize the number of people that speak the same language, in this case SBML. My work sits perfectly in this context and my hope is that it has wide application.

# References

1. Altman, R.B., et al.: Ribonucleic acid markup language (2002), http://www.smi.stanford.edu/projects/helix/riboml/
2. Bergmann, F.T., Sauro, H.M.: Sbw - a modular framework for systems biology. In: Proceedings of the 37th conference on Winter simulation. Winter Simulation Conference, pp. 1637–1645 (2006)
3. Cavaliere, M., Sedwards, S.: Modelling cellular processes using membrane systems with peripheral and integral proteins. In: Priami, C. (ed.) CMSB 2006. LNCS (LNBI), vol. 4210, pp. 108–126. Springer, Heidelberg (2006)
4. Cuellar, A.A., Lloyd, C.M., Nielsen, P.F., Bullivant, D.P., Nickerson, D.P., Hunter, P.J.: An overview of cellml 1.1, a biological model description language. Simulation 79(12), 740–747 (2003)
5. Deckard, A., Bergmann, F.T., Sauro, H.M.: Supporting the sbml layout extension. Bioinformatics 22(23), 2966–2967 (2006)
6. Eccher, C., Priami, C.: Design and implementation of a tool for translating sbml into the biochemical stochastic pi-calculus. Bioinformatics 22(24), 3075–3308 (2006)
7. Fenyo, D.: The biopolymer markup language. Bioinformatics 15(4), 339–340 (1999)
8. Finney, A.M., Hucka, M.: Systems biology markup language: Level 2 and beyond. Biochem. Soc. Trans. 31, 1472–1473 (2003)
9. Project GlassFish: The jaxb project. https://jaxb.dev.java.net/
10. Hanisch, D., Zimmer, R., Lengauer, T.: Proml - the protein markup language for specification of protein sequences, structures and families. In Silico Biology 2(3), 313–324 (2002)
11. Hedley, W.J., Nelson, M.R., Bullivant, D.P., Nielson, P.F.: A short introduction to cellml. Philos. Trans. R. Soc. Lond. A 359, 1073–1089 (2001)
12. Hermjakob, H., et al.: The hupopsiŠs molecular interaction format - a community standard for the representation of protein interaction data. Nature Biotechnol. 22(2), 177–183 (2004)

13. Hucka, M., et al.: The systems biology markup language (sbml): a medium for representation and exchange of biochemical network models. Bioinformatics 19(4), 524–531 (2003)
14. Doubletwist Inc.: Agave: architecture for genomic annotation, visualization and exchange (2001), http://www.agavexml.org
15. LabBook Inc.: Bsml (bioinformatics sequence markup language) 2.2 (2002), http://www.labbook.com/products/xmlbsml.asp
16. Joshi-Tope, G., Gillespie, M., Vastrik, I., D'Eustachio, P., Schmidt, E., de Bono, B., Jassal, B., Gopinath, G.R., Wu, G.R., Matthews, L., Lewis, S., Birney, E., Stein, L.: Reactome: a knowledgebase of biological pathways. Nucleic Acids Res. 33, D428–D432 (2005)
17. Kanehisa, M., Goto, S.: Kegg: Kyoto encyclopedia of genes and genomes. Nucleic Acids Research 28(1), 27–30 (2000)
18. Keating, S.M., Bornstein, B.J., Finney, A., Hucka, M.: Sbmltoolbox: an sbml toolbox for matlab users. Bioinformatics 22(10), 1275–1277 (2006)
19. Kholodenko, B.N.: Negative feedback and ultrasensitivity can bring about oscillations in the mitogen-activated protein kinase cascades. Eur. J. Biochem 267, 1583–1588 (2000)
20. Kleingeld, P., Brown, E.: Cosmopolitanism. In: The Stanford Encyclopedia of Philosophy, Edward N. Zalta (winter 2006)
21. Le Novère, N., Bornstein, B., Broicher, A., Courtot, M., Donizelli, M., Dharuri, H., Li, L., Sauro, H., Schilstra, M., Shapiro, B., Snoep, J.L., Hucka M.: Biomodels database: a free, centralized database of curated, published, quantitative kinetic models of biochemical and cellular systems. Nucleic Acids Research 34, D689–D691 (2006)
22. Liao, Y.M., Ghanadan, H.: The chemical markup language. Anal. Chem. 74(13), 389A-390A (2002)
23. Machné, R., Finney, A., Müller, S., Lu, J., Widder, S., Flamm, C.: The sbml ode solver library: a native api for symbolic and fast numerical analysis of reaction networks. Bioinformatics 22(11), 1406–1407 (2006)
24. McArthur, D.C.: An extensible xml schema definition for automated exchange of protein data: Proximl (protein extensible markup language) (2001), http://www.cse.ucsc.edu/douglas/proximl/
25. Mendes, P.: Gepasi: a software package for modeling the dynamics, steady states, and control of biochemical and other systems. Comput. Applic. Biosci. 9, 563–571 (1993)
26. Mi, H., Lazareva-Ulitsky, B., Loo, A., Kejariwal, R., Vandergriff, J., Rabkin, S., Guo, N., Muruganujan, A., Doremieux, O., Campbell, M.J.: The panther database of protein families, subfamilies, functions and pathways. Nucleic Acids Res. 33, D284–D288 (2005)
27. The Microsoft Research University of Trento. Centre for Computational and Systems Biology. Web page of Cyto-Sim (2006), http://www.cosbi.eu/Rpty_Soft_CytoSim.php
28. Olivier, B.G., Snoep, J.L.: Web-based modelling using jws online. Bioinformatics 20, 2143–2144 (2004)
29. Apache XML project: Xerces2 java parser 2.9.0 (2004), http://xml.apache.org/xerces2-j/
30. Rodriguez, N., Donizelli, M., Le Novère, N.: Sbmleditor: effective creation of models in the systems biology markup language (sbml). Bioinformatics 8(79) (published online March 2007)

31. Rohwer, J.M., Meadowi, N.D., Rosemani, S., Westerhoff, H.V., Postma, P.W.: Understanding glucose transport by the bacterial phosphoenolpyruvate: glycose phosphotransferase system on the basis of kinetic measurements in vitro. The Journal of Biological Chemistry 275(45), 34909–34921 (2000)
32. Schilstra, M.J., Li, L., Matthews, J., Finney, A., Hucka, M., Le Novère, N.: Cellml2sbml: conversion of cellml into sbml. Bioinformatics 22(8), 1018–1020 (2006)
33. Shapiro, B.E., Hucka, M., Finney, A., Doyle, J.: Mathsbml: a package for manipulating sbml-based biological models. Bioinformatics 20(16), 2829–2831 (2004)
34. Spellman, P.T., Miller, M.: Design and implementation of microarray gene expression markup language (mage-ml). Genome Biol. 3(9), 0046.0041–0046.0049 (2002)
35. Taylor, C.F., Paton, N.W.: A systematic approach to modeling, capturing, and disseminating proteomics experimental data. Nature Biotechnol. 21, 247–254 (2003)
36. Zhike, Z., Klipp, E.: Sbml-pet: a systems biology markup language-based parameter estimation tool. Bioinformatics 22(21), 2704–2705 (2006)

# Active Membrane Systems Without Charges and Using Only Symmetric Elementary Division Characterise P

Niall Murphy[1] and Damien Woods[2]

[1] Department of Computer Science, National University of Ireland, Maynooth, Ireland
nmurphy@cs.nuim.ie
[2] Department of Computer Science, University College Cork, Ireland
d.woods@cs.ucc.ie

**Abstract.** In this paper we introduce a variant of membrane systems with elementary division and without charges. We allow only elementary division where the resulting membranes are identical; we refer to this using the biological term symmetric division. We prove that this model characterises P and introduce logspace uniform families. This result characterises the power of a class of membrane systems that fall under the so-called P conjecture for membrane systems.

## 1 Introduction

The **P**-conjecture states that recogniser membranes systems with division rules (active membranes [6]), but without charges, characterise **P**. This was shown for a restriction of the model: without dissolution rules [4]. However, it has been shown that systems with dissolution rules and non-elementary division characterise **PSPACE** [2,9]. In this setting, using dissolution rules allows us to jump from **P** to **PSPACE**. As a step towards finding a bound (upper or lower) on systems with only elementary division rules, we propose a new restriction, and show that it has an upper bound of **P**.

Our restriction insists that the two membranes that result from an elementary division rule must be identical. This models mitosis, the biological process of cell division [1] and we refer to it using the biological term "symmetric division". We refer to division where the two resulting daughter cells are different by the biological term "asymmetric division". In nature asymmetric division occurs, for example, in stem cells as a way to achieve cell differentiation.

Since our model is uniform via polynomial time deterministic Turing machines, it trivially has a lower bound of **P**. However, we introduce logspace uniformity for this model and then prove a **P** lower bound. All recogniser membrane systems with division rules are upper bounded by **PSPACE** [9]. In this paper we show that systems with symmetric elementary division and without charges are upper bounded by **P**. From an algorithmic point of view, this result allows one to write a polynomial time algorithm that models certain membrane systems which use exponential numbers of membranes and objects.

G. Eleftherakis et al. (Eds.): WMC8 2007, LNCS 4860, pp. 367–384, 2007.
© Springer-Verlag Berlin Heidelberg 2007

## 2  Preliminaries

In this section we define membrane systems and complexity classes. These definitions are from Păun [6,7], and Sosík and Rodríguez-Patón [9]. We also introduce the notion of logspace uniformity for membrane systems. We give a **P** lower bound for the model that is the main focus of this paper.

### 2.1  Recogniser Membrane Systems

Active membranes systems are membrane systems with membrane division rules. Division rules can either only act on elementary membranes, or else on both elementary and non-elementary membranes. An elementary membrane is one which does not contain other membranes (a leaf node, in tree terminology). In Definition 1 we make a new distinction between two types of elementary division rules. When we refer to *symmetric division* ($e_s$) we mean division where the resulting two child membranes are identical. When the two child membranes are not identical we refer to the rule as being *asymmetric* ($e$).

**Definition 1.** *An active membrane system without charges using elementary division is a tuple* $\Pi = (V, H, \mu, w_1, \ldots, w_m, R)$ *where,*

1. *$m > 1$ the initial number of membranes;*
2. *$V$ is the alphabet of objects;*
3. *$H$ is the finite set of labels for the membranes;*
4. *$\mu$ is a membrane structure, consisting of $m$ membranes, labelled with elements of $H$;*
5. *$w_1, \ldots, w_m$ are strings over $V$, describing the multisets of objects placed in the $m$ regions of $\mu$.*
6. *$R$ is a finite set of developmental rules, of the following forms:*
   *(a) $[_h \; a \; \rightarrow \; v \; ]_h$,*
       *for $h \in H$, $a \in V$, $v \in V^*$*
   *(b) $a[_h \; ]_h \rightarrow [_h \; b \; ]_h$,*
       *for $h \in H$, $a, b \in V$*
   *(c) $[_h \; a \; ]_h \rightarrow [_h \; ]_h \; b$,*
       *for $h \in H$, $a, b \in V$*
   *(d) $[_h \; a \; ]_h \rightarrow b$,*
       *for $h \in H$, $a, b \in V$*
   *($e_s$) $[_h \; a \; ]_h \rightarrow [_h \; b \; ]_h \; [_h \; b \; ]_h$,*
       *for $h \in H$, $a, b \in V$*
   *(e) $[_h \; a \; ]_h \rightarrow [_h \; b \; ]_h \; [_h \; c \; ]_h$,*
       *for $h \in H$, $a, b, c \in V$.*

These rules are applied according to the following principles:

- All the rules are applied in maximally parallel manner. That is, in one step, one object of a membrane is used by at most one rule (chosen in a nondeterministic way), but any object which can evolve by one rule of any form, must evolve.

- If at the same time a membrane labelled with $h$ is divided by a rule of type $(e)$ or $(e_s)$ and there are objects in this membrane which evolve by means of rules of type $(a)$, then we suppose that first the evolution rules of type $(a)$ are used, and then the division is produced. This process takes only one step.
- The rules associated with membranes labelled with $h$ are used for membranes with that label. At one step, a membrane can be the subject of only one rule of types $(b)$-$(e)$.

In this paper we study the language recognising variant of membrane systems which solves decision problems. A distinguished region contains, at the beginning of the computation, an input — a description of an instance of a problem. The result of the computation (a solution to the instance) is "yes" if a distinguished object **yes** is expelled during the computation, otherwise the result is "no". Such a membrane system is called *deterministic* if for each input a unique sequence of configurations exists. A membrane system is called *confluent* if it always halts and, starting from the same initial configuration, it always gives the same result, either always "yes" or always "no". Therefore, given a fixed initial configuration, a confluent membrane system non-deterministically chooses from one from a number of valid configuration sequences and rule applications but all of them must lead to the same result.

## 2.2    Complexity Classes

Complexity classes have been defined for membrane systems [8]. Consider a decision problem $X$, i.e. a set of instances $\{x_1, x_2, \ldots\}$ over some finite alphabet such that to each $x_i$ there is an unique answer "yes" or "no". We consider a *family* of membrane systems to solve a decision problem if each instance of the problem is solved by some class member.

We denote by $|x_i|$ the size of any instance $x_i \in X$.

**Definition 2 (Polynomial uniform families of membrane systems).** *Let $\mathcal{D}$ be a class of membrane systems and let $f : \mathbb{N} \to \mathbb{N}$ be a total function. The class of problems solved by uniform families of membrane systems of type $\mathcal{D}$ in time $f$, denoted by* $\mathbf{MC}_{\mathcal{D}}(f)$, *contains all problems $X$ such that:*

- *There exists a uniform family of membrane systems, $\Pi_X = (\Pi_X(1); \Pi_X(2); \ldots)$ of type $\mathcal{D}$: each $\Pi_X(n)$ is constructable by a deterministic Turing machine with input $n$ and in time that is polynomial of $n$.*
- *Each $\Pi_X(n)$ is sound: $\Pi_X(n)$ starting with an input (encoded by a deterministic Turing machine in polynomial time) $x \in X$ of size $n$ expels out a distinguished object* **yes** *if an only if the answer to $x$ is "yes".*
- *Each $\Pi_X(n)$ is confluent: all computations of $\Pi_X(n)$ with the same input $x$ of size $n$ give the same result; either always "yes" or else always "no".*
- *$\Pi_X$ is $f$-efficient: $\Pi_X(n)$ always halts in at most $f(n)$ steps.*

*Polynomial Semi-uniform families of membrane systems $\Pi_X = (\Pi_X(x_1); \Pi_X(x_2); \ldots)$ whose members $\Pi_X(x_i)$ are constructable by a deterministic Turing machine with input $x_i$ in a polynomial time with respect to $|x_i|$. In this case,* for each instance of $X$ we have a special membrane system which therefore does not need an input. The resulting class of problems is denoted by $\mathbf{MC}_\mathcal{D}^S(f)$. Obviously, $\mathbf{MC}_\mathcal{D}(f) \subseteq \mathbf{MC}_\mathcal{D}^S(f)$ for a given class $\mathcal{D}$ and a complexity [3] function $f$.

We denote by

$$\mathbf{PMC}_\mathcal{D} = \bigcup_{k \in \mathbb{N}} \mathbf{MC}_\mathcal{D}(O(n^k)), \ \mathbf{PMC}_\mathcal{D}^S = \bigcup_{k \in \mathbb{N}} \mathbf{MC}_\mathcal{D}^S(O(n^k))$$

the class of problems solvable by uniform (respectively semi-uniform) families of membrane systems in polynomial time. We denote by $\mathcal{AM}$ the classes of membrane systems with active membranes. We denote by $\mathcal{EAM}$ the classes of membrane systems with active membranes and only elementary membrane division. We denote by $\mathcal{AM}_{-a}^0$ (respectively, $\mathcal{AM}_{+a}^0$) the class of all recogniser membrane systems with active membranes without charges and without asymmetric division (respectively, with asymmetric division). We denote by $\mathbf{PMC}_{\mathcal{EAM}_{-a}^0}^S$ the classes of problems solvable by semi-uniform families of membrane systems in polynomial time with no charges and only symmetric elementary division. We let poly$(n)$ be the set of polynomial complexity functions of $n$.

### 2.3   (Semi-)Uniformity Via logspace Turing Machines

In Theorem 2 we prove that $\mathbf{PMC}_{\mathcal{EAM}_{-a}^0}^S$ has a $\mathbf{P}$ upper bound. When we use (semi-)uniform families constructed in polynomial time by deterministic Turing machines we trivially have a $\mathbf{P}$ lower bound. However, to ensure that the membrane system itself is able to solve any problem in $\mathbf{P}$, and is not benefiting from

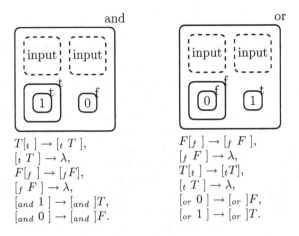

**Fig. 1.** AND and OR gadgets which can be nested together to simulate a circuit. The input is either a T, F, or a nested series of gadget membranes. A NOT gate membrane can be made with the rules $[_{not} T ] \rightarrow [_{not} ]F$, $[_{not} F ] \rightarrow [_{not} ]T$.

preprocessing by the output of the encoding Turing machine or family generating Turing machine, we restrict both of these machines to be logspace Turing machines.

In the following theorem it is understood that $\mathbf{PMC}^S_{\mathcal{EAM}^0_{-a}}$ is logspace uniform.

**Theorem 1.** $\mathbf{P} \subseteq \mathbf{PMC}^S_{\mathcal{EAM}^0_{-a}}$

*Proof.* A logspace Turing machine encodes an instance of the CIRCUIT VALUE problem (CVP) [5] as a $\mathbf{PMC}^S_{\mathcal{EAM}^0_{-a}}$ membrane system using the gadgets shown in Figure 1. The resulting membrane system directly solves the instance of CVP in polynomial time. $\qquad \square$

The main result of this paper, Theorem 2, holds for both logspace and polynomial families of $\mathbf{PMC}^S_{\mathcal{EAM}^0_{-a}}$.

# 3   An Upper Bound on $\mathbf{PMC}^S_{\mathcal{EAM}^0_{-a}}$

In this section we give an upper bound of $\mathbf{P}$ on the membrane class $\mathbf{PMC}^S_{\mathcal{EAM}^0_{-a}}$. We provide a random access machine (RAM) algorithm that simulates this class using a polynomial number of registers of polynomial length, in polynomial time. We begin with an important definition followed by informal description of our contribution.

**Definition 3 (Equivilance class of membranes).** *An equivalence class of membranes is a multiset of membranes where: each membrane shares a single parent, each has the same label, and each has identical contents. Further, only membranes without children can be elements of an equivalence class of size greater than one; each membrane with one or more children has its own equivalence class of size one.*

Throughout the paper, when we say that a membrane system has $|E|$ equivalence classes, we mean that $|E|$ is the minimum number of equivalence classes that includes all membranes of the system.

While it is possible for a computation path of $\mathbf{PMC}^S_{\mathcal{EAM}^0_{-a}}$ to use an exponential number of equivalence classes, our analysis guarantees that there is another, equally valid, computation path that uses at most a polynomial number of equivalence classes. Our algorithm finds this path in polynomial time. Moreover, via our algorithm, after a single timestep the increase in the number of equivalence classes is never greater than $|E_0||V|$, the product of the number of initial equivalence classes and the number of object types in the system. Since the system is confluent, our chosen computation path is just as valid to follow as any alternative path.

In Section 3.2 we prove that by using our algorithm:

- Type $(a)$ rules do not increase the number of equivalence classes since the rule has the same effect on each membrane of a given equivalence class.
- Type $(c)$ rules do not increase the number of equivalence classes since objects exit all child membranes for the parent membrane (which is already an equivalence class with one membrane).
- Type $(d)$ rules do not increase the number of equivalence classes since the rule is applied to all membranes in the equivalence class. The contents and child membranes are transfered to the parent (already an equivalence class).
- Type $(e_s)$ rules do not increase the number of equivalence classes, only the number of membranes in the existing equivalence classes simply increase.

Type $(b)$ rules require a more detailed explanation. In Section 3.3 we show that there is a deterministic polynomial sequential time algorithm that finds a computation path that uses only a polynomial number of equivalence classes.

Our RAM algorithm operates on a number of registers that can be thought of as a data structure (see Section 3.1). The data structure stores the state of the membrane system at each timestep. It compresses the amount of information to be stored by storing equivalence classes instead of explicitly storing all membranes. Each equivalence class contains the number of membranes in the class, a reference to each of the distinct objects in one of those membranes, and the number of copies of that distinct object. Type $(a)$ rules could therefore provide a way to create exponential space. However, we store the number of objects in binary thus we store it using space that is the logarithm of the number of objects.

Our RAM algorithm operates in a deterministic way. To introduce determinism we sort all lists of object multisets by object multiplicity, then lexicographically. We sort all equivalence classes by membrane multiplicity, then by label, and then by object. We sort all rules by rules type, matching label, matching object, and then by output object(s). The algorithm iterates through the equivalence classes and applies all rules of type $(a)$, $(c)$, $(d)$, and $(e_s)$. It then checks to see if any rules of type $(b)$ are applicable. If so, it takes each object in its sorted order and applies it to the relevant membranes in their sorted order.

**Theorem 2.** $\mathbf{PMC}^S_{\mathcal{EAM}^0_{-a}} \subseteq \mathbf{P}$

The proof is in the remainder of this section. The result holds for both logspace and polynomial time uniform membrane systems of type $\mathbf{PMC}^S_{\mathcal{EAM}^0_{-a}}$.

## 3.1  Structure of RAM Registers

Our RAM uses a number of binary registers that is a polynomial ($\mathrm{poly}(n)$) of the length $n$ of the input. The length of each register is bounded by a polynomial of $n$. For convenience our registers are grouped together in a data structure (as illustrated in Figure 2).

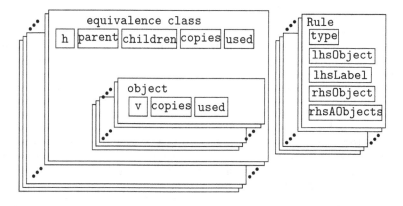

**Fig. 2.** A representation of our polynomial sized registers as a data structure

**Object registers.** For each distinct object type $v_i$, the following registers are used to encode the object in an equivalence class $e_k \in E$.

The register v represents the type of the object, $v_i \in V$ (see Definition 1). Throughout the computation, the size of the set $V$ is fixed so this register does not grow beyond its initial size.

The copies register is the multiplicity of the distinct object $v_i$ encoded in binary. At time 0 we have $|v_i|$ objects. At time 1 the worst case is that each object evolves via a type $(a)$ rule to give a number of objects that is poly($n$). This is an exponential growth function, however, since we store it using binary, the register length does not grow beyond space that is poly($n$).

The register used represents the multiplicity $v_i$ objects that have been used already in this computation step. It is always the case that used $\leq$ copies for each object type $v_i$.

**Equivalence class registers.** The following registers are used to store information about each equivalence class. To conserve space complexity we only explicitly store equivalence classes (rather than explicitly storing membranes); the number of equivalence classes is denoted $|E|$.

The register h stores the label of equivalence class $e_k$ and is an element of the set $H$ (see Definition 1). The size of register h is fixed and is bounded by poly($n$).

The register parent stores a reference to the equivalence class (a single membrane in this case) that contains this membrane. This value is bounded by the polynomial depth of the membrane structure. Since the depth of the membrane structure is fixed throughout a computation, the space required to store a parent reference is never greater than a logarithm of the depth.

The children register references all of the child equivalence classes of $e_k$ at depth one. Its size is bounded by poly($n$) via Theorem 3.

The register copies stores the number, denoted $|e_k|$, of membranes in the equivalence class. We store this number in binary. In the worst case, the number that is stored in copies doubles at each timestep (due to type $(e_s)$ rules). Since we store this number in binary we use space that is poly($n$).

The register used stores the number of membranes in the equivalence class that have been used by some rule in the current timestep and so this value is $\leq |e_k|$.

**Rules registers.** The rules registers store the rules of the membrane system; their number is bounded by the polynomial $|R|$ and is fixed for all time $t$. The rules registers can not change or grow during a computation. The type register stores if the rule is of type $(a)$, $(b)$, $(c)$, $(d)$ or $(e_s)$. The lhsObject register stores the object on the left hand side of the rule. The lhsLabel register stores the label on the left hand side of the rule. The rhsObject register stores the object on the right hand side of the rule. The rhsAObjects register stores the multiset of objects generated by the rule.

## 3.2   There Is a Computation Path That Uses Polynomially Many Equivalence Classes

In Section 3.2 we prove Theorem 3. Before proceeding to this theorem we make an important observation. Suppose we begin at an initial configuration of a recogniser membrane system. Due to non-determinism in the choice of rules and objects, after $t$ timesteps we could be in any one of a large number of possible configurations. However all computations are *confluent*. So if we are only interested in whether the computation accepts or rejects, then it does not matter which computation path we follow.

Theorem 3 asserts that after a polynomial number of timesteps, there is at least one computation path where the number of equivalence classes of a $\mathbf{PMC}^S_{\mathcal{EAM}^0_{-a}}$ system is polynomially bounded. This is shown by proving that there is a computation path where the application of each rule type $(a)$ to $(e_s)$, in a single timestep, leads to at most an additive polynomial increase in the number of equivalence classes.

**Theorem 3.** *Given an initial configuration of a $\mathbf{PMC}^S_{\mathcal{EAM}^0_{-a}}$ system $\Pi$ with $|E_0|$ equivalence classes and $|V|$ distinct object types, then there is a computation path such that at time $t \in$ poly($n$) the number of equivalence classes is $|E_t| = O(|E_0| + t|E_0||V|)$ which is poly($n$).*

*Proof. Base case:* From Definition 3, $|E_0|$ is bounded above by the (polynomial) number of membranes at time 0. Thus $|E_0| \in$ poly($n$). Each of lemmata 1 to 5 gives an upper bound on the increase in the number of equivalence classes after one timestep for rule types $(a)$ to $(e_s)$, respectively. Lemma 2 has an additive increase of $|E_0||V|$ and the other four lemmata have an increase of 0. Thus at time 1 there is a computation path where the number of equivalence classes is $|E_1| \leq |E_0| + |E_0||V|$. (From Definitions 1 and 2, $|V| \in$ poly($n$) and $|V|$ is fixed for all $t$.)

*Inductive step:* Assume that $|E_i|$, the number of equivalence classes at time $i$, is polynomial in $n$. Then, while Lemmata 1 to 5, there exists a computation path where $|E_{i+1}| \leq |E_i| + |E_0||V|$.

After $t$ timesteps we have $|E_t| = O(|E_0| + t|E_0||V|)$, which is polynomial in $n$ if $t$ is. □

The proofs of the following five lemmata assume some ordering on the set of object types $V$ and on the rules $R$. For the proof of Lemma 2, we give a specific ordering, however for the other proofs any ordering is valid.

**Lemma 1.** *Given a configuration $C_i$ of a* $\mathbf{PMC}^S_{\mathcal{EAM}^0_{-a}}$ *system with $|E|$ equivalence classes. After a single timestep, where only rules of type (a) (object evolution) are applied, there exists a configuration $C_{i+1}$ such that $C_i \vdash C_{i+1}$ and $C_{i+1}$ has $\leq |E|$ equivalence classes.*

*Proof.* If a type $(a)$ rule is applicable to an object in a membrane in equivalence class $e_k$, then the rule is also applicable in exactly the same way to all membranes in $e_k$. Due to non-determinism in the choice of rules and objects, it could be the case that the membranes in $e_k$ evolve differently. However let us assume an ordering on the object types $V$ and on the rules $R$. We apply the type $(a)$ rules to objects using this ordering. Then all membranes in an equivalence class evolve identically in a timestep, and no new equivalence classes are created. Thus there is a computation path $C_i \vdash C_{i+1}$ where there is no increase in the number of equivalence classes. □

Observe that type $(b)$ rules have the potential to increase the number of equivalence classes in one timestep by sending different object types into different membranes from the same class. For example, if objects of type $v_1$ are sent into some of the membranes in an equivalence class, and $v_2$ objects are sent into the remainder, then we increase the number of equivalence classes by 1. The following lemma gives an additive polynomial upper bound on this increase.

**Lemma 2.** *Given a configuration $C_i$ of a* $\mathbf{PMC}^S_{\mathcal{EAM}^0_{-a}}$ *system $\Pi$ with $|E|$ equivalence classes. Let $|E_0|$ be the number of equivalence classes in the initial configuration of $\Pi$. Let $|V|$ be the number of distinct object types in $\Pi$. After a single timestep, where only rules of type (b) (incoming objects) are applied, there exists a configuration $C_{i+1}$ such that $C_i \vdash C_{i+1}$ and $C_{i+1}$ has $\leq |E| + |E_0||V|$ equivalence classes.*

*Proof.* Let $e_j$ be a parent equivalence class, thus $e_j$ represents one membrane (by Definition 3). If the child membranes of $e_j$ are all parent membranes themselves, then the type $(b)$ communication rule occurs without any increase to the number of equivalence classes. The remainder of the proof is concerned with the other case, where $e_j$ contains a non-zero number of equivalence classes of elementary membranes; by the lemma statement this number is $\leq |E|$.

For the remainder of this proof let $V' \subseteq V$ be the set of distinct object types in the membrane defined by $e_j$ for which there are rules in $R$ applicable for this timestep, let $\mathbb{V}$ be the total number of objects in the membrane defined by $e_j$, let $E' \subset E$ be the set of equivalence classes that describe the children of the membrane defined by $e_j$, and let $\mathbb{M}$ be the total number of membranes that are children of the membrane defined by $e_j$ (therefore $\mathbb{M}$ is the total number of membranes in $E'$). Furthermore we assume that $E'$ is ordered by number of membranes, i.e. we let $E' = (e_1, e_2, \ldots, e_{|E'|})$ where $|e_k|$ is the number of membranes in equivalence class $e_k$ and $\forall k$, $|e_k| \leq |e_{k+1}|$. Similarly we assume that $V'$ is ordered by the number of each object type, i.e. we let $V' = (v_1, v_2, \ldots, v_{|V'|})$ where $|v_k|$ is the multiplicity of objects of type $v_k$ and $\forall\ k$, $|v_k| \leq |v_{k+1}|$. This ordering ensures that the same deterministic computation path is followed for different instances of the same input configuration. We now consider the two possible cases.

*Case 1:* $\mathbb{V} < \mathbb{M}$. Table 1 explicitly gives the proof for this case. The $\mathbb{M}$ membranes, beginning with membranes from equivalence class $e_1$, each receive one object, beginning with available objects of type $v_1$. We continue, following the above orderings on $V'$ and $E'$, until there are no more objects to communicate. Thus after these type $(b)$ rules have been applied, some number from the $E'$ has received objects, leading to the $E'$ rows in the "Range" column of Table 1. If objects of one distinct type fill up an equivalence class exactly, that class cannot be split into further equivalence classes in that time step. The "Sub-case" column captures all possible (given our ordering on $E'$ and $V'$) ways that objects can fill increasing numbers of equivalence classes. The "Increase EC" column gives the increase in the equivalence classes after one timestep each sub-case. The worst case increase is caused by no equivalence class being exclusively filled up by a distinct object, this means every distinct object communicated will create a new equivalence class. The worst case increase in the total number of equivalence classes after one timestep is $|V'|$.

*Case 2:* $\mathbb{V} \geq \mathbb{M}$. Table 2 explicitly gives the proof for this case. The $\mathbb{M}$ membranes, beginning with membranes from equivalence class $e_1$, each receive one object, beginning with available objects of type $v_1$. We continue, following the above orderings on $V'$ and $E'$, until there are no more available membranes to communicate to. Thus after these type $(b)$ rules have been applied, some number from the $V'$ have been communicated, leading to the $V'$ rows in the "Range" column of Table 2. If membranes from one equivalence class all receive objects of the same distinct type, that class cannot be split into further equivalence classes in that time step. The "Sub-case" column captures all possible (given our ordering on $E'$ and $V'$) ways that equivalence classes can be filled by increasing numbers of distinct objects. The "Increase EC" column gives the increase in the equivalence classes after one timestep each sub-case. The worst case increase is caused by no equivalence class being exclusively filled up by a distinct object, this means every distinct object communicated will create a new equivalence class. The worst case increase in the total number of equivalence classes after one timestep is $|V'| - 1$.

This procedure is iterated over all parent membranes $e_j$ where type $(b)$ rules are applicable, by Definition 3 the number of such parent membranes $\leq |E_0|$. For each parent it is the case that $|V'| \leq |V|$. Thus there is a computation path $C_i \vdash C_{i+1}$ where the increase in the number of equivalence classes is $\leq |E_0||V'| \leq |E_0||V|$. □

**Lemma 3.** *Given a configuration $C_i$ of a* $\mathbf{PMC}^S_{\mathcal{EAM}^0_{-a}}$ *system with $|E|$ equivalence classes. After a single timestep, where only rules of type (c) (outgoing objects) are applied, there exists a configuration $C_{i+1}$ such that $C_i \vdash C_{i+1}$ and $C_{i+1}$ has $\leq |E|$ equivalence classes.*

*Proof.* If a type $(c)$ rule is applicable to an object in a membrane in equivalence class $e_k$, then the rule is also applicable in exactly the same way to all membranes in $e_k$. Due to non-determinism in the choice of rules and objects it could be the case that membranes in $e_k$ eject different symbols. However lets assume an ordering on the object types $V$ and on the rules $R$. We apply the type $(c)$ rules to objects using this ordering. Then all membranes in an equivalence class evolve identically in one (each membrane ejects the same symbol), and so no new equivalence classes are created from $e_k$. The single parent of all the membranes in $e_k$ is in an equivalence class $e_j$ which, by Definition 3, contains exactly one membrane and so no new equivalence classes are created from $e_j$.

Thus there is a computation path $C_i \vdash C_{i+1}$ where there is no increase in the number of equivalence classes. □

Interestingly, dissolution is the easiest rule to handle using our approach. The following lemma actually proves something stronger than the other lemmata: dissolution *never* leads to an increase in the number of equivalence classes.

**Lemma 4.** *Given a configuration $C_i$ of a* $\mathbf{PMC}^S_{\mathcal{EAM}^0_{-a}}$ *system with $|E|$ equivalence classes. After a single timestep, where only rules of type (d) (membrane dissolution) are applied then for all $C_{i+1}$, such that $C_i \vdash C_{i+1}$, $C_{i+1}$ has $\leq |E|$ equivalence classes.*

*Proof.* If there is at least one type $(d)$ rule that is applicable to an object and a membrane in equivalence class $e_k$, then there is at least one rule that is also applicable to all membranes in $e_k$. Unlike previous proofs, we do not require an ordering on the objects and rules: all membranes in $e_k$ dissolve and equivalence class $e_k$ no longer exists. The single parent of all the membranes in $e_k$ is in an equivalence class $e_j$ which, by Definition 3, contains exactly one membrane and so no new equivalence classes are created from $e_j$.

Thus for all $C_{i+1}$, where $C_i \vdash C_{i+1}$, there is no increase in the number of equivalence classes. □

**Lemma 5.** *Given a configuration $C_i$ of a* $\mathbf{PMC}^S_{\mathcal{EAM}^0_{-a}}$ *system with $|E|$ equivalence classes. After a single timestep, where only rules of type $(e_s)$ (symmetric membrane division) are applied, there exists a configuration $C_{i+1}$ such that $C_i \vdash C_{i+1}$ and $C_{i+1}$ has $\leq |E|$ equivalence classes.*

*Proof.* If a type $(e_s)$ rule is applicable to an object and membrane in equivalence class $e_k$, then the rule is also applicable in exactly the same way to all membranes in $e_k$. Due to non-determinism in the choice of rules and objects it could be the case that membranes in $e_k$ divide using and/or creating different symbols. However lets assume an ordering on the object types $V$ and on the rules $R$. We apply the type $(e_s)$ rules to objects (and membranes) using this ordering. Then all membranes in an equivalence class evolve identically in a timestep (each membrane in $e_k$ divides using the same rule). The number of membranes in $e_k$ doubles, but since each new membrane is identical, no new equivalence classes are created from $e_k$.

Thus there is a computation path $C_i \vdash C_{i+1}$ where there is no increase in the number of equivalence classes. □

### 3.3    Polynomial Time RAM Algorithm

Here we outline a RAM algorithm that simulates the computation of any membrane system of the class $\mathbf{PMC}_{\mathcal{EAM}^0_{-a}}^S$ in polynomial time (in input length $n$). The algorithm operates on any valid initial configuration and successively applies the evolution rules of the membrane system.

The algorithm makes explicit use of the polynomial size bounded registers described in Section 3.1. It also relies on the confluent nature of recogniser membrane systems and simulates only one of the set of valid computation paths. In particular, using the results from Section 3.2, the algorithm chooses a computation path that uses polynomial space by sorting the membranes, objects and rules of a configuration.

Our `sort` function runs in polynomial time (in input length $n$) and sorts lists of

- object multisets by object multiplicity, then lexicographically.
- equivalence classes by membrane multiplicity, then by label, and then by objects.
- rules by rules type, matching label, matching object, and then by output object(s).

Since instances of $\mathbf{PMC}_{\mathcal{EAM}^0_{-a}}^S$ are constructed by polynomial time (or logspace) deterministic Turing machines they are at most polynomial size. Also, since all instances of $\mathbf{PMC}_{\mathcal{EAM}^0_{-a}}^S$ run in polynomial time, if our algorithm simulates it with a polynomial time overhead we obtain a polynomial time upper bound.

Our algorithm begins with a configuration of $\mathbf{PMC}_{\mathcal{EAM}^0_{-a}}^S$ (see Algorithm 1). The input configuration is encoded into the registers of the RAM in polynomial time. The rules of the system are sorted and the algorithm then enters a loop. At each iteration all available rules are applied, this simulates a single timestep of the membrane systems computation. The loop terminates when the system ejects a `yes` or `no` object, indicating that the computation has halted. Since all instances of $\mathbf{PMC}_{\mathcal{EAM}^0_{-a}}^S$ run in polynomial time, this loop iterates a polynomial number of times. The total time complexity for running the simulation for time $t$ is $O(t|R||E|^2|V|)$.

**Table 1.** Increase in the number of equivalence classes (EC) when $\mathbb{V} < \mathbb{M}$. This table is used in the proof of Lemma 2. In the Graphic column, a □ represents an equivalence class and a ○ represents a single membrane in an equivalence class.

| Range | Sub-cases | Increase EC | Graphic |
|---|---|---|---|
| $1 \leq \mathbb{V} < |e_1|$ | | $|V|$ | |
| $|e_1| \leq \mathbb{V} < |e_1| + |e_2|$ | $e \subset E', |e| = 1$, every EC in $e$ is filled by objects from a $v \in V'$ | $|V''| - 1$ | |
| | $\nexists e \subset E'$, every EC in $e$ is filled by objects from a $v \in V'$ | $|V'|$ | – |
| $|e_1| + |e_2| \leq \mathbb{V} < |e_1| + |e_2| + |e_3|$ | $e \subset E', |e| = 2$, every EC in $e$ is filled by objects from a $v \in V'$ | $|V''| - 2$ | – |
| | $e \subset E', |e| = 1$, every EC in $e$ is filled by objects from a $v \in V'$ | $|V''| - 1$ | – |
| | $\nexists e \subset E'$, every EC in $e$ is filled by objects from a $v \in V'$ | $|V'|$ | – |
| $|e_1| + |e_2| + |e_3| \leq \mathbb{V} < |e_1| + |e_2| + |e_3| + |e_4|$ | $e \subset E', |e| = 3$, every EC in $e$ is filled by objects from a $v \in V'$ | $|V''| - 3$ | – |
| | $e \subset E', |e| = 2$, every EC in $e$ is filled by objects from a $v \in V'$ | $|V''| - 2$ | – |
| | $e \subset E', |e| = 1$, every EC in $e$ is filled by objects from a $v \in V'$ | $|V''| - 1$ | – |
| | $\nexists e \subset E'$, every EC in $e$ is filled by objects from a $v \in V'$ | $|V'|$ | – |
| ... | ... | ... | ... |
| $\sum_{\ell=1}^{|E'|-1} |e_\ell| \leq \mathbb{V} < \sum_{\ell=1}^{|E'|} |e_\ell|$ | $e \subset E', |e| < |E'|$, every EC in $e$ is filled by objects from a $v \in V'$ | $|V''| - |E'|$ | – |
| | ... | ... | – |
| | $e \subset E', |e| = 2$, every EC in $e$ is filled by objects from a $v \in V'$ | $|V''| - 2$ | – |
| | $e \subset E', |e| = 1$, every EC in $e$ is filled by objects from a $v \in V'$ | $|V''| - 1$ | – |
| | $\nexists e \subset E'$, every EC in $e$ is filled by objects from a $v \in V'$ | $|V'|$ | – |

**Table 2.** Increase in the number of equivalence classes (EC) when $\mathbb{V} \geq \mathbb{M}$. This table is used in the proof of Lemma 2. In the Graphic column, a □ represents an equivalence class and a ○ represents a single membrane in an equivalence class.

| Range | Sub-cases | Increase EC | Graphic |
|---|---|---|---|
| $0 < \mathbb{M} \leq |v_1|$ | - | 0 | $v_1$ |
| $|v_1| < \mathbb{M} \leq |v_1| + |v_2|$ | $\forall e \in E'$, every EC in $e$ is filled by objects from a $v \in V'$ | 0 | $v_1$, $v_2$ |
| | $\forall e \in E' - \{e_{i1}\}$, every EC in $e$ is filled by objects from a $v \in V'$ | 1 | $v_1$, $v_2$ |
| $|v_1| + |v_2| < \mathbb{M} \leq |v_1| + |v_2| + |v_3|$ | $\forall e \in E'$, every EC in $e$ is filled by objects from a $v \in V'$ | 0 | - |
| | $\forall e \in E' - \{e_{i1}\}$, every EC in $e$ is filled by objects from a $v \in V'$ | 1 | - |
| | $\forall e \in E' - \{e_{i1}, e_{i2}\}$, every EC in $e$ is filled by objects from a $v \in V'$ | 2 | - |
| $|v_1| + |v_2| + |v_3| < \mathbb{M} \leq |v_1| + |v_2| + |v_3| + |v_4|$ | $\forall e \in E'$, every EC in $e$ is filled by objects from a $v \in V'$ | 0 | - |
| | $\forall e \in E' - \{e_{i1}\}$, every EC in $e$ is filled by objects from a $v \in V'$ | 1 | - |
| | $\forall e \in E' - \{e_{i1}, e_{i2}\}$, every EC in $e$ is filled by objects from a $v \in V'$ | 2 | - |
| | $\forall e \in E' - \{e_{i1}, e_{i2}, e_{i3}\}$, every EC in $e$ is filled by objects from a $v \in V'$ | 3 | - |
| $\cdots$ | $\cdots$ | $\cdots$ | - |
| $\sum_{\ell=1}^{|V'|-1} |v_\ell| < \mathbb{M} \leq \sum_{\ell=1}^{|V'|} |v_\ell|$ | $\forall e \in E'$, every EC in $e$ is filled by objects from a $v \in V'$ | 0 | - |
| | $\forall e \in E' - \{e_{i1}\}$, every EC in $e$ is filled by objects from a $v \in V'$ | 1 | - |
| | $\forall e \in E' - \{e_{i1}, e_{i2}\}$, every EC in $e$ is filled by objects from a $v \in V'$ | 2 | - |
| | $\forall e \in E' - \{e_{i1}, e_{i2}, e_{i3}\}$, every EC in $e$ is filled by objects from a $v \in V'$ | 3 | - |
| | $\cdots$ | $\cdots$ | - |
| | $\forall e \in E' - \bigcup_{k=1}^{|E'|-1} \{e_{ik}\}$, every EC in $e$ is filled by objects from a $v \in V'$ | $|V'| - 1$ | - |

---

**Algorithm 1.** The main body of the membrane simulation algorithm. The rules of the system are sorted and then applied to the configuration at each timestep until the system accepts or rejects its input.

---

**Input.** a configuration of $\mathbf{PMC}^{S}_{\mathcal{EAM}^0_{-a}}$

**Output.** The deciding configuration of the system

Initialise registers with input system;

sortedRules $\leftarrow$ sort(*rules*);

|               |                                                        |     |
|---------------|--------------------------------------------------------|-----|
| $O(t)$        | **repeat**                                             |     |
|               | /* evolve the membrane system one step                 | */  |
| $O(\|E\|)$    | **forall** equivalence_class *in* membraneSystem **do** |     |
| $O(\|R\|\|E\|\|V\|)$ | └ ApplyRules(equivalence_class);                |     |
|               | **until** yes *or* no *object is in skin membrane* ;   |     |

---

---

**Function.** ApplyRules(*equivalence_class*) Applies all applicable rules for an equivalence class for one timestep

---

**Input.** equivalence_class

**Output.** equivalence_class after one timestep of computation

b_rules $\leftarrow \emptyset$;

b_ecs $\leftarrow \emptyset$;

b_objs $\leftarrow \emptyset$;

|                     |                                                                              |
|---------------------|------------------------------------------------------------------------------|
| $O(\|R\|)$          | **forall** rule *in* sortedRules **do**                                       |
|                     | **if** rule.*label matches* equivalence_class.*label and* rule *is not type (b)* **then** |
| $O(\|V\|)$          | **forall** object *in* sortedObjects **do**                                   |
|                     | **if** *not all copies of* object *have been used* **then**                   |
|                     | **if** rule *is type (a)* **then**                                            |
| $O(\|V\|)$          | │ Apply_a_rule(equivalence_class, object, rule);                              |
|                     | **else if** rule *is type (c)* **then**                                       |
| $O(1)$              | │ Apply_c_rule(equivalence_class, object, rule);                             |
|                     | **else if** rule *is type (d)* **then**                                       |
| $O(\|V\|)$          | │ Apply_d_rule(equivalence_class, object, rule);                             |
|                     | **else if** rule *is type* $(e_s)$ **then**                                   |
| $O(1)$              | └ Apply_e_rule(equivalence_class, object, rule);                             |
|                     |                                                                              |
|                     | **if** rule *is type (b)* **then**                                            |
| $O(\|E\|)$          | **forall** child_c *in* equivalence_class **do**                             |
|                     | **if** child_c.label = rule.lhsLabel *and* object.used $\geq 1$ **then**      |
|                     | append child_c to b_ecs ;                                                     |
|                     | append object to b_objs ;                                                     |
| $O(\|V\|\|E\|)$     | └ Apply_b_rule(b_ecs, b_objs, rule)                                           |
|                     |                                                                              |
| $O(\|V\| \times \|E\|)$ | reset all used counters to 0;                                            |

**Function.** Apply_a_rule(*equivalence_class, object, rule*) applies a single type (*a*) rule to instances of an object in an equivalence class. Total time complexity $O(|V|)$.

**Input.** equivalence_class, object, rule

**Output.** equivalence_class after a type (*a*) rule on an object has been applied

$O(|V|)$    **forall** resultingObject *in* rule.outAobjects **do**

     multiplicity of resultingObject in equivalence_class + = the multiplicity of matching object − the number of object.used × the resultingObject.multiplicity ;

     used number of resultingObject in the equivalence_class + = the multiplicity of resultingObject × object.multiplicity − object.used ;

decrement object.multiplicity ;

set object.used = object.multiplicity ;

---

**Function.** Apply_c_rule(*equivalence_class, object, rule*) applies a single rule of type (*c*) to a membrane. Total time complexity $O(1)$.

**Input.** equivalence_class

**Output.** equivalence_class after a (*c*) rule have been applied

decrement object.multiplicity ;

increment object.multiplicity in equivalence_class.parent of the generated object;

increment object.used in equivalence_class.parent of the generated object;

increment equivalence_class.used ;

---

**Function.** Apply_d_rule(*equivalence_class, object, rule*). This function applies dissolution rules to an equivalence class. It calculates the total number of each object in the equivalence class and adds it to the parent. It also copies the child membranes from the dissolving membrane and adds them to the parents child list. The total time complexity is $O(|V|)$.

**Input.** equivalence_class

**Output.** equivalence_class after (*d*) rule has been applied

decrement object.multiplicity ;

increment object.multiplicity in equivalence_class.parent from the rule;

increment object.used in equivalence_class.parent from the rule;

/* move contents of the dissolved membrane to its parent     */

$O(|V|)$    **forall** move_object *in* equivalence_class *objects* **do**

     add move_object.multiplicity × equivalence_class.multiplicity to move_object.multiplicity in equivalence_class.parent ;

     add move_object.used × equivalence_class.multiplicity to move_object.used in equivalence_class.parent ;

     move_object.multiplicity ← 0;

     move_object.used ← 0;

equivalence_class.parent.children ← equivalence_class.parent.children ∪ equivalence_class.children ;

equivalence_class.multiplicity ← 0;

equivalence_class ← ∅;

---

**Function.** `Apply_es_rule`(*equivalence_class, object, rule*). Applies a single rule of type ($e_s$) to a membrane. Total time complexity $O(1)$.

---

**Input.** equivalence_class
**Output.** equivalence_class after ($e_s$) rule has been applied
decrement object.multiplicity ;
increment object.multiplicity from the rule;
increment object.used from the rule;
increment equivalence_class.used ;
equivalence_class.multiplicity ← equivalence_class.multiplicity × 2;

---

**Function.** `Apply_b_rules`(*b_equivalence_classes, b_objects, b_rules*). Total time complexity $O(|V||E|)$.

---

**Input.** membrane
**Output.** membrane after ($b$) rules have been applied
b_objects_sorted ← `sort`(b_objects);
b_equivalence_classes_sorted ← `sort`(b_equivalence_classes);
$O(|V|)$     **forall** object *in* b_objects_sorted **do**
$O(|E|)$         **forall** equivalence_class *in* b_equivalence_classes_sorted **do**
           **if** object.multiplicity $<$ equivalence_class.multiplicity **then**
             copy equivalence_class to new_equiv_class ;
             subtract object.multiplicity from new_equiv_class.multiplicity ;
             equivalence_class.multiplicity ← object.multiplicity ;
             equivalence_class.used ← equivalence_class.multiplicity ;
             increment equivalence_class.object.multiplicity ;
             increment equivalence_class.object.used ;
           **else if** object.multiplicity $\geq$ equivalence_class.multiplicity **then**
             increment equivalence_class.object.multiplicity ;
             increment equivalence_class.object.used ;
             equivalence_class.used ← equivalence_class.multiplicity ;
             subtract equivalence_class.multiplicity from object.multiplicity ;

---

## 4   Conclusion

We have given a **P** upper bound on the computational power of one of a number of membrane systems that fall under the so-called **P**-conjecture. In particular we consider a variant of membrane systems that allows only symmetric devision. This variant can easily generate an exponential number of membranes and objects in polynomial time. We restricted the uniformity condition to logspace, making the **P** lower bound more meaningful. Our technique relies on being able to find computation paths that use only polynomial space in polynomial time. It seems that this technique is not naïvely applicable to the case of asymmetric division: it is possible to find examples where all computation paths are forced to use an exponential number of equivalence classes.

Furthermore the result seems interesting since before before now, all models without dissolution rules were upper bounded by **P** and all those with dissolution rules characterised **PSPACE**. This result shows that despite having dissolution rules, by using only symmetric elementary division we restrict the system so that it does not create exponential space on all computation paths in polynomial time.

## Acknowledgements

Niall Murphy is supported by the Irish Research Council for Science, Engineering and Technology. Damien Woods is supported by Science Foundation Ireland grant number 04/IN3/1524. We give a special thanks to Thomas J. Naughton for interesting comments and ideas.

## References

1. Alberts, B., Johnson, A., Lewis, J., Raff, M., Roberts, K., Walter, P.: Molecular Biology of the Cell, 4th edn. Garland Science, New York (2002)
2. Alhazov, A., Pérez-Jiménez, M.J.: Uniform solution to QSAT using polarization-less active membranes. In: Gutiérrez-Naranjo, M.A., Păun, G., Riscos-Núñez, A., Romero-Campero, F.J. (eds.) Fourth Brainstorming Week on Membrane Computing, Sevilla, January 30-February 3, 2006, vol. I, pp. 29–40. Fénix Editora (2006)
3. Balcázar, J.L., Diaz, J., Gabarró, J.: Structural complexity I, 2nd edn. Springer, Berlin (1988)
4. Gutiérrez-Naranjo, M.A., Pérez-Jiménez, M.J., Riscos-Núñez, A., Romero-Campero, F.J.: Computational efficiency of dissolution rules in membrane systems. International Journal of Computer Mathematics 83(7), 593–611 (2006)
5. Ladner, R.E.: The circuit value problem is log space complete for P. SIGACT News 7(1), 18–20 (1975)
6. Păun, G.: P Systems with active membranes: Attacking NP-Complete problems. Journal of Automata, Languages and Combinatorics 6(1), 75–90 (2001)
7. Păun, G.: Membrane Computing. An Introduction. Springer, Berlin (2002)
8. Pérez-Jiménez, M.J., Romero-Jiménez, A., Sancho-Caparrini, F.: Complexity classes in models of cellular computing with membranes. Natural Computing 2(3), 265–285 (2003)
9. Sosík, P., Rodríguez-Patón, A.: Membrane computing and complexity theory: A characterization of PSPACE. Journal of Computer and System Sciences 73(1), 137–152 (2007)

# Balancing Performance, Flexibility, and Scalability in a Parallel Computing Platform for Membrane Computing Applications

Van Nguyen, David Kearney, and Gianpaolo Gioiosa

School of Computer and Information Science
University of South Australia
{Van.Nguyen,David.Kearney,Gianpaolo.Gioiosa}@unisa.edu.au

**Abstract.** It is an open question whether it is feasible to develop a parallel computing platform for membrane computing applications that significantly outperforms equivalent sequential computing platforms while still achieving acceptable flexibility and scalability. To move closer to an answer to this question, we have investigated a novel approach to the development of a parallel computing platform for membrane computing applications that has the potential to deliver a good balance between performance, flexibility and scalability. This approach involves the use of reconfigurable hardware and an intelligent software component that is able to configure the hardware to suit the specific properties of the membrane computing model to be executed. We have already developed a prototype computing platform called Reconfig-P based on the approach. Reconfig-P is the first computing platform of its type to implement parallelism at both the system and region levels. In this paper, we describe the functionality of the intelligent software component responsible for hardware configuration in Reconfig-P, and perform an empirical analysis of the performance, flexibility and scalability of Reconfig-P. The empirical results suggest that the implementation approach on which Reconfig-P is based is a viable means of attaining a good balance between performance, flexibility and scalability.

## 1 Introduction

To exploit the performance advantage of the large-scale parallelism of membrane computing models, it is necessary to execute them on a parallel computing platform. However, the use of a parallel computing platform instead of a sequential computing platform often comes at the cost of reduced flexibility and scalability.

The parallel computing platforms for membrane computing applications that predate the research described in this paper [1, 6, 7] do not achieve a good balance between performance, flexibility and scalability. Even so, because research in this area is in its early stages, it is still an open question whether it is feasible to develop a parallel computing platform for membrane computing applications that significantly outperforms equivalent sequential computing platforms while still achieving acceptable flexibility and scalability. To move closer to an answer

G. Eleftherakis et al. (Eds.): WMC8 2007, LNCS 4860, pp. 385–413, 2007.
© Springer-Verlag Berlin Heidelberg 2007

to this question, it is important to investigate the viability of implementation approaches that have the potential to deliver a good balance between performance, flexibility and scalability.

The research presented in this paper involves an investigation of a novel approach to the development of a parallel computing platform for membrane computing applications. This approach involves the use of reconfigurable hardware and an intelligent software component that is able to configure the hardware to suit the specific properties of the membrane computing model to be executed. We have developed a prototype computing platform called Reconfig-P based on the approach. In a companion paper [4], we describe the hardware implementation of membrane computing that is at the foundation of Reconfig-P, and present a theoretical analysis of its performance. In this paper, we describe the functionality of the intelligent software component responsible for hardware configuration in Reconfig-P, and perform an empirical analysis of the performance, flexibility and scalability of Reconfig-P.

## 2   Background

In this section, we introduce key concepts and previous research associated with parallel computing platforms for membrane computing applications.

### 2.1   Membrane Computing and Its Applications

Membrane computing [5] investigates models of computation inspired by certain structural and functional features of biological cells, especially features that arise because of the presence and activity of biological membranes. We call membrane computing models *P system models* and their instances *P systems*.

Following is a definition of an example P system model. All P systems $\Pi$ that instantiate the model have all the fundamental features of a P system plus two common additional features (catalysts and reaction rule priorities). We call the model the *core P system model*.

$$\Pi = (V, T, C, \mu, w_1, ..., w_m, (R_1, \rho_1), ..., (R_m, \rho_m)), \text{ where}$$

- $V$ is an alphabet that contains labels for all the *types of objects* in the system;
- $T \subseteq V$ is the *output alphabet*, which contains labels for all the types of objects that are relevant to the determination of the system output;
- $C \subseteq V - T$ is the alphabet that contains labels for all the *types of catalysts*, which are the types of objects whose multiplicities cannot change through the application of a reaction rule;
- $\mu$ is a hierarchical membrane structure consisting of $m$ membranes, with the membranes (and hence the regions defined by the membranes) injectively labeled by the elements of a given set $H$ of $m$ labels (in this paper, $H = \{1, 2, ..., m\}$);
- each $w_i, 1 \leq i \leq m$, is a string over $V$ that represents the *multiset of objects* contained in region $i$ of $\mu$ in the initial configuration of the system;

- each $R_i, 1 \leq i \leq m$, is a finite *set of reaction rules* over $V$ associated with the region $i$ of $\mu$;
- a *reaction rule* is a pair $(r, p)$, written in the form $r \to p$, where $r$ is a string over $V$ representing a multiset of reactant objects and $p$ is a string over $\{a_{\mathrm{here}}, a_{\mathrm{out}}, a_{\mathrm{in}} \mid a \in V\}$ representing a multiset of product objects, each of which either (a) stays in the region to which the rule is associated (the subscript 'here' is usually omitted), (b) travels 'out' into the region that immediately contains the region to which the rule is associated, or (c) travels 'in' to one of the regions that is immediately contained by the region to which the rule is associated; and
- each $\rho_i$ is a partial-order relation over $R_i$ which defines the *relative priorities of the reaction rules* in $R_i$.

P system models have been applied in a variety of domains [2]. Most existing applications of membrane computing are targeted at the modeling and simulation of biological systems.

## 2.2 Quality Attributes of Computing Platforms for Membrane Computing Applications

Performance, flexibility and scalability are three of the most important quality attributes for a computing platform for membrane computing applications. Ensuring that a computing platform has all three of these attributes to an acceptable degree is a challenge, because a factor that promotes one of the attributes can sometimes demote another one of the attributes. In this section, we define the attributes of performance, flexibility and scalability in the context of a computing platform for membrane computing applications, explain the significance of these attributes, and indicate the connections that exist between them.

**Performance.** By the performance of a computing platform for membrane computing applications we mean the speed at which it executes P systems; that is, the amount of useful processing it performs per unit time. A suitable measure of the amount of useful processing performed is the number of reaction rule applications performed. Thus the performance of a computing platform for membrane computing applications can be measured in reaction rule applications per unit time.

**Flexibility.** By the flexibility of a computing platform for membrane computing applications we mean the extent to which it can support the execution of a wide range of P systems. Thus a flexible computing platform for membrane computing applications must be able to adapt to the specific properties of the P system to be executed. The greater the flexibility of the computing platform, the greater the diversity among the P systems in the class of P systems that the computing platform is able to execute.

**Scalability.** By the scalability of a computing platform for membrane computing applications we mean the extent to which increases in the size of the P system to be executed do not lead to a reduction in the ability of the computing platform to perform its functions or a reduction in the performance of the computing platform. We take the size of a P system as being largely determined by the number of regions and the number of reaction rules it contains.

**Connections between performance, flexibility and scalability.** The performance of a computing platform for membrane computing applications can be increased by tailoring its implementation to the specific properties of the P systems it is intended to execute. However, the greater the diversity of these P systems, the more difficult it is to efficiently tailor the implementation to their specific properties. Therefore, increasing the performance of the computing platform is likely to come at the cost of reduced flexibility, while increasing the flexibility of the computing platform is likely to come at the cost of reduced performance.

Increasing the flexibility of a computing platform for membrane computing applications involves supporting additional P system features. Naturally, this usually requires the implementation of additional data structures and algorithms. In software-based computing platforms, the implementation of additional data structures is likely to come at the cost of increased memory consumption. In hardware-based computing platforms, the implementation of additional data structures comes at the cost of increased hardware resource consumption, as does the implementation of additional algorithms. Therefore, given that memory resources and hardware resources are limited, implementing additional P system features reduces the maximum size of the P systems that a computing platform for membrane computing applications can execute. Thus increasing the flexibility of a computing platform for membrane computing applications is likely to come at the cost of reduced scalability, while increasing the scalability of such a computing platform is likely to come at the cost of reduced flexibility.

### 2.3   Types of Computing Platforms

We identify three major types of computing platforms: sequential computing platforms, software-based parallel computing platforms and hardware-based parallel computing platforms.

*Sequential computing platforms* are typically based on a software-programmed microprocessor. When such a microprocessor is used, the execution hardware is abstracted by the instruction set architecture, which provides a set of specific instructions that the microprocessor can process to perform computations. This is a very flexible computing solution since it is possible to change the functionality of the computing platform simply by modifying its software — there is no need to modify the hardware configuration. As a result of this flexibility, the same fixed hardware can be used for many applications. However, the flexibility comes at the cost of lower performance. As each instruction needs to be sequentially fetched from memory and decoded before being executed, there is a high execution overhead associated with each individual operation. Furthermore, only one instruction can be executed at a time.

*Software-based parallel computing platforms* are typically based on a cluster of software-programmed microprocessors. Because the microprocessors execute in parallel, software-based parallel computing platforms can significantly outperform sequential computing platforms for many applications. The microprocessors synchronize their activities by using shared memory or by sending messages to each other (often over a network). Such synchronization can be very time consuming, and therefore can hinder performance significantly. Increasing the performance of a software-based parallel computing platform involves increasing the amount of parallelism and therefore requires the inclusion of additional microprocessors. However, as the number of microprocessors increases, the overheads associated with synchronization increase substantially (unless the overall algorithm executed by the computing platform can be neatly partitioned into separate procedures that are largely independent of each other). This fact limits the scalability of software-based parallel computing platforms.

*Hardware-based parallel computing platforms* execute algorithms that have been directly implemented in hardware. The hardware platform implements the algorithm in terms of the parallel activities of a certain number of processors that are spatially, rather than temporally, related. The ability to use parallel processors brings a potentially very significant improvement in execution time performance. However, the use of the spatial dimension means that the number of processors, and therefore the class of algorithms, that can be implemented on the platform is constrained by the amount of hardware resources available on the platform. In one approach, an application-specific integrated circuit (ASIC) is used. The design of an ASIC is tailored to a specific algorithm. As a consequence, ASICs usually achieve a higher performance than software-programmed microprocessors when executing the algorithm for which they were designed. However, with this higher performance comes reduced flexibility: as the implemented algorithm is fabricated on a silicon chip, it cannot be altered without creating another chip. In another approach, reconfigurable hardware is used. Unlike ASICs, reconfigurable hardware can be modified. Therefore, by using reconfigurable hardware, it is possible to improve on the performance of software-based computing platforms while retaining some of their flexibility. A field-programmable gate array (FPGA) is a type of reconfigurable hardware device. An FPGA consists of a matrix of logic blocks which are connected by means of a network of wires. The logic blocks at the periphery of the matrix can perform I/O operations. The functionality of the logic blocks and the connections between them can be modified by loading configuration data from a host computer. In this way, any custom digital circuit can be mapped onto the FPGA, thereby enabling it to execute a variety of applications. Such digital circuits are specified in hardware description languages. A very popular hardware description language is VHDL. VHDL allows circuits to be specified either in terms of a structural description of the circuit or in terms of low-level algorithmic behaviors of the circuit. Another popular hardware description language is Handel-C. Unlike VHDL, Handel-C does not support the specification of the structural features of a hardware circuit. However, having a syntax similar to that of the C programming language,

Handel-C allows algorithms to be specified at a very abstract level, and therefore eases the process of designing a circuit for an application.

## 2.4   Existing Parallel Computing Platforms for Membrane Computing Applications

In this section, we provide a brief survey of existing parallel computing platforms for membrane computing applications.

**Software-based parallel computing platforms.** Two research groups have created prototypes of software-based parallel computing platforms for membrane computing applications. Ciobanu and Guo [1] have implemented a simulation of P systems on a Linux cluster using C++ and a library of functions for message-passing parallel computation called the Message Passing Interface (MPI), while Syropoulos and colleagues [7] have implemented a distributed simulation of P systems using Java Remote Method Invocation (RMI). We discuss Ciobanu and Guo's computing platform below.

*Ciobanu and Guo's computing platform.* Ciobanu and Guo's computing platform is a software program written in C++ that is designed to run on a cluster of computers. The communication mechanism for the computing platform is implemented using MPI. In its prototype form, the computing platform consists of a Linux cluster, in which each node has two 1.4GHz Intel Pentium III CPUs and 1GB of memory, and the nodes are connected by gigabit Ethernet.

Ciobanu and Guo's computing platform supports the execution of a class of P systems that is very similar to the class of P systems that instantiate the core P system model. That is, the computing platform implements most of the basic features of P systems, but does not implement additional features such as membrane creation and dissolution. In the computing platform, each region of a P system is modeled as a separate computational process. Such a process implements the application of the reaction rules in its corresponding region. The processes for the regions in the P system execute in parallel. Communication and synchronization between regions is implemented using MPI.

As the threads for the reaction rules in a region execute on the same node in the cluster, and there are only two processors per node, it would seem that it is impossible for the computing platform to achieve region-level parallelism for anything other than small P systems. To achieve region-level parallelism for larger P systems, it would be necessary to increase the number of processors in a node from two to a number at least equal to the number of reaction rules in the region corresponding to that node. Thus without the inclusion of additional nodes, the computing platform cannot be said to implement region-level parallelism, although it does implement system-level parallelism.

Ciobanu and Guo indicate that the major problem with their computing platform from the point of view of performance is the overhead associated with communication and cooperation between regions. Such communication and cooperation consumes most of the total execution time.

Ciobanu and Guo do not evaluate the scalability of their computing platform. However, it is clear that the scalability of the computing platform is limited to a large extent by the nature of a cluster-based implementation approach. For example, to execute P systems with a large number of regions, the computing platform would have to include a large number of nodes, since there is a one-to-one correspondence between regions and nodes. As a consequence, there would be very significant overheads associated with communication and synchronization between regions, and this would have an adverse impact on the performance of the computing platform.

As it implements only a basic P system model, Ciobanu and Guo's computing platform is not capable of executing P systems that have additional features such as symport and antiport rules. This detracts from its flexibility. Nevertheless, since the existing implementation is expressed at a level of abstraction at which the high-level features of a P system are apparent, it seems very feasible that the computing platform could be extended to support additional P system features.

**Hardware-based parallel computing platforms.** A few researchers have designed digital circuits for particular aspects of P systems (e.g., see [3]). However, to the best of our knowledge, only Petreska and Teuscher [6] have implemented a hardware-based computing platform for membrane computing applications. We discuss Petreska and Teuscher's computing platform below.

*Petreska and Teuscher's computing platform.* Petreska and Teuscher [6] have developed a full implementation of a particular P system model on reconfigurable hardware. This P system model is similar to the core P system model, except that it also includes the feature of membrane creation and dissolution. The hardware architecture for the specific P system to be executed, which is specified in structural VHDL, is elegant in that it contains only one type of high-level hardware component (a universal component) and interconnections between components of this type.

Petreska and Teuscher have demonstrated the feasibility of implementing some of the important features of membrane computing on reconfigurable hardware. Nevertheless, their computing platform has four main limitations.

First, the computing platform does not exploit the performance advantages of the membrane computing paradigm. This is primarily because it does not implement parallelism at the region level (i.e., the reaction rules in a region are applied sequentially). Achieving region-level parallelism requires the implementation of a scheme for the resolution of conflicts that arise when different reaction rules compete for or produce the same types of objects in the same region at the same time. It is difficult to implement such a scheme efficiently in hardware, especially when a low-level hardware description language is used, and this is perhaps a major reason why Petreska and Teuscher did not attempt to do so.

Second, the computing platform is inflexible. As the computing platform uses only one type of high-level hardware component and connects components of this type to build hardware architectures in a fixed manner, the extent to which the hardware architecture for a P system can be tailored to the specific characteristics of the P system is limited.

Third, the computing platform is not extensible. As it is specified at the hardware level in a low-level hardware description language, adding support for additional P system features would require redesigning the hardware for the computing platform directly. This is likely in most cases to be a difficult and time-consuming task, given the dependence of the computing platform on the design of a single universal hardware component. Thus there is limited opportunity to improve the flexibility of the computing platform.

Fourth, the computing platform has limited scalability. As there is only a limited ability to tailor the hardware architecture to the specific characteristics of the P system to be executed, the hardware architecture often includes many redundant hardware components. These redundant components unnecessarily consume hardware resources.

As it implements membrane creation and dissolution in addition to the basic P system features included in the core P system model, Petreska and Teuscher's computing platform can execute a wider range of P systems than Ciobanu and Guo's computing platform. So, in this respect, it is more flexible than Ciobanu and Guo's computing platform.

# 3   A Proposed Implementation Approach

In this section, we specify the key features of the novel implementation approach on which Reconfig-P is based, and explain why this implementation approach has the potential to deliver a good balance between performance, flexibility and scalability.

## 3.1   Key Features of the Implementation Approach

The implementation approach on which Reconfig-P is based involves

- use of a reconfigurable hardware platform,
- generation of a customized digital circuit for each P system to be executed, and
- use of a hardware description language that allows digital circuits to be specified at a level of abstraction similar to the level of abstraction at which a general-purpose procedural software programming language (such as C) allows algorithms to be specified.

In the approach, a software component of the computing platform is responsible for analyzing the structural and behavioral features of the P system to be executed and producing a hardware description for the P system that is tailored to these features. When determining the hardware description for the P system, the software component aims to maximize performance and minimize hardware resource consumption.

## 3.2   Potential of the Implementation Approach

The use of reconfigurable hardware opens up the possibility of generating custom digital circuits for P systems. The ability to generate a custom circuit for the P system to be executed makes it possible to design this circuit according to the specific structural and behavioral features of the P system, and therefore facilitates the design of circuits that exhibit good performance and economical hardware usage. Therefore the implementation approach facilitates the development of a computing platform that exhibits good performance and economical hardware usage. For example, because the number of reaction rules in the P system to be executed is known before it is executed, the circuit for the P system can be designed in such a way that it includes exactly that number of processing units to implement the reaction rules. Without the possibility of generating a custom circuit, the circuit for the P system would have to include a fixed number of processing units for reaction rules, and therefore would often include redundant hardware components. Also, because it is possible by inspection of the definitions of the reaction rules in a P system to determine for any two regions in the P system whether it is possible for objects to traverse between these regions, the circuit for a P system can be designed in such a way that the logic that implements object traversal is included only for those inter-region connections over which object traversal is possible.

The fact that digital circuits are specified at a level of abstraction similar to that at which a general-purpose procedural software programming language specifies algorithms, rather than at a level of abstraction that reveals the structure or low-level algorithmic behavior of the circuits, makes it more feasible to develop a software component that is able to flexibly adapt to the specific features of the P system to be executed when generating a circuit for that P system. The greater the ability of the software component to flexibly adapt to the specific features of the P system to be executed, the greater the range of P systems for which it is capable of generating circuits that exhibit good performance and economical hardware usage. Therefore the implementation approach facilitates the development of a computing platform that exhibits good flexibility. For example, as mentioned in Section 2.4, implementing parallelism at the region level of a P system requires resolving conflicts that may occur when different reaction rules update the same multiplicity values. If a low-level hardware description language were used, it would be very difficult to resolve such conflicts in an efficient manner. The use of a high-level hardware description language makes it more feasible that a solution to the conflict resolution problem can be found.

Because it involves the use of a hardware description language that is incapable of expressing the low-level structure and behavior of digital circuits, the implementation approach limits the extent to which low-level optimizations of circuits can be carried out. However, it is unlikely that the benefits of customization and flexibility mentioned above could be achieved if a low-level hardware description language were used.

The above considerations suggest that the implementation approach has the potential to deliver a good balance between performance, flexibility and scalability in a parallel computing platform for membrane computing applications.

# 4   Description of Reconfig-P

In this section, we describe Reconfig-P, our prototype hardware-based parallel computing platform for membrane computing applications based on the implementation approach specified in Section 3. Being the first hardware-based computing platform to implement parallelism at both the system and region levels, Reconfig-P advances the state-of-the-art in hardware implementations of membrane computing. First, we specify the functional requirements of Reconfig-P. Second, we provide an overview of the major components of Reconfig-P and the roles of these components in the execution of applications. Third, we provide an overview of the functionality of P Builder, the software component of Reconfig-P that is responsible for generating customized hardware representations for P systems. Finally, we describe how P Builder represents the fundamental structural and behavioral features of P systems in hardware.

## 4.1   Functional Requirements

Reconfig-P is required to execute P systems that instantiate the core P system model on reconfigurable hardware. In addition, to facilitate testing of P system designs, Reconfig-P is required to enable the user to execute a P system in software, and view the configuration-by-configuration evolution of the P system, before generating a hardware circuit for the P system.

It is not a strict requirement that Reconfig-P implement the nondeterminism of P systems.

## 4.2   System Overview

Figure 1 shows the major components of Reconfig-P and the roles of these components in the execution of a P system.

(1) The user begins using Reconfig-P by writing a *P system specification*. This specification defines a P system that is described in terms of the *core P system model*. (2) The *hardware source code generator* called *P Builder* (which is hidden from the user) processes the input information. (3) P Builder analyses the P system specification, and then generates Handel-C source code that implements a customized hardware representation for the P system. (4) The user can choose to (a) execute the source code in hardware, or (b) simulate the execution of the source code in software. (5) The ability to generate *simulation source code* enables users to examine their P system design before building a corresponding hardware circuit. (6) The *simulation instance* (specified by a DLL file) is executed on a *host* computer. The host computer invokes the simulation feature provided by the Celoxica *DK Design Suite* to allow users to (a) view the evolution of their P system one configuration at a time, or (b) return the output of the simulation in an *output file*. (7) The generation of *hardware execution source code* allows the user, once they have finalized the design of their P system, to build a hardware circuit for the P system. (8) The hardware execution source code is then synthesized into a hardware circuit. A *hardware execution instance* (specified by a bitstream) can then be executed on a *reconfigurable hardware*

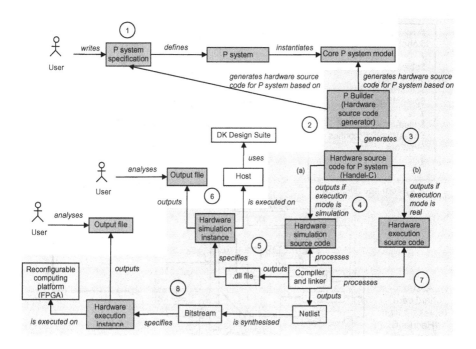

**Fig. 1.** An overview of Reconfig-P. The shaded region covers the components of Reconfig-P that are transparent to the user.

*platform* (an FPGA). The FPGA communicates with the host computer via a PCI bus. The output of the execution instance is stored in an *output file*, which can then be analyzed by the user. Much of the process of executing a P system is transparent to the user. The shaded region in Figure 1 covers the components of Reconfig-P that are transparent to the user.

### 4.3   Overview of P Builder

P Builder is responsible for implementing the hardware reconfiguration capability of Reconfig-P. It generates customized Handel-C source code for a P system based on the specific characteristics of the P system.

P Builder interprets a simple declarative language which is used by the user to specify P systems. More specifically, P Builder supports the execution of a P system by

1. converting a text representation of the P system (the P system specification) into software objects (written in Java);
2. converting the object representation into Handel-C source code that may be regarded as an abstract hardware representation; and then
3. converting the Handel-C source code into a hardware circuit (by invoking Xilinx tools) or initiating a software simulation of the hardware circuit specified by the Handel-C code (by invoking the DK Design Suite).

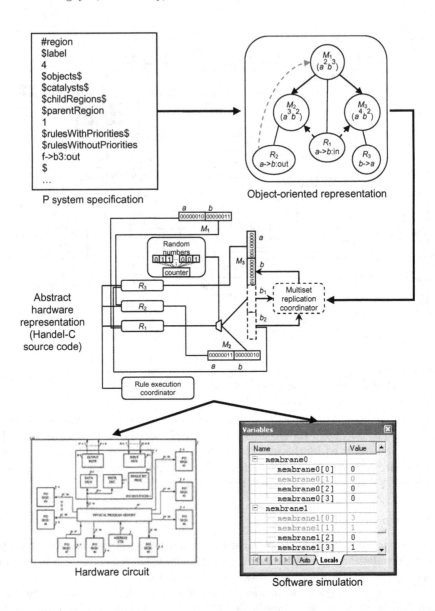

**Fig. 2.** P Builder converts the text specification of a P system into an executable hardware circuit or a software simulation of such a hardware circuit. In the intermediate stages of the conversion process, the P system is represented as a set of software objects and then as a set of abstract hardware components (specified in Handel-C).

Since P Builder is hidden from the user, the mechanics of the conversion process it performs are transparent to the user. The conversion process is illustrated in Figure 2.

## 4.4   How P Builder Generates Abstract Hardware Representations for P Systems

In [4], we describe the hardware implementation of membrane computing that is at the foundation of Reconfig-P. In this section, we describe how P Builder generates the Handel-C source code that specifies the hardware implementation for a specific P system.

Although at the software level P Builder represents all features of the P system to be executed, at the hardware level it represents only those aspects that are needed to compute the evolution of the P system. This is done to save hardware resources. The essential aspects include: (a) multisets of objects in regions, (b) inter-region containment relationships, (c) reaction rules, (d) application of reaction rules, and (e) synchronization of reaction rules.

**Multisets of objects in regions.** The multiset of objects in a region, when represented in software, has two key attributes: the labels for the object types and the current multiplicity values of these object types in the region. Instead of representing both attributes in hardware, P Builder represents the multiplicity values as the values stored in an array of registers. Each register in the array is mapped to an object type in the alphabet of the P system (stored in software). For example, in Figure 4 the multisets for regions $M_1$, $M_2$ and $M_3$ are each represented as arrays containing three registers for the storage of the current multiplicity value in the relevant region of object types $a$, $b$ and $c$, respectively. P Builder assigns to the array registers a fixed bitwidth, which can be defined by the user (the default bitwidth is 8 bits). For example, the multisets for regions $M_1$, $M_2$ and $M_3$ in Figure 4 could be declared in Handel-C as shown in Figure 3.

```
unsigned int 8 multiset1[3] = {0b00000010, 0b00000001, 0b00000011};
unsigned int 8 multiset2[3] = {0b00000011, 0b00000000, 0b00000010};
unsigned int 8 multiset3[3] = {0b00000100, 0b00000000, 0b00000010};
```

**Fig. 3.** Handel-C code that P Builder could generate to represent the multisets in Figure 4

Note that the array for each region contains one register for each object type in the alphabet of the P system, regardless of which object types the reaction rules in the region consume or produce. This is to accommodate the traversal of objects between regions.

**Inter-region containment relationships.** To represent the inter-region containment relationships essential to computing the evolution of a P system, P Builder connects each rule processing unit (see below) to the array that represents the multiset of objects in the region to which it belongs, as well as to the arrays for all the regions to which it is possible that objects controlled by the rule traverse. For example, in Figure 4, rule $R_1$ is connected to the array for its accommodating membrane $M_1$, as well as to the arrays for $M_1$'s child regions $M_2$ and $M_3$ (since it is possible for the execution of $R_1$ to result in the traversal of

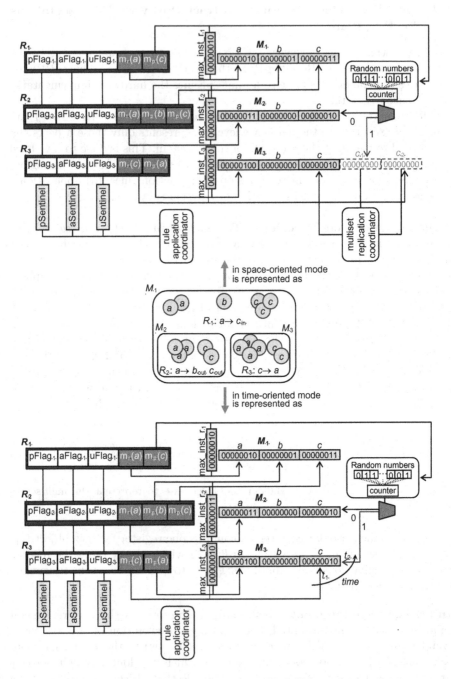

**Fig. 4.** The abstract hardware representations that P Builder generates for an example P system when Reconfig-P executes in time-oriented mode and in space-oriented mode

$b$ objects to $M_2$ and $M_3$). P Builder achieves this by including in the generated Handel-C code for the relevant rule processing unit a reference to the array that represents that multiset.

**Reaction rules.** To implement parallelism at both the system and region levels, P Builder generates parallel processing units for all reaction rules in all regions of the P system to be executed. These processing units are all connected to a global clock. Each processing unit is implemented in Handel-C as a potentially infinite while loop (see Figure 6, for example). All the rule processing units are placed in a **par** block.[1]

Each rule processing unit contains references to data relevant to its application and synchronization with other reaction rules, as well as logic that accomplishes its application. As shown in Figure 4, a rule processing unit includes references to registers that store the multiplicity information recorded in the definition of the reaction rule, three 1-bit registers called **pFlag**, **aFlag** and **uFlag**, and one 8-bit register called **max_inst_rx** (where x is the label of the rule). The flags are used by the *rule application coordinator* (a processing unit that executes in parallel with the rule processing units) to synchronize the execution of the rule processing units.[2] The **max_inst_rx** register for a reaction rule is used to store the value of the maximum number of instances of the rule that can be applied in the current transition.

**Application of reaction rules.** During a transition of a P system, all reaction rules are executed in parallel. Unless appropriate measures are taken, implementing this parallelism would lead to situations in which a rule processing unit updates the multiset of objects in its region before one or more of the other rule processing in the region have finished acquiring objects from the multiset.

To prevent such situations from occurring, P Builder separates the application of a reaction rule into two phases: a *preparation phase.* and an *updating phase.*

*Code for the preparation phase.* The block of Handel-C code in a rule processing unit that implements the preparation phase of a reaction rule $r$ (called **prepBlock**) implements the calculation of max-instances$_r$, the maximum number of instances of $r$ that can be applied in the current transition of the P system (according to the maximally parallel reaction rule application property of P systems). More specifically, the code implements the division of the current multiplicity value of each required object type by the number of objects of that type required for the application of one instance of $r$, and then the calculation of the minimum of the series of results thus obtained, which is equivalent to max-instances$_r$. The operation of determining the minimum ratio can be represented as a binary tree in which each node corresponds to the execution of a MIN operation and executing the MIN operation at the root node gives the value of the minimum ratio.

---

[1] The blocks of code included in a **par** block are executed in parallel.

[2] See [4, Section 3.1.4] for a more detailed discussion of how synchronization is implemented in Reconfig-P.

---

**procedure** generatePreparationCode (*r*: reaction rule)

---

*m*: region to which *r* belongs

*reactants* (*r*): list of the reactant (including catalyst) object types in *r*

*reactants* (*r*, *i*): the object type at the $i^{th}$ position of *reactants* (*r*)

*multiplicity* (*o*, *m*): the multiplicity of object type *o* in region *m* in the current transition of the P system

*amtToConsume* (*o*, *m*, *r*): the amount of objects of type *o* in region *m* to be consumed by reaction rule *r* in the current transition of the P system

*amtConsumed* (*o*, *m*, *r*): the amount of objects of type *o* in region *m* consumed by reaction rule *r* in the current transition of the P system

*evalAmtConsumedExpr* (*o*, *m*): string that represents an arithmetical expression that is used to evaluate the amount of object type *o* in region *m* already consumed in the current transition of the P system

*maxInst* (*r*): the maximum number of instances of *r* that can be applied in the current transition of the P system

*prepBlock*: (initially empty) Handel-C block of code for the preparation phase of *r*

*instCalcBlock*: (initially empty) Handel-C block of code for the calculation of *maxInst* (*r*)

*delaysBlock*: (initially empty) Handel-C block of code consisting of delay statements

**if** *r* has been assigned a priority

    *noOfDelays* ← 0

    **for** each reaction rule *s* in the same region as *r* with a higher priority than *r*

        *noOfDelays* ← *noOfDelays* + $\lceil \log_2(\text{size of } reactants\ (s))\rceil$

    Append *noOfDelays* many delay statements to *delaysBlock*

    Append *delaysBlock* to the current contents of *prepBlock*

    Obtain the value for *amtConsumed* (*reactants* (*r*, *i*), *m*, *r*) from *evalAmtConsumedExpr* (*reactants* (*r*, *i*), *m*)

**else**

    *amtConsumed* (*reactants* (*r*, *i*), *m*, *r*) ← 0

*n*: number of elements in *reactants* (*r*)

$o_i$: abbreviation for '*reactants* (*r*, *i*)'

**if** $n = 1$

    Append to *instCalcBlock* Handel-C code that evaluates *maxInst* (*r*) according to the following formula:

        *maxInst* (*r*) = (*multiplicity* ($o_1$, *m*) − *amtConsumed* ($o_1$, *m*)) / *amtToConsume* ($o_1$, *m*, *r*)

**else**

    Append to *instCalcBlock* Handel-C code that evaluates in as parallel a manner as possible the minimum value in the set {(*multiplicity* ($o_1$, *m*) − *amtConsumed* ($o_1$, *m*)) / *amtToConsume* ($o_1$, *m*, *r*), (*multiplicity* ($o_2$, *m*) − *amtConsumed* ($o_2$, *m*)) / *amtToConsume* ($o_2$, *m*, *r*) , ..., (*multiplicity* ($o_n$, *m*) − *amtConsumed* ($o_n$, *m*)) / *amtToConsume* ($o_n$, *m*, *r*)}

    **if** *r* has been assigned a priority

        **for** *i* = 0 to *n*

            *evalAmtConsumedExpr* ($o_i$, *m*) ← Handel-C code that evaluates *maxInst* (*r*) ∗ *amtToConsume* ($o_i$, *m*, *r*) + the result of evaluating *evalAmtConsumedExpr* ($o_i$, *m*)

Append *instCalcBlock* to the current contents of *prepBlock*

---

**Fig. 5.** The procedure that P Builder follows when generating the Handel-C code for the preparation phase of a reaction rule

MIN values are calculated in as parallel a manner as possible. In the preparation phase, parallelism is achieved at both the system and region levels. However, recall that in the core P system model, reaction rules associated with the same region may have relative priorities. If this is the case, P Builder ensures that the rule processing units with relative priorities determine their maximum number of instances one after the other according to their relative priorities. P Builder achieves this by calculating for each reaction rule the number of clock cycles taken by all the reaction rules with higher priorities to calculate their respective max-instances$_r$ values, and then inserting an equivalent number of `delay` statements into the Handel-C code for the rule processing unit for the reaction rule (above the code that calculates the max-instances$_r$ value for the rule).[3] P Builder also generates Handel-C code that evaluates the amount of objects of each type remaining in the region (after reaction rules with a higher priority are executed in a maximally parallel manner) which is to be executed before the code that determines max-instances$_r$. In this version of Reconfig-P, there is the constraint that reaction rules associated with the same region that attempt to obtain the same type of object must be assigned relative priorities. The procedure `generatePreparationCode` in Figure 5 illustrates how P Builder generates Handel-C code for the preparation phase of a reaction rule. Figure 6 shows example Handel-C code for the preparation phases of two reaction rules that belong to the same region.

*Code for the updating phase.* The block of Handel-C code in a rule processing unit that implements the updating phase of a reaction rule is called `updatingBlock`. In the updating phase of a reaction rule, if the rule is applicable, every instance of the rule is applied. P Builder implements the checking of the applicability of a reaction rule by generating a condition statement for its `updatingBlock` based on its max-instances$_r$ value. If its max-instances$_r$ value is greater than 0, the rest of the code in the relevant rule processing unit that implements the updating phase for the rule is executed; that is, the combined effect of the application of the instances of the rule is brought about. The relevant rule processing unit decreases/increases certain multiplicity values in certain multiset arrays according to the type, amount and source/destination of the objects consumed/produced by the instances of the rule. If the rule includes 'in' target directives and there are multiple child regions, P Builder creates and associates a *random number module* to the relevant rule processing unit. The random number module contains an array of numbers and a counter that is used to retrieve a particular number at random (see Figure 4). Each number refers to a particular child region. The numbers are stored in read-only memory instead of in registers in order to save hardware resources. The procedure `generateUpdatingCodeForAnObjectType` in Figure 7 describes the way in which P Builder generates Handel-C code that implements the updating of the multiplicity of an object type in a region. Figure 9 shows some example Handel-C code for the updating phase of a reaction rule.

---

[3] A `delay` statement specifies the execution of an empty process that takes one clock cycle to complete.

```
//Multiset in region: a^p b^q c^r d^s
//Rules in region: Rule 1 and Rule 2
//priority (Rule 1) > priority (Rule 2)
par
{
  ...
  //Rule 1: a^a1 b^b1 c^c1 -> d^d1
  while(1)
  {
    ...
    par
    {
      ...
      seq
      {
        tempMin = MIN(p/a_1, q/b_1);
        max_instances_r_1 = MIN(tempMin,
         r/c_1);
      }
    }
  }
```

```
  //Rule 2: a^a2 -> b_in^b2
  while(1)
  {
    ...
    par
    {
      ...
      seq
      {
        delay;
        delay;
        max_instances_r_2 =
          (p - max_instances_r_1*a_1)/a_2;
      }
    }
  }
  ...
}
```

**Fig. 6.** Example Handel-C source code for the preparation phases of two reaction rules that belong to the same region

A significant problem associated with implementing the updating phase of a transition of a P system in hardware is dealing with resource conflicts that arise because different rules need to update the multiplicity for the same type of object in the same clock cycle. When generating Handel-C code, P Builder overcomes the problem by providing two alternative strategies for conflict resolution: the *time-oriented strategy* and the *space-oriented strategy*. The current version of P Builder requires users to select the conflict resolution strategy to be used.[4] In both strategies, P Builder first determines the potential conflicts between reaction rules through the construction of a *conflict matrix* for the P system to be executed. Each row of a conflict matrix for a P system is a quadruple ($p$, $q$, $r$, $s$), where $p$ is an object type in the alphabet of the P system, $q$ is a region in the P system, $r$ is the set of reaction rules whose application results in the consumption and/or production of objects of type $p$ in $q$, and $s$ — called the *conflict degree* of ($p$, $q$) — is the size of $r$. There is a row for every pair ($p$, $q$). P Builder constructs the conflict matrix before run-time, and then generates the Handel-C code for the P system in such a way that all rule processing units can execute independently without any possibility of writing to the same register at the same time.

We now describe how P Builder generates Handel-C code for the updating phase of a reaction rule when (a) the time-oriented strategy is used, and (b) the space-oriented strategy is used.

[4] In a future version of P Builder, we plan to implement automatic selection of the most appropriate conflict resolution strategy based on the specific characteristics of the P system to be executed.

*Time-oriented conflict resolution.* In the Handel-C code that P Builder generates when the time-oriented conflict resolution strategy is used, all conflicting update operations for a certain type of object in a certain region are rolled out over different clock cycles. This is done in such a way that throughout the whole P system any update operations or sub-operations that can execute in parallel will execute in parallel, and all the update operations or sub-operations with conflicts will execute as early as possible. For example, in Figure 4, since without conflict resolution the multiplicity of object type $b$ in region $M_3$ would be updated by both rule processing unit $R_1$ and rule processing unit $R_3$ in the same clock cycle, the two operations of updating the register that stores the multiplicity of object type $b$ will be done in two consecutive clock cycles, with the update operation in the first clock cycle occurring in parallel with the rest of the update operations occurring in the P system. P Builder achieves the necessary interleaving among conflicting operations by inserting the appropriate number of `delay` statements in the appropriate places in the Handel-C code for the rule processing units. In this example, the updating phase for the whole P system takes two clock cycles to complete, since the maximum conflict degree in the conflict matrix for the P system is two. In principle, the time taken to complete the updating phase for a P system is not affected by the number of reaction rules in the P system, but is determined by the maximum conflict degree in the conflict matrix for the P system.

*Space-oriented conflict resolution.* In the space-oriented conflict resolution strategy, if $n$ reaction rules need to update the multiplicity value for the same type of object in the same region, then $n$ copies are made of the register that stores the multiplicity value. The processing units for the conflicting reaction rules are assigned one copy register each, and in the updating phase write to their respective copy registers. For example, in Figure 4, since the multiplicity of object type $b$ in region $M_3$ might be updated by both rule processing unit $R_1$ and rule processing unit $R_3$ in the same clock cycle, two copies $b_1$ and $b_2$ are made of that part of the multiset array for $M_3$ that represents the current multiplicity of object type $b$ in $M_3$. $R_1$ and $R_3$ are assigned one copy each — $R_1$ is able to update only $b_1$, and $R_3$ is able to update only $b_2$. So if both processing units need to update the multiplicity of object type $b$ in $M_3$ in the same clock cycle, they can do so without any conflict, because they update different parts of the multiset array. To accomplish the updating of the original register, P Builder generates a processing unit called a *multiset replication coordinator* (see Figure 4). After the end of a transition, and before the beginning of the next transition, the multiset replication coordinator reads the values stored in the copy registers and then updates accordingly (in a single clock cycle) that part of the multiset array that represents the multiplicity of the object type whose multiplicity value is being updated.

The space-oriented strategy offers a potentially significant performance advantage over the time-oriented strategy. As already mentioned, when the time-oriented strategy is used, the time taken to complete the entire updating phase for a P system depends on the maximum conflict degree in the conflict matrix

---

**procedure** generateUpdatingCodeForAnObjectType (*o*: object type to be updated, *m*: region to which *o* belongs, *r*: reaction rule)

---

*objectBlock*: (initially empty) string to store the Handel-C code that is to be returned by this procedure

*amtToConsume* (*o, m, r*): the amount of objects of type *o* in region *m* to be consumed by one instance of reaction rule *r* in the current transition of the P system

**if** *o* is a reactant in *r*

    **call** getDelayStatements **with** *o* and *m* **returning** *noOfDelays*

    **call** getRegisterToUpdate **with** *o* and *m* **returning** *registerToUpdate*

    Append to *objectBlock* (a) *noOfDelays* sequential delay statements and (b) Handel-C code that decreases the value stored at *registerToUpdate* by *amtToConsume* (*o, m, r*) * *maxInst* (*r*)

**else if** *o* is a product in *r*

    **if** direction of *o* in *r* is 'out'

        *p*: parent region of *m*

        **call** getDelayStatements **with** *o* and *p* **returning** *noOfDelays*

        **call** getRegisterToUpdate **with** *o* and *p* **returning** *registerToUpdate*

        Append to *objectBlock* (a) *noOfDelays* sequential delay statements and (b) Handel-C code that increases the value stored at *registerToUpdate* by *amtToProduce* (*o, p, r*) * *maxInst* (*r*)

    **else if** direction of *o* in *r* is 'in'

        **if** *m* has exactly one child region $c_1$

            **call** getDelayStatements **with** *o* and $c_1$ **returning** *noOfDelays*

            **call** getRegisterToUpdate **with** *o* and $c_1$ **returning** *registerToUpdate*

            Append to *objectBlock* (a) *noOfDelays* sequential delay statements and (b) Handel-C code that increases the value stored at *registerToUpdate* by *amtToProduce* ($c_1$, *m, r*) * *maxInst* (*r*)

        **else if** *m* has more than one child region

            Create the beginning of a Handel-C switch statement, with the expression of the switch containing the variable that stores the random number returned by the random number module associated with *m*

            **for** each child region *c* of *m*

                Create a Handel-C case block, where the condition of the case is that the random number returned by the random number module associated with *m* is equal to the label of *c*

                **call** getDelayStatements **with** *o* and *c* **returning** *noOfDelays*

                **call** getRegisterToUpdate **with** *o* and *c* **returning** *registerToUpdate*

                Insert into the case block (a) *noOfDelays* sequential delay statements and (b) Handel-C code that increases the value stored at *registerToUpdate* by *amtToProduce* (*o, c, r*) * *maxInst* (*r*)

                Insert the case block into the switch block

            End the switch block

            Append the switch block to *objectBlock*

    **else**

        **call** getDelayStatements **with** *o* and *m* **returning** *noOfDelays*

        **call** getRegisterToUpdate **with** *o* and *m* **returning** *registerToUpdate*

        Append to *objectBlock* (a) *noOfDelays* sequential delay statements and (b) Handel-C code that increases the value stored at *registerToUpdate* by *amtToProduce* (*o, m, r*) * *maxInst* (*r*)

**return** *objectBlock*

**Fig. 7.** The procedure that P Builder follows when generating Handel-C code that implements the updating of the multiplicity of an object type in a region. (See also the associated procedures **getDelayStatements** and **getRegisterToUpdate** in Figure 8.)

---

**procedure** getDelayStatements (*o*: object type, *m*: region)

---

*noOfDelays*: the number of delay statements to be inserted into a rule processing unit
for the purpose of conflict resolution

*currentIndex* (*o*, *m*): index that refers to the copy register in *objectCopies* (*o*, *m*) to which
the current update is to be made

*initDegree* (*o*, *m*): the degree of conflict associated with (*o*, *m*) initially to be processed

**if** *initDeg* (*o*, *m*) = 1 **or** *initDeg* (*o*, *m*) = 0 **or** the mode is space-oriented
   *noOfDelays* ← 0
**else**
   *noOfDelays* ← *initDeg* (*o*, *m*) − *currDeg* (*o*, *m*)
   *currDeg* (*o*, *m*) ← *currDeg*(*o*, *m*) − 1

**return** *noOfDelays*

---

**procedure** getRegisterToUpdate (*o*: object type, *m*: region)

---

*registerToUpdate*: the register to which the update of the multiplicity of object type *o*
in region *m* in the current transition should be made

*currentIndex* (*o*, *m*): index that refers to the copy register in *objectCopies* (*o*, *m*) to which
the current update is to be made

*objectCopies* (*o*, *m*): the array containing all copy registers for object *o* in region *m*

**if** the mode is time-oriented
   *registerToUpdate* ← multiplicity (*o*, *m*)
**else if** the mode is space-oriented
   *registerToUpdate* ← the copy register at the index *currIndex* (*o*, *m*) of *objectCopies* (*o*, *m*)
   *currIndex* (*o*, *m*) ← *currIndex* (*o*, *m*) + 1

**return** *registerToUpdate*

---

**Fig. 8.** The procedures invoked by the procedure `generateUpdatingCode-ForAnObjectType` shown in Figure 7

for the P system, and therefore is sensitive to the existence of conflicts between reaction rules. When the space-oriented strategy is used, on the other hand, the time taken to complete the entire updating phase is fixed at only two clock cycles. However, the performance advantage of the space-oriented strategy comes at the cost of increased hardware resource usage because of the need to replicate registers and implement special processing units for the coordination of this replication.

**Synchronization of reaction rules.** As mentioned above, each rule processing unit is associated with three 1-bit flags: a `pFlag`, a `uFlag` and an `aFlag`. The `pFlag` records whether the rule processing unit has finished its preparation phase, the `uFlag` records whether the rule processing unit has finished its updating phase, and the `aFlag` records whether the reaction rule implemented by the rule processing unit is applicable in the current transition. All the flags of a given type are stored in a single array.

```
//Multiset in region: a^p b^q c^r d^s
//Rules in region: Rule 1 and Rule 2
//priority (Rule 1) > priority (Rule 2)
//Rule 1: a^{a1} b^{b1} c^{c1} → d^{d1}        //Rule 1: a^{a1} b^{b1} c^{c1} → d^{d1}
while(1)                                       while(1)
{                                              {
    ...                                            ...

    par                                            par
    {                                              {
      multiset0[0] = p - max_instances_r_1*a_1;      multiset0_a_copy1 =
      multiset0[1] = q - max_instances_r_1*b_1;        - max_instances_r1*a_1;
      multiset0[2] = r - max_instances_r_1*c_1;      multiset0[1] = q - max_instances_r_1*b_1;
      multiset0[3] = s + max_instances_r_1*d_1;      multiset0[2] = r - max_instances_r_1*c_1;
    }                                                multiset0[3] = s + max_instances_r_1*d_1;
    ...                                            }
}                                                  ...

                                               }
//Rule 2: a^{a2} → b_{in}^{b2}
while(1)                                       //Rule 2: a^{a2} → b_{in}^{b2}
{                                              while(1)
    ...                                        {
                                                   ...
    par
    {                                              par
      seq                                          {
      {                                              multiset0_a_copy2 =
        delay;                                         - max_instances_r_2*a_2;
        multiset0[0] = p                             ...
        - max_instances_r_1*a_1                    }
        - max_instances_r_2*a_2;                   ...
      }                                          }
      ...
    }
    ...
}
```

        (a)                                            (b)

**Fig. 9.** Example Handel-C code for the updating phase of a reaction rule using (a) the time-oriented conflict resolution strategy and (b) the space-oriented conflict resolution strategy

The flags associated with the rule processing units are used by the *rule application coordinator*, a processing unit that executes in parallel with the other processing units in Reconfig-P, to synchronize the execution of the rule processing units both within and across transitions (see Figure 4). The rule application coordinator uses three 1-bit registers called pSentinel, aSentinel and uSentinel to manage system-level synchronization. pSentinel records whether all the rule processing units in the P system have finished their respective preparation phases (and hence are ready to progress to their respective updating phases), aSentinel records whether at least one reaction rule in the P system being executed is applicable in the current transition (and hence whether the P system should halt or progress to a new transition), and uSentinel records whether all the rule processing units in the P system have finished their respective updating phases (and hence whether the P system can progress to the next transition). At every clock cycle, the rule application coordinator checks the status of the rule processing units in the system, and sets the values of the sentinels to reflect

**Table 1.** Details of the P systems used in the experiments

| P system | n | Regions | Horizontal cascading | | | Vertical cascading | | | Horizontal and vertical cascading | | |
|---|---|---|---|---|---|---|---|---|---|---|---|
| | | | Rules | C | k | Rules | C | k | Rules | C | k |
| 1 | 1 | 4 | 11 | 27 | 7 | 11 | 27 | 7 | 11 | 27 | 7 |
| 2 | 2 | 7 | 21 | 53 | 7 | 22 | 54 | 7 | 22 | 54 | 7 |
| 3 | 4 | 13 | 41 | 97 | 7 | 44 | 108 | 7 | 42 | 106 | 7 |
| 4 | 8 | 25 | 81 | 193 | 9 | 88 | 216 | 7 | 83 | 211 | 7 |
| 5 | 16 | 49 | 161 | 377 | 17 | 176 | 432 | 7 | 165 | 415 | 7 |

this status. Also at every clock cycle, the rule application coordinator reads the current values of the sentinels and takes any action required to ensure that the system is properly synchronized.

## 5 Evaluation of Reconfig-P

We have conducted a series of experiments to investigate the performance and hardware resource usage of Reconfig-P.[5] In this section, we present the results of these experiments, and evaluate the performance, flexibility and scalability of Reconfig-P in light of these results.

### 5.1 Details of Experiments

Table 1 shows the P systems that were executed in the experiments. Each P system was constructed by first taking $n$ copies of the basic P subsystem shown at the top-right of Figure 10, then cascading these copies in a horizontal, vertical or horizontal and vertical manner (as shown in Figure 10), and finally placing the copies into the region shown at the top-left of Figure 10. The value of $n$ is a measure of the size of the constructed P system; the larger the value of $n$, the larger the P system. Thus in the experiments a series of P systems of different sizes and different structures were executed.

Table 1 lists a $C$ value for each P system. The $C$ value for a P system is a measure of the amount of conflict that exists between reaction rules in the P system. More specifically, $C$ is the sum of the conflict degrees of all pairs $(p, q)$ for the P system, where $p$ is an object type, $q$ is a region and the conflict degree of the pair is greater than 1.

In the experiments, the logic depth reduction feature of Reconfig-P was not used (this is the default setting).

---

[5] To facilitate a more precise comparison of the performance of Reconfig-P in the space-oriented and time-oriented modes than that presented in the pre-proceedings version of this paper, a more precise method of measuring execution times has been adopted. This method involves, for example, ignoring the time taken to initialize the FPGA.

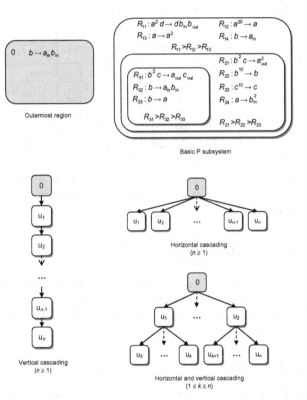

**Fig. 10.** Each P system used in the experiments was constructed by first cascading $n$ copies of a basic P subsystem in a horizontal, vertical or horizontal and vertical manner, and then placing these copies into an outermost region.

The target circuit for hardware executions was the Xilinx Virtex-II XC2V6000-FF11 52-4, and the Handel-C code for the P systems was synthesized, placed and routed using Xilinx tools. The computing platform for sequential executions was a 1.73GHz Intel Pentium M processor with 2GB of memory.

## 5.2   Results of Experiments

The graph at the top of Figure 11 shows the experimental results related to the performance of Reconfig-P, both when it executes in time-oriented mode and when it executes in space-oriented mode. It also shows, for the sake of comparison, the corresponding results for a software-based sequential computing platform (i.e., a Java simulator for the core P system model).

The graph at the bottom of Figure 11 shows the experimental results related to the hardware resource usage of Reconfig-P, both when it executes in time-oriented mode and when it executes in space-oriented mode. We use the number of LUTs (lookup tables) on the circuit generated for a P system as the measure of the hardware resource usage of Reconfig-P for that P system. We also record the percentage of the LUTs available on the FPGA that is used by the circuit, because this percentage provides an indication of the extent to which current FPGA technology meets the hardware resource requirements of Reconfig-P.

## 5.3  Evaluation of the Performance of Reconfig-P

In evaluating the performance of Reconfig-P, we make the following observations about the performance results shown in the graph at the top of Figure 11:

- Reconfig-P executes P systems significantly faster than the software-based sequential computing platform (from 131 to 3192 times faster, and on average 770 times faster). The larger the P system that is executed, the greater the extent to which Reconfig-P outperforms the sequential computing platform. This is as expected, because larger P systems have more regions and more reaction rules and therefore more opportunity for parallelism at both the system and region levels. The highest performance achieved by Reconfig-P is approximately 700 million reaction rule applications per second.

- The results show that, in general, the rate of increase in the performance of Reconfig-P in both the space-oriented and time-oriented modes as the number of reaction rules in the P system it executes increases is closer to being linear than constant. This is a good result, because it indicates that as the size of the P system to be executed increases, Reconfig-P is able to take advantage of the increased opportunities for parallelism.

- Reconfig-P performed better in space-oriented mode than in time-oriented mode in all the experiments. The performance in space-oriented mode is roughly twice as good as the performance in time-oriented mode. However, the performance difference between the two modes is more pronounced than usual when the largest P system generated using horizontal cascading is executed. This is a consequence of the fact that the $k$ value for this P system is much higher than for the other P systems. Increasing $k$ increases the number of clock cycles per transition when the time-oriented conflict resolution strategy is used, but does not affect the number of clock cycles per transition when the space-oriented conflict resolution strategy is used. Although not verified in the experiments, it is possible that if $C$ is large and $k$ is not too large (e.g., if a large number of conflicts are evenly distributed among regions), the performance in time-oriented mode would be closer to or perhaps better than the performance in space-oriented mode. This is because (a) a large $C$ value results in significant hardware resource usage (due to multiset replication) and a relatively low clock rate (due to greater logic depths associated with implementing the coordination of multiset replication) when the space-oriented mode is used, and (b) if $k$ is not too large, the performance degradation due to sensitivity to $k$ in the time-oriented mode may be offset by the performance degradation due to a relatively low clock rate in the space-oriented mode.

In summary, the experimental results indicate that Reconfig-P achieves very good performance.

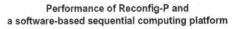

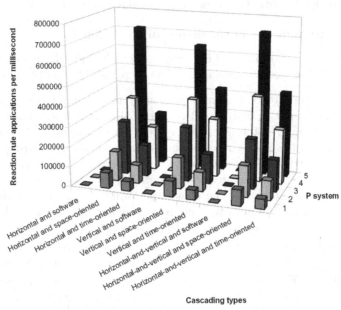

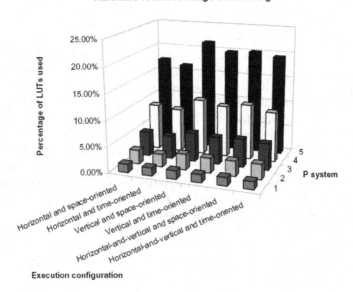

**Fig. 11.** Experimental results for the performance of Reconfig-P and a sequential computing platform and for the hardware resource usage of Reconfig-P

## 5.4   Evaluation of the Scalability of Reconfig-P

By showing how the performance of Reconfig-P changes as the size of the P system it executes changes, the performance results shown in the graph at the top of Figure 11 form the basis of an evaluation of the scalability of Reconfig-P. The hardware resource usage results shown in the graph at the bottom of Figure 11 also provide insight into the scalability of Reconfig-P, because they indicate the extent to which Reconfig-P can support the execution of large P systems.

In evaluating the scalability of Reconfig-P, we make the following observations about the results shown in Figure 11:

- Due to the nature of the membrane computing paradigm, the ideal limit of the scalability of Reconfig-P is that its performance increases linearly with respect to the number of reaction rules in the P system that it executes. If Reconfig-P were not scalable, its performance would remain constant or decrease as the number of reaction rules in the P system that it executes increases. As observed in Section 5.3, the rate of increase in the performance of Reconfig-P in both the space-oriented and time-oriented modes as the number of reaction rules in the P system that it executes increases is closer to being linear than constant. This suggests that there are no problems of scale in the hardware design (e.g., nonlinearly growing logic depths in certain parts of the hardware circuit that would reduce the clock rate of the FPGA). However, note that the scalability of Reconfig-P is more limited if (a) it executes in time-oriented mode, and (b) the number of conflicts per region increases at a linear (or close to linear) rate with respect to the size of the P system that it executes.
- The hardware resource usage of Reconfig-P scales linearly with respect to the size of the P system executed (i.e., with respect to $n$). This is as good as can reasonably be expected, and indicates that Reconfig-P is scalable with respect to hardware resource usage.
- The type of cascading employed in the construction of the P system that is executed has little effect on hardware resource usage.
- For all P systems, Reconfig-P uses less than 22% of the LUTs available on the FPGA. Given that the largest P system has 49 regions and 176 reaction rules, this is an impressive result. Not only does it strongly suggest that current FPGA technology meets the hardware resource requirements of Reconfig-P, it also indicates that it would be feasible to extend Reconfig-P to support P system features not covered by the core P system model.
- Reconfig-P uses only slightly more hardware resources in space-oriented mode than in time-oriented mode. This suggests that, at least for the P systems executed in the experiments, multiset replication has only a relatively small effect on hardware usage. Indeed, even for P systems with $C > 400$ and therefore with more than 400 copies of multiplicity values, the hardware resources consumed by Reconfig-P to store, access and coordinate these copies is relatively small.

In summary, the experimental results related to performance clearly demonstrate the scalability of Reconfig-P. These results are backed up by the results related to hardware resource usage, which indicate that Reconfig-P makes economical use of hardware resources, and therefore is scalable with respect to hardware resource usage.

### 5.5 Evaluation of the Flexibility of Reconfig-P

In its current prototype form, Reconfig-P supports the basic P system features covered by the core P system model. Therefore it is not able to execute P systems that include additional features such as structured objects and membrane permeability. This counts against its flexibility. However, there is good reason to believe that Reconfig-P can be extended to support additional P system features. As we have observed, Reconfig-P exhibits exceptionally economic hardware resource usage: for the P systems used in the experiments, more than 75% of the available hardware resources are left unused. Thus there is ample space on the FPGA for the inclusion of additional data structures and logic required for the implementation of additional features. Furthermore, the fact that Reconfig-P is implemented in a high-level hardware description language should ease the process of incorporating additional features into the existing implementation.

## 6   Conclusion

By developing Reconfig-P, we have demonstrated that it is possible to efficiently implement both the system-level and region-level parallelism of P systems on reconfigurable hardware and thereby achieve significant performance gains.

Theoretical results (presented in [4]) demonstrate that the parallel algorithm executed by Reconfig-P is significantly faster than the sequential algorithm used in sequential implementations of membrane computing. Empirical results show that for a variety of P systems Reconfig-P achieves very good performance while making economical use of hardware resources. And there is good reason to believe that Reconfig-P can be extended in the future to support additional P system features. Therefore, there is strong evidence that the implementation approach on which Reconfig-P is based is a viable means of attaining a good balance between performance, flexibility and scalability in a parallel computing platform for membrane computing applications.

## References

1. Ciobanu, G., Guo, W.: P Systems Running on a Cluster of Computers. In: Martín-Vide, C., Mauri, G., Păun, G., Rozenberg, G., Salomaa, A. (eds.) Membrane Computing. LNCS, vol. 2933, pp. 123–139. Springer, Heidelberg (2004)
2. Ciobanu, G., Păun, G., Pérez-Jiménez, M.J. (eds.): Applications of Membrane Computing. Springer, Heidelberg (2006)

3. Fernandez, L., Martinez, V.J., Arroyo, F., Mingo, L.F.: A Hardware Circuit for Selecting Active Rules in Transition P Systems. In: Pre-proceedings of the First International Workshop on Theory and Application of P Systems, Timisoara, Romania, September 26–27, 2005, pp. 45–48 (2005)
4. Nguyen, V., Kearney, D., Gioiosa, G.: An Implementation of Membrane Computing using Reconfigurable Hardware. Computing and Informatics (to appear)
5. Păun, G.: Membrane Computing: An Introduction. Springer, Heidelberg (2002)
6. Petreska, B., Teuscher, C.: A Reconfigurable Hardware Membrane System. In: Martín-Vide, C., Mauri, G., Păun, G., Rozenberg, G., Salomaa, A. (eds.) Membrane Computing. LNCS, vol. 2933, pp. 269–285. Springer, Heidelberg (2004)
7. Syropoulos, A., Mamatas, E.G., Allilomes, P.C., Sotiriades, K.T.: A Distributed Simulation of Transition P Systems. In: Martín-Vide, C., Mauri, G., Păun, G., Rozenberg, G., Salomaa, A. (eds.) Membrane Computing. LNCS, vol. 2933, pp. 357–368. Springer, Heidelberg (2004)

# On Flip-Flop Membrane Systems with Proteins

Andrei Păun[1,2,3] and Alfonso Rodríguez-Patón[3]

[1] Department of Computer Science/IfM, Louisiana Tech University
P.O. Box 10348, Ruston, LA 71272, USA
apaun@latech.edu
[2] National Institute Research and Development for Biological Sciences,
Splaiul Independenţei nr. 296, Sector 6, 060031 Bucharest,
[3] Universidad Politécnica de Madrid - UPM, Facultad de Informática
Campus de Montegancedo S/N, Boadilla del Monte, 28660 Madrid, Spain
arpaton@fi.upm.es

**Abstract.** We consider once again the membrane systems with proteins on membranes. This model is bridging the membrane systems and brane calculi areas together, thus it is interesting to study it in more depth. We improve previous results in the area and also define a new variant of these systems based on time as the output of the computation. The new model allows (due to its flexibility) even stronger improvements with respect to the number of proteins needed to perform the computation.

## 1 Introduction

We continue the work on a membrane systems model combining membrane systems and brane calculi as introduced in [14]. In brane calculi introduced in [5], one works only with objects – called proteins – placed on membranes, while the evolution is based on membrane handling operations, such as exocytosis, phagocytosis, etc. In the membrane computing area we have rules associated with each region defined by a membrane, and in the recent years the rules in membrane computing have been considered mainly to work on symbol objects rather than other structures such as strings. The extension considered in [14] and in [15] was to have both types of rules (both at the level of the region delimited by membranes and also at the level of membrane controlled by a protein). The reason for considering both extensions was that in biology, many reactions taking place in the compartments of living cells are controlled/catalysed by the proteins embedded in the membranes bilayer. For instance, it is estimated that in the animal cells, the proteins constitute about 50% of the mass of the membranes, the rest being lipids and small amounts of carbohydrates. There are several types of such proteins embedded in the membrane of the cell; one simple classification places these proteins into two classes, that of integral proteins (they "work" both inside the membrane as well as in the region outside the membrane), and that of peripheral proteins (they can only work in one region of the cell) – see [1].

In the present paper we continue the discussion in the direction of membrane systems with proteins, but we extend the model to have also a "more natural" output of the computation with ideas from [8].

G. Eleftherakis et al. (Eds.): WMC8 2007, LNCS 4860, pp. 414–427, 2007.
© Springer-Verlag Berlin Heidelberg 2007

Briefly, the systems that we consider in this paper extend the original definition by using the paradigm of time as the output of a computation as previously introduced in [6] and [8]. The idea originates in [17] as Problem W; the novelty is that instead of the "standard" way to output, like the multiplicities of objects found at the end of the computation in a distinguished membrane as it was defined in the model from [14] and in [15], it seems more "natural" to consider certain *events* (i.e., configurations) that may occur during a computation and to relate the output of such a computation with the time interval between such distinguished configurations. Our system will compute a set of numbers similarly with the case of "normal" symport/antiport systems as defined in [14], but the benefit of the current setting is that the computation and the observance of the output are now close to the biology and to the tools used for cell biology (fluorescence microscopy, FACS).

## 2    The Types of Rules in the System

In what follows we assume that the reader is familiar with membrane computing basic elements, e.g., from [16] and from [19], as well as with basic elements of computability, so that we only mention here a few notations we use. The rules based on proteins on membranes were described in detail in [14], and we refer the interested reader to that publication and to [15] for further details.

As usual, we represent multisets of objects from a given alphabet $V$ by strings from $V^*$, and the membrane structures by expressions of correctly matching labeled parentheses. The family of recursively enumerable sets of natural numbers is denoted by $NRE$.

In the P systems which we consider below, we use two types of objects, *proteins* and usual *objects*; the former are placed **on** the membranes, the latter are placed **in** the regions delimited by membranes. The fact that a protein $p$ is on a membrane (with label) $i$ is written in the form $[_i p|$. Both the regions of a membrane structure and the membranes can contain multisets of objects and of proteins, respectively.

We consider the following types of rules for handling the objects and the proteins; in all of them, $a, b, c, d$ are objects, $p$ is a protein, and $i$ is a label ("cp" stands for "change protein"), where $p, p'$ are two proteins (possibly equal; if $p = p'$, then the rules of the type $cp$ become rules of the type $res$; i.e., restricted):

| Type | Rule | Effect (besides changing also the protein) |
|------|------|--------------------------------------------|
| 1cp | $[_i p|a \rightarrow [_i p'|b$ | |
|  | $a[_i p| \rightarrow b[_i p'|$ | modify an object, but not move |
| 2cp | $[_i p|a \rightarrow a[_i p'|$ | |
|  | $a[_i p| \rightarrow [_i p'|a$ | move one object unmodified |
| 3cp | $[_i p|a \rightarrow b[_i p'|$ | |
|  | $a[_i p| \rightarrow [_i p'|b$ | modify and move one object |
| 4cp | $a[_i p|b \rightarrow b[_i p'|a$ | interchange two objects |
| 5cp | $a[_i p|b \rightarrow c[_i p'|d$ | interchange and modify two objects |

An intermediate case between *res* and *cp* can be that of changing proteins in a restricted manner, by allowing at most two states for each protein, $p, \bar{p}$, and the rules working either in a *res* manner (without changing the protein), or changing it from $p$ to $\bar{p}$ and back (like in the case of bistable catalysts). Rules with such flip-flop proteins are denoted by $nff, n = 1, 2, 3, 4, 5$ (note that in this case we allow both rules which do not change the protein and rules which switch from $p$ to $\bar{p}$ and back).

Both in the case of rules of type $ff$ and of type $cp$ we can ask that the proteins are always moved in another state (from $p$ into $\bar{p}$ and vice versa for $ff$). Such rules are said to be of *pure* $ff$ or $cp$ type, and we indicate the use of pure $ff$ or $cp$ rules by writing $ffp$ and $cpp$, respectively.

We can use these rules in devices defined in the same way as the symport/antiport P systems (hence with the environment containing objects, in arbitrarily many copies each – we need such a supply of objects, because we cannot create objects in the system), where also the proteins present on each membrane are mentioned.

That is, a *P system with proteins on membranes* is a device of the form

$$\Pi = (O, P, \mu, w_1/z_1, \ldots, w_m/z_m, E, R_1, \ldots, R_m, i_o),$$

where:

1. $m$ is the degree of the system (the number of membranes);
2. $O$ is the set of objects;
3. $P$ is the set of proteins (with $O \cap P = \emptyset$);
4. $\mu$ is the membrane structure;
5. $w_1, \ldots, w_m$ are the (strings representing the) multisets of objects present in the $m$ regions of the membrane structure $\mu$;
6. $z_1, \ldots, z_m$ are the multisets of proteins present on the $m$ membranes of $\mu$;
7. $E \subseteq O$ is the set of objects present in the environment (in an arbitrarily large number of copies each);
8. $R_1, \ldots, R_m$ are finite sets of rules associated with the $m$ membranes of $\mu$;
9. $i_o$ is the output membrane, an elementary membrane from $\mu$.

The rules can be of the forms specified above, and they are used in a nondeterministic maximally parallel way: in each step, a maximal multiset of rules is used, that is, no rule can be applied to the objects and the proteins which remain unused by the chosen multiset. As usual, each object and each protein can be involved in the application of only one rule, but the membranes are not considered as involved in the rule applications, hence the same membrane can appear in any number of rules at the same time.

If, at one step, two or more rules can be applied to the same objects and proteins, then only one rule will be non-deterministically chosen. At each step, a P system is characterized by a configuration consisting of all multisets of objects and proteins present in the corresponding membranes (we ignore the structure $\mu$, which will not be changed, and the objects from the environment). For example, $C = w_1/z_1, \ldots, w_m/z_m$ is the initial configuration, given by the definition of the P system. By applying the rules in a non-deterministic maximally parallel manner, we obtain transitions between the configurations of the system. A finite sequence of configurations is called computation. A computation halts if it reaches a configuration where no rule can be applied to the existing objects and proteins.

Only halting computations are considered successful, thus a non-halting computation will yield no result. With a halting computation we associate a result, in the form of the multiplicity of objects present in region $i_o$ in the halting configuration. We denote by $N(\Pi)$ the set of numbers computed in this way by a given system $\Pi$. (A generalization would be to distinguish the objects and to consider vectors of natural numbers as the result of a computation, but we do not examine this case here.)

We denote, in the usual way, by $NOP_m(pro_r; list\text{-}of\text{-}types\text{-}of\text{-}rules)$ the family of sets of numbers $N(\Pi)$ generated by systems $\Pi$ with at most $m$ membranes, using rules as specified in the list-of-types-of-rules, and with at most $r$ proteins present on a membrane. When parameters $m$ or $r$ are not bounded, we use $*$ as a subscript.

The new definition introduced by the current paper is the addition of time to the above model, in brief, *P system with proteins on membranes and time* is a device of the form

$$\Pi = (O, P, \mu, w_1/z_1, \ldots, w_m/z_m, E, R_1, \ldots, R_m, C_{start}, C_{stop}),$$

where:

1. $m$, $O$, $P$, $\mu$, $w_1, \ldots, w_m$, $z_1, \ldots, z_m$, $E$, $R_1, \ldots, R_m$ are as defined above;
2. $C_{start}$, $C_{stop}$ are regular subsets of $(O^*)^m$, describing configurations of $\Pi$. We will use a regular language over $O \cup \{\$\}$ to describe them, the special symbol $\$ \notin O$ being used as a marker between the configurations[1] in the different regions of the system. More details are given in [8] and [12].

As an example for the $C_{start}$ and $C_{stop}$ configurations, let us give the following restriction[2] $C = b^3 d^7 (O - \{a, b, d\})^*$ for a single membrane (the proofs obtained below need only one membrane, thus we can simplify the notation by not using the symbol $\$$). This means that in the region delimited by the only membrane in the system, the configuration $C$ is satisfied if and only if we do not have any symbol of type $a$, we must have exactly 3 symbols of type $b$ and exactly 7 symbols of type $d$. Any other symbol not mentioned is not restricted, e.g. we can have any number of symbols of type $c$.

We emphasize the fact that in the definition of $\Pi$ we assume that $C_{start}$ and $C_{stop}$ are regular. Other, more restrictive, cases can be of interest but we do not discuss them here.

---

[1] We express by these configurations restrictions that need to be satisfied by each of the current multisets in their respective regions so that the overall configuration can be satisfied.

[2] $C$ can be written also in the following form $C = (a^0 b^3 d^7)$.

We can now denote the systems (defined as above) based on time with $NTOP_m$ $(pro_r; list\text{-}of\text{-}types\text{-}of\text{-}rules)$ the family of sets of numbers $N(\Pi)$ generated by systems $\Pi$ with at most $m$ membranes, using rules as specified in the list-of-types-of-rules, and with at most $r$ proteins present on a membrane. When parameters are not bounded we replace them by $*$.

## 3   Register Machines

In the proofs from the next sections we will use register machines as devices characterizing $NRE$, hence the Turing computability.

Informally speaking, a register machine consists of a specified number of registers (counters) which can hold any natural number, and which are handled according to a program consisting of labeled instructions; the registers can be increased or decreased by 1 – the decreasing being possible only if a register holds a number greater than or equal to 1 (we say that it is non-empty) –, and checked whether they are non-empty.

Formally, a (non-deterministic) *register machine* is a device as follows: $M = (m, B, l_0, l_h, R)$, where $m \geq 1$ is the number of counters, $B$ is the (finite) set of instruction labels, $l_0$ is the initial label, $l_h$ is the halting label, and $R$ is the finite set of instructions labeled (hence uniquely identified) by elements from $B$ ($R$ is also called the *program* of the machine). The labeled instructions are of the following forms:

- $l_1 : (\text{ADD}(r), l_2, l_3)$, $1 \leq r \leq m$   (add 1 to register $r$ and then jump in a non-deterministic way to one of the instructions with labels $l_2$, $l_3$),
- $l_1 : (\text{SUB}(r), l_2, l_3)$, $1 \leq r \leq m$   (if register $r$ is not empty, then subtract 1 from it and go to the instruction with label $l_2$, otherwise go to the instruction with label $l_3$),
- $l_h : \text{HALT}$   (the halt instruction, which can only have the label $l_h$).

We say that a register machine has no ADD instructions looping to the same label (or *without direct loops*) if there are no instructions of the form $l_1 : (\text{ADD}(r), l_1, l_2)$ or $l_1 : (\text{ADD}(r), l_2, l_1)$ in $R$. For instance, an instruction of the form $l_1 : (\text{ADD}(r), l_1, l_2)$ can be replaced by the following instructions, where $l_1'$ is a new label: $l_1 : (\text{ADD}(r), l_1', l_2)$, $l_1' : (\text{ADD}(r), l_1, l_2)$. The generated set of numbers is not changed.

A register machine generates a natural number in the following manner: we start computing with all $m$ registers being empty, with the instruction labeled by $l_0$; if the computation reaches the instruction $l_h : \text{HALT}$ (we say that it halts), then the values of register 1 is the number generated by the computation. The set of numbers computed by $M$ in this way is denoted by $N(M)$. It is known (see [?]) that non-deterministic register machines with three registers generate exactly the family $NRE$, of Turing computable sets of numbers. Moreover, without loss of generality, we may assume that in the halting configuration all registers except the first one, where the result of the computation is stored are empty.

## 4 Previous Results

In [14] the following results were proved:

**Theorem 1.**

$$NOP_1(pro_2; 2cpp) = NRE. \quad \textit{(Theorem 5.1 in [14])}$$
$$NOP_1(pro_*; 3ffp) = NRE. \quad \textit{(Theorem 5.2 in [14])}$$
$$NOP_1(pro_2; 2res, 4cpp) = NRE. \quad \textit{(Theorem 6.1 in [14])}$$
$$NOP_1(pro_2; 2res, 1cpp) = NRE. \quad \textit{(Theorem 6.2 in [14])}$$
$$NOP_1(pro_*; 1res, 2ffp) = NRE. \quad \textit{(Theorem 6.3 in [14])}$$

As an extension of the work reported in [14], a significant amount of energy was devoted to the flip-flopping variant of these membrane systems in [9]. S.N Krishna was able to prove [9] several results improving Theorem 5.2, and Theorem 6.3 from [14]:

**Theorem 2.**

$$NOP_1(pro_7; 3ffp) = NRE. \quad \textit{(Theorem 1 in [9])}$$
$$NOP_1(pro_7; 2ffp, 4ffp) = NRE. \quad \textit{(Theorem 2 in [9])}$$
$$NOP_1(pro_7; 2ffp, 5ffp) = NRE. \quad \textit{(Corollary 3 in [9])}$$
$$NOP_1(pro_{10}; 1res, 2ffp) = NRE. \quad \textit{(Theorem 4 in [9])}$$
$$NOP_1(pro_7; 1ffp, 2ffp) = NRE. \quad \textit{(Theorem 6 in [9])}$$
$$NOP_1(pro_9; 1ffp, 2res) = NRE. \quad \textit{(Theorem 7 in [9])}$$
$$NOP_1(pro_9; 2ffp, 3res) = NRE. \quad \textit{(Theorem 9 in [9])}$$
$$NOP_1(pro_8; 1ffp, 3res) = NRE. \quad \textit{(Theorem 10 in [9])}$$
$$NOP_1(pro_9; 3res, 4ffp) = NRE. \quad \textit{(Theorem 11 in [9])}$$
$$NOP_1(pro_8; 2ffp, 5res) = NRE. \quad \textit{(Theorem 13 in [9])}$$

A close reading of the theorems mentioned above will yield some improvements that are given in the following section.

## 5 New Results

We start this section by first discussing the results from [9] which we mentioned in Theorem 2. The main idea in all the proofs reported in [9] was to simulate register machines (it is known that such devices with 3 registers are universal). The novelty of the proof technique in [9] was to consider for all ADD instructions associated with a particular register a single protein, similarly we use one protein for all the SUB instructions associated with a specific register. Thus in the proofs of the results mentioned in Theorem 2 we will have 6 proteins used for the simulation of the instructions in the register machine, (both ADD and

SUB instructions for the 3 registers in the machine) the other(s) protein(s) being needed mainly for the test with zero processing in the simulation of SUB instructions.

The main observation that we want to make at this point is the fact that register machines with three registers out of which one (the output register) is non-decreasing are still universal, thus all the results from [9] are better by one protein without any major changes in their proofs. This is due to the fact that we only need two proteins to simulate the SUB instructions, and also the proof technique allows for such a modification. Subsequently, the following results were shown in [9]:

**Theorem 3.**

$$NOP_1(pro_6; 3ffp) = NRE. \quad \textit{(Theorem 1 in [9])}$$

$$NOP_1(pro_6; 2ffp, 4ffp) = NRE. \quad \textit{(Theorem 2 in [9])}$$

$$NOP_1(pro_6; 2ffp, 5ffp) = NRE. \quad \textit{(Corollary 3 in [9])}$$

$$NOP_1(pro_9; 1res, 2ffp) = NRE. \quad \textit{(Theorem 4 in [9])}$$

$$NOP_1(pro_6; 1ffp, 2ffp) = NRE. \quad \textit{(Theorem 6 in [9])}$$

$$NOP_1(pro_8; 1ffp, 2res) = NRE. \quad \textit{(Theorem 7 in [9])}$$

$$NOP_1(pro_8; 2ffp, 3res) = NRE. \quad \textit{(Theorem 9 in [9])}$$

$$NOP_1(pro_7; 1ffp, 3res) = NRE. \quad \textit{(Theorem 10 in [9])}$$

$$NOP_1(pro_8; 3res, 4ffp) = NRE. \quad \textit{(Theorem 11 in [9])}$$

$$NOP_1(pro_7; 2ffp, 5res) = NRE. \quad \textit{(Theorem 13 in [9])}$$

We will proceed now to consider the same framework, but with the extra feature of the output based on time. We show that we can improve the result from Theorem 11 from [9]:

**Theorem 4.** $NRE = NTOP_1(pro_7, 3res, 4ffp)$.

*Proof.* We consider a register machine $M = (m, B, l_0, l_h, R)$ and we construct the system

$$\Pi = (O, P, [_1 \; ]_1, \{l_0, b\}/P, E, R_1, C_{start}, C_{stop})$$

with the following components:

$$O = \{a_r, a'_r \mid 1 \le r \le 3\} \cup \{i, i', l_i, l'_i, l''_i, l'''_i, l^{iv}_i, L_i, L'_i \mid 0 \le i \le h\}$$
$$\cup \{o, o_1, o_2, b, h, \dagger\}.$$

$$E = \{a_r, a'_r \mid 1 \le r \le 3\} \cup \{i \mid 0 \le i \le h\} \cup \{o\}.$$

$$P = \{p_1, p_2, p_3, s_2, s_3, p, t\}.$$

$C_{start} = l''_h(O - \{l''_h, \dagger\})^*$, in other words, $l''_h$ appears exactly once and there are no copies of $\dagger$ in the membrane, and the rest of the symbols can appear in any multiplicity as they are ignored.

$C_{stop} = (O - \{a_1\})^*$, in this case $a_1$ does not appear in the membrane.

The proteins $p$ and $t$ are of the type 3res while all the others are of the type 4ffp. Proteins $p$ and $p_i$ are used in the simulation of ADD instructions of register $i$, proteins $p$, $t$ and $s_i$ are used in the simulation of SUB instructions of register $i$, and protein $p$, $t$, $s_2$ and $s_3$ are used in the simulation of the instructions for counting or termination.

The system has the following rules in $R_1$:

For an **ADD instruction** $l_1 : (\mathtt{ADD}(r), l_2, l_3) \in R$, we consider the rules as shown in Table 1.

**Table 1.** Steps for ADD instruction for Theorem 4

| Step | Rules | Type | Environ. | Membrane |
|---|---|---|---|---|
| 1 | $a'_r[_1 p_r \mid l_1 \to l_1[_1 p'_r \mid a'_r$ | 4ffp | $El_1$ | $ba'_r$ |
| 2 or | $a_r[_1 p'_r \mid a'_r \to a'_r[_1 p_r \mid a_r$ and $l_1[_1 p \mid \to [_1 p \mid l_2$ | 4ffp, 3res | $E$ | $bl_2 a_r$ |
| 2 | $a_r[_1 p'_r \mid a'_r \to a'_r[_1 p_r \mid a_r$ and $l_1[_1 p \mid \to [_1 p \mid l_3$ | 4ffp, 3res | $E$ | $bl_3 a_r$ |

We simulate the work of the ADD instruction in two steps. First we send out the current instruction label $l_1$ and bring in a copy of the (padding) symbol $a'_r$ using the protein $p_r$. Next we simultaneously apply the rules to replace $a'_r$ with $a_r$ using the protein $p'_r$ and we bring in the next instruction label $l_2$ or $l_3$ according to the currently simulated rule $l_1$. Of course, $l_1$ uniquely identifies which rule was simulated, thus there is no ambiguity about which symbols $l_i$ are able to enter the membrane at this time. Let us now consider the case of the SUB instructions:

For a **SUB instruction** $l_1 : (\mathtt{SUB}(r), l_2, l_3) \in R$ we consider the rules as shown in Table 2.

We simulate the work of the SUB instruction in several steps (eight if the register is not empty and ten if it is empty). We first send out the current label as $l'_1$ using the protein $p$. At the next step the symbol $l'_1$ is brought in as $l''_1$. Next we exchange $l''_1$ and $o$ using the protein $s_r$ (the protein $s_r$ is moved in its primed version of the flip-flop). We can now apply two rules in parallel and bring in $l'''_1$ as $l'''_1$ while sending out $o$ as $o_1$. Next, $l'''_1$ is sent out as $l^{iv}_1$ while we bring in $o_1$ as $o_2$ in parallel.

In this moment our system will perform the checking of the contents of the register $r$. If the register is not empty, then $l^{iv}_1$ will enter the membrane, decreasing the register and at the same time another marker $o_2$ is sent outside as $o$ to help identify the correct case later. At the next stage $l^{iv}_1$ will be sent out as $2'$ using protein $p$. Finally $2'$ will return as the next instruction label to be brought in (in this case $l_2$ as the register is not empty). If $l^{iv}_1$ comes back in the membrane through the protein $t$ instead of $s'_r$, we will have a wrong computation. In this case we can send out $o_2$ as symbol $\dagger$ in parallel using the protein $p$ (as this is the only channel available at this time to $o_2$, $t$ being used by $l^{iv}_1$). Next we can bring in a copy of the symbol $\dagger$ into the membrane. The application of this rule will never satisfy the starting configuration; hence, we will not be able to use the time counter.

**Table 2.** Steps for SUB instruction for Theorem 4

| Step | Rules | Type | Environ. | Membrane |
|---|---|---|---|---|
| 1 | $[_1p \mid l_1 \to l'_1[_1p \mid$ | 3res | $El'_1$ | $ba_r$ |
| 2 | $l'_1[_1p \mid \to [_1p \mid l''_1$ | 3res | $E$ | $bl''_1a_r$ |
| 3 | $o[_1s_r \mid l''_1 \to l''_1[_1s'_r \mid o$ | 4ffp | $El''_1$ | $boa_r$ |
| 4 | $l''_1[_1p \mid \to [_1p \mid l'''_1$ and $[_1t \mid o \to o_1[_1t\mid$ | 3res, 3res | $Eo_1$ | $bl'''_1a_r$ |
| 5 | $[_1p \mid l'''_1 \to l^{iv}_1[_1p \mid$ and $o_1[_1t \mid \to [_1t \mid o_2$ | 3res, 3res | $El^{iv}_1$ | $bo_2a_r$ |
| | Register r is non-empty | | | |
| 6 | $l^{iv}_1[_1s'_r \mid a_r \to a_r[_1s_r \mid l^{iv}_1$ and $[_1t \mid o_2 \to o[_1t\mid$ | 4ffp, 3res | $Eoa_r$ | $bl^{iv}_1$ |
| 7 | $[_1p \mid l^{iv}_1 \to 2'[_1p \mid$ | 3res | $E2'$ | $b$ |
| 8 | $2'[_1p \mid \to [_1p \mid l_2$ | 3res | $E$ | $bl_2$ |
| | Wrong computation | | | |
| 6 | $l^{iv}_1[_1t \mid \to [_1t \mid L'_3$ and $[_1p \mid o_2 \to \dagger[_1p \mid$ | 3res, 3res | $E\dagger$ | $bL'3a_r$ |
| 7 | $\dagger[_1t \mid \to [_1t \mid \dagger$ | 3res | $E$ | $b \dagger a_r$ |
| | Register r is empty | | | |
| 6 | $[_1t \mid o_2 \to o[_1t\mid$ | 3res | $Eo$ | $b$ |
| 7 | $l^{iv}_1[_1t \mid \to [_1t \mid L'_3$ | 3res | $E$ | $bL'_3$ |
| 8 | $3[_1s'_r \mid L'_3 \to L'_3[_1s_r \mid 3$ | 4ffp | $EL'_3$ | $b3$ |
| 9 | $[_1p \mid 3 \to 3'[_1p \mid$ | 3res | $E3'$ | $b$ |
| 10 | $3'[_1p \mid ]ra[_1p \mid l_3$ | 3res | $E$ | $bl_3$ |

If the register is empty, after step 5 we have $l^{iv}_1$ in the environment and $o_2$ in the membrane, and the protein associated with the subtract rule for the register $r$ $(s_r)$ is primed. At this moment $l^{iv}_1$ cannot enter the membrane through the protein $s'_r$ as there are no $a_r$ objects in the membranes with which it must be exchanged. There are two choices: either $l^{iv}_1$ enters the membrane through $t$ (and we get the wrong computation case as above) or $t$ is used by $o_2$, and then $l^{iv}_1$ sits one step in the environment. At the next step we have the "branching point": rather than exchanging with $a_r$ (which will be present in the membrane in the case when the register is not empty), $l^{iv}_1$ comes into the membrane as $L'_3$ through $t$. Next we use the protein $s'_r$ to exchange $L'_3$ and 3, and then send out 3 as $3'$ using protein $p$. Now we bring in $3'$ as the next instruction to be simulated $l_3$.

**Terminating/counting work.** It is clear that at the end of the simulation, if the register machine has reached the final state, we will have the halting instruction symbol in the membrane along with one copy of the symbol $b$ and multiple copies of the three different objects associated with their respective registers. At that time we will have the computed value encoded as the multiplicity of the object $a_1$ that is associated with the output register. We will also have in the system the label of the halting instruction, $l_h$; thus, the rule $([_1p \mid l_h \to l'_h[_1p \mid)$ can be applied only when the simulation is performed correctly. At the next step, using the protein $s_2$ we exchange $l'_h$ and $b$.

The terminating/counting work is done by the rules as shown in Table 3.

Next we apply two rules in parallel and bring in $b$ as $l''_h$ while sending out $l'_h$ as $l''_h$, satisfying the $C_{start}$ configuration. One can note that if there are no copies of $a_1$ in the membrane, then also the configuration $C_{stop}$ is satisfied at the same time, thus our system would compute the value zero in that case. Next we exchange $h$ from the environment with $l''_h$ and $l''_h$ from the environment with $a_1$ until we reach the stopping configuration. For any other value encoded in the multiplicity of $a_1$ it will take exactly the same number of steps to push the number of copies of object $a_1$ from the membrane.                        □

**Table 3.** Steps for terminating/counting instructions for Theorem 4

| Step | Rules | Type | Environ. | Membrane |
|------|-------|------|----------|----------|
| 1 | $[_1p \mid l_h \to l'_h[_1p \mid$ | 3res | $El'_h$ | $ba_1^{n_1}a_2^{n_2}a_3^{n_3}$ |
| 2 | $l'_h[_1s_2 \mid b \to b[_1s'_2 \mid l'_h$ | 4ffp | $Eb$ | $l'_ha_1^{n_1}a_2^{n_2}a_3^{n_3}$ |
| 3 | $b[_1p \mid \to [_1p \mid l''_h$ and $[_1t \mid l'_h \to l''_h[_1t \mid$ | 3res | $El''_h$ | $l''_ha_1^{n_1}a_2^{n_2}a_3^{n_3}$ |
| 4 | $h[_1s_2 \mid l''_h \to l''_h[_1s'_2 \mid h$ or $h[_1s'_2 \mid l''_h \to l''_h[_1s_2 \mid h$ | 4ffp | $El''_ha_1$ | $l''_ha_1^{n_1}a_2^{n_2}a_3^{n_3}h$ |
| | and $l''_h[_1s_3 \mid a_1 \to a_1[_1s'_3 \mid l''_h$ or | 4ffp | | |
| | or $l''_h[_1s'_3 \mid a_1 \to a_1[_1s_3 \mid l''_h$ | 4ffp | | |

An interesting observation is the fact that the object $b$ is used for the counting at the end of the computation. If one considers the same construct for membrane systems with proteins as defined in [14] (the "classical" systems with the output the multiplicity of objects in the membrane), then our construction is still valid even in the case of systems without time, thus we have the following theorem also proven:

**Theorem 5.** $NRE = NOP_1(pro_7, 3res, 4ffp)$.

The theorem above is valid as one can restrict the register machine to be simulated (without loss of generality) to the case when the machine halts with the non-output registers empty.

Thus it can be seen that we are able to improve the result shown in Theorem 10 in [9] both for systems based on multiplicity output and also for systems based on time. The next result improves significantly Theorem 11 from [9], in this case for systems based on time, and later one we will discuss also about the non-timed systems.

**Theorem 6.** $NRE = NTOP_1(pro_3, 2ffp, 5res)$.

*Proof.* We consider a register machine $M = (m, B, l_0, l_h, R)$ and we construct the system

$$\Pi = (O, P, [_1 \ ]_1, \{l_0, b, e\}/P, E, R_1, C_{start}, C_{stop})$$

with the following components:

$$O = \{l_i, l'_i \mid 0 \le i \le h\} \cup \{a_1, a_2, a_3, b, o, y\}.$$
$$E = \{a_1, a_2, a_3, o\}.$$
$$P = \{p, q, s\}.$$

$C_{start} = (O - \{b\})^*$, in other words, there are no copies of $b$ in the membrane, and the rest of the symbols can appear in any multiplicity as they are ignored.

$C_{stop} = (O - \{a_1\})^*$, in this case $a_1$ does not appear in the membrane.

Protein $q$ is of type 5res while all the others are of the type 2ffp. Proteins $p$ and $q$ are used in the simulation of the ADD instruction, proteins $q$ and $s$ are used in the simulation of the SUB instruction, and protein $q$ is used in the simulation of the instructions for counting or termination.

The system has the following rules in $R_1$:

For an **ADD instruction** $l_1 : (\text{ADD}(r), l_2, l_3) \in R$, we consider the rules as shown in Table 4.

**Table 4.** Steps for ADD instruction for Theorem 6

| Step | Rules | Type | Environment | Membrane |
|------|-------|------|-------------|----------|
| 1 | $a_r[_1q \mid l_1 \to l'_1[_1q \mid a_r$ | 5res | $El'_1$ | $bea_r$ |
| 2 | $l'_1[_1q \mid e \to e[_1q \mid l_2$ | 5res | $Ee$ | $bl_2a_r$ |
| 2 | $l'_1[_1q \mid e \to e[_1q \mid l_3$ | 5res | $Ee$ | $bl_3a_r$ |
| 3 | $e[_1p \mid \to [_1p' \mid e$ or $e[_1p' \mid \to [_1p \mid e$ | 2ffp, 2ffp | $E$ | $bea_r$ |

We simulate the work of the ADD instruction in two steps. First we send out the current instruction label $l_1$ as $l'_1$ and bring in a copy of the symbol $a_r$ using the protein $q$. Next we apply the rule to send out $e$ using the protein $q$ and we bring $l'_1$ in as the new instruction label. To simulate the non-deterministic behavior of these machines we have two rules that do the same job, the only difference being the next instruction label being brought back in the system. It is clear that the simulation of the ADD instruction is performed correctly. The work is finished in this case by the rule $(e[_1p \mid \to [_1p' \mid e)$ or $(e[_1p' \mid \to [_1p \mid e)$.

For a **SUB instruction** $l_1 : (\text{SUB}(r), l_2, l_3) \in R$ we consider the rules as shown in Table 5.

**Table 5.** Steps for SUB instruction for Theorem 6

| Step | Rules | Type | Environment | Membrane |
|------|-------|------|-------------|----------|
| 1 | $[_1s \mid l_1 \to l_1[_1s' \mid$ | 2ffp | $El_1$ | $bea_r$ |
| | Register r is non-empty | | | |
| 2 | $o[_1s' \mid \to [_1s \mid o$ **and** $l_1[_1q \mid a_r \to a_r[_1q \mid l'_1$ | 2ffp, 5res | $Ea_r$ | $beol'_1$ |
| 3 | $o[_1q \mid l'_1 \to l'_1[_1q \mid l_2$ | 5res | $El'_1$ | $beol_2$ |
| 4 | $l'_1[_1q \mid o \to o[_1q \mid y$ | 5res | $Eo$ | $bey$ |
| | Register r is empty | | | |
| 2 | $o[_1s' \mid \to [_1s \mid o$ | 2ffp | $El_1$ | $beo$ |
| 3 | $l_1[_1q \mid o \to o[_1q \mid l_3$ | 5res | $Eo$ | $bel_3$ |

We simulate the work of the SUB instruction in several steps (four if the register is not empty and three if it is empty). At step 1 we first send out the current label $l_1$ using the protein $s$. If the register is not empty, at step 2, $l_1$ will enter the membrane, decreasing the register and at the same time the symbol $o$ is brought in. At the next stage (step 3) $l'_1$ will be sent out using protein $q$, and $o$ will return as the next instruction label to be brought in (in this case $l_2$ as the register is not empty). Finally $l'_1$ will return as the symbol $y$ while sending out $o$, so that no extra copies of $o$ are left in the membrane so that future SUB simulations will be performed correctly. The symbols $y$ will accumulate in the membrane.

In the case when the register to be decremented is empty, we perform the same initial step, sending out the current label using the protein $s$. This time $l_1$ cannot enter the membrane at the step 2 as there is no $a_r$ in the membrane to help bring it in. So $l_1$ will wait for one step in the environment. $o$ is entering the membrane at step 2, so at the step 3 $l_1$ can now come into the membrane through $q$ and is changed into the label of the next instruction to be simulated $l_3$.

The **terminating/counting work** stage is done by the rules as shown in Table 6.

**Table 6.** Steps for terminating/counting instructions for Theorem 6

| Step | Rules | Type | Environment | Membrane |
|------|-------|------|-------------|----------|
| 1 | $o[_1q \mid l_h \to l_h[_1q \mid y$ | 5res | $El_h$ | $bea_1^{n_1}a_2^{n_2}a_3^{n_3}$ |
| 2 | $l_h[_1q \mid b \to l'_h[_1q \mid y$ | 5res | $Ebl'_h$ | $ea_1^{n_1}a_2^{n_2}a_3^{n_3}$ |
| 3 | $l'_h[_1q \mid a_1 \to l'_h[_1q \mid y$ | 5res | $El'_h a_1$ | $ea_1^{n_1}a_2^{n_2}a_3^{n_3}$ |

It is clear that at the end of the simulation, if the register machine has reached the final state, we will have the halting instruction symbol in the system membrane, along with one copy of the symbol $b$ and multiple copies of the three different objects associated with the respective registers and the symbol $y$. At that time we will have the computed value encoded as the multiplicity of the object $a_1$ that is associated with the output register. We will also have in the system the label of the halting instruction, $l_h$, thus the rule $(o[_1q \mid l_h \to l_h[_1q \mid y)$ can be applied only when the simulation is performed correctly. At the next step, using the protein $q$ we bring in $l_h$ as $y$ while sending out $b$ as $l'_h$, satisfying the $C_{start}$ configuration. One can note that if there are no copies of $a_1$ in the membrane, then also the configuration $C_{stop}$ is satisfied at the same time, thus our system would compute the value zero in that case. Next we bring in $l'_h$ as $y$ while sending out $a_1$ as $l'_h$ until we reach the stopping configuration. For any other value encoded in the multiplicity of $a_1$ it will take exactly the same number of steps to push the $a_1$-s out of the membrane.                                    □

Thus it can be seen that by using time as the output, we are able to improve the result shown in Theorem 13 from [9], where seven proteins were required for universality, as opposed to the three used in the above proof.

If one wants to still restrict the discussion to only the case of the non-timed systems, with the price of one protein we can remove the objects $y$ and $e$ from the membrane (by first modifying them into some other symbols such as $y'$ and $o'$ and then expelling them to the environment). In this way it is easy to see that our proof for Theorem 6 leads to the following theorem:

**Theorem 7.** $NRE = NOP_1(pro_4, 2ffp, 5res)$.

# 6    Final Remarks

We have shown that previous results about membrane systems with proteins on membranes can be improved in what concerns the number of proteins. We have also extended the model to have the output encoded as the time between two configurations and this has lead to a significant improvement as opposed to the previous results reported in [9]. Additional similar improvements are under investigation.

## Acknowledgments

A. Păun gratefully acknowledges the support in part by LA BoR RSC grant LEQSF (2004-07)-RD-A-23 and NSF Grants IMR-0414903 and CCF-0523572. We acknowledge the significant improvements to the paper suggested by the anonymous referees.

## References

1. Alberts, B., Johnson, A., Lewis, J., Raff, M., Roberts, K., Walter, P.: Molecular Biology of the Cell, 4th edn. Garland Science, New York (2002)
2. Alhazov, A., Freund, R., Rogozhin, Y.: Some Optimal Results on Symport/Antiport P Systems with Minimal Cooperation. In: Gutiérrez-Naranjo, M.A., et al. (eds.) Cellular Computing (Complexity Aspects), ESF PESC Exploratory Workshop, pp. 23–36. Fénix Editora, Sevilla (2005)
3. Alhazov, A., Freund, R., Rogozhin, Y.: Computational Power of Symport / Antiport: History, Advances and Open Problems. In: Freund, R., Păun, G., Rozenberg, G., Salomaa, A. (eds.) WMC 2005. LNCS, vol. 3850, pp. 1–30. Springer, Heidelberg (2006)
4. Bernardini, F., Păun, A.: Universality of Minimal Symport/Antiport: Five Membranes Suffice. In: Martín-Vide, C., Mauri, G., Păun, G., Rozenberg, G., Salomaa, A. (eds.) Membrane Computing. LNCS, vol. 2933, pp. 43–54. Springer, Heidelberg (2004)
5. Cardelli, L.: Brane Calculi – Interactions of Biological Membranes. In: Danos, V., Schachter, V. (eds.) CMSB 2004. LNCS (LNBI), vol. 3082, pp. 257–280. Springer, Heidelberg (2005)
6. Cavaliere, M., Freund, R., Păun, G.: Event–Related Outputs of Computations in P Systems. In: Gutiérrez-Naranjo, M.A., et al. (eds.) Cellular Computing (Complexity Aspects), ESF PESC Exploratory Workshop, pp. 107–122. Fénix Editora, Sevilla (2005)

7. Freund, R., Păun, A.: Membrane Systems with Symport/Antiport: Universality Results. In: Păun, G., Rozenberg, G., Salomaa, A., Zandron, C. (eds.) Membrane Computing. LNCS, vol. 2597, pp. 270–287. Springer, Heidelberg (2003)
8. Ibarra, O.H., Păun, A.: Counting Time in Computing with Cells. In: Carbone, A., Pierce, N.A. (eds.) DNA Computing. LNCS, vol. 3892, pp. 112–128. Springer, Heidelberg (2006)
9. Krishna, S.N.: Combining Brane Calculus and Membrane Computing. In: BIC-TA 2006. Proc. Bio-Inspired Computing – Theory and Applications Conf, Wuhan, China (September 2006), Membrane Computing Section and Journal of Automata Languages and Combinatorics (in press)
10. Minsky, M.L.: Recursive Unsolvability of Post's Problem of "Tag" and Other Topics in Theory of Turing Machines. Annals of Mathematics 74, 437–455 (1961)
11. Minsky, M.L.: Computation: Finite and Infinite Machines. Prentice Hall, Englewood Cliffs, New Jersey (1967)
12. Nagda, H., Păun, A., Rodríguez-Patón, A.: P Systems with Symport/Antiport and Time. In: Hoogeboom, H.J., Păun, G., Rozenberg, G., Salomaa, A. (eds.) WMC 2006. LNCS, vol. 4361, pp. 429–442. Springer, Heidelberg (2006)
13. Păun, A., Păun, G.: The Power of Communication: P Systems with Symport/Antiport. New Generation Computing 20(3), 295–306 (2002)
14. Păun, A., Popa, B.: P Systems with Proteins on Membranes. Fundamenta Informaticae 72(4), 467–483 (2006)
15. Păun, A., Popa, B.: P Systems with Proteins on Membranes and Membrane Division. In: Ibarra, O.H., Dang, Z. (eds.) DLT 2006. LNCS, vol. 4036, pp. 292–303. Springer, Heidelberg (2006)
16. Păun, G.: Membrane Computing – An Introduction. Springer, Heidelberg (2002)
17. Păun, G.: Further Twenty-six Open Problems on Membrane Computing. In: The Third Brainstorming Meeting on Membrane Computing, Sevilla, Spain (February 2005)
18. Rozenberg, G., Salomaa, A. (eds.): Handbook of Formal Languages. 3 volumes, Springer, Berlin (1997)
19. The P Systems Website: http://psystems.disco.unimib.it

# Characterizing Membrane Structures Through Multiset Tree Automata

José M. Sempere and Damián López

Departamento de Sistemas Informáticos y Computación
Universidad Politécnica de Valencia
Camino de Vera s/n 46017 Valencia, Spain
{jsempere,dlopez}@dsic.upv.es

**Abstract.** The relation between the membrane structures of P systems and an extension of tree automata which introduces multisets in the transition function has been proposed in previous works. Here we propose two features of tree automata which have been previously studied (namely, reversibility and local testability) in order to extend them to multiset tree automata. The characterization of these families will introduce a new characterization of membrane structures defined by the set of rules used for membrane creation and deletion.

## 1 Introduction

The relation between membrane structures and tree languages has been explored in previous works. So, Freund et al. [4] proved that P systems are able to generate recursively enumerable sets of trees through their membrane structures. Other works have focused on extending the definition of finite tree automata in order to take into account the membrane structures generated by P systems. For instance, in [13], the authors propose an extension of tree automata, namely multiset tree automata, in order to recognize membrane structures. In [7], this model is used to calculate editing distances between membrane structures. Later, a method to infer multiset tree automata from membrane observations was presented in [14].

In this work we introduce two new families of multiset tree automata, by using previous results taken from tree language theory. We propose a formal definition of reversible multiset tree automata and local testable multiset tree automata. These features have been widely studied in previous works [6,8].

The structure of this work is as follows: first we give basic definitions and notation for tree languages, P systems and multiset tree automata and we define the new families of multiset tree automata. Finally, we give some guidelines for future research.

## 2 Notation and Definitions

In the sequel we provide some concepts from formal language theory, membrane systems, and multiset processing. We suggest the books [12], [10] and [2] to the reader.

G. Eleftherakis et al. (Eds.): WMC8 2007, LNCS 4860, pp. 428–437, 2007.
© Springer-Verlag Berlin Heidelberg 2007

First, we will provide some definitions from multiset theory as exposed in [15].

**Definition 1.** *Let $D$ be a set. A multiset over $D$ is a pair $\langle D, f \rangle$ where $f : D \longrightarrow \mathbb{N}$ is a function. We say that $A$ is* empty *if for all $a \in D$, $f(a) = 0$.*

**Definition 2.** *Suppose that $A = \langle D, f \rangle$ and $B = \langle D, g \rangle$ are two multisets. The removal of multiset $B$ from $A$, denoted by $A \ominus B$, is the multiset $C = \langle D, h \rangle$ where for all $a \in D$ $h(a) = max(f(a) - g(a), 0)$. Their sum, denoted by $A \oplus B$, is the multiset $C = \langle D, h \rangle$, where for all $a \in D$ $h(a) = f(a) + g(a)$.*

*Then, we say that $A = B$ if the multiset $(A \ominus B) \oplus (B \ominus A)$ is empty.*

The size of any multiset $M$, denoted by $|M|$ will be the number of elements that it contains. We are specially interested in the class of multisets that we call *bounded multisets*. They are multisets that hold the property that the sum of all the elements is bounded by a constant $n$. Formally, we denote by $\mathcal{M}_n(D)$ the set of all multisets $\langle D, f \rangle$ such that $\sum_{a \in D} f(a) = n$.

A concept that is quite useful to work with sets and multisets is the *Parikh mapping*. Formally, a Parikh mapping can be viewed as the application $\Psi : D^* \rightarrow \mathbb{N}^n$ where $D = \{d_1, d_2, \cdots, d_n\}$. Given an element $x \in D^*$ we define $\Psi(x) = (\#_{d_1}(x), \cdots, \#_{d_n}(x))$ where $\#_{d_j}(x)$ denotes the number of occurrences of $d_j$ in $x$.

We introduce now basic concepts from membrane systems taken from [10]. A general P system of degree $m$ is a construct

$$\Pi = (V, T, C, \mu, w_1, \cdots, w_m, (R_1, \rho_1), \cdots, (R_m, \rho_m), i_0), \text{ where:}$$

- $V$ is an alphabet (the *objects*)
- $T \subseteq V$ (the *output alphabet*)
- $C \subseteq V$, $C \cap T = \emptyset$ (the *catalysts*)
- $\mu$ is a membrane structure consisting of $m$ membranes
- $w_i$, $1 \leq i \leq m$, is a string representing a multiset over $V$ associated with the region $i$
- $R_i$, $1 \leq i \leq m$, is a finite set of *evolution rules* over $V$ associated with the $i$th region and $\rho_i$ is a partial order relation over $R_i$ specifying a *priority*.
  An evolution rule is a pair $(u, v)$ (or $u \rightarrow v$) where $u$ is a string over $V$ and $v = v'$ or $v = v'\delta$ where $v'$ is a string over

$$\{a_{here}, a_{out}, a_{in_j} \mid a \in V, 1 \leq j \leq m\}$$

  and $\delta$ is a special symbol not in $V$ (it defines the *membrane dissolving action*).
  From now on, we will denote the set *tar* by $\{here, out, in_k : 1 \leq k \leq m\}$.
- $i_0$ is a number between 1 and $m$ and it specifies the *output* membrane of $\Pi$ (in the case that it equals to $\infty$ the output is read outside the system).

The language generated by $\Pi$ in external mode ($i_0 = \infty$) is denoted by $L(\Pi)$ and it is defined as the set of strings that can be defined by collecting the objects that leave the system by arranging them in the leaving order (if several objects leave the system at the same time then permutations are allowed). The set of vector numbers that represent the objects in the output membrane $i_0$ will be denoted by $N(\Pi)$. Obviously, both sets $L(\Pi)$ and $N(\Pi)$ are defined only for *halting computations*.

One of the multiple variations of P systems is related to the creation, division and modification of membrane structures. There have been several works in which these variants have been proposed (see, for example, [1,9,10,11]).

In the following, we enumerate some kind of rules which are able to modify the membrane structure:

1. 2-division: $[_h a]_h \rightarrow [_{h'} b]_{h'} [_{h''} c]_{h''}$
2. Creation: $a \rightarrow [_h b]_h$
3. Dissolving: $[_h a]_h \rightarrow b$

The power of P systems with the previous operations and other ones (*exocytosis*, *endocytosis*, etc.) has been widely studied in the membrane computing area.

Now, we will introduce some concepts from tree languages and automata as exposed in [3,5]. First, let a *ranked alphabet* be the association of an alphabet $V$ together with a finite relation $r$ in $V \times \mathbb{N}$. We denote by $V_n$ the subset $\{\sigma \in V \mid (\sigma, n) \in r\}$.

The set $V^T$ of trees over $V$, is defined inductively as follows:

$a \in V^T$ for every $a \in V_0$

$\sigma(t_1, ..., t_n) \in V^T$ whenever $\sigma \in V_n$ and $t_1, ..., t_n \in V^T$, $(n > 0)$

and let a *tree language* over $V$ be defined as a subset of $V^T$.

Given the tuple $l = \langle 1, 2, ..., k \rangle$ we will denote the set of permutations of $l$ by $perm(l)$. Let $t = \sigma(t_1, ..., t_n)$ be a tree over $V^T$. We denote the set of permutations of $t$ at first level by $perm_1(t)$. Formally, $perm_1(t) = \{\sigma(t_{i_1}, ..., t_{i_n}) \mid \langle i_1, i_2, ..., i_n \rangle \in perm(\langle 1, 2, ..., n \rangle)\}$.

Let $\mathbb{N}^*$ be the set of finite strings of natural numbers, separated by dots, formed using the catenation as the composition rule and the empty word $\lambda$ as the identity. Let the prefix relation $\leq$ in $\mathbb{N}^*$ be defined by the condition that $u \leq v$ if and only if $u \cdot w = v$ for some $w \in \mathbb{N}^*$ ($u, v \in \mathbb{N}^*$). A finite subset $D$ of $\mathbb{N}^*$ is called a *tree domain* if:

$$u \leq v \text{ where } v \in D \text{ implies } u \in D, \text{ and}$$
$$u \cdot i \in D \text{ whenever } u \cdot j \in D \ (1 \leq i \leq j)$$

Each tree domain $D$ could be seen as an unlabeled tree whose nodes correspond to the elements of $D$ where the hierarchy relation is the prefix order. Thus, each tree $t$ over $V$ can be seen as an application $t : D \rightarrow V$. The set $D$ is called the *domain of the tree* $t$, and denoted by $dom(t)$. The elements of the tree domain $dom(t)$ are called *positions* or *nodes* of the tree $t$. We denote by $t(x)$ the label of a given node $x$ in $dom(t)$.

Let the level of $x \in dom(t)$ be $|x|$. Intuitively, the level of a node measures its distance from the root of the tree. Then, we can define the depth of a tree $t$ as

$depth(t) = max\{|x| : x \in dom(t)\}$. In the same way, for any tree $t$, we denote the size of the tree by $|t|$ and the set of subtrees of $t$ (denoted with $Sub(t)$) as follows:

$$Sub(a) = \{a\} \text{ for all } a \in V_0$$

$$Sub(t) = \{t\} \cup \bigcup_{i=1,\dots,n} Sub(t_i) \text{ for } t = \sigma(t_1, \dots, t_n) \ (n > 0)$$

Given a tree $t = \sigma(t_1, \dots, t_n)$, the root of $t$ will be denoted as $root(t)$ and defined as $root(t) = \sigma$. If $t = a$ then $root(t) = a$. The successors of a tree $t = \sigma(t_1, \dots, t_n)$ will be defined as $H^t = \langle root(t_1), \dots, root(t_n) \rangle$. Finally, $leaves(t)$ will denote the set of leaves of the tree $t$.

**Definition 3.** *A finite deterministic tree automaton is defined by the tuple $A = (Q, V, \delta, F)$ where $Q$ is a finite set of states; $V$ is a ranked alphabet with $m$ as the maximum integer in the relation $r$, $Q \cap V = \emptyset$; $F \subseteq Q$ is the set of final states and $\delta = \bigcup_{i:V_i \neq \emptyset} \delta_i$ is a set of transitions defined as follows:*

$$\delta_n : (V_n \times (Q \cup V_0)^n) \to Q \qquad n = 1, \dots, m$$
$$\delta_0(a) = a \qquad \forall a \in V_0$$

Given the state $q \in Q$, we define the *ancestors* of the state $q$, denoted by $Ant(q)$, as the set of strings

$$Ant(q) = \{p_1 \cdots p_n \mid p_i \in Q \cup V_0 \wedge \delta_n(\sigma, p_1, \dots, p_n) = q\}$$

From now on, we will refer to finite deterministic tree automata simply as *tree automata*. We suggest [3,5] for other definitions on tree automata.

The transition function $\delta$ is extended to a function $\delta : V^T \to Q \cup V_0$ on trees as follows:

$$\delta(a) = a \text{ for any } a \in V_0$$
$$\delta(t) = \delta_n(\sigma, \delta(t_1), \dots, \delta(t_n)) \text{ for } t = \sigma(t_1, \dots, t_n) \ (n > 0)$$

Note that the symbol $\delta$ denotes both the set of transition functions of the automaton and the extension of these functions to operate on trees. In addition, you can observe that the tree automaton $A$ cannot accept any tree of depth zero.

Given a finite set of trees $T$, let the *subtree automaton* for $T$ be defined as $AB_T = (Q, V, \delta, F)$, where:

$$Q = Sub(T)$$
$$F = T$$
$$\delta_n(\sigma, u_1, \dots, u_n) = \sigma(u_1, \dots, u_n) \qquad \sigma(u_1, \dots, u_n) \in Q$$
$$\delta_0(a) = a \qquad a \in V_0$$

Let \$ be a new symbol in $V_0$, and $V_{\$}^T$ the set of trees $(V \cup \{\$\})^T$ where each tree contains \$ only once. We will name the node with label \$ as *link point* when necessary. Given $s \in V_{\$}^T$ and $t \in V^T$, the operation $s\#t$ is defined as:

$$s\#t(x) = \begin{cases} s(x) & \text{if } x \in dom(s), \ s(x) \neq \$ \\ t(z) & \text{if } x = yz, \ s(y) = \$, \ y \in dom(s) \end{cases}$$

therefore, given $t, s \in V^T$, let the tree quotient $(t^{-1}s)$ be defined as

$$t^{-1}s = \begin{cases} r \in V_{\$}^T & : \ s = r\#t \text{ if } t \in V^T - V_0 \\ t & \text{if } t \in V_0 \end{cases}$$

This quotient can be extended to consider set of trees $T \subseteq V^T$ as:

$$t^{-1}T = \{t^{-1}s \mid s \in T\}$$

For any $k \geq 0$, let the $k$-root of a tree $t$ be defined as follows:

$$root_k(t) = \begin{cases} t, & \text{if } depth(t) < k \\ t' : t'(x) = t(x), \ x \in dom(t) \wedge |x| \leq k, & \text{otherwise} \end{cases}$$

## 3   Multiset Tree Automata and Mirrored Trees

We extend now over multisets some definitions of tree automata and tree languages. We will introduce the concept of multiset tree automata and then we will characterize the set of trees that it accepts.

Given any tree automaton $A = (Q, V, \delta, F)$ and $\delta_n(\sigma, p_1, p_2, \ldots, p_n) \in \delta$, we can associate to $\delta_n$ the multiset $\langle Q \cup V_0, f \rangle \in \mathcal{M}_n(Q \cup V_0)$ where $f$ is defined by $\Psi(p_1 p_2 \ldots p_n)$. The multiset defined in such way will be denoted by $M_\Psi(\delta_n)$. Alternatively, we can define $M_\Psi(\delta_n)$ as $M_\Psi(p_1) \oplus M_\Psi(p_2) \oplus \cdots \oplus M_\Psi(p_n)$ where $\forall 1 \leq i \leq n \ M_\Psi(p_i) \in \mathcal{M}_1(Q \cup V_0)$. Observe that if $\delta_n(\sigma, p_1, p_2, \ldots, p_n) \in \delta$, $\delta'_n(\sigma, p'_1, p'_2, \ldots, p'_n) \in \delta$ and $M_\Psi(\delta_n) = M_\Psi(\delta'_n)$ then $\delta_n$ and $\delta'_n$ are defined over the same set of states and symbols but in different order (that is the multiset induced by $\langle p_1, p_2, \cdots, p_n \rangle$ equals the one induced by $\langle p'_1 p'_2 \ldots p'_n \rangle$).

Now, we can define a *multiset tree automaton* that performs a bottom-up parsing as in the tree automaton case.

**Definition 4.** *A multiset tree automaton is defined by the tuple* $MA = (Q, V, \delta, F)$, *where $Q$ is a finite set of states, $V$ is a ranked alphabet with maxarity$(V) = n$, $Q \cap V = \emptyset$, $F \subseteq Q$ is a set of final states and $\delta$ is a set of transitions defined as follows:*

$$\delta = \bigcup_{\substack{1 \leq i \leq n \\ i : V_i \neq \emptyset}} \delta_i$$

$$\delta_i : (V_i \times \mathcal{M}_i(Q \cup V_0)) \rightarrow \mathcal{P}(\mathcal{M}_1(Q)) \qquad i = 1, \ldots, n$$
$$\delta_0(a) = M_\Psi(a) \in \mathcal{M}_1(Q \cup V_0) \qquad \forall a \in V_0$$

We can observe that every tree automaton $A$ defines a multiset tree automaton $MA$ as follows

**Definition 5.** *Let $A = (Q, V, \delta, F)$ be a tree automaton. The multiset tree automaton induced by $A$ is defined by the tuple $MA = (Q, V, \delta', F)$ where each $\delta'$ is defined as follows: $M_\Psi(r) \in \delta'_n(\sigma, M)$ if $\delta_n(\sigma, p_1, ..., p_n) = r$ and $M_\Psi(\delta_n) = M$.*

Observe that, in the general case, the multiset tree automaton induced by $A$ is non deterministic.

As in the case of tree automata, $\delta'$ could also be extended to operate on trees. Here, the automaton carries out a bottom-up parsing where the tuples of states and/or symbols are transformed by using the Parikh mapping $\Psi$ to obtain the multisets in $\mathcal{M}_n(Q \cup V_0)$. If the analysis is completed and $\delta'$ returns a multiset with at least one final state, the input tree is accepted. So, $\delta'$ can be extended as follows

$$\delta'(a) = M_\Psi(a) \text{ for any } a \in V_0$$
$$\delta'(t) = \{M \in \delta'_n(\sigma, M_1 \oplus \cdots \oplus M_n) \mid M_i \in \delta'(t_i) 1 \le i \le n\}$$
$$\text{for } t = \sigma(t_1, \ldots, t_n) \ (n > 0)$$

Formally, every multiset tree automaton $MA$ accepts the following language

$$L(MA) = \{t \in V^T \mid M_\Psi(q) \in \delta'(t), q \in F\}$$

Another extension which will be useful is the one related to the ancestors of every state. So, we define $Ant_\Psi(q) = \{M \mid M_\Psi(q) \in \delta_n(\sigma, M)\}$.

**Theorem 1.** *(Sempere and López, [13]) Let $A = (Q, V, \delta, F)$ be a tree automaton, $MA = (Q, V, \delta', F)$ be the multiset tree automaton induced by $A$ and $t = \sigma(t_1, \ldots, t_n) \in V^T$. If $\delta(t) = q$ then $M_\Psi(q) \in \delta'(t)$.*

**Corollary 1.** *(Sempere and López, [13]) Let $A = (Q, V, \delta, F)$ be a tree automaton and $MA = (Q, V, \delta', F)$ be the multiset tree automaton induced by $A$. If $t \in L(A)$ then $t \in L(MA)$.*

We introduce the concept of *mirroring* in tree structures as exposed in [13]. Informally speaking, two trees will be related by mirroring if some permutations at the structural level hold. We propose a definition that relates all the trees with this mirroring property.

**Definition 6.** *Let $t$ and $s$ be two trees from $V^T$. We will say that $t$ and $s$ are mirror equivalent, denoted by $t \bowtie s$, if one of the following conditions holds:*

1. $t = s = a \in V_0$
2. $t \in perm_1(s)$
3. $t = \sigma(t_1, \ldots, t_n)$, $s = \sigma(s_1, \ldots, s_n)$ and there exists $\langle s^1, s^2, \ldots, s^k \rangle$
   $\in perm(\langle s_1, s_2, \ldots, s_n \rangle)$ such that $\forall 1 \leq i \leq n \; t_i \bowtie s^i$

**Theorem 2.** *(Sempere and López, [13]) Let $A = (Q, V, \delta, F)$ be a tree automaton, $t = \sigma(t_1, \ldots, t_n) \in V^T$ and $s = \sigma(s_1, \ldots, s_n) \in V^T$. Let $MA = (Q, V, \delta', F)$ be the multiset tree automaton induced by $A$. If $t \bowtie s$ then $\delta'(t) = \delta'(s)$.*

**Corollary 2.** *(Sempere and López, [13]) Let $A = (Q, V, \delta, F)$ be a tree automaton, $MA = (Q, V, \delta', F)$ the multiset tree automaton induced by $A$ and $t \in V^T$. If $t \in L(MA)$ then, for any $s \in V^T$ such that $t \bowtie s$, $s \in L(MA)$.*

The last results were useful to propose an algorithm to determine whether two trees are mirror equivalent or not [13]. So, given two trees $s$ and $t$, we can establish in time $\mathcal{O}((min\{|t|, |s|\})^2)$ if $t \bowtie s$.

## 4    $k$-Testable in the Strict Sense (k-TSS) Multiset Tree Languages

In this section, we will define a new class of multiset tree languages. The definitions related to multiset tree automata come from the relation between mirrored trees and multiset tree automata which we have established in the previous section. So, whenever we refer to multiset tree languages we are taking under our consideration the set of (mirrored) trees accepted by multiset tree automata.

We refer to [6] for more details about reversibility and local testability in tree languages.

First, we define $k$-TSS multiset tree languages for any $k \geq 2$.

**Definition 7.** *Let $T \subseteq V^T$ and the integer value $k \geq 2$. $T$ is a $k$-TSS multiset tree language if and only if, given whatever two trees $u_1, u_2 \in V^T$ such that $root_{k-1}(u_1) = root_{k-1}(u_2)$, $u_1^{-1}T \neq \emptyset$ and $u_2^{-1}T \neq \emptyset$ implies that $u_1^{-1}T = u_2^{-1}T$.*

Any multiset tree automaton as in the definition given before will be named a $k$-TSS multiset tree automaton. Given $A$ a tree automaton, $t_1, t_2$ valid contexts over $V_\$^T$, as an extension of a result concerning $k$-TSS tree languages, we give the following definition:

**Definition 8.** *Let $A$ be a multiset tree automaton over $V_\$^T$. Let $u_1, u_2 \in V^T$ be two trees such that $root_{k-1}(u_1) = root_{k-1}(u_2)$ and $t_1 \# u_1, t_2 \# u_2 \in L(A)$ for some valid contexts $t_1$ and $t_2$. If $A$ is a $k$-TSS mirror tree automaton then $\delta(u_1) = \delta(u_2)$.*

We can give the following characterization of such automata.

**Corollary 3.** *Let $A$ be a $k$-TSS multiset tree automaton. There does not exist two distinct states $q_1, q_2$ such that $root_k(q_1) \cap root_k(q_2) \neq \emptyset$.*

The previous result can be easily deduced from the definition of $k$-TSS multiset tree automata and the definitions given in section 2 about tree automata and tree languages.

*Example 1.* Consider the multiset tree automaton with transitions:

$$\delta(\sigma, aa) = q_1, \qquad \delta(\sigma, a) = q_2,$$
$$\delta(\sigma, aq_2) = q_2, \qquad \delta(\sigma, q_1q_1) = q_1,$$
$$\delta(\sigma, aq_2q_1) = q_3 \in F.$$

Note that the multiset tree language accepted by the automaton is $k$-TSS for any $k \geq 2$. Note also that the following one does not have the $k$-TSS condition for any $k \geq 2$:

$$\delta(\sigma, aa) = q_1, \qquad \delta(\sigma, bb) = q_2,$$
$$\delta(\sigma, q_2q_2) = q_2, \qquad \delta(\sigma, q_1q_1) = q_1,$$
$$\delta(\sigma, q_2q_1) = q_3 \in F,$$

because the states $q_1$ and $q_2$ (and $q_3$) share a common $k$-root.

## 5   Reversible Multiset Tree Automata

We also extend a previous result concerning $k$-reversible tree languages (for any $k \geq 0$) to give the following definition.

**Definition 9.** *Let $T \subseteq V^T$ and the integer value $k \geq 0$. $T$ is a $k$-reversible multiset tree language if and only if, given whatever two trees $u_1, u_2 \in V^T$ such that $root_{k-1}(u_1) = root_{k-1}(u_2)$, whenever there exists a context $t \in V_\$^T$ such that both $u_1 \# t, u_2 \# t \in T$, then $u_1^{-1}T = u_2^{-1}T$.*

**Definition 10.** *Let $A$ be a multiset tree automaton over $V_\$^T$. Let $p_1, p_2 \in Q$ be two states such that $root_k(L(p_1)) \cap root_k(L(p_2)) \neq \emptyset$. $A$ is order $k$ reset free if the automaton does not contain two transitions such that*

$$\delta(\sigma, q_1q_2 \ldots q_n p_1) = \delta(\sigma, q_1q_2 \ldots q_n p_2)$$

*where $q_i \in Q$, $1 \leq i \leq n$.*

**Definition 11.** *Let $A$ be a multiset tree automaton. $A$ is $k$-reversible if $A$ is order $k$ reset free and for any two distinct final states $f_1$ and $f_2$ the condition $root_k(L(f_1)) \cap root_k(L(f_2)) = \emptyset$ is fulfilled.*

*Example 2.* Consider the multiset tree automaton with transitions:

$$\delta(\sigma, aa) = q_1, \qquad \delta(\sigma, a) = q_2,$$
$$\delta(\sigma, q_2q_2) = q_2, \qquad \delta(\sigma, aaq_1) = q_1,$$
$$\delta(\sigma, q_1q_1) = q_3 \in F, \quad \delta(\sigma, q_2q_1) = q_3 \in F.$$

The multiset tree language accepted by this automaton is $k$-reversible and it is also an example of non $k$-TSS multiset tree language.

Finally, we can relate the two families of multiset tree languages that we have previously defined with the following result.

**Theorem 3.** *Let $T \subseteq V^T$ and an integer value $k \geq 2$. If $T$ is $k$-TSS then $T$ is $(k-1)$-reversible.*

*Proof.* Let $t\#t_1$ and $t\#t_2$ belong to $T$, with $t \in V_{\$}^T$ and $root_k(t_1) = root_k(t_2)$. Trivially, $t_1^{-1}T \neq \emptyset$ and $t_2^{-1}T \neq \emptyset$. If $T$ is a $k$-TSS tree language, then by previous definitions, $t_1^{-1}T = t_2^{-1}T$, and also $T$ is $(k-1)$-reversible. $\square$

## 6   From Transitions to Membrane Structures

Once we have formally defined the two classes of multiset tree automata, we will translate their characteristics in terms of membrane structures. First we will give a meaning to the concept of $root_k(t)$. Observe that in a membrane structure $t$, which is represented by a set of mirrored trees $\{t' \mid t \bowtie t'\}$, the meaning of $root_k(t)$ is established by taking into account the (sub)structure of the membranes from the top region up to a depth of length $k$. Another concept that we have managed before is the operator $\#$. Observe that this is related to membrane creation of P systems. So, we can go from membrane configuration $t$ to $t\#s$ by creating a new membrane structure $s$ in a predefined region of $t$ (established by $\#$).

So, $k$ testability implies that, whenever we take two membrane (sub)structures of the P system, $u_1$ and $u_2$, if they share a common substructure of order $k-1$ then $u_1$ appears in a membrane configuration if and only if it can be substituted by $u_2$ to give a different membrane configuration.

On the other hand, $k$-reversibility implies that whenever two membrane structures $u_1$ and $u_2$ share the same substructure up to length $k-1$, if $u_1\#t$ and $u_2\#t$ are substructures of valid configurations of the P system, then $u_1\#s$ is a substructure of a valid configuration of the P system if and only if so is $u_2\#s$.

## 7   Conclusions and Future Work

We have introduced two new families of multiset tree languages. These classes have characterized the membrane structures defined by P systems. We think that other classes of tree languages will imply new classes of membrane structures. So, all the theory that has been previously established on tree languages can enrich the way in which we look up to the membrane structures.

In addition, there is another way to explore the relation between the membrane structures of P systems and the languages that they can accept or generate. So, a natural question arises: How is affected the structure of the language by the structure of the membranes? This issue will be explored in future works.

## Acknowledgements

Work supported by the Spanish Generalitat Valenciana under contract GV06/068.

# References

1. Alhazov, A., Ishdorj, T.O.: Membrane operations in P systems with active membranes. In: Proc. Second Brainstorming Week on Membrane Computing. TR 01/04 of RGNC, pp. 37–44, Sevilla University (2004)
2. Calude, C.S., Păun, G., Rozenberg, G., Salomaa, A. (eds.): Multiset Processing. LNCS, vol. 2235. Springer, Heidelberg (2001)
3. Comon, H., Dauchet, M., Gilleron, R., Jacquemard, F., Lugiez, D., Tison, S., Tommasi, M.: Tree automata techniques and applications (1997) (release October 1, 2002), available on: http://www.grappa.univ-lille3.fr/tata
4. Freund, R., Oswald, M., Păun, A.: P Systems Generating Trees. In: Mauri, G., Păun, G., Pérez-Jiménez, M.J., Rozenberg, G., Salomaa, A. (eds.) WMC 2004. LNCS, vol. 3365, pp. 309–319. Springer, Heidelberg (2005)
5. Gécseg, F., Steinby, M.: Handbook of Formal Languages, ch. Tree languages, vol. 3, pp. 1–69. Springer, Heidelberg (1997)
6. López, D.: Inferencia de lenguajes de árboles. PhD Thesis DSIC, Universidad Politécnica de Valencia (2003)
7. López, D., Sempere, J.M.: Editing Distances between Membrane Structures. In: Freund, R., Păun, G., Rozenberg, G., Salomaa, A. (eds.) WMC 2005. LNCS, vol. 3850, pp. 326–341. Springer, Heidelberg (2006)
8. López, D., Sempere, J.M., García, P.: Inference of reversible tree languages. IEEE Transactions on Systems, Man and Cybernetics Part B: Cybernetics 34(4), 1658–1665 (2004)
9. Păun, A.: On P systems with active membranes. In: UMC 2000. Proc. of the First Conference on Unconventional Models of Computation, pp. 187–201 (2000)
10. Păun, G.: Membrane Computing. An Introduction. Springer, Heidelberg (2002)
11. Păun, G., Suzuki, Y., Tanaka, H., Yokomori, T.: On the power of membrane division on P systems. Theoretical Computer Science 324(1), 61–85 (2004)
12. Rozenberg, G., Salomaa, A. (eds.): Handbook of Formal Languages, vol. 1. Springer, Heidelberg (1997)
13. Sempere, J.M., López, D.: Recognizing membrane structures with tree automata. In: Gutiérrez Naranjo, M.A., Riscos-Núñez, A., Romero-Campero, F.J., Sburlan, D. (eds.) 3rd Brainstorming Week on Membrane Computing 2005. RGNC Report 01/2005 Research Group on Natural Computing, Sevilla University, pp. 305–316. Fénix Editora (2005)
14. Sempere, J.M., López, D.: Identifying P Rules from Membrane Structures with an Error-Correcting Approach. In: Hoogeboom, H.J., Păun, G., Rozenberg, G., Salomaa, A. (eds.) WMC 2006. LNCS, vol. 4361, pp. 507–520. Springer, Heidelberg (2006)
15. Syropoulos, A.: Mathematics of Multisets. In: Calude, C.S., Pun, G., Rozenberg, G., Salomaa, A. (eds.) Multiset Processing. LNCS, vol. 2235, pp. 347–358. Springer, Heidelberg (2001)

# $OPERAS_{CC}$: An Instance of a Formal Framework for MAS Modeling Based on Population P Systems

Ioanna Stamatopoulou[1], Petros Kefalas[2], and Marian Gheorghe[3]

[1] South-East European Research Centre, Thessaloniki, Greece
istamatopoulou@seerc.org
[2] Department of Computer Science, CITY College, Thessaloniki, Greece
kefalas@city.academic.gr
[3] Department of Computer Science, University of Sheffield, UK
M.Gheorghe@dcs.shef.ac.uk

**Abstract.** Swarm-based systems are biology-inspired systems which can be directly mapped to multi-agent systems (MAS), possessing characteristics such as local control over the decisions taken by the agents and a highly dynamic structure which continuously changes. This class of MAS is of a particular interest because it exhibits emergent behavior through self-organization and finds itself applicable to a wide range of domains. In this paper, we present $OPERAS$, an open formal framework that facilitates modeling of MAS, we describe how a particular instance of this framework, namely $OPERAS_{CC}$, could employ existing biological computation systems, such as population P systems, and demonstrate how the resulting method can be used to formally model a swarm-based system of autonomous spacecrafts.

## 1 Introduction

Lately, there has been an increasing interest toward biological and biology-inspired systems. From the smallest living elements, the cells, and how they form tissues in organisms to entire ecosystems and how they evolve, there is growing investigation on ways of specifying such systems. The intention is to create software that mimics the behavior of their biological counterparts. Examples of biological systems of interest also include insect colonies (of ants, termites, bees etc.), flocks of birds, tumors growth—the list is endless. The understanding of how nature deals with various situations has inspired a number of problem solving techniques [1] that are applicable to a wide range of situations that had been puzzling computer scientists for decades. Swarm intelligence [2,3], ant colony optimization techniques [4] for example, has been successfully applied to robotics [5], network routing [6,7] and data mining [8] and has inspired agent-based modeling platforms [9].

The promising feature is that these systems can be directly mapped to multi-agent systems (MAS) by considering each entity as an agent, with its own behavioral rules, knowledge, decision making mechanisms and means of communication with the other entities and with the environment. The overall system's

G. Eleftherakis et al. (Eds.): WMC8 2007, LNCS 4860, pp. 438–452, 2007.

behavior is merely the result of the agents' individual actions, the interactions among them and between them and the environment. This also points to the issue of self-organization and how collective behavioral patterns emerge as a consequence of individuals' local interactions in the lack of knowledge of the entire environment or global control.

An additional modeling key aspect of MAS has not received much attention so far; it is the dynamic nature of MAS and how their structure is constantly reconfigured. By *structure* we imply (i) the changing number of agents in a MAS, and (ii) either their physical placement in the environment or, more generally, the structure that is dictated by the communication channels among them. Most modeling methodologies assume a fixed, static structure that is not realistic since in a dynamic MAS, communication between two agents may need to be established or ceased at any point and also new agents may appear in the system while existing ones may be removed. One additional issue that the inherent dynamic nature of these systems raises has to do with distinguishing between the modeling of the individual agents (behavior) and the rules that govern the communication and evolution of the collective MAS (control). By 'control' we do not imply central control, as this would cancel any notion of self-organization. Rather, we refer to the part of the agent that takes care of non-behavioral issues. A modeling method that allows such a distinction, would greatly assist the modeler by breaking down the work into two separate and independent activities, modeling the behavior and modeling the control.

Population P systems (PPS) with active membranes [10], a class of variants of P Systems [11] are membrane structures composed of membranes configured in an arbitrary graph and naturally possess the trait of reconfiguring their own structure through rules that restructure the graph and allow membranes to divide and die. Inspired by this appealing characteristic, in this paper we present a formal framework, called $OPERAS$, that facilitates the development of dynamic MAS of the nature of many biology and biology-inspired systems. The next section introduces $OPERAS$ formal definition, while section 3 presents an instance of this framework, namely $OPERAS_{CC}$ which utilizes population P systems in order to model MAS. A brief description of a representative case study dealing with a swarm-based system follows in Section 4 which also deals with the formal model for the case problem in question. Finally, Section 5 discusses issues arising from our attempt and concludes the paper.

# 2   OPERAS: Formal Modelling of MAS

In an attempt to formally model each individual agent as well as the dynamic behavior of the overall system, we need a formal method that is capable of rigorously describing all the essential aspects, i.e. knowledge, behavior, communication and dynamics. It is also important that the level of abstraction imposed by a formal method is appropriate enough to lead toward the implementation of a system. New computation approaches as well as programming paradigms inspired by biological processes in living cells, introduce concurrency as well as neatly tackle the dynamic structure of multi-component systems (P systems, brane calculus, Gamma, Cham, MGS) [11,12,13]. In agent-oriented software engineering, there have been several attempts to use formal methods, each one focusing on different aspects of agent systems development [14,15,16,17,18]. Other formal methods, such as $\pi$-calculus, mobile ambients and P systems with mobile membranes [19,20,21], successfully deal with the dynamic nature of systems and concurrency of processes but lack intuitiveness when it comes to the modeling of an individual agent (lack of primitives and more complex data structures). An interesting comparison of various formal methods for the verification of emergent behaviors in swarm-based systems is reported in [22].

### 2.1 OPERAS Definition

We start this section by providing the definition of a model for a dynamic MAS in its general form. A *multi-agent system* can be defined by the tuple $(O, P, E, R, A, S)$ containing:

- a set of reconfiguration rules, $O$, that define how the system structure evolves by applying appropriate reconfiguration operators;
- a set of percepts, $P$, for the agents;
- the environment's model / initial configuration, $E$;
- a relation, $R$, that defines the existing communication channels;
- a set of participating agents, $A$, and
- a set of definitions of types of agents, $S$, that may be present in the system.

More particularly:

- the rules in $O$ are of the form *condition* $\Rightarrow$ *action* where *condition* refers to the computational state of agents and *action* involves the application of one or more of the operators that create/remove a communication channel between agents or introduce/remove an agent into/from the system;
- $P$ is the distributed union of the sets of percepts of all participating agents;
- $R : A \times A$ with $(A_i, A_j) \in R$, $A_i, A_j \in A$ meaning that agent $A_i$ may send messages to agent $A_j$;
- $A = \{A_1, \ldots A_n\}$ where $A_i$ is a particular agent defined in terms of its individual behavior and its local mechanism for controlling reconfiguration;
- $S_k = (Behaviour_k, Control_k) \in S, k \in Types$ where $Types$ is the set of identifiers of the types of agents, $Behaviour_k$ is the part of the agent that deals with its individual behavior and $Control_k$ is the local mechanism for controlling reconfiguration; each participating agent $A_i$ of type $k$ in $A$ is a particular instance of a type of agent: $A_i = (Beh_k, Ctrl_k)_i$.

### 2.2 OPERAS as an Open Framework

The general underlying idea is that an agent model consists of two parts, its behavior and its control. The behavior of an agent can be modeled by a formal method with its computation being driven by percepts from the environment.

The control can be modeled by a set of reconfiguration rules which given the computation states of agents can change the structure of the system. The MAS structure is determined through the relation that defines the communication between the agents. The set of participating agents are instances of agent types that may participate in the system. This deals with the fact that an agent may be present at one instance of the system but disappear at another or that a new agent comes into play during the evolution of the MAS. This assumes that all agent types that may participate in the system should be known in advance.

There are still some open issues which, however, make the $OPERAS$ approach a framework rather than a formal method. These are: (i) Which are the formal methods that can be used in order to model the behavior? (ii) Which are the formal methods that can to use in order to model the control? (iii) Could the methods in (i) and (ii) be different? (iv) Should the agents' behavior models communicate directly with other agents' behavior models? (v) Should the agents' control models communicate with other agents' control models? (vi) Could communication be established implicitly through percepts of the environment? (vii) Which method chosen from (i) or from (ii) drives the computation of the resulting system? There is no unique answer to these questions but the modeling solution will depend on the choice of formal methods which are considered suitable to model either behavior or control.

It is therefore implied that there are several options which could instantiate $OPERAS$ into concrete modeling methods. Regarding the modeling of each type of agent $S_k$, there are more than one options to choose from in order to specify its behavioral part and the same applies for its control mechanism. We have long experimented with various formal methods, such as X-machines with its communicating counterpart and Population P Systems with active cells. In this paper we present an instance of the framework that employs ideas from the latter, using a PPS to model the behavior and the control part of the agent.

## 3   $OPERAS_{CC}$

### 3.1   Population P Systems with Active Cells

A *population P system* [10] is a collection of different types of cells evolving according to specific rules and capable of exchanging biological / chemical substances with their neighboring cells (Fig. 1). More particularly, a PPS with active cells [10] is defined as $\mathcal{P} = (V, K, \gamma, \alpha, w_E, C_1, C_2, \ldots, C_n, R)$ where:

- $V$ is a finite alphabet of symbols called objects;
- $K$ is a finite alphabet of symbols, which define different types of cells;
- $\gamma = (\{1, 2, \ldots n\}, A)$, with $A \subseteq \{\{i, j\} \mid 1 \leq i \neq j \leq n\}$, is a finite undirected graph;
- $\alpha$ is a finite set of bond-making rules of the form $(t, x_1; x_2, p)$, with $x_1, x_2 \in V^*$, and $t, p \in K$ meaning that in the presence of objects $x_1$ and $x_2$ inside two cells of type $t$ and $p$ respectively, a bond is created between the two cells;
- $w_E \in V^*$ is a finite multi-set of objects initially assigned to the environment;

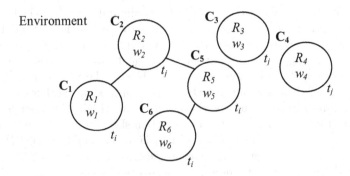

**Fig. 1.** An abstract example of a population P system; $C_i$: cells, $R_i$: sets of rules related to cells; $w_i$: multi-sets of objects associated to the cells

- $C_i = (w_i, t_i)$, for each $1 \leq i \leq n$, with $w_i \in V^*$ a finite multi-set of objects, and $t_i \in K$ the type of cell $i$;
- $R$ is a finite set of rules dealing with communication, object transformation, cell differentiation, cell division and cell death.

All rules present in the PPS are identified by a unique identifier, $r$. More particularly:

*Communication rules* are of the form $r : (a; b, in)_t$, $r : (a; b, enter)_t$, $r : (b, exit)_t$, for $a \in V \cup \{\lambda\}$, $b \in V$, $t \in K$, where $\lambda$ is the empty string, and allow the moving of objects between neighboring cells or a cell and the environment according to the cell type and the existing bonds among the cells. The first rule means that in the presence of an object $a$ inside a cell of type $t$ an object $b$ can be obtained by a neighboring cell non-deterministically chosen. The second rule is similar to the first with the exception that object $b$ is not obtained by a neighboring cell but by the environment. Lastly, the third rule denotes that if object $b$ is present it can be expelled out to the environment.

*Transformation rules* are of the form $r : (a \rightarrow b)_t$, for $a \in V$, $b \in V^+$, $t \in K$, where $V^+$ is the set of non-empty strings over $V$, meaning that an object $a$ is replaced by an object $b$ within a cell of type $t$.

*Cell differentiation rules* are of the form $r : (a)_t \rightarrow (b)_p$, with $a, b \in V$, $t, p \in K$ meaning that consumption of an object $a$ inside a cell of type $t$ changes the cell, making it become of type $p$. All existing objects remain the same besides $a$ which is replaced by $b$.

*Cell division rules* are of the form $r : (a)_t \rightarrow (b)_t (c)_t$, with $a, b, c \in V$, $t \in K$. A cell of type $t$ containing an object $a$ is divided into two cells of the same type. One of the new cell has $a$ replaced by $b$ while the other by $c$. All other objects of the originating cell appear in both new cells.

*Cell death rules* are of the form $r : (a)_t \rightarrow \dagger$, with $a \in V$, $t \in K$ meaning that an object $a$ inside a cell of type $t$ causes the removal of the cell from the system.

PPS provide a straightforward way for dealing with the change of a system's structure and this is the reason why we have chosen them to define an instance of the $OPERAS$ framework, namely $OPERAS_{CC}$.

## 3.2   Definition of $OPERAS_{CC}$

In $OPERAS_{CC}$, each agent (behavior) is modeled as a PPS cell, and has a membrane wrapped around it, that is responsible for taking care of structure reconfiguration issues (control). In essence, this may be considered as a usual PPS in which each cell is virtually divided in two regions, inner (for behavior) and outer (for control), that deal with different sets of objects and have different kinds of rules that may be applied to them. An abstract example of an $OPERAS_{CC}$ model consisting of two agents is depicted in Fig. 2.

Additionally, when using a PPS for modeling purposes, we consider all objects to be attribute-value pairs of the form $att : v$ so that it is clear to which characteristic of the agent an object corresponds to.

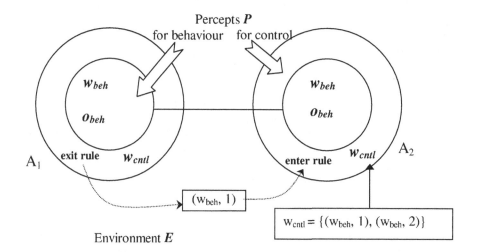

**Fig. 2.** An abstract example of a $OPERAS_{CC}$ consisting of two agents

A MAS in $OPERAS_{CC}$ is defined as the tuple $(O, P, E, R, (A_1, \ldots A_n), S)$ (in correspondence to $P = (R, V, w_E, \gamma, (C_1, \ldots C_n), k)$ of a PPS) where:

- $A_i = (w_{beh}, w_{ctrl}, t)$, $w_{beh}$ being the objects of the agent behavior cell, $w_{ant}$ the objects of the control cell (these objects possibly hold information about the $w_{beh}$ objects (computation states) of neighboring agent cells) and $t \in k$ the type of the cell;
- $O = O_A \cup O_C$.
  - The rules in $O_A$ (to be applied only by the behavior cells on the $w_{beh}$ objects) are the transformation rules of a PPS that rewrite the objects, as well as the communications rules that move objects between cells that

are linked with a bond (both kinds of rules do not affect the structure of the system).

- The rules in $O_C$ (to be applied only by the control cells on the $w_{ctrl}$ objects) are the birth, death, differentiation and bond-making rules of a PPS (the kinds of rules that affect the structure of the system) as well as environment communication rules (receiving/sending objects from/to the environment) so that there is indirect communication between the control cells.

- $P = P_A \cup P_C$, the set of percepts of all participating agents where $P_A$ is the set of inputs perceived by the behavior cells and $P_C$ is the set of inputs perceived by the control cells.
- $E$ is the set of objects assigned to the environment holding information about the computation states of all the participating agents;
- $R$ is the finite undirected graph that defines the communication links between the behavior cells;
- $S$ is the set of possible types of cells.

It should be noted that although the agent descriptions' set $A$ appears fifth in $OPERAS$ definition tuple, from a practical perspective it is the first element being defined; the other tuple elements and their form are naturally dependent on the particular method(s) chosen to define the behavioral and control part of the agents.

In every computation cycle:

- In all the cells modeling the behavior of the agent, all applicable object rules in $O_A$ (transformation and communication) are applied;
- All control cells expel in the environment the $w_{beh}$ objects (computation states of behavior cells) along with the cell identity;
- All control cells import the computation states, $w_{beh}$, of neighboring agents;
- All rules in $O_C$ (bond-making, birth, death, differentiation) are triggered in the control cells, (if applicable) reconfiguring the structure of the system.

Since the model follows the computation rules of a PPS system, the overall system's computation is synchronous. Asynchronous computation may be achieved with the use of other methods for modeling the agents' behavior and/or control. In [23] we present another instance of the framework, namely $OPERAS_{XC}$, which uses X-machines for the behavioral part of the agent and membranes wrapped around the machines for the control part, and apply it on the same swarm-based system that we present hereafter. Because in that version of the framework computation is driven by the computation of the participating X-machines, overall computation is asynchronous.

## 4  $OPERAS_{CC}$ for a Swarm-Based System

### 4.1  Autonomous Spacecrafts for Asteroid Exploration

A representative example of a system which clearly possesses all the aforementioned characteristics of a dynamic MAS is the NASA Autonomous

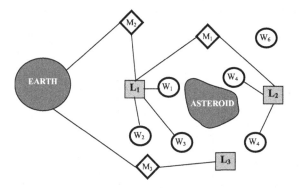

**Fig. 3.** An instance of the ANTS mission, $L$: Leader, $W$: Worker, $M$: Messenger

Nano-Technology Swarm (ANTS) system [22]. The NASA ANTS project aims at the development of a mission for the exploration of space asteroids with the use of different kinds of unmanned spacecrafts. Each each spacecraft is considered as an autonomous agent and the successful exploration of an asteroid depends on the overall behavior of the mission, which emerges as a result of self-organization. We chose this case study because relevant work on the particular project included research on and comparison of a number of formal methods [24,22].

The ANTS mission uses of three kinds of unmanned spacecrafts: $L_i$, leaders, $W_i$, workers and $M_i$, messengers (Fig. 3). The leaders are the spacecrafts that are aware of the goals of the mission and have a non-complete model of the environment. Their role is to coordinate the actions of the spacecrafts that are under their command but by no means should they be considered to be a central controlling mechanism as all spacecrafts' behavior is autonomous. Depending on its goals, a leader creates a team consisting of a number of workers and at least one messengers. Workers and messengers are assigned to a leader upon request by (i) another leader, if they are not necessary for the fulfilment of its goals, or (ii) earth (if existing spacecrafts are not sufficient in number to cover current needs, new spacecrafts are allocated to the mission).

A worker is a spacecraft with a specialized instrument able, upon request from its leader, to take measurements from an asteroid while flying by it. It also possesses a mechanism for analyzing the gathered data and sending the analysis results back to its leader in order for them to be evaluated. This in turn might update the view of the leader, i.e. its model of the environment, as well as its future goals.

The messengers, finally, are the spacecrafts that coordinate communication among workers, leaders and the control center on earth. While each messenger is under the command of one leader, it may also assist in the communication of other leaders if its positioning allows it and conditions demand it.

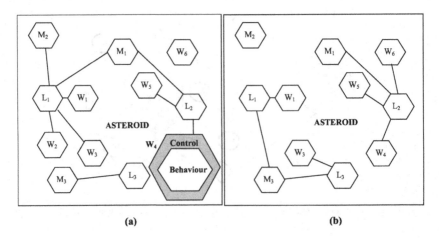

**Fig. 4.** (a) An instance of the MAS structure corresponding to Fig. 3 with an *OPERAS* agent ($W_4$) consisting of separate Behavior and Control components. (b) A change in the structure of MAS after possible events ($W_2$ aborting, $L_1$ employing $W_6$ etc.).

What applies to all types of spacecrafts is that in the case that there is a malfunctioning problem, their superiors are being notified. If the damage is irreparable they need to abort the mission while on the opposite case they may "heal" and return back to normal operation.

### 4.2   The $OPERAS_{CC}$ Approach to the ANTS Mission

The swarm-based system in the ANTS mission can be directly mapped into the *OPERAS* framework (Fig. 4). A number of agents of three different types (workers, $W$, leaders, $L$, and messengers, $M$) compose the MAS system. System configuration is highly dynamic due to its nature and unforeseen situations that may come up during the mission.

**Leader: Formal Modeling of Behavior in $OPERAS_{CC}$.** For the modeling of the leader agent, one has to identify the internal states of the agent, its knowledge as well as the inputs it is capable of perceiving, so that they are represented as objects of the corresponding PPS.

The state of a leader can be either one of the three: *Processing* for an leader that is fully operational, *Malfunctioning* for one that its facing problems and *Aborting* for one that is either facing irreparable problems or has been commanded by the control center on earth to abort the mission.

The knowledge of the agent consists of the objects presented in Table 1 along with their description. Similarly, the leader type of cell will be able to also perceive other objects representing input from the environment or from other agents. The most prominent ones are summarized in Table 2.

**Table 1.** Objects representing the knowledge of the Leader agent

| Object | Description |
|---|---|
| *status* | The current operational state of the leader |
| *existingWorkers* | The set of IDs and statuses of the workers under its command |
| *existingMsgs* | The set of IDs and statuses of the messengers under its command |
| *results* | The set containing analysis results it has gathered |
| *model* | The current model of the agent's surroundings |
| *goals* | The agent's goals |

**Table 2.** Objects representing the percepts of the Leader agent

| Object | Description |
|---|---|
| *abrt* | A request from the control center that the agent should abort the mission |
| *worker* | A new worker that joins the team under the leaders command |
| *messenger* | A new messenger that joins the team under the leaders command |
| *requestForWorker* | A request for a worker, made by another leader (so that the worker is reallocated) |
| *requestForMsg* | A request for a messenger, made by another leader (so that the messenger is reallocated) |
| *message* | An object representing a message sent by another agent |

Indicatively, two of the operations that a leader may perform in the form of transformation rules follow. The rule representing the joining of a worker $w_i$ to the leader's team of *Workers* is specified as:

$workerJoining$ :
$(status : processing\ worker : w_i\ existingWorkers : Workers$
$\quad \rightarrow status : processing\ existingWorkers : \{w_i\} \cup Workers)_L$

The newly allocated worker $w_i$ may be received by another leader with the use of a communication rule of the form:

$receiveWorker : (message : canSendYouAWorker\ ;\ worker : w_i,\ in)_L$

which assumes that a $canSendYouAWorker$ message has been previously sent by the other leader informing that it is willing to reallocate one of its workers.

Similarly, the rule representing the reallocation of the messenger $m_i$ to another leader is:

$reAllocatingMessenger$ :
$(status : processing\ percept : requestForMsg\ existingMsgs : Messengers$
$\quad \rightarrow status : processing\ existingMsgs : Messengers\backslash\{m_i\})_L,$
$\quad \text{if } isMessengerNeeded(m_i) == false$

**Table 3.** Objects representing the knowledge of the Worker agent

| Object | Description |
|---|---|
| *status* | The current operational state of the worker |
| *myLeader* | the identity of its commanding leader, |
| *teamWorkers* | The set of other co-workers belonging to the same team |
| *teamMsgs* | The set of messengers belonging to the same team |
| *Target* | The target asteroid |
| *Data* | The set of data collected from the asteroid |
| *Results* | The set of the data analysis results |

**Worker: Formal Modeling of Behavior in $OPERAS_{CC}$.** Similarly for a worker agent, the internal states in which a it may be in are *Measuring*, when taking measurements from an asteroid, *Analysing*, when analyzing the measurements in order to send results to its leader, *Idle*, *Malfunctioning* and *Aborting*. The knowledge of the agent consists of the objects presented in Table 3 along with their description.

The worker type of cell will also be able to also perceive other objects that represent either environmental stimuli or messages from other agents. Indicative ones are being summarized in Table4.

Indicatively, some of the operations that a worker may perform in the form of transformation rules follow. The rule representing the measurements' analysis mechanism of the worker is:

*analysingData* :
$(status : Analysing \ data : Data \rightarrow status : Idle \ results : Results)_W$

The rule that informs a worker that it is being reallocated to another leader is:

*reAllocating* :
$(status : Idle \ myLeader : Leader \ reassignedTo : NewLeader$
$\rightarrow status : Idle \ myLeader : NewLeader)_W$

**Table 4.** Objects representing the percepts of the Worker agent

| Object | Description |
|---|---|
| *abrt* | A request from the control center that the agent should abort the mission |
| *reassignedTo* | The identifier of the new leader the worker is being reassigned to |
| *data* | The set of measurements taken from the asteroid |

## 4.3    Formal Modeling of Control in $OPERAS_{CC}$

According to $OPERAS_{CC}$, for the definition of the given system as a dynamic MAS, we need to assume an initial configuration. To keep the size restricted for demonstrative purposes, let us consider an initial configuration that includes one leader $L_1$, one messenger $M_1$ and two workers $W_1, W_2$. According to $OPERAS_{CC}$ the above system would be defined as follows.

The set $O$ contains all the aforementioned transformation rules that model the agents' behavior as well as the reconfiguration rules (birth, death and bond-making) regarding (i) the generation of a new worker when the control center on earth decides it should join the mission, (ii) the destruction (i.e. removal from the system) of any kind of agent in the case it must abort the mission, (iii) the establishment of a communication channel between a leader and all members of its team. More particularly $O$ additionally contains the following rules.

The following birth rules create a new worker $w_i$ or messenger $m_i$ under the command of a leader $L_i$ when the leader has received the corresponding messages (objects $earthSendsWorker$ and $earthSendsMsg$) from the control center on earth.

$newWorkerFromEarth$ :
$(status : Processing\ earthSendsWorker : w_i\ existingWorkers : Workers)_{L_i}$
$\quad \rightarrow (status : Processing\ existingWorkers : Workers \cup \{w_i\})_{L_i}$
$\quad (status : Idle\ myLeader : L_i)_{W_i}$

$newMessengerFromEarth$ :
$(status : Processing\ earthSendsMsg : m_i\ existingMsgs : Messengers)_{L_i}$
$\quad \rightarrow (status : Processing\ existingMsgs : Messenger \cup \{m_i\})_{L_i}$
$\quad (status : Idle\ myLeader : L_i)_{M_i}$

Inputs, such as $earthSendsMsg : m_i$, from the environment are perceived with the use of communication rules of the form:

$receiveInput :\ (\varepsilon\ ;\ earthSendsMsg : m_i,\ enter)_L$

The death rule below removes agent instances that have aborted the mission from the model ($t \in S$ stands for any type of agent).

$abortion : (status : aborting)_t \rightarrow \dagger$

Finally, the following bond-making rules ensure the creation of a communication bond between a leader agent ($\varepsilon$ stands for the empty multi-set, i.e. no object is necessary) and any messenger or worker that belongs to this leader's team.

$workerBondMaking : (L_i\ \varepsilon\ ;\ myLeader : L_i\ W)$

$messengerBondMaking : (L_i\ \varepsilon\ ;\ myLeader : L_i\ M)$

The set $P$ contains all objects recognized by the Population P System.

Regarding the environment $E$, it should initially contain objects representing the initial percepts for all agents.

Since in the assumed initial configuration we consider to have one group of spacecrafts under the command of one leader, all agents should be in communication with all others and so:

$$R = \{(L_1, W_1), (L_1, W_2), (W_1, W_2), (M_1, L_1), (M_1, W_1), (M_1, W_2)\}$$

Finally, the set $S$ that contains the agent types is: $S = \{L, W, M\}$.

## 5    Conclusions and Further Work

We presented $OPERAS$ with which one can model multi-agent systems that exhibit dynamic structure, emergent and self-organization behavior. The contributions of $OPERAS$ can be summarized in the following:

- A formal framework for MAS modeling.
- The behavior and the control of an agent are separate components which imply distinct modeling mental activities.
- Flexibility on the choice of formal methods to utilize and option to combine different formal methods.

It is because of this distinct separation between behavior and control that $OPERAS$ provides this flexibility of choosing different methods for modeling these two aspects; while some methods are better at capturing the internal states, knowledge and actions of an agent, others focusing on the dynamic aspect of a MAS are more suitable for capturing the control mechanisms.

In this paper, we employed population P systems with active cells to define $OPERAS_{CC}$, an instance of the general framework. We presented the $OPERAS_{CC}$ model of a swarm-based system of a number of autonomous spacecrafts. It could easily be spotted that an $OPERAS_{CC}$ model does resemble (as a final outcome) a model which could be developed if one used population P systems with active cells from scratch [25]. However, in the current context we have the following advantages:

- PPS can be viewed as a special case for $OPERAS$.
- The distinction of modeling behavior and control as separate, offers the ability to deal with transformation/communication rules separately from cell birth/division/death and bond making, with implications both at theoretical as well as practical level.
- Practically, during the modeling phase, one can find advantages and drawbacks at any of the behavior or control component and switch to another formal method for this component if this is desirable.

As far as the last point is concerned, we verified our initial findings [25] in which it was stated that modeling the behavior of an agent with PPS rewrite and communication rules may be rather cumbersome. Especially in this rather complex case study, although the modeling of the control is absolutely straightforward, we had difficulties to establish the necessary peer to peer communication between agents by employing just the communication rules. That gave us the opportunity to consider alternatives. For example, we have experimented with communicating X-machines which have a number of advantages in terms of modeling the behavior of an agent. The resulting model, $OPERAS_{XC}$ [23], seems to ease the modeling process in complex MAS. It is worth noticing that none of the two formal methods (X-machines and population P systems) by itself could successfully (or at least intuitively) model a MAS [26,27]. This is true for other formal methods too, which means the current framework gives the opportunity to combine those methods that are best suited to either of the two modeling tasks.

We would like to continue the investigation of how $OPERAS$ could employ other formal methods that might be suitable for this purpose. In the near future, we will focus on theoretical aspects of the framework. Towards this direction, we are also currently working on various types of transformations that could prove its power for formal modeling as well as address legacy issues concerned with correctness.

Finally, efforts will also be directed towards enhancing existing animation tools on population P systems in order to come up with a new version of the tool that will be able animate $OPERAS_{CC}$ specified models. More particularly, the PPS-System [28] is a tool that generates Prolog executable code from population P systems models written in a particular notation. Future work will involve extending the notation and the system in order to integrate the necessary $OPERAS$ features, allowing us to gain a deeper understanding of the modeling issues involved with $OPERAS_{CC}$ and helping us investigate the practicability of our approach.

# References

1. Gheorghe, M. (ed.): Molecular Computational Models: Unconventional Approaches. Idea Publishing Inc. (2005)
2. Kennedy, J., Eberhart, R.C.: Particle swarm optimization. In: Proceedings of IEEE Intern. Conference on Neural Networks, Piscataway, NJ, pp. 1942–1948 (1995)
3. Kennedy, J., Eberhart, R.C., Shi, Y.: Swarm intelligence. Morgan Kaufmann Publishers, San Francisco (2001)
4. Dorigo, M., Maniezzo, V., Colorni, A.: The ant system: Optimisation by a colony of co-operating agents. IEEE Trans. on Systems, Man and Cybernetics 26, 1–13 (1996)
5. Dorigo, M., Trianni, V., Sahin, E., Gross, R., Labella, T.H., Baldassarre, G., Nolfi, S., Deneubourg, J.L., Mondada, F.: Evolving self-organizing behavior. Autonomous Robots 17, 223–245 (2004)
6. White, T., Pagurek, B.: Towards multi-swarm problem solving in networks. In: Proceedings of the 3rd Intern. Conference on Multi Agent Systems, p. 333 (1998)
7. Di Caro, G., Dorigo, M.: Mobile agents for adaptive routing. In: Proceedings of the 31st Hawaii International Conference on Systems (1998)
8. Abraham, A., Grosan, C., Ramos, V. (eds.): Swarm Intelligence in Data Mining. Studies in Computational Intelligence, vol. 34. Springer, Heidelberg (2006)
9. Minar, N., Burkhart, R., Langton, C., Askenazi, M.: The swarm simulation system: a toolkit for building multi-agent simulations. Working paper 96-06-042, Santa Fe Institute, Santa Fe (1996)
10. Bernandini, F., Gheorghe, M.: Population P Systems. Journal of Universal Computer Science 10, 509–539 (2004)
11. Păun, G.: Computing with membranes. Journal of Computer and System Sciences 61, 108–143 (2000), also circulated as a TUCS report since 1998
12. Banatre, J., Le Metayer, D.: The gamma model and its discipline of programming. Science of Computer Programming 15, 55–77 (1990)

13. Berry, G., Boudol, G.: The chemical abstract machine. Journal of Theoretical Computer Science 96, 217–248 (1992)
14. d'Inverno, M., Kinny, D., Luck, M., Wooldridge, M.: A formal specification of dMARS. In: Rao, A., Singh, M.P., Wooldridge, M.J. (eds.) ATAL 1997. LNCS, vol. 1365, pp. 155–176. Springer, Heidelberg (1998)
15. Rosenschein, S.R., Kaebling, L.P.: A situated view of representation and control. Artificial Intelligence 73, 149–173 (1995)
16. Brazier, F., Dunin-Keplicz, B., Jennings, N., Treur, J.: Formal specification of multiagent systems: a real-world case. In: ICMAS 1995. Proceedings of International Conference on Multi-Agent Systems, pp. 25–32. MIT Press, Cambridge (1995)
17. Benerecetti, M., Giunchiglia, F., Serafini, L.: A model-checking algorithm for multiagent systems. In: Rao, A.S., Singh, M.P., Müller, J.P. (eds.) ATAL 1998. LNCS (LNAI), vol. 1555, pp. 163–176. Springer, Heidelberg (1999)
18. Fisher, M., Wooldridge, M.: On the formal specification and verification of multiagent systems. Intern. Journal of Cooperating Information Systems 6, 37–65 (1997)
19. Krishna, S.N., Păun, G.: P systems with mobile membranes. Natural Computing: an international journal 4, 255–274 (2005)
20. Cardelli, L., Gordon, A.D.: Mobile ambients. In: Nivat, M. (ed.) ETAPS 1998 and FOSSACS 1998. LNCS, vol. 1378, pp. 140–155. Springer, Heidelberg (1998)
21. Milner, R., Parrow, J., Walker, D.: A calculus of mobile processes, i. Information and Computation 100, 1–40 (1992)
22. Rouf, C., Vanderbilt, A., Truszkowski, W., Rash, J., Hinchey, M.: Verification of NASA emergent systems. In: ICECCS 2004. Proceedings of the 9th IEEE International Conference on Engineering Complex Computer Systems, pp. 231–238 (2004)
23. Stamatopoulou, I., Kefalas, P., Gheorghe, M.: OPERAS for space: Formal modelling of autonomous spacecrafts. In: Papatheodorou, T., Christodoulakis, D., Karanikolas, N. (eds.) Current Trends in Informatics. PCI 2007. Proceedings of the 11th Panhellenic Conference in Informatics, Patras, Greece, May 18-20, 2007, vol. B, pp. 69–78 (2007)
24. Rouff, C., Vanderbilt, A., Hinchey, M., Truszkowski, W., Rash, J.: Properties of a formal method for prediction of emergent behaviors in swarm-based systems. In: SEFM 2004. Proceedings of the Second International Conference on Software Engineering and Formal Methods, pp. 24–33 (2004)
25. Stamatopoulou, I., Gheorghe, M., Kefalas, P.: Modelling dynamic configuration of biology-inspired multi-agent systems with Communicating X-machines and Population P Systems. In: Mauri, G., Păun, G., Pérez-Jiménez, M.J., Rozenberg, G., Salomaa, A. (eds.) WMC 2004. LNCS, vol. 3365, pp. 389–401. Springer, Heidelberg (2005)
26. Kefalas, P., Stamatopoulou, I., Gheorghe, M.: A formal modelling framework for developing multi-agent systems with dynamic structure and behaviour. In: Pěchouček, M., Petta, P., Varga, L.Z. (eds.) CEEMAS 2005. LNCS (LNAI), vol. 3690, pp. 122–131. Springer, Heidelberg (2005)
27. Stamatopoulou, I., Kefalas, P., Gheorghe, M.: Modelling the dynamic structure of biological state-based systems. BioSystems 87, 142–149 (2007)
28. Stamatopoulou, I., Kefalas, P., Eleftherakis, G., Gheorghe, M.: A modelling language and tool for Population P Systems. In: Proceedings of the 10th Panhellenic Conference in Informatics, November 11-13, 2005, Volos, Greece (2005)

# Author Index

# Lecture Notes in Computer Science

Sublibrary 1: Theoretical Computer Science and General Issues

For information about Vols. 1– 4525
please contact your bookseller or Springer

Vol. 4684: L. Kang, Y. Liu, S. Zeng (Eds.), Evolvable Systems: From Biology to Hardware. XIV, 446 pages. 2007.

Vol. 4683: L. Kang, Y. Liu, S. Zeng (Eds.), Advances in Computation and Intelligence. XVII, 663 pages. 2007.

Vol. 4681: D.-S. Huang, L. Heutte, M. Loog (Eds.), Advanced Intelligent Computing Theories and Applications. XXVI, 1379 pages. 2007.

Vol. 4672: K. Li, C. Jesshope, H. Jin, J.-L. Gaudiot (Eds.), Network and Parallel Computing. XVIII, 558 pages. 2007.

Vol. 4671: V.E. Malyshkin (Ed.), Parallel Computing Technologies. XIV, 635 pages. 2007.

Vol. 4669: J.M. de Sá, L.A. Alexandre, W. Duch, D. Mandic (Eds.), Artificial Neural Networks – ICANN 2007, Part II. XXXI, 990 pages. 2007.

Vol. 4668: J.M. de Sá, L.A. Alexandre, W. Duch, D. Mandic (Eds.), Artificial Neural Networks – ICANN 2007, Part I. XXXI, 978 pages. 2007.

Vol. 4666: M.E. Davies, C.J. James, S.A. Abdallah, M.D. Plumbley (Eds.), Independent Component Analysis and Blind Signal Separation. XIX, 847 pages. 2007.

Vol. 4665: J. Hromkovič, R. Královič, M. Nunkesser, P. Widmayer (Eds.), Stochastic Algorithms: Foundations and Applications. X, 167 pages. 2007.

Vol. 4664: J. Durand-Lose, M. Margenstern (Eds.), Machines, Computations, and Universality. X, 325 pages. 2007.

Vol. 4661: U. Montanari, D. Sannella, R. Bruni (Eds.), Trustworthy Global Computing. X, 339 pages. 2007.

Vol. 4649: V. Diekert, M.V. Volkov, A. Voronkov (Eds.), Computer Science – Theory and Applications. XIII, 420 pages. 2007.

Vol. 4647: R. Martin, M.A. Sabin, J.R. Winkler (Eds.), Mathematics of Surfaces XII. IX, 509 pages. 2007.

Vol. 4646: J. Duparc, T.A. Henzinger (Eds.), Computer Science Logic. XIV, 600 pages. 2007.

Vol. 4644: N. Azémard, L. Svensson (Eds.), Integrated Circuit and System Design. XIV, 583 pages. 2007.

Vol. 4641: A.-M. Kermarrec, L. Bougé, T. Priol (Eds.), Euro-Par 2007 Parallel Processing. XXVII, 974 pages. 2007.

Vol. 4639: E. Csuhaj-Varjú, Z. Ésik (Eds.), Fundamentals of Computation Theory. XIV, 508 pages. 2007.

Vol. 4638: T. Stützle, M. Birattari, H. H. Hoos (Eds.), Engineering Stochastic Local Search Algorithms. X, 223 pages. 2007.

Vol. 4630: H.J. van den Herik, P. Ciancarini, H.H.L.M.(J.) Donkers (Eds.), Computers and Games. XII, 283 pages. 2007.

Vol. 4628: L.N. de Castro, F.J. Von Zuben, H. Knidel (Eds.), Artificial Immune Systems. XII, 438 pages. 2007.

Vol. 4627: M. Charikar, K. Jansen, O. Reingold, J.D.P. Rolim (Eds.), Approximation, Randomization, and Combinatorial Optimization. XII, 626 pages. 2007.

Vol. 4624: T. Mossakowski, U. Montanari, M. Haveraaen (Eds.), Algebra and Coalgebra in Computer Science. XI, 463 pages. 2007.

Vol. 4623: M. Collard (Ed.), Ontologies-Based Databases and Information Systems. X, 153 pages. 2007.

Vol. 4621: D. Wagner, R. Wattenhofer (Eds.), Algorithms for Sensor and Ad Hoc Networks. XIII, 415 pages. 2007.

Vol. 4619: F. Dehne, J.-R. Sack, N. Zeh (Eds.), Algorithms and Data Structures. XVI, 662 pages. 2007.

Vol. 4618: S.G. Akl, C.S. Calude, M.J. Dinneen, G. Rozenberg, H.T. Wareham (Eds.), Unconventional Computation. X, 243 pages. 2007.

Vol. 4616: A.W.M. Dress, Y. Xu, B. Zhu (Eds.), Combinatorial Optimization and Applications. XI, 390 pages. 2007.

Vol. 4614: B. Chen, M. Paterson, G. Zhang (Eds.), Combinatorics, Algorithms, Probabilistic and Experimental Methodologies. XII, 530 pages. 2007.

Vol. 4613: F.P. Preparata, Q. Fang (Eds.), Frontiers in Algorithmics. XI, 348 pages. 2007.

Vol. 4600: H. Comon-Lundh, C. Kirchner, H. Kirchner (Eds.), Rewriting, Computation and Proof. XVI, 273 pages. 2007.

Vol. 4599: S. Vassiliadis, M. Bereković, T.D. Hämäläinen (Eds.), Embedded Computer Systems: Architectures, Modeling, and Simulation. XVIII, 466 pages. 2007.

Vol. 4598: G. Lin (Ed.), Computing and Combinatorics. XII, 570 pages. 2007.

Vol. 4596: L. Arge, C. Cachin, T. Jurdziński, A. Tarlecki (Eds.), Automata, Languages and Programming. XVII, 953 pages. 2007.

Vol. 4595: D. Bošnački, S. Edelkamp (Eds.), Model Checking Software. X, 285 pages. 2007.

Vol. 4590: W. Damm, H. Hermanns (Eds.), Computer Aided Verification. XV, 562 pages. 2007.

Vol. 4588: T. Harju, J. Karhumäki, A. Lepistö (Eds.), Developments in Language Theory. XI, 423 pages. 2007.

Vol. 4583: S.R. Della Rocca (Ed.), Typed Lambda Calculi and Applications. X, 397 pages. 2007.

Vol. 4580: B. Ma, K. Zhang (Eds.), Combinatorial Pattern Matching. XII, 366 pages. 2007.

Vol. 4576: D. Leivant, R. de Queiroz (Eds.), Logic, Language, Information and Computation. X, 363 pages. 2007.

Vol. 4547: C. Carlet, B. Sunar (Eds.), Arithmetic of Finite Fields. XI, 355 pages. 2007.

Vol. 4546: J. Kleijn, A. Yakovlev (Eds.), Petri Nets and Other Models of Concurrency – ICATPN 2007. XI, 515 pages. 2007.

Vol. 4545: H. Anai, K. Horimoto, T. Kutsia (Eds.), Algebraic Biology. XIII, 379 pages. 2007.

Vol. 4533: F. Baader (Ed.), Term Rewriting and Applications. XII, 419 pages. 2007.

Vol. 4528: J. Mira, J.R. Álvarez (Eds.), Nature Inspired Problem-Solving Methods in Knowledge Engineering, Part II. XXII, 650 pages. 2007.

Vol. 4527: J. Mira, J.R. Álvarez (Eds.), Bio-inspired Modeling of Cognitive Tasks, Part I. XXII, 630 pages. 2007.